Marine Corrosion and Cathodic Protection

Marine Corrosion and Cathodic Protection

Chris Googan

CRC Press
Taylor & Francis Group
Boca Raton London New York

CRC Press is an imprint of the
Taylor & Francis Group, an **informa** business

First edition published 2022
by CRC Press
2 Park Square, Milton Park, Abingdon, Oxon, OX14 4RN

and by CRC Press
6000 Broken Sound Parkway NW, Suite 300, Boca Raton, FL 33487-2742

CRC Press is an imprint of Informa UK Limited

British Library Cataloguing-in-Publication Data
A catalogue record for this book is available from the British Library

Library of Congress Cataloging-in-Publication Data
Names: Googan, C. G. (Christopher George), 1952- author.
Title: Marine corrosion and cathodic protection / Chris Googan.
Description: First edition. | Abingdon, Oxon ; Boca Raton : CRC Press, 2022. |
Includes bibliographical references and index.
Identifiers: LCCN 2021040846 (print) | LCCN 2021040847 (ebook) |
ISBN 9781032105819 (hbk) | ISBN 9781032105826 (pbk) | ISBN 9781003216070 (ebk)
Subjects: LCSH: Offshore structures—Protection. | Corrosion and
anti-corrosives. | Cathodic protection. | Seawater corrosion.
Classification: LCC TC1670 .G66 2022 (print) | LCC TC1670 (ebook) |
DDC 627/.98—dc23/eng/20211117
LC record available at https://lccn.loc.gov/2021040846
LC ebook record available at https://lccn.loc.gov/2021040847

ISBN: 978-1-032-10581-9 (hbk)
ISBN: 978-1-032-10582-6 (pbk)
ISBN: 978-1-003-21607-0 (ebk)

DOI: 10.1201/9781003216070

Typeset in Sabon
by codeMantra

To my wife Sue

Contents

Preface

WHY THIS BOOK?

This book is intended for a range of users. I hope that even seasoned cathodic protection (CP) campaigners will find it useful. But it is mainly directed at readers looking to learn about marine CP from scratch, or who have previously gained a nodding acquaintance and wish to delve deeper.

There are few CP specialists who do not have a copy of John Morgan's *Cathodic Protection* or Von Baeckman and Schwenk's multi-authored *Handbook of Cathodic Corrosion Protection* on their bookshelves. These remain important works, I recommend both. Nevertheless, the most recent edition of each is over 30 years old. A cursory look at the CP standards that have been issued in the interim (see below) tells us that, although the fundamentals remain unchanged, some things have moved on.

Both of these classic works cover the gamut of CP applications: marine, onshore and concrete reinforcement. However, as you will discover, there is more than enough material to explore in a single book; without venturing out of the seawater and into the soil, or onto a reinforced concrete structure. I make no apology for leaving these other spheres of CP alone. It is relevant to focus on marine CP because, after a period of stasis, new areas – such as offshore wind, applications in deeper waters and the need to extend the lives of ageing offshore assets – are gaining traction. All of these areas of CP activity are buoyed by the relatively recent ascendancy of computer modelling.

WHAT IS IN IT?

I was tempted to get all the theory out of the way first. However, that is not the way anybody, myself included, actually develops their understanding of CP. The acquisition of practical experience and theoretical knowledge progress hand-in-hand. Interestingly, I have worked on CP one way and another for over 40 years, but I only got around to tackling some elements of the theory when I was obliged to think how I would present them here.

With this in mind, Chapter 1 starts with a review of the corrosion of steel in seawater. This is obviously relevant since the vast majority of marine CP is targeted at protecting immersed steelwork in the form of ships' hulls, pipelines, fixed and moveable offshore structures, harbours, jetties and so on. However, you may regard this chapter as optional. If your only interest in CP is because a construction contract or legislation requires it, then the type and extent of corrosion that would otherwise occur may not be relevant to you.

The content of Chapter 2 is clear from its title: "CP Basics". It presents sufficient theory to enable you to tackle run-of-the-mill CP designs for simple offshore structures, or to review designs prepared by others. This leads to Chapter 3 "Designing According to the Codes" which, by the way of some examples, illustrates how to create simple sacrificial anode CP designs using industry recognised codes. This chapter also introduces the topic of the competency levels necessary to carry out a CP design. Competency is a topic that reappears at various points throughout the book.

So far so good. We now know enough to design a basic CP system and understand, conceptually at least, how it works. However, by this point, we should also be aware that a great deal remains to be covered. For example, we have not yet touched upon impressed current systems, nor have we dealt with numerous other applications such as ships and floating structures, offshore wind farm foundations, internal spaces and so on. Furthermore, a thinking reader may be wondering: what are the origins of the guideline design parameters, and why do they vary between the codes?

These questions can only be realistically approached with a deeper understanding of how CP works. In keeping with the way corrosion science is taught, we start with thermodynamics (Chapter 4). In essence, this examines whether or not a given metal can corrode. Again, you have a choice. You are welcome to take the theory on board. Alternatively, you can assume that, since steel is observed to corrode in seawater, thermodynamics can offer nothing else but confirmation of that observation.

Things get more interesting in Chapter 5. Here, we are not concerned with the reality that iron or steel can corrode. We are interested in how quickly or, hopefully, how slowly that process will be. This requires an understanding of what controls electrochemical reaction rates. This brings us to the topic of "electrode kinetics". Importantly, it is our examination of kinetics that provides the more relevant insights into how CP actually does work.

As our understanding develops, it emerges that the key CP parameters are potential and current. Chapter 6 is dedicated to the pivotal concept of potential; and the question: what change in potential do we need for protection? Then, since that change is brought about by the application of current, Chapter 7 investigates what the magnitude of this current should be. It explains why it varies in different seawater exposure conditions and, importantly, why it changes over time.

Up to this point, we have only considered carbon steel because steel is the first-choice material for structures and equipment in seawater. However, for various engineering reasons, other alloys find their way into seawater service. In Chapter 8, therefore, we take a detour and survey some of the other alloys used in seawater service and their CP parameters. Some of these alloys are fully resistant to seawater corrosion. Others are not. However, the distinction is often unimportant because they are frequently connected to carbon steel, and thence into its CP system.

In many applications, such as ships and pipelines, CP is used in conjunction with protective coatings. Indeed, the coating is generally referred to as the primary defence against corrosion, and CP is regarded as secondary. A CP practitioner, therefore, needs to know something about protective coatings, even if it is only enough to be able to cut through some of the coating industry's jargon. To this end, Chapter 9 presents a layman's guide to the coating materials and systems most usually employed on subsea structures and pipelines.

We return to the subject of applying CP in Chapters 10 and 11. Chapter 10 focusses on sacrificial anodes, where we track the evolution of the bespoke alloys of aluminium, zinc and, to a lesser extent, magnesium. We look at how the alloy composition affects the anode performance. We also take the opportunity to explore the anode resistance equations that we used to calculate anode current outputs in Chapter 2. Chapter 10 also covers some topics such as achieving rapid polarisation or applying CP at limited potentials.

Chapter 11 provides our first substantive encounter with impressed current cathodic protection (ICCP) systems. We outline the basics of the systems: power supplies, cables and, in particular, anodes. This serves as a background to our further encounters with ICCP systems in subsequent chapters.

In many engineering applications, the loss of steel thickness by corrosion is not the life-limiting factor. The durability of many offshore structures is instead dictated by fatigue. Accordingly, Chapter 12 studies the fatigue process of steel and examines how that process is exacerbated by corrosion; and how the resulting corrosion fatigue is mitigated by CP. It also examines other interactions of CP with the mechanical properties of metals. In particular, we address the role of hydrogen, which is produced by CP, in causing embrittlement and cracking in some higher strength steels and corrosion-resistant alloys.

Chapter 13 gives a brief history of the evolution of CP on fixed offshore structures and examines examples of both sacrificial and ICCP systems. We then examine the protection of wind turbine foundation structures. This, in turn, leads to a discussion on "retrofit" CP installations. That is installing CP on structures where the original system has either failed prematurely, or else the required service life has been stretched beyond the original intent. This chapter also looks at CP applications on other types of massive fixed structures such as harbours and jetties.

Chapter 14 takes us to the CP of subsea pipelines. These are generally simpler to protect than the more complex offshore structures because they are universally very well coated. However, as we shall see, the CP of offshore pipelines is not without its own challenges.

There is no technical difference between steel that finds itself part of a fixed structure, and steel on a floating structure such as a ship. However, there are sufficient differences between the approaches to, and methodologies of, CP that we deal with ships and other floating structures separately in Chapter 15.

So far, we have only dealt with the protection of structures in open seawater. This position is reversed in Chapter 16 ("Internal CP") where we explore examples of seawater contained within a structure: ships' ballast tanks and the internals of offshore wind turbine foundations. Having found our way into internal spaces, we then look at instances where CP can be, and is, applied for the internal protection of seawater-filled piping and equipment such as heat exchangers.

The computational modelling of CP systems has probably generated more technical papers in the 21st century than any other aspect of the subject. We turn to this in Chapter 17. There is no doubt that modelling can be a powerful tool for visualising and solving CP problems. However, it is also necessary to counsel against the overselling of its benefits. As with so much of CP engineering, a critical faculty is a useful asset when it comes to modelling.

We conclude in Chapter 18 with a discussion on CP system management. Here we review the information that we need from our system, how we might obtain it, how modelling might help us interpret it and what practical use we can make of it.

Acknowledgements

I can take no personal credit for the developments in CP charted in this book. I have occasionally chipped in with very minor original contributions. For the most part, however, I have been just a journeyman practitioner, but with a keen – and sometimes cynical – interest. This book is built on the achievements, expertise and occasional mishaps of others: from Nobel Laureates to lesser mortals. I have endeavoured to distil that part of their wisdom that resides in their published writings into these pages. I am grateful to all of the hundreds of authors whose work is referenced. I also apologise to the many others whose publications do not get a mention. This is often for no other reason than lack of space. I am also indebted to the unsung heroes who give so much of their time on committees drafting the CP codes that support this book.

It may be invidious to single out any individuals by name. However, I feel that certain experts deserve special mention. I must start with Vic Ashworth who first explained CP to me when I was a student in Manchester. He then acted as my academic supervisor for both my MSc dissertation and PhD thesis. We subsequently spent 17 years as colleagues in Global Corrosion Consultants. Robin Jacob, another former Global colleague, has the distinction of the longest time-span between citations in this book (45 years). A special mention is due to the late Dave Scantlebury, who co-supervised my post-graduate research, and with whom I collaborated on protective coatings during my 5 years with International Paint.

Of course, although I have leaned on the efforts of others, it is only my name on the cover. So, any errors are down to me.

Chris Googan
Much Wenlock, England

Author

Dr Chris Googan is an AMPP Accredited Corrosion Specialist. After graduating in chemistry, and spending several years in the steel and copper industries, he gained an MSc in corrosion science (1977) and a PhD in cathodic protection (1979). After 5 years in a corrosion role in the protective coatings industry, he became a freelance corrosion engineering consultant. His experience of corrosion, materials, coatings and cathodic protection has been gained in over 40 countries.

Units, abbreviations and symbols

For the most part, this book adopts SI units, but with occasional departure. For example, the SI unit for corrosion rate should be nm/s (nanometres per second). This is very difficult to picture compared to the more rational unit of mm/year. Similarly, quantities that were originally reported in imperial units have been converted to their SI equivalents. For example, the US customary unit for corrosion rates is "mpy" (mils per year). Confusingly, a "mil" in this instance does not stand for "millimetre", but for a "thousandth of an inch".

Elsewhere, we try to stick with IUPAC chemical nomenclature, even if it does require "sulphate" to be spelled "sulfate". We also use the common names for some chemicals. Where possible, we refer to alloys by their UNS designations.

The following abbreviations are also used in this book. Elsewhere, symbols are defined after the equation or formula in which they appear.

ABBREVIATIONS

3LPE	3-layer polyethylene
3LPP	3-layer polypropylene
AC	alternating current
AISI	American Iron and Steel Institute
ALWC	accelerated low water corrosion
AMPP	Association for Materials Protection and Performance
ASTM	ASTM International (formerly: American Society for Testing and Materials)
AUV	autonomous underwater vehicle
BASIC	Beginner's all-purpose symbolic instruction code
bcc	body-centred cubic
BCE	before the common era
bct	body-centred tetragonal
BEM	boundary element method (computer modelling technique)
BHRA	The British Hydromechanics Research Association (now BHR Group)
BoD	basis of design
CA	corrosion allowance
CAD	computer-aided design
CD	current density
CE	common era
CEOCOR	Western European Committee on Corrosion and Protection of Conduits
COIPM	Comité International Permanent pour Protection des Matériaux en Milieu Marin
CP	cathodic protection
CRA	corrosion-resistant alloy

CSE	copper sulfate electrode
CSP	chlorosulfonated polyethylene (electrical insulation)
CTE	coal tar epoxy (paint)
CTE	coal tar enamel (pipeline coating)
DC	direct current
DFT	dry film thickness
DNV	Det Norske Veritas
DNVGL	Det Norske Veritas Gemanischer Lloyd
DSS	duplex stainless steel
DVM	digital voltmeter
EFC	European Federation of Corrosion
EIS	electrochemical impedance spectroscopy
EMF	electromotive force
EPR	ethylene propylene rubber
FBE	fusion bonded epoxy
fcc	face-centred cubic
FDM	finite difference method (computer modelling technique)
FEM	finite element method (computer modelling technique)
FiGS	field gradient sensor
FJC	field joint coating
FPSO	floating production, storage and offloading (vessel)
GRP	glass-reinforced polymer (or polyester)
HAT	highest astronomical tide
HE	hydrogen embrittlement
HMS	His Majesty's Ship
HISC	hydrogen-induced stress cracking
HSC	hydrogen stress cracking
ICCP	impressed current cathodic protection
ILI	in-line inspection
IMCA	International Marine Contractors Association
IMO	International Maritime Organisation
ISGOTT	International Safety Guide for Oil Tankers and Terminals
ISO	International Organization for Standardisation
IUPAC	International Union of Pure and Applied Chemistry
LAT	lowest astronomical tide
LCIA	London Court of International Arbitration
LID	Lynn and Inner Dowsing (wind farm)
mdd	milligrammes per square decimetre per day
MDFT	minimum dry film thickness
MHWS	mean high water springs
MLWS	mean low water springs
MIC	microbiologically influenced corrosion
MMO	mixed metal oxide
MWL	mean water level
NAB	nickel aluminium bronze
NACE	NACE International (formerly National Association of Corrosion Engineers) now AMPP
NDFT	nominal dry film thickness
NORSOK	Norsk Sokkels Konkuranseposisjon (Standards developed by the Norwegian Technology Centre)

NPCA	Norwegian Pollution Control Authority
PE	polyethylene
PP	polypropylene
PRE_N	pitting resistance equivalent number
PTFE	poly-tetrafluoro-ethylene
PVC	polyvinyl chloride
PVDF	polyvinylidene fluoride
PSA	Petroleum Safety Authority (Norway)
RA	reduction in area (of a tensile test specimen)
RCP	resistor controlled cathodic protection
RIB	rigid inflatable boat
RMS	Royal Mail Steamer
RNA	ribonucleic acid
S-N	stress vs number of cycles (fatigue curve)
SCC	stress corrosion cracking
SCE	saturated calomel electrode
SCF	stress concentration factor
SDSS	superduplex stainless steel
SEM	scanning electron microscope
SGL	sales gas line
SHE	standard hydrogen electrode
SI	Système international d'unités
SMSS	super martensitic stainless steel
SMYS	specified minimum yield stress
SRB	sulfate reducing bacteria
SRP	sulfate reducing prokaryotes
T/R	transformer–rectifier
TSA	thermally sprayed aluminium
UNS	unified numbering system (for alloys)
USC	United States customary (units)
USS	United States Ship
VOC	volatile organic compound
WBT	water ballast tank
w/w	weight/weight
XLPE	cross-linked polyethylene
ZRA	zero resistance ammeter

Cathodic protection codes

Like all engineering activities, CP is governed by codes, which I use as a catch-all term embracing: standards, specifications, recommended practices and guidelines. I list below the codes that we refer to in this book. Codes not primarily concerned with CP are listed with references at the end of each chapter. Where a code is described simply by its publisher and its number, then the reference is to the edition current at the time of writing (June 2021). By the time you are reading this, at least some of these codes will have been revised and reissued.

In addition to the current codes, there is much that we can learn about the development of CP from historical codes that have now been withdrawn or superseded. In this book, where I refer to an earlier version of a code, I add the year of issue.

Codes current at the time of writing

AMPP

In January 2021, NACE International (formerly the National Association of Corrosion Engineers) merged with The Society for Protective Coatings (formerly the Steel Structures Painting Council - SSPC) to form the Association for Materials Protection and Performance (AMPP).

1. NACE SP0169 *Control of External Corrosion on Underground or Submerged Metallic Piping Systems (2013).*
2. NACE SP-0176 *Corrosion Control of Steel, Fixed Offshore Platforms Associated with Petroleum Production (2007).*
3. NACE SP 0387 *Metallurgical and Inspection Requirements for Cast Galvanic Anodes for Offshore Operations (2019).*
4. NACE TM 0190-98 *Impressed Current Laboratory Testing of Aluminum Alloy Anodes 2017.*

ASTM

5. ASTM B418 *Standard Specification for Cast and Wrought Galvanic Zinc Anodes (2021).*

BS

6. BS 3900 Part F10: *Method of test for paints: Determination of resistance to cathodic disbonding of coating for use in marine environments 1985.*

DNV/DNVGL

The Norwegian Organization Det Norske Veritas (DNV) has been a dominant force in setting, and policing, standards for offshore engineering. There are numerous references to DNV in this book. DNV merged with Germanischer Lloyd (GL) in 2013 to become

DNVGL. It changed its name back to DNV in March 2021. References to DNV should be read as including DNVGL for documents issued between 2014 and 2020.

7. DNV-CG-0288 *Corrosion protection of ships 2017.*
8. DNV-RP-B101 *Corrosion Protection of Floating Production and Storage Units 2019.*
9. DNV-RP-B401 *Cathodic protection design 2021.*
10. DNV-RP-F103 *Cathodic protection of submarine pipelines 2016.*
11. DNV-RP-0416 *Corrosion protection for wind turbines 2016.*

European Standards

12. EN 12473 *General principles of cathodic protection in seawater 2014.*
13. EN 12495 *Cathodic protection for fixed steel offshore structures 2000.*
14. EN 12499 *Internal cathodic protection of metallic structures 2003.*
15. EN 13173 *Cathodic protection for offshore floating structures 2001.*
16. EN 13509 *Cathodic protection measurement techniques 2003.*
17. EN 16222 *Cathodic Protection of ship hulls 2012.*
18. EN 50162 *Protection against corrosion by stray currents from direct current systems 2005.*

ISO

19. ISO 13174 *Cathodic protection of harbour installations 2012.*
20. ISO 15589-2 *Petroleum and natural gas industries – Cathodic protection of pipeline transportation systems Part 2 Offshore pipelines 2012 (revision expected 2021).*
21. ISO 20313 *Ships and marine technology - Cathodic protection of ships 2018.*
22. ISO 24656 *Cathodic protection of offshore wind structures (expected 2021)*

NORSOK

23. M-503 Rev 4 *Common Requirements Cathodic Protection 2016 (plus 2018 corrigendum).*

US Military Specifications

24. MIL-A-18001 K *Anodes, Sacrificial Zinc Alloy* (1991 amended 2007).
25. MIL-DTL-24779D(SH) *Detail Specification, Anodes Sacrificial, Aluminum Alloy* (2015).

Superseded or withdrawn codes (in date order)

To help us trace the historical development of CP, we occasionally make reference to earlier editions of the codes listed above. For the same reason, we also occasionally refer to the following codes that have now been withdrawn.

1. NACE RP0169 *Control of External Corrosion on Underground or Submerged Metallic Piping Systems.* (1969). This Recommended Practice (RP) has been subject to numerous revisions, before being re-issued as SP 0169 (SP=Standard Practice).
2. NACE RP0675 Control of External Corrosion on Offshore Steel Pipelines (1988).
3. BS CP 1021 *Code of Practice – cathodic protection* 1973 (superseded by BS7361-1).
4. DNV-TNA-703 *Cathodic Protection Evaluation* 1981 (superseded by various editions of DNV/DNVGL-RP-B401 and DNV/DNVGL-RP-F103).
5. BS 7361-1 (1991) *Cathodic Protection Part 1: Code of practice for land and marine applications* (replaced by multiple EN standards).
6. EN 13174 *Cathodic Protection of Harbour Installations* 2001 (Re-issued as an ISO in 2012).
7. EN 12474 *Cathodic protection of submarine pipelines* 2001. (Still current at the time of writing, but effectively superseded by ISO 15589-2).
8. DNV-RP-F112 *Design of duplex stainless steel subsea equipment exposed to cathodic protection* 2008. (The 2017 edition has dropped "*cathodic protection*" from its title).

Chapter 1

The marine corrosion of steel

A lawyer calls

Some time ago, a lawyer called me and asked: does seawater corrode steel? That simple, and some would say obvious, question led to my being formally instructed to provide an expert opinion to a court. My unsurprising opinion was: yes, seawater does indeed corrode steel.

I am now free to admit that my opinion was as useless as it was obvious. The relevant question should not have been: *is seawater corrosive to steel*? It should have been: *how corrosive is it*? This question could usefully have been followed up with: *what forms does the corrosion take? What corrosion rates can we expect? Do these rates vary in different locations and circumstances*? Even if these were answered, there would remain an even more important unasked question: *does the corrosion actually matter?*

We explore these questions in this chapter. Then, in the rest of this book, we will look at how we can control or manage this corrosion using cathodic protection (CP).

1.1 THE CORROSION OF STEEL IN SEAWATER

1.1.1 How much do we know?

Steel is by far the most commonly used metal employed for construction, and a good deal of that construction finds its way into the sea which covers about 70% of the planet's surface. On that basis, we might expect the corrosion of steel in seawater to have been studied very extensively indeed. In reality, there has been less research in the area than we might have imagined. This apparent lack of study does not signify an absence of interest. Rather, since it has always been well known that seawater rusts steel, the relevant lines of research have been directed at technologies to protect steel, or else at finding alternative alloys that resist marine corrosion.

1.1.2 Why does steel corrode?

About three and a half millennia ago, in what is now Turkey, the Bronze Age gave way to the Iron Age. That was pivotal for the subsequent destiny of mankind in general. On a less imposing scale, early in the 21st century, it prompted this book.

Since the dawn of the Iron Age, humanity has known that iron and steel corrode. The concept of rust is embedded in popular culture. The Bible tells us ... *lay not up for yourselves treasures upon earth, where moth and rust doth corrupt* [2]; and Shakespeare's Othello warns Brabantio and Roderigo to... *keep up your bright swords, for the dew will rust them* [1]. Somewhat less archaically, Neil Young and Crazy Horse entitled their 1979 album... *Rust Never Sleeps.*

DOI: 10.1201/9781003216070-1

The reason that rusting of iron and steel, which is an example of corrosion, occurs emerges from the fact that iron, the principal ingredient of steel, is found in the earth's crust in the form of chemical compounds. These compounds, referred to as ores, include: oxides, sulfides, carbonates and so on. Extracting the metal from its ore, and converting it to a useful engineering material, requires a considerable energy input. This is clear to anyone who has spent time close to an operating blast furnace.

Here we touch on thermodynamics, a scientific discipline to which we return in Chapter 4. The laws of thermodynamics essentially tell us that if we push Nature, then Nature will push back. If we heat up water, Nature cools it down again. The same laws say that where we have put energy into extracting a metal from its ore, Nature will release that energy by converting the extracted metal into chemical compounds resembling the ores from which it was extracted. This process of spontaneous chemical change of the metal is corrosion. We may regard it as completing the cycle that began when the metal was first extracted from its ore (Figure 1.1).

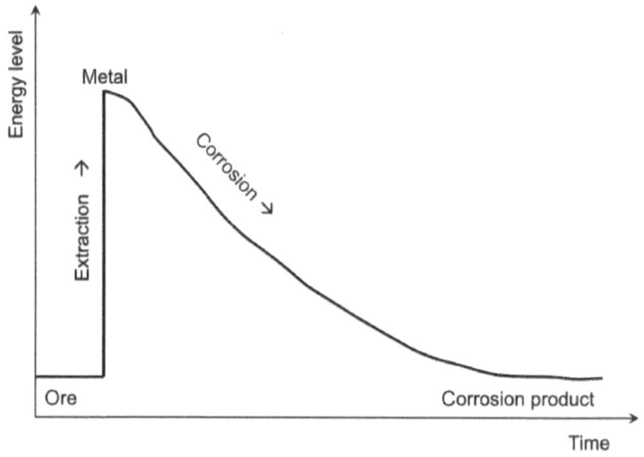

Figure 1.1 The corrosion cycle.

Practically, all of the metals relevant to engineering will corrode, sooner or later, in natural environments. The exceptions are the so-called "noble" metals, of which gold is the best-known example. When prospectors pan for gold in stream sediments, they spot it because of its metallic lustre. Unlike most metals, it occurs naturally in the metallic state, not as an ore. Since it has never needed to be won from its ore by the input of energy, it does not naturally revert to a chemical compound. It does not corrode.

1.1.3 How does corrosion happen?

1.1.3.1 A definition

Before we talk about how corrosion occurs, we need a working definition of the process. There are plenty of candidates from which to choose. My dictionary defines the noun "corrosion" as… *the act or process of eating or wasting away*. It further explains that the word derives from the Latin verb "rodere" meaning to gnaw. This Latin root also gives us the word *rodent*. The prefix "*cor*" adds the sense that this gnawing away is intensive.

Descriptive as this dictionary definition may be, it does not tell us why corrosion occurs. For that, we would be better to use a more scientific definition: *corrosion is the passage of a metal into a chemically combined state.*

This description is more satisfactory because it tells us that we are dealing with a chemical process that converts a metal, with all its desirable mechanical properties, into a chemical compound, with mechanical attributes little better than a soggy biscuit.

1.1.3.2 A school corrosion experiment

I recall a simple experiment that was taught in my school chemistry class. An iron nail was placed in each of two test tubes. Untreated tap water was added to one tube, which was then left open to the air. The second tube was filled with water that had been boiled to expel the dissolved air. That tube was promptly stoppered with a rubber bung. The result of this experiment was apparent when we returned to the lab the following week. The nail in the open tube had turned reddish brown due to the formation of rust. The nail in the stoppered tube remained unchanged – at least as far as we could see.

The conclusion from this experiment was that the rusting of iron requires not only water but also the air dissolved in that water.

The science teacher then explained that the rusting was a chemical reaction between iron and oxygen that took place in the water. We would then have copied equation 1.1 into our notebooks. This shows that the iron, given the chemical symbol Fe (from the Latin *ferrum*), reacts with water (H_2O) and oxygen (O_2) to form ferrous hydroxide ($Fe(OH)_2$). In other words, the iron passes into the chemically combined state, and so, by our definition, it corrodes.

$$2Fe + 2H_2O + O_2 \rightarrow 2Fe(OH)_2 \tag{1.1}$$

A complex series of further reactions will then take place whereby the essentially colourless, gelatinous ferrous hydroxide is converted into the complex mixture of reddish-brown hydrated oxides that we recognise as rust. However, these subsequent reactions have no bearing on the essential fact that our piece of iron has passed into the chemically combined state, and its mechanical value has been lost.

Equation 1.1 is a statement of chemical balance. It tells us what we start with (the reactants: iron, water and air), and the reaction product we end with (ferrous hydroxide). Crucially, it tells us nothing about how the reaction proceeds from reactants to product. To address this question, we need to stray from the domain of the chemist into that of the electrochemist.

1.1.3.3 Some electrochemistry

1.1.3.3.1 Why do we need this?

Electrochemistry is often regarded as a slightly weird subject, described using obscure words and symbols, many of which have their origins in ancient Greek. Very little of the subject is included in a typical school chemistry syllabus. For many engineers involved in CP, the schoolroom was their last contact with chemistry. Unsurprisingly, therefore, many seasoned CP engineers keep their heads down when electrochemical jargon finds its way into the discussion.

This is a pity. CP is an application of electrochemistry. So, we really do need to know at least a little bit about the subject. That is not to say, of course, that we can only work in CP if we possess an advanced degree in electrochemistry. However, a modicum of electrochemical nous is occasionally useful. With that in mind, in this book we introduce some very basic electrochemistry. We then up the level from time to time where a deeper understanding is useful.

1.1.3.3.2 Electrochemical half-reactions

To understand the corrosion process, it is helpful to examine separately the changes taking place on the surface of the metal, and in the adjacent environment. We can do this by breaking equation 1.1 down into two half-reactions: one for the metal and the other for the environment. Equation 1.2 illustrates iron atoms being converted into electrically charged atoms called ferrous ions (Fe^{2+}). Unlike iron atoms, ferrous ions are soluble in water. This conversion of the metal to its soluble ion releases two electrons (e^-) from each iron atom dissolved. The iron changes from a valence state of zero in the metal to a valence state of +2 as the ferrous ion. Chemists call such an increase in positive valence: *oxidation*.

However, because the process involves the release of electrons, the electrochemist claims it and applies the electrochemical term: *anodic*.

$$2Fe \rightarrow 2Fe^{2+} + 4e^- \tag{1.2}$$

The half-reaction undergone by the environment is indicated in equation 1.3. The water and oxygen combine with electrons to form hydroxyl ions (OH^-). This represents a decrease in positive valence and is termed *reduction* by chemists. Again, because this reaction involves the consumption of electrons, it is claimed by the electrochemist who gives it the term: *cathodic*.

$$2H_2O + O_2 + 4e^- \rightarrow 4OH^- \tag{1.3}$$

We can see that if we add the anodic and cathodic equations, and cancel out the electrons common to both, we return to equation 1.1. This is a roundabout way of saying that the rusting of iron is a chemical process that proceeds via an electrochemical mechanism.

Incidentally, there is an important philosophical point that needs to be made here. Just because we can conjure up a pair of electrochemical equations does not actually prove that corrosion is an electrochemical process. That proof lies in a body of 19th- and 20th-century scientific research that is too extensive to describe here.

Before moving on, we should mention that the words anodic and cathodic derive from the two words "anode" and "cathode". These words were coined, from the ancient Greek for "ascent" and "descent", by the Cambridge priest and polymath William Whewell[1] in 1834. He did so at the request of Michael Faraday, whom we shall meet again in this book. Whewell also introduced other electrochemical terms such as "ion", "dielectric" and "electrode" to the English language.

1.1.3.3.3 The corrosion cell

On a naturally corroding metal, the anodes and cathodes coexist on the surface forming innumerable submicroscopic corrosion cells. These cells are not fixed in space or time. At any one instant, one location might act as an anode. For example, this could be a submicroscopic site where the orientation of the metal's crystalline structure leaves an atom standing proud from the adjacent surface. This atom is more readily dissolved than neighbouring surface atoms that are more firmly bound into the lattice. At some other location, perhaps where an intermetallic particle protrudes at the surface, the conditions favour the cathodic reaction. When steel is corroding in seawater, the situation is usually highly dynamic, with

[1] Apparently, he pronounced his surname "hewell", the first "W" being silent.

individual locations, separated by submicroscopic distances: sometimes being anodic in character, and sometimes cathodic.

This process can be exceedingly rapid when viewed on an atomic scale. Consider a piece of steel with an area of a postage stamp corroding in natural seawater. As we will see later, this might be corroding at a rate of about 0.1 mm/year. This might seem undramatic. However, it would be losing about ten trillion atoms every second by anodic reactions on its surface. On the same surface, about five trillion oxygen molecules will be reduced every second by the corresponding cathodic reaction.

A highly stylised depiction of this situation is the corrosion cell shown in Figure 1.2. Oxygen and water are reduced to hydroxyl ions (OH^-) on that part of the surface acting, at that instant, as a cathode. The process consumes electrons liberated by the anodic oxidation of iron atoms at adjacent anodic sites. The current in the metal is carried by electrons flowing from anodes to cathodes. Metals are very good electronic conductors. This electronic current is balanced by the flow of current in the electrolyte (seawater). Unlike in metals, electrons do not exist freely in aqueous solutions such as seawater. Here, this balancing current is carried by ions; and so is referred to as the ionic, or electrolytic, current. Negatively charged ions, termed anions (e.g., Cl^- and OH^-), migrate towards the anodes. Positively charged ions, termed cations (e.g., Na^+ and H^+), migrate towards the cathodes.

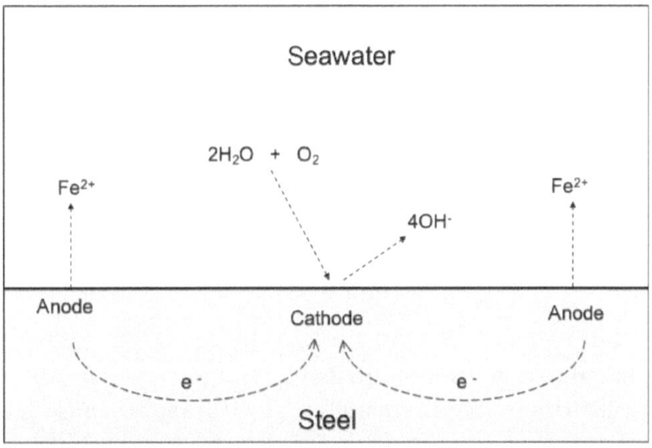

Figure 1.2 Idealised corrosion cell.

In the example of iron (or steel) freely corroding in seawater, there are innumerable anode and cathode sites on the metal surface at any instant. These sites are separated by submicroscopic distances. In other words, a corroding metal is a "mixed" electrode supporting both anodic and cathodic reactions.

1.1.3.3.4 The corrosion current

As mentioned, the issue of interest is not whether steel will corrode. That much is inevitable. What matters to us is: *how fast it will corrode.*

Corrosion rates are usually expressed in units relevant to engineering structures, such as mm/year. However, the existence of corrosion cells opens up an alternative electrochemical way of measuring corrosion. For every iron atom that passes into the chemically combined state, two electrons are released (equation 1.2). This means that a rate of dissolution of iron atoms can be expressed as a rate of passage of electrons: in other words, as an electric current.

It follows that we can relate the quantity of iron dissolved to the quantity of electric charge. This is Michael Faraday's first law of electrolysis, which states: *the mass of any substance deposited or liberated at any electrode is directly proportional to the quantity of electricity passed.*

His second law tells us that the mass of an element separated by the passage of the electric current is directly proportional to its atomic mass divided by a (usually small) integer, which turns out to be the number of electrons released per atom.

$$M = \frac{ZQ}{nF} \tag{1.4}$$

where

M = mass of metal dissolved (gramme-equivalents)
Z = atomic weight of the metal
Q = quantity of electric charge passed (C)[2]
n = number of electrons per atom oxidised
F = Faraday's constant (96487 C/gramme-equivalent)

Simplifying equation 1.2 to 1.5, we see that $n = 2$. Since iron has an atomic mass of 55.85, the passage of 96487 C of electricity is equivalent to the dissolution of about 27.9 g of iron.

$$Fe \rightarrow Fe^{2+} + 2e^{-} \tag{1.5}$$

This means that, in addition to expressing a corrosion rate in terms of thickness loss, as understood by the engineer, we can also express it in terms of an electric current: a quantity more familiar to the electrochemist.

The two expressions are directly related by Faraday's law of electrochemical equivalence. Let us assume that we have a piece of iron corroding at 0.1 mm/year. If we consider an area of 1 m^2:

- the volume of metal lost in a year is $10^{-4} m^3$,
- since iron has a density of (approximately) $7.9 \times 10^3 kg/m^3$, the loss of mass is 0.79 kg,
- applying Faraday's second law (equation 1.4) tells us that for 0.79 kg of iron to dissolve requires the passage of $790\ g \div (27.9\ g \times 96487\ C) = 2.73 \times 10^6 C$,
- then dividing this figure by the number of seconds in a year ($\sim 31.56 \times 10^6$) tells us that the current is ~0.087 A.

Chemistry is a subject that deals with elements. So, in this book, we discuss the chemistry, and electrochemistry, of iron. Engineers, on the other hand, do not construct with pure iron which is relatively soft. They use stronger materials comprising iron alloyed with small amounts of carbon and various other elements. These materials are known collectively as steels or carbon steels. Fortunately for our discussions, the corrosion behaviour of steels with a low alloying content is very similar to that of iron. This means that the above result for Faraday's law calculation for iron applies, to a good enough approximation, to steel.

Obviously, the rate of mass loss from the surface, and therefore, the magnitude of the current, is directly proportional to the area of metal exposed. This means that a corrosion rate of 0.1 mm/year for iron (or steel) is equivalent to a current of 0.087 A/m^2 of surface. We refer

[2] The SI unit for electrical charge is the coulomb C. It is equivalent to the charge passed when 1 amp flows for 1 second (1 C = 1 As). Many workers use the As, partly to avoid confusing C with Celcius.

to the quantity of current per unit area as the current density. It is conventionally given the symbol "i". The value that is equivalent to the corrosion rate is termed the corrosion current density (i_{corr}). We explore i_{corr} in more detail in Chapter 5.

1.1.3.3.5 Potential and voltage

Even the most basic discussion about electricity will involve both current and voltage, and the relationship between the two. The same applies to electrochemistry. Having talked about the relevance of current, we must now introduce the topic of voltage. This parameter, which is more usually referred to as "potential" in electrochemistry and corrosion science, is of pivotal importance in the theory and application of CP. For this reason, we devote a considerable amount of time to the meaning and the measurement of potential at various points in this book. At this early stage, we can ease ourselves in gently. All we need to appreciate for now is that a potential difference exists between a metal such as steel and the seawater in which it immersed. Furthermore, we can make a measurement of this potential difference (or voltage) using a reference half-cell and a voltmeter as illustrated in Figure 1.3.

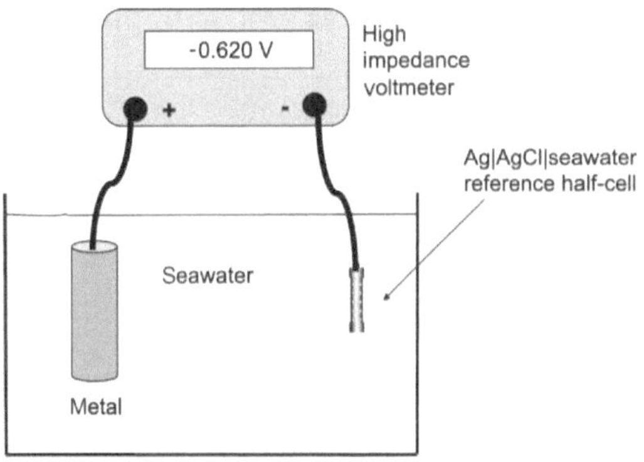

Figure 1.3 Potential measurement.

In this diagram, we see a piece of steel connected to the positive terminal of the voltmeter. The negative terminal is connected to a reference. In this illustration, it is a silver–silver chloride–seawater half-cell.[3] Elsewhere in this book, we write it using the IUPAC[4] terminology "Ag|AgCl|seawater". This is the type of reference most commonly used in marine CP. We will deal with its workings in Chapter 18. For now, all we need to do is understand that it exhibits a fixed potential with respect to the seawater. This means that we can use it as a reference against which we can measure the (unknown and variable) potential of the piece of metal (for example, steel).

In our illustration, the potential of the steel is 620 mV more negative than the Ag|AgCl|seawater reference: hence the "–" sign on the voltmeter display.

The potential we measure is not fixed. Importantly, it varies from metal to metal. As a generalisation, metals that are more corrosion resistant will have more positive potentials. That is, their measured potentials will usually be smaller negative values than that of steel. For any given metal-environment combination, the potential will vary over time, reflecting the progress of the various electrochemical corrosion processes taking place on the surface.

We will have much more to say about potential as this book unfolds.

[3] Frequently, but incorrectly, referred to as a "silver chloride electrode".

[4] IUPAC: International Union of Pure and Applied Chemistry.

1.1.3.4 Aerated seawater

The corrosion current density (i_{corr}) cannot be measured directly. There is no way of inserting the two terminals of an ammeter between anodes and cathodes separated by submicroscopic distances on the surface of a corroding piece of steel.

Nevertheless, there are various electrochemical techniques, collectively termed polarisation experiments, which examine the relationship between current and potential in electrochemical systems. We can then apply electrochemical theory to deduce, at least approximately, the value of i_{corr} quickly and conveniently. This has the benefit that corrosion rates can be determined in the laboratory without the need to carry out time-consuming weight loss experiments. We will return to polarisation experiments in Chapter 5.

For example, the simple school science experiment which we described earlier, which shows that oxygen is needed for corrosion, can be explored further. Electrochemical experiments show that the rate of corrosion is directly related to the rate of diffusion of the dissolved oxygen molecules through the environment to the steel surface. This is illustrated by the work of Ashworth [15] who conducted polarisation experiments on rotating steel electrodes in synthetic seawater. Some of his results are reproduced in Table 1.1.

Table 1.1 Corrosion rates predicted from electrochemical experiments

Experiments in seawater 9 mg/L O_2 at 7°C		
	Corrosion rate	
Linear velocity (m/s)	*mA/m²*	*mm/year*
0	102	0.12
0.3	118	0.138
1	153	0.18
4	307	0.36

From [15].

Ashworth's experiments showed that, in still seawater saturated with air at 7°C (9 mg/L dissolved oxygen), the corrosion rate of a clean steel surface, expressed as a current density, is ~102 mA/m². By Faraday's law, this is equivalent to a metal loss rate of ~0.12 mm/year. This figure is in the same ballpark as the range of corrosion rates reported from short-term site testing which we discuss below.

Ashworth's experiments also demonstrate that, as the velocity of the seawater over the steel surface increases, so too does the corrosion rate. This reflects the fact that the rate of diffusion of dissolved oxygen from the bulk solution to the steel surface increases with increases in the fluid velocity.

Furthermore, for any fluid temperature and hydrodynamic flow condition, the rate of oxygen diffusion to the surface, and therefore the rate of corrosion, is directly proportional to the concentration of dissolved oxygen in the seawater. Hence, Ashworth was able to write-in a set of predicted corrosion rates for different levels of dissolved oxygen. Some of these predictions, which only apply for the initial corrosion rate of clean, polished steel are reproduced in Table 1.2.

1.1.3.5 Deaerated seawater

Seawater usually contains dissolved oxygen to a greater or lesser extent. However, there are some circumstances under which seawater becomes fully deaerated, or anaerobic. The most obvious location where we might expect to find deaerated seawater is in seabed sediments.

Table 1.2 Corrosion rates from electrochemical experiments

Predictions in static seawater at 7°C		
	Corrosion rate	
mg/L O_2	*mA/m²*	*mm/year*
6	68	0.08
8	91	0.11
10	114	0.13

From [15].

Usually, a corrosion engineering assessment will ignore the rate of general corrosion in fully deaerated seawater. For example, DNV-RP-B401, of which much more will be said throughout this book, states... *Completely closed water flooded compartments... without access to external air will not normally require CP.* However, that is not to say that the corrosion rate will actually be zero.

It is informative to evaluate even the low corrosion rates anticipated in deaerated seawater. In the absence of dissolved oxygen, the only cathodic process available to drive the corrosion reaction is the reduction of hydrogen ions to form hydrogen gas (equation 1.6).

$$2H^+ + 2e^- \rightarrow H_2 \tag{1.6}$$

This is the familiar process that causes the rapid corrosion of steel and many other metals, in a mineral acid such as hydrochloric acid (HCl). It is not usually considered relevant to steel in deaerated seawater because the pH of natural seawater is typically slightly above 8. This means that the hydrogen ion concentration in seawater is about 10^8–10^9 times lower than in HCl. So, the "acid" corrosion rate should also be proportionally lower.

I do not know of any literature reports dealing with measurements of corrosion rates in oxygen-free seawater. A natural exposure experiment would inevitably take a very long time to produce a measurable weight loss. However, a study by Noor and Al-Moubaraki [36], on the corrosion of steel in varying strengths of HCl, affords an opportunity to extrapolate to seawater pH values. The result of this extrapolation is shown in Figure 1.4.

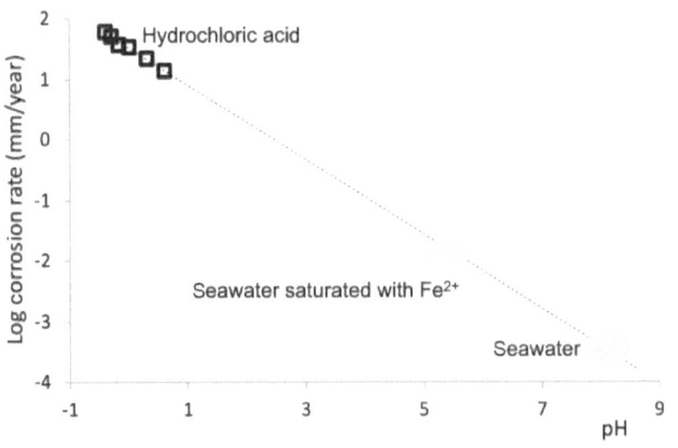

Figure 1.4 Extrapolation of steel corrosion rates in HCl to deaerated seawater.

We must always be careful when extrapolating data on a log-log plot. This is particularly so when the experimental data are bunched at one end, and the extrapolation is required to span numerous orders of magnitude. Any linear extrapolation also adopts the untested assumption that the corrosion mechanism does not change. Nevertheless, if we accept these assumptions and extrapolate the graph to the pH of seawater (about pH 8), the result obtained from Figure 1.4 suggests a corrosion rate for steel of ~0.3 μm/year in deaerated seawater. This seems intuitively reasonable. This equates to only 1 mm of metal loss over a period equivalent to that which separates the present day from the Trojan Wars.

However, the real rate is almost certainly higher. The reason for this is that, although the natural pH of seawater is ~8, the corrosion process produces ferrous ions (equation 1.5). These then enter into equilibrium with water to produce ferrous hydroxide ($Fe(OH)_2$). As can be inferred from equation 1.7, the process produces hydrogen ions (H^+) and therefore lowers the pH; the final equilibrium value of which is determined by the solubility product of ferrous hydroxide and by the pH buffering capacity of the seawater. It has been observed that the pH near the interface of steel corroding in seawater shifts from ~8 to ~5.

$$Fe^{2+} + 2H_2O \rightarrow Fe(OH)_2 + 2H^+ \quad (1.7)$$

Thus, the corrosion process of steel in deaerated seawater is, to some extent, autocatalytic. From Figure 1.4, we see that this shift in pH would be expected to increase the corrosion rate from ~0.3 μm/year by a factor of about 30 to ~0.01 mm/year. This predicted rate of about 1 mm per century is negligible from practically every corrosion engineering perspective.

Unfortunately, however, it is not always possible to ignore corrosion threats in fully deaerated seawater. To understand why, we need to understand microbiological influences on corrosion. We will return to this in Section 1.3. First, we need to examine the limitations of the basic science we have been examining.

1.1.4 What doesn't the basic science tell us?

Our idealised model of the corrosion cell in Figure 1.2 indicates iron atoms oxidising and then disappearing into the environment as ferrous ions. This is useful, but it is also very limited as a description of what goes on when steel corrodes in seawater.

For example, the experimental results summarised in Tables 1.1 or 1.2 relate to clean, polished steel surfaces. However, clean polished steel is never used in offshore engineering. If it were, it would very soon become brown rusty steel.

In a similar vein, Figure 1.2 depicts the metal as a specimen of pure iron. As we have pointed out, we engineer with steels. So far, we have ignored the alloying elements, because the corrosion chemistry of steel is approximately the corrosion chemistry of iron. Nevertheless, we need to consider the extent to which our approximation holds true; and to what extent, alloying elements or impurities in steel are likely to influence its corrosion behaviour.

Likewise, the "seawater" in the corrosion cell is depicted as a conducting electrolyte, the only role of which is to deliver the cathodic reactant (dissolved oxygen) to the surface, and to transport away the dissolved ferrous ions. There is no indication of the roles that the other constituents dissolved in the seawater might play in the corrosion process.

Finally, the concept of a corrosion cell considers only inorganic chemistry and electrochemistry. It pays no heed to the fact that natural seawater is a medium with varying levels of organisms. Rather than a solution of salts, it might more properly be thought of as a biological soup of varied, and largely unknown, ingredients.

Thus, although the basic science gives us a sound explanation of why steel corrodes, and can even point us towards a ballpark figure for the corrosion rate, it falls well short of providing a comprehensive picture of the corrosion process.

If we extrapolate this conceptual image from the atomic scale to real life, we might interpret the simple corrosion cell as indicating that the surface of the steel will simply corrode away uniformly at a constant rate. In practice, however, we are aware that the reality is more subtle. Yes: the corrosion does cause loss of metal from the surface; but, as a matter of experience, it also causes the build-up of corrosion product scales (rust). However, the nature of these corrosion products, and their effect on the subsequent corrosion, depend on a raft of metallurgical, chemical, biological and other environmental factors. Our simple corrosion cell model does not help us here.

What this means, in practice, is that we cannot theorise our way to a corrosion rate, or corrosion pattern, for steel in seawater. In the end, we need to either carry out tests or refer to test work published by others.

1.2 CORROSION RATES

1.2.1 Laboratory tests

1.2.1.1 Weight loss tests

If we want to know how fast steel will corrode in seawater then, in principle, there is a simple laboratory experiment to be done. We take a steel specimen. We weigh it and place it in a sample of real or artificial seawater for a fixed period of time. We then remove it, clean it and then reweigh it. Since we know the surface area of the specimen and the time of exposure, we can determine the corrosion rate in terms of mass per unit area of steel lost. Traditionally, such information was expressed in units of $mg/dm^2/day$ (mdd).

Since we also know the approximate density of steel ($\sim 7.9 \times 10^3 kg/m^3$), we can convert this measured rate of mass loss into an average rate of penetration. This is more usefully expressed in units of mm/year (or µm/year).

1.2.1.2 The importance of the "Blank" weight loss measurements

In the above description of a weight loss test, I mentioned cleaning the specimens. This is not a trivial exercise, particularly where it means removing an adherent corrosion product scale such as rust. Numerous specimen cleaning techniques have been used, ranging from rigorous scrubbing, or even abrading, under running water to chemical cleaning, for example with inhibited hydrochloric acid. These cleaning activities will themselves cause some corrosion of steel. Unless it is properly accounted for, any metal loss during cleaning will lead to an exaggeration of the determined corrosion rate.

Furthermore, the relative effect of this exaggeration is more pronounced if the exposure tests themselves are of short duration, or if the corrosion rate being studied is actually very small. ASTM G1 [49] provides useful information about how to prepare, clean and evaluate corrosion test specimens.

1.2.1.3 Limitations of laboratory testing

There have been numerous such seawater corrosion tests. However, they have invariably been of short duration and often conducted using polished specimens immersed in artificial seawater. It also transpires that many test programmes have inexcusably omitted the relevant blank tests. In the event, the effect of these short-term tests has been to exaggerate the long-term corrosion rates. Furthermore, the majority of the laboratory tests have failed to mimic true seawater immersion conditions, particularly with respect to microbiological activity.

It follows that the most reliable way we can examine the corrosion rate of steel in seawater is either to carry out long-term testing in the sea or to assess the actual corrosion experienced by unprotected steel structures.

1.2.2 Seawater immersion tests

Various test programmes have been carried out since the 1940s. Typically, these have involved pre-weighed specimen panels suspended from rafts or jetties, at relatively shallow depths, in coastal locations. After a predetermined interval, the panels have been retrieved, cleaned and re-weighed. From the measured weight loss, the average general corrosion rate is determined. For example, Chandler and Hudson [18] quote rates of corrosion for nine grades of mild steel panel fully immersed in the sea off Plymouth (UK) for 203 days. The average general penetration rates fell within the range of 0.136–0.158 mm/year. This permitted the conclusion that, for all practical purposes, the rate of corrosion of carbon steel is independent of the material grade.

On the other hand, it was established relatively early that the rate of corrosion does depend on the time of exposure. It is initially high but then decreases with prolonged exposure. Rowlands [19] cites British Admiralty test results which are reproduced in Figure 1.5. These results show quite high initial rates which then settle down to an approximately constant value of ~0.11 mm/year. This emphasises the importance of long-term testing. Rowlands results also align with the work of Blekkenhorst et al. [11] who measured average corrosion rates of 0.11–0.12 mm/year at depths of 45 and 90 m in the North Sea. They also align with more recent work by Grolleau et al. [41] who carried out laboratory immersion tests in water from Cherbourg Harbour and found average corrosion rates of 0.09–0.12 mm/year in 1-year tests under ambient conditions (8°C–19°C).

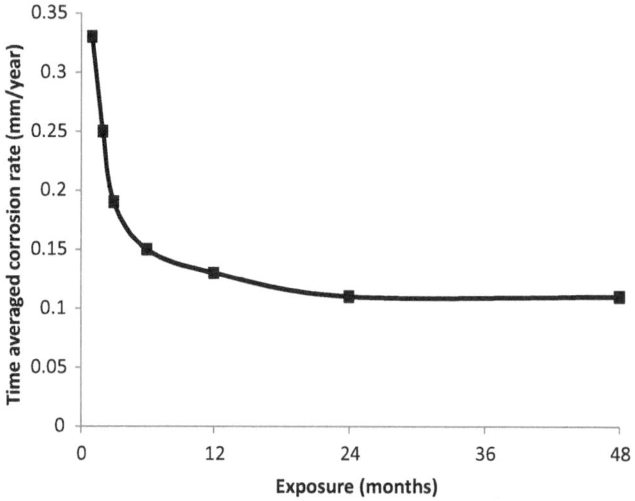

Figure 1.5 British Admiralty data – corrosion of steel in seawater [19].

In 1969 Southwell [6] published the results of the US Navy's long-term (16 years) immersion testing carried out in the Pacific Ocean offshore Panama. Like other such test programmes, the main focus of the work was on the performance of corrosion resistant alloys. However, fortunately for our purposes, carbon steel panels were also included. He reported that the early corrosion rate[5], determined in the first two years, was quite variable; but thereafter it settled down to a near linear value ~0.07 mm/year. He also noted that early pitting penetration was 2.5–5 times deeper than general average corrosion thickness loss. However, by about year 8, all pits were deepening at about the same rate as general corrosion.

The following year, Boyd and Fink [8] of Battelle Columbus Laboratories published a literature review of the corrosion of metals in marine environments. Their review was updated and republished in 1977. They observed that, in panel tests in seawater at widely different marine locations, average rates of penetration for steel... *were found to range from 0.05 to 0.125 mm/year.*

Subsequently, a collaborative international programme of long-term exposure tests in open seawater was carried out under the sponsorship of ASTM. The results were first published in 1988 by Kirk and Pikul [13]. It is worth quoting verbatim from their 1988 report.

> *Although carbon and structural steels corrode over 100% of the surface exposed, the result is not uniform thickness loss....*
>
> *Average corrosion rates for carbon steel in this program were the same after 6 months and 1 year (168 and 172 μm/year) and decreased to 117 μm/year after 3 years... This is the commonly accepted rate throughout the industry for static exposure in relatively quiescent seawater...*
>
> *An increase in the maximum depth of attack with increased exposure time... appears constant with time...*

An extension of the same test programme to 5 years resulted in a further slight decrease in the average corrosion rate from 117 to 107 μm/year [20]. Interestingly, and unusually, the authors of the latter report recorded that they also carried out blank tests to enable them to compensate for any weight loss arising from the cleaning process (pickling in inhibited hydrochloric acid). In that case, because the test panels were quite large, and the duration had been lengthy, the recorded weight loss had been substantial. They made measurements on balances with a discrimination of ±100 mg. Because of this, the report authors were able to report that... *No detectable mass loss was noted on control specimens in any pickling treatment.* However, this does not mean that blank measurements, as described by ASTM G1 are unnecessary, had they been making such measurements of much lower corrosion losses on laboratory balances, with typical resolutions of ±0.1 mg, it is certain that a weight loss due to the cleaning would have been observed.

The corrosion rate results from the ASTM programme broadly agree with the reports of Boyd and Fink, Hudson, and Chandler and Rowlands quoted above. They also agree with the earlier work by LaQue [3] who reported 0.13 mm/y and by Schmitt and Phelps who reported rates falling over time from 0.12 (1.5-year test) to 0.07 mm/year (8.5-year test) [7].

Taken together, the results are in line with the much-used, albeit intentionally conservative, ballpark figure of 0.12 mm/year [15] for static seawater taken from the North Sea. It may also be noted that an EFC/NACE joint publication [44] tells us that... *in quiescent seawater, the general corrosion rate of carbon steel and cast iron is ~0.1* mm/year.

As noted, the values derived from panel weight loss experiments represent average rates of corrosion. Locally, the measured rates of corrosion penetration were found to be much

[5] In common with early corrosion work in the USA, corrosion rates were originally reported in units of mpy (mils per year), where 1 mil is 1/1000th of an inch. Here, and elsewhere, I have been converted the reported units to mm/year.

higher. For example, Table 6 of the Kirk and Pikul report [13] gives the following results of the maximum penetration observed:

Exposure period	*Maximum thickness loss range*
6 months	0.16–0.95 mm
1 year	0.22 mm to complete perforation
3 years	0.92 mm to complete perforation

The panels showing localised perforation through the full 6 mm (¼ inch) steel thickness had been exposed at the Hawaii site. This was the only site where the panels were exposed horizontally, and with... *deposits such as sand and silt settling out of the water and onto the top surface.* Subsequent work by the US Navy [28] also observed that corrosion was more aggressive on horizontally orientated coupons, an observation that was attributed to microbiological influences (see Section 1.3).

Thus, the consensus of all studies carried out in near-surface seawater, irrespective of the geographical location, is that the long-term corrosion rate of carbon steel immersed in seawater falls in the range of 0.10–0.15 mm/year. This is an average which includes the contribution to the weight loss of more rapid, but more localised, pitting.

1.2.2.1 Effect of temperature

The Kirk and Pikul data include tests carried out in 14 locations around the world, spanning a range of seawater temperatures. Fortunately, their report provides sufficient detail to enable us to extract some useful information about the variation of corrosion rate with temperature. For example, Figure 1.6 shows the average corrosion rates reported from all sites after 3 years exposure as a function of mean temperature. The results indicate a weak positive dependence (indicated by the trend line), but any temperature trend is largely obscured by the scatter in the data.

This is consistent with the work of Grolleau et al. [40] who measured the rate of corrosion in laboratory trials in natural seawater. In 1-year exposures, they found that increasing the temperature from ambient (8°C–19°C) to 30°C resulted in the mean corrosion rate increasing from 0.09–0.12 mm/year to 0.15–0.21 mm/year.

These data do not seriously conflict with the widely held view that the corrosion rate of steel in seawater around the world is more uniform than would be expected just from considering the effect of temperature on chemical reaction rates. For example, Figure 1.6 suggests, but does not confirm, that the natural tendency for increasing temperature to increase chemical reaction rates is offset by other factors. As we will discuss elsewhere, these mitigating factors are the reduction in oxygen solubility as the temperature increases and the increased tendency for protective scale formation in warmer waters. Elsewhere, Phull [39] has reported seasonal variations in corrosion rate at a single site. This shows a correlation between corrosion rate and temperature similar to that shown in Figure 1.6. It also confirms that the oxygen content of the water reduces as it becomes warmer.

Melchers [24] has carried out a more detailed analysis of the relationship between corrosion rate and seawater temperature than indicated in Figure 1.6. Whilst acknowledging a high degree of scatter in the data, he observed an increase in corrosion rate with seawater temperature as the temperature increased from 0°C to 10°C. This was followed by a drop in the corrosion rate over a few degrees before the onset of... *an almost linear rise in corrosion with seawater temperature.*

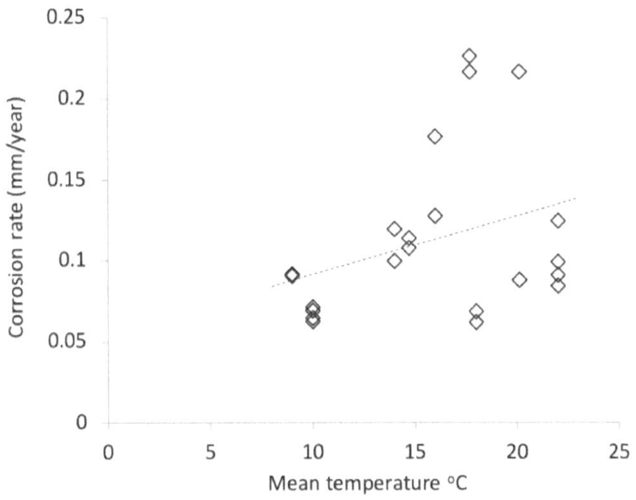

Figure 1.6 Corrosion rate of steel in seawater (3 years immersion) [13].

Recently, Larché et al. [50] have measured low-temperature corrosion rates in seawater at a harbour in the Arctic using electrical resistance probes. These are very sensitive devices in which the metal loss is not measured as a weight reduction, but as in increase in electrical resistance of a thin shim of metal (usually steel). They recorded an "initial" corrosion rate of 0.37 mm/year near the surface and 0.21 mm/year near the seabed. Whilst these rates seem high, it should be appreciated that the duration of the testing was very short (6 days). As has already been demonstrated, corrosion rates of steel in seawater reduce with time.

1.2.2.2 Effect of water depth

Most field studies to measure corrosion rates have been carried out in near-surface coastal waters. Going slightly deeper, Blekkenhorst et al. [11] reported in 1986 that, for two sites in the North Sea, corrosion rates... *at the site of 90 m depth were somewhat lower than those at the 45 m site.* However, even by the time of that testing, offshore hydrocarbon exploration and production was taking place in much deeper waters. The Cognac platform had been installed in 320 m water depth since 1977. It follows that we might usefully consider how the seawater depth affects its corrosivity to carbon steel.

Generally speaking, deeper waters are colder, of lower pH and lower in dissolved oxygen than the near-surface strata. Dexter and Culberson [10] have explained that the world's deep oceans are of a relatively constant temperature (<5°C). This is due to circulation emanating from the poles. The lowering of the dissolved oxygen content with depth is due to the biochemical oxidation of organic matter. This in turn produces CO_2 which, being acidic, lowers the pH. The extent to which these parameters alter with depth is dependent on the geographical location and the seasons. For a North Atlantic test location, the dissolved oxygen content reduces from ~7 to ~5 mg/L over the first 1000 m of depth. For a North Pacific location, of comparable latitude, the reduction is from ~9 to ~1.5 mg/L over the same depth. Interestingly, at both sites, the oxygen levels then increase at depths beyond ~1 km. This is believed to reflect the circulation of cold oxygenated water from polar regions.

In addition to this variation between the Atlantic and the Pacific oceans, there is also a variation within the North Atlantic Area. For example, a NACE study [37], whilst generally supporting Dexter and Culberson's observations of reducing oxygen levels with depth (down to 1000 m), shows higher levels of oxygen in the Norwegian Sea than at another NW Atlantic site. Moreover, compared to other sites, there is comparatively little reduction in the level of dissolved oxygen with depth offshore Norway, with oxygen levels only reducing from ~9 to ~8 mg/L over a depth of 500 m.

It was also found that the pH values dropped in a similar profile. At the Pacific location the pH value fell from ~8.25 at the surface to 7.60 at ~1000 m, while at the North Atlantic test location it fell less dramatically from ~8.25 to ~7.96 over the same depth interval. This emphasises that, in addition to the lateral variation in seawater properties around the world, we also need to take account of variations with depth.

Thus, based on the known reduction in temperature and dissolved oxygen content, we would expect corrosion rates to reduce with increasing depth. The limited available test data support this. For example, Rowlands [19] quotes the results of some work undertaken by US Naval laboratories in the Pacific Ocean in the mid-1960s (see Table 1.3). He does not tell us the duration of the exposure, but the value of 0.127 mm/year for near-surface conditions implies that it was at least 12 months. Based on these few results, it would appear that corrosion rates generally decrease with increasing depth. However, there also appears to be an increase in corrosion rate at very considerable depth, at least at the location where this testing took place.

Table 1.3 Effect of depth on the corrosion rate of steel

Depth (m)	*General corrosion rate (mm/year)*
0	0.127
704	0.043
1600	0.023
1700	0.020
2050	0.058

From [19].

That there has been comparatively little site work on the corrosion rate of carbon steel as a function of depth is not surprising. Such testing is inevitably expensive, and there is little obvious practical reward to be gained. It is universally accepted that carbon steel needs to be protected from corrosion when employed in seawater. The more onerous engineering requirements of deep water will only increase the criticality of those protection measures. Thus, there is limited practical interest in attempting to measure the corrosion rate of unprotected steel.

1.2.3 Information from existing structures

There are two main classes of structure where it is possible to gain further information on the natural corrosion rates of steel in seawater: shipwrecks and harbour piling.

1.2.3.1 Shipwrecks

In principle, shipwrecks offer the opportunity to examine corrosion that has been progressing for many decades, or even longer. For example, Rémazeilles et al. [38] reported on iron ingots found in an ancient shipwreck. These ingots remain recognisable after 2000 years at a depth of 12 m in the Mediterranean Sea. Unfortunately, the fact that a hunk of iron remains "recognisable" does not tell us how much metal has actually corroded away.

This points to a general problem in this respect. Despite shipwrecks adventitiously, and often tragically, constituting long-term corrosion test sites, there has been comparatively little systematic study on the progress of corrosion over the decades they have been on the seabed.

Although we usually have a good idea about the length of time the steelwork has been exposed, it is seldom easy to determine how much metal has been lost by corrosion. This makes the estimation of long-term corrosion rates from shipwrecks very difficult.

1.2.3.1.1 RMS Titanic

Arguably the most famous of all wrecks is that of the Royal Mail Steamer (RMS) *Titanic* which has lain on the North Atlantic sea bed, about 960 km SE of Newfoundland, since 15 April 1912. That she remained identifiable, at least up until 2020, means that a substantial part of her ¾-inch (~19 mm) thick hull plating had remained intact, despite being exposed to seawater from both sides for more than a century. This tells us that, to date, the average rate of corrosion must have been notably less than 0.1 mm/year, although we cannot gauge the contribution the paint coating has made to restraining the corrosion, particularly in the early decades. However, the depth of the wreck of the Titanic (3784 m) is considerably greater than the structures of interest to most CP designers.

In 2010, the website www.livescience.com published a somewhat alarmist article titled: *New Species of Rust-Eating Bacteria Destroying the Titanic.* The article goes on to state that... *the RMS Titanic may soon be lost, thanks to a newly discovered rust-eating bacteria* (sic). This claim arises from the observation of "rusticles" (agglomerations similar to an icicle or stalactite, containing up to 36% iron). The rusticles also contain a newly identified bacterium named *Halomonas titanicae.*

The article, which hyperbolically describes the newly classified microbe as an iron-oxide-munching bacterium, goes on to warn that the wreck will shortly disappear as a result of its rapaciousness. The same theme is pursued by Mann [43] who states... *Unfortunately, due to rusticles consumption, the Titanic cannot be preserved for ever as an underwater heritage site.* A subsequent dive in 2020 has revealed that the disintegration of the wreck is continuing. Reading the articles published online, this seems to have engendered both dismay and surprise.

I have not researched rusticles, so I cannot give an authoritative opinion on this publication. However, we can make the point that bacteria do not physically eat anything. More important, we can also question whether or not the articles' authors have made the classic blunder of confusing correlation with cause.

It is true that there are iron-rich bacteria-infested accretions on the hull. However, as far as I am aware, no one has taken the trouble to determine whether *Halomonas titanicae* is accelerating the corrosion of the steel, or is simply thriving because the natural corrosion enriches the dissolved iron content of the seawater near its surface. These microbes are identified as eating the rust, not causing it. Until, or unless, this is resolved we should keep an open mind about the role of *Halomonas titanicae.*

However, irrespective of whether or not he is correct about the role of rusticles in corrosion, Mann is certainly correct that the *Titanic* will not last forever.

Nothing does.

1.2.3.1.2 Wrecks posing pollution risks

Michel et al. [33] reported on the compilation of a database containing details of 8569[6] potentially polluting wrecks around the world. About 75% of these date from World War II. The authors point out that it is difficult to estimate how much oil is contained in these wrecks, but offer a view that the cumulative figure lies between 2.5×10^6 and 20.4×10^6

[6] The survey excludes tankers of less than 150 tonnes and other vessels of less than 400 tonnes. It also excludes all wrecks known now to be non-polluting.

tonnes. Some of these wrecks have been leaking oil since they were sunk: for example, the battleship *HMS Royal Oak*, which was sunk at Scapa Flow in 1939, and the US Navy oil tanker *USS Mississinewa*, which was sunk in 1944 in Micronesia. Others have yet to leak any oil, as far as we know. The authors point out that, as all these vessels age, there will be... *added concern that corrosion will lead to increased oil discharges.*

The Norwegian Pollution Control Authority (NPCA) has examined shipwrecks in Northern European waters [16]. It has compiled a register of 2100 wrecks of over 100 tonnes gross weight that sank in Norwegian territorial waters between 1914 and 1992. Of these, 30 were considered to present a *considerable* pollution risk. Fifteen of those were given a rough inspection by divers or remote-operated vehicles in 1993. NPCA reported that... *steel plates in the shell bulkheads and tank tops do not appear to have rusted more than 1–3 mm in the course of 50 years.*

Care is required in assessing the outcome of "rough" inspections, particularly since the report does not provide information on how the steel thickness measurements were made. Superficially, the results indicate that the average corrosion rates have been in the range of 0.02–0.06 mm/year.

However, there has apparently been no consideration of the role of the paint systems in providing some degree of corrosion protection, at least during the early years of submersion. It might well be that the progressive deterioration of the paint explains NPCA's comment that... *experience indicates that the speed of corrosion increases with the passage of time:* otherwise, this view contradicts evidence from long-term testing programmes.

An interesting, albeit only qualitative, comment in the NPCA report is that... *wrecks lying near the surface do not seem to have rusted to a greater degree than those lying at a depth of 50–60 m.*

In dealing with the problem of potentially polluting wrecks, the key decision has usually been whether or not it is necessary to carry out the costly, and inherently risky, operation to salvage the oil. This decision is invariably based on an environmental risk assessment which considers the consequences of the oil escaping. Thus, the assessment is driven by the quantity of oil believed to remain in the wreck and the sensitivity of the location. There have been no reported instances where the rate of corrosion of the hull has been considered in the risk assessment.

1.2.3.1.3 USS Arizona

The *USS Arizona* was sunk during the Japanese attack on Pearl Harbor in December 1941. She lies in 9.1 m of water and 7.6 m of mud, and was declared a National Historic Landmark by the US Government in 1989. In view of her heritage status, she has been subject to more intensive study than most wrecks. In 2002, coupons were trepanned from the hull and the steel thickness measured. The remaining steel thicknesses have been referenced to the original thickness values shown on the construction drawings. From this, it was concluded that the average corrosion rate over 60 years has been 0.08 mm/year in seawater, and 0.03 mm/year in the mud [35].

It may be noted that these results are in the same ballpark as the values reported by the NPCA for wrecks in much cooler waters.

Circumspection needs to be exercised when interpreting the *USS Arizona* data in terms of structures at depths elsewhere in the world. Being in shallow, tropical waters she is covered in hard biofouling (termed *concretion*). Elsewhere, structures inevitably become fouled, particularly at depths of less than about 30 m, but this fouling is rarely so dense that it is referred to as *concretion.*

A further difficulty in interpreting the corrosion rates reported for the *USS Arizona* is that there is no explicit consideration of the possibility that areas of the sampled coupons may have suffered corrosion between her launch (1917) and her sinking. Conversely, there is no consideration of the benefit the paint layers, in place in 1941, would have exerted on restraining the onset of corrosion.

1.2.3.1.4 Other World War II wrecks

There are other publications reporting surveys of World War II wrecks. However, they all suffer the problem of presenting no hard data on the rate or extent of the corrosion. For example, Overfield [30] considers six wrecks at various depths in the Gulf of Mexico. Unfortunately, the author's perspective is that of a marine archaeologist, not a corrosionist. Corrosion is assessed solely in terms of the observation of rusticles, which are assumed to be the cause of the deterioration. My view is that the rusticles are non-quantitative confirmation of the obvious reality that corrosion is taking place.

1.2.3.2 Harbour piling

The other potentially plentiful source of corrosion rate data is harbour piling. Historically, harbour piling has been installed without either coating or CP. It has simply been accepted that, providing steel of adequate thickness is employed, even allowing for corrosion, a long service life (up to 120 years) can be anticipated. However, it should be stressed that this habit of using unprotected piling in harbours is coming under scrutiny due to a phenomenon known as accelerated low-water corrosion (ALWC). We return to this in Section 1.3.2.

For now, it is sufficient to note that, in the UK for example, thickness measurements on aged harbour piling were distilled into corrosion rate guidelines (summarised in Table 1.4) in BS 6349-1 [22] in 2000.

Table 1.4 BS 6349-1 (2000) guidelines for UK harbour piling

	Corrosion rate (mm/year per side)	
Zone	*Average*	*Upper limit*
Atmospheric	0.04	0.10
Splash (>MHWS)	0.08	0.17
Tidal (MLWS to MHWS)	0.04	0.10
Intertidal Low Water (LAT-0.5 m to MLWS)	0.08	0.17
Continuous immersion	0.04	0.13
Seabed		0.015

1.2.3.3 Seabed burial

There have been comparatively few studies on the corrosion rates of steel in seabed sediments. Wang et al. [32] have reviewed data published prior to 2004. They refer to 3-year test results where the measured (average) corrosion rate in sand was typically 0.06 mm/year; and in mud, it varied between 0.03 and 0.06 mm/year. The details of the testing are not given in the paper. However, it is reasonable to assume that traditional weight loss experiments were carried out. This means that the figures represent average values, with some areas of the surface being more attacked than others. This would be consistent with the observations of marine corrosion testing more generally. In 2012, Budiea et al. [46] provided a more detailed report, referencing ASTM G1, of similar experiments. In these shorter-term experiments, four steel coupons were buried (~1 m depth) for 6 months at three sites on the Malaysian coast. They estimated average corrosion rates of between ~0.06 and 0.16 mm/year, with only one of the three sites showing rates above 0.1 mm/year. The authors endeavoured to link the higher observed rate by reference to measured pH, temperature and conductivity; but without success. Interestingly, however, they did not consider the microbiological dimension – a topic to which we return in Section 1.3 below.

1.2.3.4 Intertidal and splash zones

This book is about marine CP. In this chapter, we are examining the corrosion of carbon steel in seawater which, as we shall see, can be controlled by CP. It is, therefore, important to recognise at the outset that CP does not work above the waterline. Thus, to complete our review of marine corrosion rates of carbon steel, we need to say only a few words about corrosion in the splash and intertidal zones because CP is ineffective here. If the corrosion rates in these zones are too high, the only option is to change the material from steel to a more corrosion-resistant alloy or to use a high-performance protective coating. In the majority of instances, the latter approach is adopted.

1.2.4 How do we use corrosion rate information?

Although there is no such thing as "the corrosion rate of steel in seawater", there is a range of long-term rates that can be inferred from data available from planned experiments and opportunistic observations of existing structures. Where codes, such as BS 6349-1 [22], offer guideline corrosion rates, these generally reflect the upper-bound values derived from long-term observational data.

From the practical engineering standpoint, the natural rate of corrosion in seawater may, or may not, present a problem. For example, an average overall corrosion rate of 0.1 mm/year only amounts to a 3 mm metal loss (on average) over a 30-year life. In many situations, this could be catered for by the addition of a 3 mm corrosion allowance (CA) to the thickness of the steel plate. On this basis, no additional corrosion control would be warranted. Indeed, this approach is not uncommon. As discussed above, harbour construction often employs unprotected steel sheet piling, the thickness of which implicitly incorporates a CA.

However, it is not always possible, or wise, to rely solely on the CA. In the first place, we have to remember that the corrosion of steel is not uniform. Figures for average corrosion rates subsume higher rates of local penetration that are inevitable when the corrosion pattern is uneven. Whereas a localised increased rate of corrosion penetration might be tolerable on jetty piling, it would be unacceptable on a high-pressure gas pipeline.

Similarly, loss of steel thickness is not necessarily the only consideration when it comes to deciding to control the corrosion of a steel structure. Rusting roughens the surface. This would be intolerable on the hull of a ship since the resulting hydrodynamic drag would reduce speed and increase fuel consumption.

It also has to be borne in mind that metal loss, either uniform or localised, is not necessarily the life-determining degradation mechanism for a marine installation. Very often, that turns out to be fatigue or corrosion fatigue. We return to this in Chapter 12.

1.3 THE MICROBIOLOGICAL DIMENSION

From the corrosion point of view, it is tempting to think of the sea as a large body of salty water containing sufficient dissolved oxygen to give us something of a corrosion problem. Fortunately, for the existence of life on this planet, there is a lot more to the story than that.

The oceans and their contiguous seas are alive with organisms ranging in size from microscopic plankton to blue whales. They also contain the organic by-products of this teeming inventory of life. So, rather than thinking of seawater as just a salt solution, corrosionists would perhaps do better to think of it as a biologically active soup. This is exemplified by the observation that just about any surface immersed in seawater for more than an hour becomes fully covered in an (initially invisible) biofilm.

These biofilms include the cells of living marine micro-organisms. The cells are frequently embedded in a self-produced polymeric substance, sometimes (incorrectly) described as slime. Over time, the biofilm will come to include not only the growing cells but also their metabolic by-products and the decaying matter of dead organisms.

As with any real-world biological experiment, the biofilm is both complex and variable. Its nature and constituent populations of micro-organisms are neither predictable nor controllable. This biological dimension of seawater has to be factored into any consideration of corrosion, and into our thoughts about cathodic protection. This is a tall order.

1.3.1 Clean seawater

The corrosion rate data from shipwrecks, harbour piling and long-term seawater immersion testing must have included any local contributions due to microbiological activity. Given that the reported rates are generally not particularly high, we can conclude that marine micro-organisms do not exert a dramatic effect on the overall corrosion of steel under normal conditions in seawater.

This trite statement is amplified by Melchers who, in a series of publications with various co-workers [25,29,45,47,48], has reviewed published corrosion rate data from test programmes. He has also analysed opportunistic data from some steel structures that have been exposed to seawater without any protection in the form of coatings or CP. The outcome of this work is a proposed model of the seawater corrosion of steel that follows the pattern shown diagrammatically as a cumulative corrosion versus time curve in Figure 1.7. The curve comprises four stages.

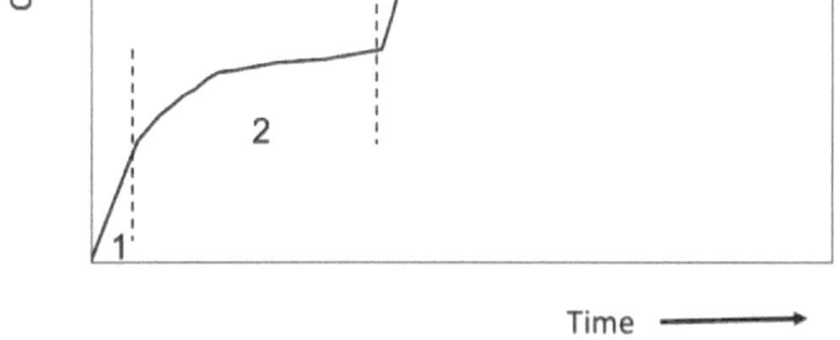

Figure 1.7 Model of marine corrosion of steel. (Adapted from Melchers [25].)

Stage 1 shows an initial linear relationship between the corrosion loss and time. During this stage, the rate of corrosion is governed by the rate of diffusion of dissolved oxygen to the steel surface. Initially, the corrosion rate is high. (See, for example, Ashworth's experiments on clean steel described in Section 1.1.3.4.) This linear relationship between corrosion and time is relatively short-lived. Over time, the corrosion rate reduces.

Stage 2 shows this reduction. The shape of the graph appears parabolic. This is due to the accumulated adherent corrosion product limiting the rate of diffusion of dissolved oxygen. As the corrosion product scale thickens, the corrosion rate reduces further. However, it never reduces to zero. Rather, it approaches the very low rates expected in fully deaerated seawater that we discussed above. If the testing were carried out in synthetic (i.e. non-biological) seawater, then this attenuation of the corrosion rate would persist, and the long-term corrosion rate would be very low indeed. However, in real seawater, the rate of corrosion kicks off again.

Stage 3 shows a sudden increase in the rate of corrosion. This is due to the effects of micro-organisms on the corrosion process: a phenomenon that falls under the heading of microbiologically influenced corrosion (MIC). We explore this topic further below. For a while, the rate of corrosion is even higher than that of freshly immersed clean steel.

Stage 4 shows that, ultimately, the very rapid rate of MIC settles down to a near steady-state value. During this stage, the corrosion rate is controlled, in part at least, by the rate at which nutrients for the micro-organisms can diffuse through the corrosion product scale.

It needs to be stressed that this model represents an average, and that individual data points are highly scattered due to the variability of the seawater's physical, chemical and biological parameters. Notwithstanding this scatter of the data, however, the model does appear reasonable. Significantly, even though there is a pronounced contribution from MIC throughout the exposure life, the overall long-term rate of corrosion is always less than the initial value determined on clean steel in short-term tests. Furthermore, indirect support for the Melchers model is provided by Grolleau et al. [41] who observed that chlorination of the seawater reduced the average corrosion rate, and the maximum pitting rate, compared with tests carried out in unchlorinated natural seawater. The role of chlorination is to restrict the activity of micro-organisms.

1.3.2 Slightly polluted seawater

Generally speaking, around the developed world, the quality of the water in coastal harbours has improved over recent decades as tighter controls have been brought in regarding discharges from berthed ships and effluent release into local water courses. Nevertheless, coastal seawater is often more nutrient-rich than pelagic zones. This is usually due to agricultural run-off or incomplete treatment of discharged sewage. Thus, in many parts of the world, the seawater in harbours, although free of gross pollution, contains higher concentrations of nutrients for micro-organisms than further offshore. There appears to be a correlation, and probably a causal relationship, between the development of these microbiologically fertile conditions and the emergence of ALWC since the mid- to late 1980s. Such has been the upsurge in interest in this "new" corrosion phenomenon that the abbreviation ALWC soon became firmly established in the lexicon of corrosion engineers and many civil engineers responsible for harbour installations.

There is some evidence that ALWC was first identified in the 1960s. At the time it was described as "concentrated" corrosion. However, it was not firmly documented in UK waters until the 1980s. In passing, it also merits mentioning that Humble [4] reported relatively high levels of corrosion around the mean low water level as long ago as 1949. This suggests that ALWC might not be as recent a phenomenon as some workers claim.

It has been found on the seaward faces of unprotected sheet steel piling in harbours. Surveys show that, where ALWC arises, it is typically in a band about 0.5 m deep centred at an elevation about 0.3 m above the lowest astronomical tide (LAT). The phenomenon appears as deep pits clustered in patches which are often about 0.15 m wide.

Maximum annual penetration rates are stated to be in the order of 0.4–1.0 mm/year. This is several times the upper rate advised by BS 6349-1 for that zone (refer to Table 1.4). Piles 12 mm thick have been found to have perforated within 20 years of installation. Gubner and Beech [21], who suggest that the phenomenon only became newsworthy after about 1984, cite a 16-year survey of 71 structures in Japan in which the maximum corrosion rate was 0.93 mm/year; adding that... *the highest corrosion rates were reported at, or just below, the mean low water level.* This maximum rate is echoed by Christie [23], who dates the onset of ALWC issues in the UK from the early 1980s. He gives a figure of 1 mm/year. Adopting a civil engineering perspective, he points out that 10 mm thick piling plate, which should have a nominal life of 50 years, would have its life reduced to about 10 years.

Perhaps the most comprehensive study of the rates of ALWC derives from a 30-year study of harbour piling carried out by the US military [31]. This describes the appearance of ALWC... *bright orange patches overlying a black sludge.* The measured long-term corrosion rates were up to 0.8 mm/year; a figure that compares well with the estimates in the 1 mm/year ballpark obtained from shorter test durations.

The bright orange colour is invariably associated with iron corrosion products formed under aerobic conditions, whereas the black sludge is symptomatic of anaerobic conditions. There is little doubt that ALWC is a form of MIC, and it further seems that it involves some form of "cooperation" between aerobic micro-organisms in the outer (orange) layer and anaerobic micro-organisms in the inner layer. Beyond that, however, there has been no successful mechanistic investigation. This is not surprising since, to date, it has not been reproduced in the laboratory.

A 2005 CIRIA guideline [34], intended for harbour owners, operators and maintenance engineers, advises that ALWC can be effectively controlled by coatings and/or CP. It also advises that... *in the absence of an accurate predictive model, it is prudent to assume that ALWC will occur at all locations.* The latter advice may, however, be conservative since ALWC has not been found, for example, in the Baltic Sea [42].

1.3.3 Heavily polluted seawater and sediments

To a first approximation, the corrosion situation in unpolluted seabed mud is the same as that in deaerated seawater. So, we might reasonably expect corrosion rates of no more than 0.01 mm/year. However, we have already seen that there is a microbiological influence on corrosion in natural seawater (see Figure 1.7). As we have also observed, this influence can become more pronounced in the case of nutrient-rich seawater in harbours, giving rise to ALWC. Furthermore, this MIC can be even more pronounced in seawater that is so polluted that it is deaerated. Examples of such situations include: underneath the sludge that accumulates in ships' ballast tanks and in polluted seabed sediments. Under these circumstances, the dominant corrosion threat is not the low rate of general corrosion associated with deaerated seawater. It is the relatively high rate of localised pitting due to MIC. In polluted seawater environments, this MIC invariably involves sulfate reducing organisms.

1.3.3.1 Sulfate reducing micro-organisms

For decades, the sulfate reducing species involved in MIC were taken to be sulfate reducing *bacteria* (SRB), and the abbreviation SRB has become part of every corrosion engineer's vocabulary. Indeed, there was a tendency among corrosionists to regard SRB as a single type of organism, whereas, in fact, the term encompasses a very large number of species. Furthermore, in recent years, the position has become more complicated because RNA studies have demonstrated that many of these organisms are not bacteria. They are examples of much more primitive life forms classified as *archaea*.

It seems that corrosionists now have to unlearn the abbreviation SRB and replace it with SRP. The "P" stands for prokaryote. I am told by workers in the field that this is the collective term for simple organisms, such as bacteria and archaea, that do not have a cell nucleus.

Irrespective of how we choose to classify these sulfate reducing microbes, they are naturally present in aerated seawater, typically at a concentration of one cell or spore per litre. However, they remain inactive since they are strictly anaerobic. They can only metabolise in the absence of dissolved oxygen. This is a strong indication that they evolved in the vicinity of sulphurous volcanic vents in the earth's primitive oceans, long before there was oxygen in our atmosphere. They may well be direct linear descendants of the very first life forms to appear on the planet.

Although the oceans are generally oxygenated to a greater or lesser degree, circumstances can arise that produce anaerobic conditions. An example is when a volume of seawater stagnates such that it does not receive convective replenishment by oxygenated water, or the photosynthetic species are denied access to light. Under these conditions, anaerobic conditions can develop relatively quickly. Examples of where anaerobic seawater can be found are in seabed sediments and in flooded closed compartments such as ships' seawater ballast tanks. In recent years, the internal spaces of foundations for offshore wind turbine generator towers have also been designed with the intention of developing anaerobic conditions.

Such environments pose a prima facie risk of MIC. However, even the confirmed presence of SRP does not inevitably mean that MIC is in progress. Moreover, even the presence of some MIC does not necessarily constitute a corrosion problem.

1.3.3.2 MIC mechanisms

It needs to be stressed at the outset that, although MIC involves micro-organisms, the actual corrosion mechanism remains electrochemical. It is not biological.

The microbes do not chomp away at the steel in the way cartoon bugs, in adverts for toothpaste, are represented as attacking tooth enamel. Rather, they modify the environment at the steel surface and, in so doing, influence the corrosion. In principle, that influence could be to either stimulate or inhibit the corrosion process.

1.3.3.3 MIC morphology and rates

MIC of carbon steel is highly localised. It manifests itself in the form of pits, the locations of which coincide with the seemingly random locations of microbial colonies containing active SRP. An example from a ship's ballast tank is shown in Figure 1.8.

There is very little attack on the areas surrounding the pits. The practical implication of this is that MIC, if unmitigated, would have profound practical consequences for a pressurised steel pipeline because it presents the threat of loss of containment. However, it might be tolerable in the context of structural steelwork.

Although the risk of MIC always has to be assumed for seabed sediments, the rate of expected corrosion, should it become established, remains open to question. Based on surveys of damage in ships' bilges and, less relevantly for us, inside oil pipelines, a penetration rate of 2 mm/year is reckoned to be an upper bound value in marine environments at ambient temperatures [40]. This aligns reasonably with King's opinion [9] that, in seabed mud... *microbiological activity can increase the corrosion rate of steel by up to 15 times.*

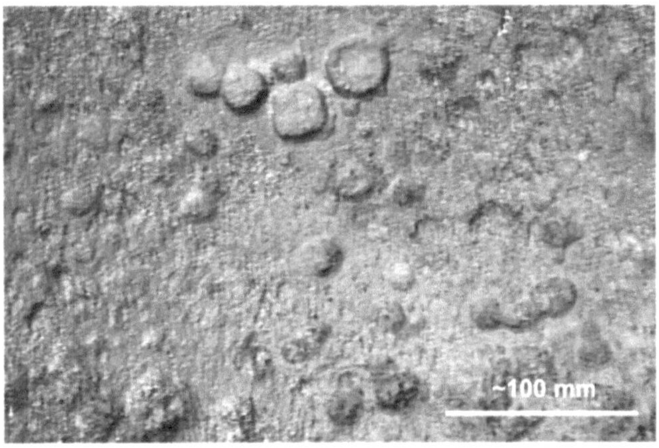

Figure 1.8 Example of MIC pitting in a ship's water ballast tank.

At first sight, a figure of 2 mm/year is at odds with a brief 1966 article by Copenhagen [5] who reported MIC of ship's bottom plate where pits had grown to a depth of 6 mm within 2 years. However, Copenhagen's figures relate to bilge plates near the engine room, so temperatures will have been higher than in seabed sediments.

Nevertheless, there is always a problem with trying to estimate rates of MIC in real structures such as ships. That problem is that we can never be sure when the MIC initiated. In the case of MIC in a ship's ballast tank, such as shown in Figure 1.8, there is a temptation to assume that the newly discovered MIC started soon after the previous class survey, on the basis that it was not reported at that survey. This is the basis of some MIC rates reported in the literature. For example, Campbell et al. [14] claimed rates of 8 mm in 12 months on this basis. However, they also pointed out that... *any corrosion damage is not readily visible due to its localised nature.* To my mind, the latter comment points to the very strong possibility that the developing MIC was missed during the vessel's previous class surveys. Having been involved in some such surveys this would not surprise me. The activities are, after all, surveys. They are not inspections.

It is also worth mentioning that MIC has been the topic of considerable laboratory research, often carried out by enthusiastic PhD students. However, although this work has done a great deal to elucidate the mechanisms that might be involved, the maximum rates of corrosion achieved in controlled laboratory experiments have always been considerably lower than the maximum rates that have been claimed for real structures. By way of example, Beech et al. [17] used electrochemical techniques to measure MIC rates in the laboratory. The maximum figure they obtained was 0.18 mm/year. This was despite the fact that they had previously reported 1.34 mm/year in harbour piling.

In the 1990s, I was involved in three marine insurance claims that were settled on the basis of claimed MIC rates in the order of 10 mm/year. The affected structures were a ship, a jetty and a semi-submersible drilling rig. All were located, or operated predominantly, in tropical locations; and the latter two were heavily fouled. In each case, the insurers paid up, accepting the claimants' reported corrosion rates, and agreeing that they exceeded the *normal wear and tear* that was excluded by the wording of the policies. It has to be admitted, however, that the evidence put forward in support of the ~10 mm/year allegation was, to say the least, flimsy in every case.

Irrespective of the uncertainties involved in obtaining real-field data for MIC rates, we also need to understand that the maximum rate, whatever that might be, is unlikely to be permanently sustainable. It is in the nature of MIC that corrosion initiates and then propagates, sometimes quite rapidly, but then the rate subsides. This may be due to a shortfall of nutrients for the microbial population or to the build-up of their metabolic waste. Under some circumstances, once activity subsides, waste products diffuse away from, and nutrients diffuse back into, the biomass. The corrosion cycle then resumes. This cyclic-type behaviour is ascribed to the occasional observation of a "terraced" appearance in some MIC pits.

1.4 THE FORMS OF CORROSION

1.4.1 General corrosion

Books on corrosion identify various morphologies of corrosion damage. Most agree that the rusting of steel, in seawater or other environments, is a case of general or uniform corrosion. That is, the corrosion damage penetrates the steel more-or-less uniformly over the surface. Beyond that, however, it is hard to find two books that agree on how to describe the various forms that can arise for different metals in different environments and under different circumstances of exposure. For example, one source might categorise MIC as a standalone form of corrosion, another might simply treat it as one of a number of corrosion situations that can result in the morphology of pitting. I am not setting out to rewrite those corrosion textbooks here. Rather, I will simply consider the forms of corrosion that can occur, or which some workers believe can occur, when carbon steel is exposed to seawater.

1.4.2 Galvanic corrosion

1.4.2.1 Classic example

In 1763, the British Admiralty discovered that using iron nails to fix copper sheeting to the hulls of its ships was not a good idea. The first test application was on the frigate *HMS Alarm.* When her hull was examined after two years of service, it was observed that the nails were "much rotted", the sheathing had become detached from the hull in many places, and in some areas it had completely disappeared. The ensuing report concluded that iron should not be allowed in direct contact with copper in seawater [12].

As far as we know, the British Royal Navy put no name to the phenomenon of corrosion between two dissimilar metals that it had unwittingly discovered. While *HMS Alarm* was patrolling in the West Indies, a young Italian named Luigi Galvani began his career as a lecturer of anatomy at Bologna University. Twenty-five years later, he discovered that a (dead) frog's muscle could be made to twitch by touching an iron wire to the muscle, a copper wire to the nerve and then touching the two wires together.

It is for this reason that, since the British Navy had not coined the obvious name (bimetallic corrosion), the type of problem experienced by *HMS Alarm* became known as "galvanic corrosion".

1.4.2.2 Why does it happen?

Our simple corrosion cell in Figure 1.2 implies a featureless interface between the steel and the seawater. By implication, the myriad individual anodic and cathodic sites are unrestricted in their movement across the surface. At one instant, a submicroscopic area might be anodic, and at another, it might be cathodic. According to this model, the overall effect is that iron atoms are lost randomly over the surface, with the cumulative effect that the corrosion appears uniform.

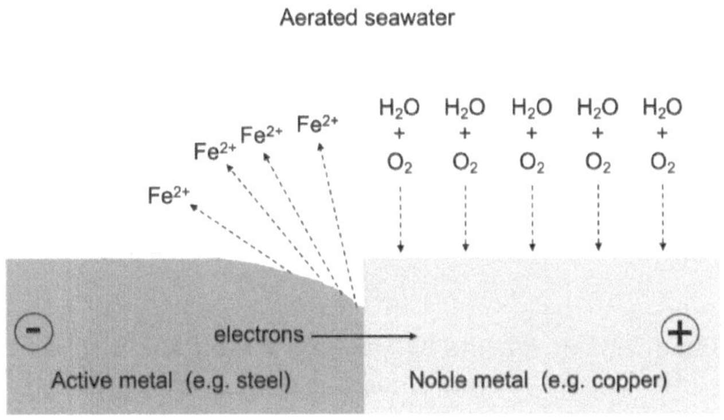

Figure 1.9 Schematic of galvanic corrosion.

However, when we electrically connect carbon steel to another metal such as copper and immerse them both in seawater, this uniform surface distribution of anodes and cathodes no longer occurs. There tends to be a separation, such that the cathodic processes occur predominantly on the metal with the more positive potential (in this case the copper)[7]; and the anodic processes are favoured on the metal with the more negative potential (in this case the steel). Furthermore, not only does this connection of carbon steel to copper, also referred to as "coupling", mean that the corrosion is focussed on the steel. It also leads to the rate of the anodic dissolution reaction on the steel increasing, and that on the copper decreasing. We explain the electrochemical reasons for this in Chapter 7.

The effect of this galvanic, or bimetallic, coupling is illustrated schematically in Figure 1.9.

1.4.2.3 What are the risk factors?

There are many instances in engineering where different metals are coupled together, but most situations do not result in a galvanic corrosion problem.

In the first place, for galvanic corrosion to occur, both metals need to be in electrical contact with each other and immersed in the same continuous body of conducting liquid (seawater in this case). This applied in the case of the copper sheathing and iron nails on *HMS Alarm*, but it would not apply, for example, where different piping materials carrying a non-conducting fluid are coupled together in an above-ground facility.

Second, the environment itself needs to be corrosive. Taking the case illustrated in Figure 1.9 as an example, if there was no oxygen dissolved in the seawater, there would be virtually no corrosion and no galvanic corrosion problem. This is why domestic central heating systems can use a combination of steel radiators and copper piping. These systems are, or at least should be, sealed. Once the finite quantity of oxygen, present in the water when the system is filled, is consumed by corrosion, subsequent corrosion, galvanic or otherwise, becomes negligible.

Thirdly, the relative areas of anode (steel) and cathode (copper) are crucial. Where the cathode area is relatively small in comparison to that of the anode, any galvanic stimulation of the corrosion is relatively small and is spread over a relatively large area of anode. The corrosion situation may be regarded as "safe". Conversely, where the anode presents a relatively small surface area compared to that of the cathode, the relative galvanic stimulation of the corrosion is enhanced. Furthermore, this stimulated corrosion is focussed on the small area of anode. This makes the rate of corrosion of the anode very high indeed. It is the

[7] In the jargon of corrosion science, the metal with the more positive potential is said to be more "noble".

"unsafe" manifestation of galvanic corrosion. This was the problem encountered by *HMS Alarm*. The surface area of the copper sheathing was very much larger than that of the iron nails, so the nails corroded very rapidly.

Finally, if the coupling of two metals, and the likelihood of some galvanic corrosion, is inevitable for engineering reasons then it is wise to ensure that the more critical component is the cathode. Again, this rule was unknown when *HMS Alarm* put out to sea. The critical components in the wooden hull's protection system were the nails that fixed the sheathing to the hull.

1.4.3 Pitting

There have been some lively arguments between corrosion scientists as to whether or not steel suffers pitting corrosion in seawater. Such a division of opinion is not that surprising when we contrast the classic view, namely that the corrosion of steel in seawater is an example of general corrosion, with the appearance of the corroded surface that we see, for example, as a result of MIC in Figure 1.8.

It is important here to draw a distinction between the language of the corrosion scientist and the corrosion engineer or corrosion inspector. To a corrosion scientist, pitting corrosion can only occur on metals that are normally protected by a passive oxide film. The most common example of these are the stainless steels and aluminium. We deal with these metals in more detail in Chapter 8. For now, we simply need to be aware that, where the passive film breaks down locally, a galvanic cell develops between the small area of metal exposed at the film breakdown site and the majority of the surface which retains a coherent passive film. The metal exposed at the film breakdown is anodic, and the oxide covered surface is cathodic. Galvanic corrosion current will flow between these two areas. Furthermore, as we discussed above, the high cathode-to-anode area ratio means that the anodic dissolution will be very rapid. For this reason, where a passive metal undergoes pitting corrosion, the individual pit propagates into the metal very rapidly in comparison with its rate of lateral growth. Such pits are described as having a high aspect ratio. Figure 1.10 shows an example of pitting that had developed on an unprotected stainless steel ship's propeller boss in static seawater during the period that the ship was under construction.

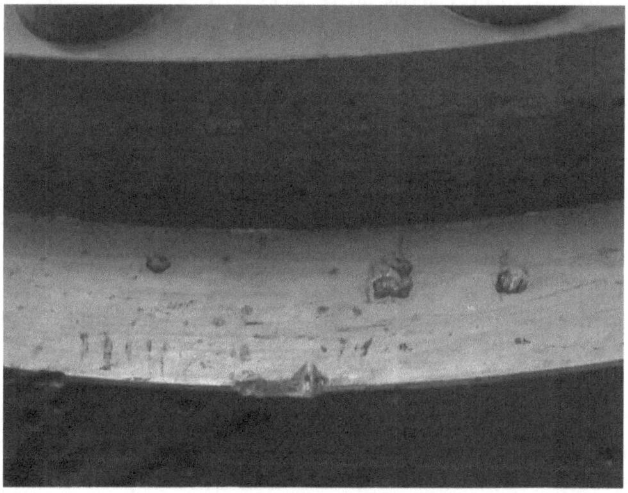

Figure 1.10 Example of a pit in a passive metal.

Carbon steel cannot form a passive film in the chloride-rich seawater environment. It follows that it does not exhibit this type of high aspect ratio pitting, so many corrosion scientists would simply say that it does not undergo pitting corrosion. The reality, however, is more subtle. This is illustrated by Melchers [26,27] who points out that he uses the unqualified term "pitting" to mean deeper local areas of penetration.

For a start, if we take a polished piece of steel, immerse it in seawater for some hours, and then examine the surface under a microscope, we observe small corrosion indentations. We would refer to these as "micro-pits". They may be regarded as a manifestation of galvanic corrosion. As we know, steel is not a pure metal. It is an alloy of iron and carbon plus a number of other minor elements that either have been added intentionally or else are present as impurities. Some of these impurities combine chemically to form inclusion phases in the metal. Where these precipitates are exposed at the polished surface, they will often be cathodic to the continuous iron-based phase. This results in selective dissolution around the inclusion, undermining it until it is lost from the surface, creating the "micro-pit".

Micro-pitting is only of academic interest. Polished steel is not exposed to seawater in any engineering application. Furthermore, micro-pits are self-cancelling. Once the cathodic inclusion is lost, the galvanic impetus for the micro-pit at that location ceases.

Where researchers refer to "pits" that are observable to the naked eye, they invariably mean "macro-pits" which are relatively wide, shallow depressions. That is, they have a low aspect ratio. They reflect the natural variation of scaling and fouling that inevitably occurs during any marine exposure. In the extreme, they can reflect the distribution of accumulations of organisms implicated in MIC (as seen in Figure 1.8).

In summary, carbon steel does not undergo "pitting corrosion" in seawater, but the appearance of the corrosion can often be described, with some justification, as exhibiting shallow pitting.

1.4.4 Crevice corrosion

Crevice corrosion is a phenomenon that affects alloys such as stainless steel or aluminium (we also cover this in Chapter 8). There is no reason to expect carbon steel to undergo crevice corrosion in seawater. Despite this, I have met corrosion engineers who assume that crevice corrosion is possible, or even inevitable, in carbon steel whenever there is a crevice present. The only crevice corrosion testing of carbon steel in seawater of which I am aware was carried out by Blekkenhorst et al. [10]. They reported that unalloyed carbon steel... *did not show any sign of crevice corrosion in any of the exposures.* The exposure durations in question were up to 4 years.

Case closed.

1.4.5 Fatigue and corrosion fatigue

In practice, even if corrosion were not controlled, comparatively few structures would fail because the steel had thinned to the extent that it could no longer sustain the deadweight load. In most instances, failure would come earlier as a result of fatigue or corrosion fatigue. We will say no more about this subject here. It features heavily when we discuss the mechanical aspects of CP in Chapter 12.

1.4.6 Other forms of corrosion

The above does not exhaust all of the forms of corrosion that appear in the basic corrosion textbooks. However, it does cover all of the forms of corrosion that carbon steel might normally encounter in the marine environment, and which we might expect to counteract with CP. There are forms of corrosion that simply do not occur on carbon steel in seawater, so we do not need to worry about them. These include stress corrosion cracking, intergranular corrosion, and selective dealloying. Similarly, there are forms of corrosion that could occur but are rare in well-engineered systems and would be difficult to control by CP in any event. The latter include erosion-corrosion and cavitation.

REFERENCES

1. W. Shakespeare, *Othello* Act 1 Scene 2 (1603).
2. *The Bible*, King James Edition: Matthew 6:19 (1611).
3. F. LaQue, *Corrosion Handbook* (ed. H.H. Uhlig) Wiley (New York) 1948, p. 383.
4. H.A. Humble, *The cathodic protection of steel piling in sea water.* Corrosion **5** (2), 292 (1949).
5. W.J. Copenhagen, *Accelerated corrosion of ship's bottom plate.* British Corrosion Journal **1** (11), 344 (1966).
6. C.R. Southwell, *The corrosion rates of structural metals in sea water, fresh water and tropical atmospheres.* Corrosion Science **9**, 179 (1969).
7. R.J. Schmitt and E.H. Phelps, *Corrosion performance of constructional steels in marine applications.* Journal of Metals 22 (3), 47 (1970).
8. W.K. Boyd and F.W. Fink, *Corrosion of Metals in Marine Environments.* Metal and Ceramics Information Center Report No. MCIC-78-37, 1978.
9. R.A. King, *Prediction of corrosiveness of sea bed sediments.* Materials Performance **19** (1), 39 (1980).
10. S.C. Dexter and C. Culberson, *Global variability of natural sea water.* Materials Performance 19 (9), 16 (1980).
11. F. Blekkenhorst, et al., *Development of high strength low alloy steels for marine applications Part 1: Results of long term exposure tests on commercially available and experimental steels.* British Corrosion Journal **21** (3), 163 (1986).
12. K.R. Trethewey and J. Chamberlain, Chapter 1, in *Corrosion for Students of Science and Engineering.* Longmans (Harlow) 1988. [ISBN 0-582-45089-6].
13. W.W. Kirk and S.J. Pikul, *Seawater corrosivity around the world: results from three years of testing*, in *Corrosion in Natural Waters* (ed. C.H. Baloun) ASTM STP 1086 (1988).
14. S.A. Campbell, R.A. Scannell and F.C. Walsh, *Microbially-Assisted Pitting Corrosion of Ships Hull Plate.* Proc. UK Corrosion/88.
15. V. Ashworth, *Cathodic Protection: Theory and Practice*, Eds. V. Ashworth and C.G. Googan. Ellis Horwood Ltd. (Chichester) 1993.
16. K. Idaas, *Norwegian Pollution Control Authority Work on Shipwrecks* (1993). Downloadable from www.iosc.org.
17. I.B. Beech et al., *Microbially Influenced Corrosion of Carbon Steel in Marine Environments.* Proc. UK Corrosion/93.
18. K.A. Chandler and J.C. Hudson, *Corrosion*, Vol. 1, 3rd edition, Eds. L.L. Shreir, R.A. Jarman and G.T. Burstein, Butterworth–Heinemann (Oxford) 1994, p. 3:15.
19. J.C. Rowlands, *Corrosion*, Vol. 1, 3rd edition. Eds. L.L. Shreir, R.A. Jarman and G.T. Burstein, Butterworth–Heinemann (Oxford) 1994, p. 2.69.
20. B.S. Phull, S.J. Pikul and R.M. Kain, *Seawater corrosivity around the world: results from five years of testing*, in *Corrosion Testing Natural Waters* (ed. M. Kain and W.T. Young) ASTM STP 1300 (1997).

21. R. Gubner and I. Beech, *Statistical Assessment of the Risk of Biocorrosion in Tidal Waters.* Paper 184 CORROSION/99
22. BS 6349-1, *Maritime structures — Part 1: Code of practice for general criteria* (2000).
23. J. Christie, *Concentrated Corrosion on Berths and Jetties – Accelerated Low Water Corrosion,* Proc. UK Corrosion, 2001.
24. R.E. Melchers, *Effect of temperature on the marine immersion corrosion of carbon steels.* Corrosion **58** (9), 768 (2002).
25. R.E. Melchers, *Modeling of marine immersion corrosion for mild and low-alloy steels – part 1: Phenomenological model.* Corrosion **59** (4), 319, 2003.
26. R.E. Melchers, *Pitting corrosion of mild steel in marine immersion environment—Part 1: Maximum pit depth.* Corrosion **60** (9), 824 (2004).
27. R.E. Melchers, *Pitting corrosion of mild steel in marine immersion environment—Part 2: Variability of maximum pit depth.* Corrosion **60** (10), 937 (2004).
28. J.S. Lee, et al., *An evaluation of carbon steel corrosion under stagnant seawater conditions,* Paper 04595 CORROSION/2004.
29. R.E. Melchers, *Effect of nutrient- based pollution on the corrosion of mild steel in immersion conditions.* Corrosion **61** (3), 237 (2005).
30. M.L. Overfield, *Corrosion of Deep Gulf Shipwrecks of World War II.* Proceedings of *International Oil Spill Conference*, 2005.
31. A. Kumar and L.D. Stephenson, *Accelerated Low Water Corrosion of Steel Pilings in Seawater,* Paper 221 CORROSION/2005.
32. X. Wang, et al., *Corrosion of steel structures in sea-bed sediment.* Bulletin of Materials Science **28** (2), 81 (2005).
33. J. Michel, *Potentially Polluting Wrecks in Marine Waters.* 2005 *International Oil Spill* Conference www.iosc.org.
34. J.E. Breakell, et al., *Management of accelerated low water corrosion in maritime structures.* CIRIA Report C634 CIRIA (London) 2005.
35. D.L. Johnson, et al., *Corrosion of steel shipwreck in the marine environment USS Arizona.* Materials Performance **35** (10), 40 (2006).
36. E.A. Noor and A.H. Al-Moubaraki, *Corrosion behaviour of mild steel in hydrochloric acid solutions.* International Journal of Electrochemical Science **3**, 806–818 (2008).
37. Publication 7L19, *Cathodic Protection Design Considerations for Deep Water Structures,* NACE, updated (2009).
38. C. Rémazeilles, et al., *Mechanisms of long-term anaerobic corrosion of iron archaeological artefacts in seawater.* Corrosion Science **62**, 2932 (2009).
39. B. Phull, *Marine Corrosion*, Chapter 18, in *Shreir's Corrosion*, Vol. 2, Eds. J.A. Richardson, et al. Elsevier (Amsterdam) 2010.
40. J.D.F. Stott, *Corrosion in microbial environments*, Chapter 20, in *Shreir's Corrosion*, Vol. 2, Eds. J.A. Richardson, et al. Elsevier (Amsterdam) 2010.
41. A.-M. Grolleau, et al., *Electrochemical Characterization of low carbon steels During long Term Immersion in Natural Seawater,* Paper 10395 Corrosion/2010.
42. H. Wall and L. Wadsö, *Sheet Pile Corrosion in Swedish Harbours – An inventory of Corrosion Surveys along the Swedish Coast*, Paper 4420 Eurocorr 2011.
43. H. Mann, *New Bacterium Species Discovered on RMS Titanic Rusticles*, UNESCO Scientific Colloquium on *Factors Impacting the Underwater Heritage*, Brussels, 2011.
44. R. Francis, in *The Corrosion Performance of Metals in the Marine Environment: A Basic Guide*, p. 5, Eds. C. Powell and R. Francis) EFC/NACE Publication No. 63. Maney Publishing (London) 2012.
45. I.A. Chaves, and R.E. Melchers, *External corrosion of carbon steel pipeline weld zones.* Proceedings 22nd International *Offshore and Polar Engineering Conference* Rhodes, 2012.
46. A.M.A. Budiea, N. Yahaya and N.M. Nor, *Corrosion of API X70 Steel Due to Near Shore Sediment.* International Journal of Civil and Environmental Engineering **12** (3), 84 (2012).

47. I.A. Chaves and R.E. Melchers, *Long term corrosion of marine steel piling welds.* Corrosion Engineering Science and Technology **48** (6), 469 (2013).
48. R.E. Melchers and R.J. Jeffrey, *Long-term corrosion of steels and steel piling in seawaters with elevated nutrient concentration*, Paper 7199 Eurocorr 2014.
49. ASTM G1, *Standard Practice for Preparing, Cleaning, and Evaluating Corrosion Test Specimens* (original publication 1990) 2017.
50. N. Larché, et al., *Cathodic protection in arctic conditions.* Materials Performance **57** (8), 26 (2018).

Chapter 2

Cathodic protection basics

Sir Humphry and the law of unintended consequences

As we mentioned in Chapter 1, the British Royal Navy first used copper sheets to clad wooden hulls in the 1760s. It was discovered in 1763 that the iron nails used to fix the sheets to the hull of *HMS Alarm* became "much rotted". This was solved by ensuring that the iron and copper did not come into direct contact [2]. Nevertheless, there remained the problem that the copper itself still corroded. Seeking a solution, in 1822 the Navy Board turned to the President of the Royal Society: Sir Humphry Davy. A little over a year later, Sir Humphry was standing before members of the Society delivering the results of his investigations [1].

He opened by informing his audience that his findings... *promise to illustrate some obscure parts of electro-chemical science.* He advanced the crucial proposition that, because... *copper is a metal only weakly positive in the electrochemical scale... if it could be rendered slightly negative, the corroding action of seawater upon it would be nul.* This led him to the insightful view that... *a very feeble chemical reaction would be destroyed by a very feeble electrical force.*

Davy's well-attended presentation signalled the start of cathodic protection.

We should bear in mind that Davy was not an electrochemist. That job description was not yet in use. He referred to himself as a natural philosopher. Furthermore, even though he can literally be called the father of cathodic protection, he never used the word "cathodic". As we have already mentioned, the words *anode* and *cathode* were coined by William Whewell in 1834; 5 years after Davy had died in Switzerland.

Nevertheless, despite not being an electrochemist, Davy designed a CP system based on zinc blocks. In the first actual example of what we now know as CP, these were fitted to copper sheathing on the hull of the warship *HMS Samarang*. After a couple of years on patrol, it was observed that there had been very little wastage of the copper. Davy's zinc blocks, which the Royal Navy referred to as "Davy's protectors" well into the 20th century, had evidently done their job.

This makes the point that it is not always necessary to have a detailed understanding of electrochemistry in order to design a functional CP system. We will show how to do this in the next chapter. Here, we will simply provide enough coverage of the fundamental principles to get us going. We will delve more deeply into the science, as the need arises, later in the book.

But, before we begin, we should be aware of an interesting historical footnote to Davy's story. The CP system on *HMS Samarang* worked entirely as intended. The corrosion of the copper was effectively controlled. Unfortunately, however, the exercise was a singular failure from a practical point of view. The main benefit of the copper sheathing was to prevent the growth of marine fouling. That antifouling process actually requires the copper to corrode and produce toxic copper ions which prevent the settlement of marine micro-organisms onto the surface. Because Davy's protectors prevented that corrosion, by the time Samarang

DOI: 10.1201/9781003216070-2

returned to Portsmouth, her hull was heavily fouled. Her speed and fighting capability were badly compromised: a classic example of the Law of Unintended Consequences.

The Admiralty abandoned the use of Davy's protectors. Following its unchanging habit, the British press turned on a former hero. Davy and his invention were ridiculed. He never really recovered from this fall from grace. He subsequently produced little work. Before the decade was out, he was in his grave in Switzerland. Whether or not CP hastened his demise must remain a matter of conjecture.

2.1 A THEORETICAL EXPERIMENT

We discussed the basics of corrosion of steel in seawater in Chapter 1 where we developed the idea that, although corrosion is a chemical process, it proceeds via an electrochemical mechanism. This led us to develop a simple picture of a corrosion cell. There is an obvious implication: the fact that corrosion is electrochemical in nature suggests that it should be controllable by an electrochemical technique.

We can now permit ourselves the luxury of carrying out a "thought experiment" to consider how this electrochemical control might be brought about. We can imagine a situation where we have the capability of interfering with the corrosion process by making the system more electrically positive (or more negative). We can then theorise about what would happen, and we can do this without the inconvenience of having to venture into the laboratory. All that we need to do is consider the established principle that, if we try and impose any change on nature, then nature will react to try to oppose our applied change.

2.1.1 Removing electrons

Let us start by removing electrons from the system. The beauty of doing this as a thought experiment is that, at this stage, we do not need to trouble ourselves over the practicalities of how we actually do this in practice. We simply assume that it can be done and are then content just to consider the consequences.

It has been known since Millikan's Nobel Prize–winning oil drop experiments in 1909, for which he won the Nobel Prize for physics in 1923, that the charge on the electron is negative. It follows that our theoretical removal of electrons amounts to removing negative charges. If we remove negative charge, the steel would tend to accumulate a surplus positive charge.

We can now reintroduce the term potential which we met briefly in Chapter 1. At its simplest, the potential is a parameter that tells us how positive, or negative, the steel is with respect to the electrolyte (in this case seawater). We will have much more to say about potential in later chapters. For now, on the basis of this simplistic interpretation, we can say that removing negatively charged electrons will make the potential of the steel more positive.

We should also recall the anodic electrochemical half-reaction for the corrosion of iron that we introduced in Chapter 1.

$$Fe \rightarrow Fe^{2+} + 2e^- \tag{2.1}$$

So, if we remove electrons (from the right-hand side of the equation), the system will react against our imposed change. It seeks to replace those that are lost. It does this by increasing the rate at which iron atoms are converted into Fe^{2+} ions. In other words, abstracting electrons makes the potential more positive and increases the rate of corrosion. This is an interesting, but undesirable, outcome. In fact, our iron would be "much rotted" as were the nails on *HMS Alarm*.

2.1.2 Adding electrons

We can modify our theoretical experiment easily enough. We can throw an imaginary switch and add electrons instead of removing them. Again, we can consider what the implications of this action might be without worrying about any practical details of how we might do it. From the reasoning developed above, we see that adding negatively charged electrons will make the potential of the steel more negative with respect to its environment.

We now picture the addition of electrons (to the right-hand side of equation 2.1). Again, following the same logic, the system reacts to oppose what it sees as an excess of electrons. We add electrons, so the system compensates by reducing the rate at which they are produced. This can only be achieved by decreasing the rate at which iron atoms are converted to Fe^{2+} ions.

In other words, adding electrons to the metal makes the potential more negative and decreases the rate of corrosion. This is both interesting and desirable.

We know that the corrosion process involves both anodic and cathodic reactions. It follows that we also need to consider what will happen to the cathodic processes when we add electrons.

In this case, the cathodic reaction is the reduction of dissolved oxygen (equation 2.2) which we also met in Chapter 1.

$$2H_2O + O_2 + 4e^- \rightarrow 4OH^- \tag{2.2}$$

If we add electrons to the left-hand side of equation and we follow the above reasoning, we would conclude that the cathodic reaction rate will tend to speed up. The system sees a surplus of electrons and reacts to reduce that surplus by consuming them. Providing there is enough oxygen dissolved in the water, the excess electrons are consumed by increasing the rate of reduction of the dissolved oxygen to produce hydroxyl ions.

We can summarise the outcome of our thought experiment in Table 2.1.

Table 2.1 "Thought" experiment: electrical interventions

Electrons	*Removed*	*Added*
Effect on potential of the metal	→Positive	→Negative
Effect on anodic process	Faster	Slower
Effect on cathodic process	Slower	Faster
Effect on corrosion rate	Increased	**Reduced**

2.2 A SIMPLE MODEL

2.2.1 How does cathodic protection work?

Our thought experiment does not produce a working CP system. To explore how we can do that we should first refer back to our simple model of the corrosion cell (Figure 2.1). In the case of a freely corroding metal, the rate of electron consumption by the cathodic process must equal the rate of electron release by the anodic process. We know this because, as a matter of experience, corroding metals do not charge up. We do not suffer an electric shock if we accidently touch some rusty iron.

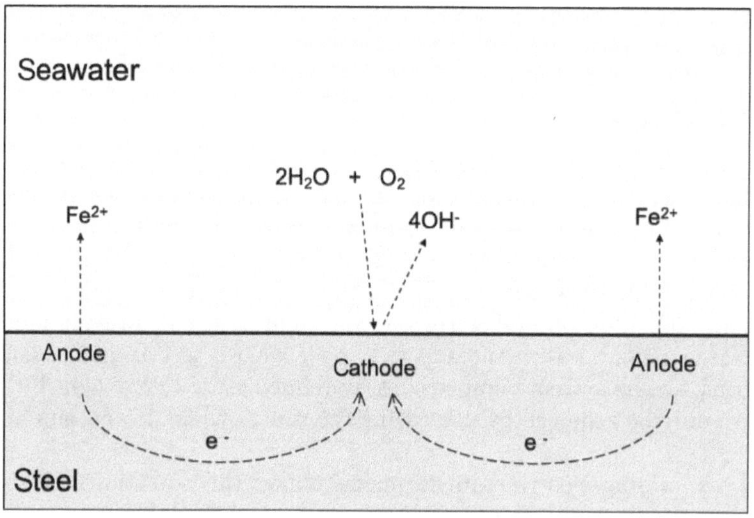

Figure 2.1 Simple corrosion cell.

We appear to have ignored this physical reality in our thought experiment. That is a problem that we need to return to in a little while. But, for now, we can stick with our concept. This means that we can picture the addition of electrons to the system as indicated in Figure 2.2. We have created an external source of electrons. These electrons interact with the electrochemical processes occurring on the surface of the steel.

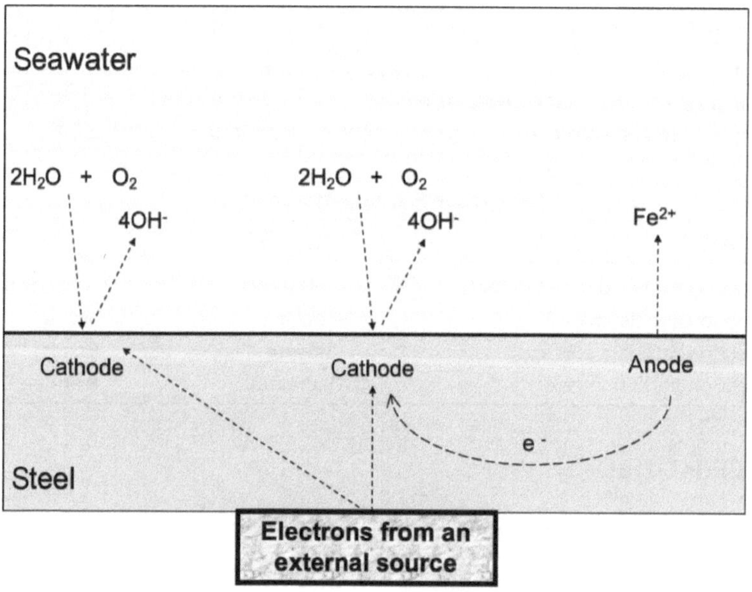

Figure 2.2 Partial cathodic protection.

This theoretical picture, or model, indicates that some of the current demand needed to cathodically reduce the dissolved oxygen on the metal surface is now satisfied by the external electrons. Hence, fewer electrons need to be provided by the anodic dissolution of iron atoms. Corrosion consequently slows down.

We do not need to stop there. We are, after all, conducting a thought experiment. We have no problem increasing the rate at which electrons can be supplied to our metal. In this model, we can progressively open the tap and pump electrons as fast as we wish.

In principle, there is no reason why we cannot achieve the situation illustrated in Figure 2.3. Now, the rate of electron delivery is such that all of the cathodic reduction of dissolved oxygen is sustained by the externally supplied electrons. There is no requirement for iron atoms to be lost anodically from the surface. Iron no longer passes into the chemically combined state, so corrosion has ceased. The entire surface behaves as a cathode.

It is fully cathodically protected.

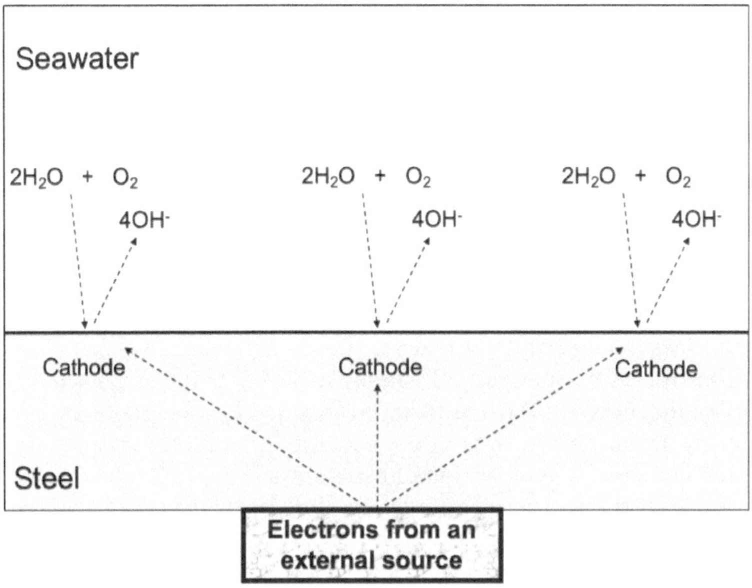

Figure 2.3 Full cathodic protection.

2.2.2 Implementation

We must be clear that, just because we can conjure up a thought experiment, and illustrate it with a couple of simple diagrams, does not mean that we have proved anything about the mechanism of CP. It will emerge later that there are subtleties to this picture. Indeed, there are some CP practitioners who would raise fundamental challenges to the basic concept of our thought experiment. We will return to these subtleties and challenges throughout this book. Nevertheless, this simple picture is enough to get us up and running.

If we are going to make CP work for us, we need to consider how we physically implement the concepts depicted in Figures 2.2 and 2.3. This raises three practical questions:

1. Since the act of injecting electrons makes the steel more negative, we should ask ourselves: how negative do we have to go to control the corrosion? This brings us to the question of potential: what is it and how do we measure it?
2. Given that we need to deliver electrical current to the steel structure, there is an obvious question: how much current do we need?

3. In practice, how do we provide the electrons needed to satisfy the cathodic demand of the steel surface we wish to protect? The supply of electrons is, of course, an electric current. This current has to be driven against the resistance of the electrical circuit formed by the structure and the seawater environment. This requires electrical power.

The topics of potential, current and power form the core of this book. Potential is outlined in 2.4 and is the subject of more detailed consideration in Chapter 6. We touch upon the question of current briefly in 2.5, and consider it in much more detail in Chapter 7. Similarly, in 2.6, we introduce the two options for supplying this power: sacrificial anodes and impressed current. These options are covered in more detail in Chapters 10 and 11, respectively. Taken together, these three topics are what this book is all about.

However, before we can develop our thinking on these, we need to clear up a minor point about how electrical engineers and electrochemists might describe the flow of electricity in CP circuits.

2.3 THE TWO VIEWS OF CURRENT FLOW

The science of electricity had been developing since the middle of the 18th century. Benjamin Franklin had coined the expressions "positive" and "negative" to describe the charges involved. Subsequently, as the concept of electric current developed, it was taken as a matter of convention that current flowed from positive to negative poles.

As we have already mentioned, it was not until 1909 that Millikan demonstrated that the electron carried a negative charge. Of course, this does not mean that the scientific work on electricity prior to 1909 was wrong. It simply meant that the established convention of describing *electrical current* as flowing from positive to negative required an appreciation that the flow of *electrical charges* was in the opposite direction. Fortunately, this subtlety has not held back humanity's exploitation of electricity.

As a result of our thought experiment, we concluded that we could achieve CP by causing electrons to flow into the metal. This is the electrochemists' perspective, since they are generally interested in where electrons are going, and what they do when they get there. In practice, however, there are probably more people with an electrical, rather than electrochemical, background working in CP. Workers with this background are more inclined to view CP as being achieved by draining current from the metal. Indeed, the point of electrical connection to a cathodically protected structure is termed the "drain point" for this reason.

Both views of CP are correct. Indeed, as we will see at various points in this book, it is sometimes convenient to visualise the process of CP in terms of the flow of electrons; and sometimes, it helps to wear an electrical engineering hat and consider the conventional (positive to negative) flow of electric current.

2.4 POTENTIAL

We have made the point that reducing the rate of corrosion by CP involves making the potential of the metal more negative with respect to its environment. This prompts three further obvious questions:

- What do we mean by *potential*?
- How do we measure it?
- How negative does the potential need to be to confirm that we have achieved protection?

We introduced a simple view of potential, and a means of measuring it, in Chapter 1. It's time for a little more theory.

2.4.1 What do we mean by "Potential"?

It is convenient to consider an idealised, artificial situation. Let's imagine that we insert a metal (M) into water. We can assume that it is energetically favourable for some of the atoms at the surface of the metal to enter the water forming a solution of M^{2+} ions (equation 2.3). This is energetically favourable because the M^{2+} ion, which is stabilised by the water molecules surrounding it, is in a lower energy state than the M atoms in the metal. In other words, the dissolution of the metal is energetically favourable from a chemical point of view because it releases energy.

$$M \rightarrow M^{2+} + 2e^- \qquad (2.3)$$

For each atom that ionises and enters solution, two electrons are left in the metal. Since electrons are negatively charged, and M^{2+} ions are positively charged, the metal becomes slightly more negative in its overall charge, and the solution becomes slightly more positively charged.

In this model, the dissolution cannot go on indefinitely. Although the ionisation of M to M^{2+} is chemically advantageous, there is an electrical disadvantage involved in separating the positive ions from the negative electrons. This is because opposite charges are naturally attracted to each other. As the concentration of M^{2+} builds up in the solution, the magnitude of this energy disadvantage increases. A point is reached where it becomes energetically favourable for some of the M^{2+} ions to electrodeposit back onto the surface of M. This is indicated by equation 2.4, where the direction of the arrow has been reversed to show ions are returning to the metallic state.

$$M \leftarrow M^{2+} + 2e^- \qquad (2.4)$$

Taking this idea a little further, we can see that a situation will be reached in which chemical energy advantage gained by the dissolution is equal in magnitude to the electrical energy disadvantage arising from the separation of positive and negative charges. At this point, the rate of the (forward) dissolution reaction will be equal to the rate of the (reverse) electrodeposition process. We can then say that the process is in equilibrium. This is indicated in equation 2.5, where the $\Leftrightarrow$ symbol signifies an equilibrium state in which there is no nett change.

$$M \Leftrightarrow M^{2+} + 2e^- \qquad (2.5)$$

We can picture this equilibrium situation in Figure 2.4. This allows us to introduce the term potential or, more precisely, the electrical potential energy. Electrochemists habitually resort to Greek for terminology and symbols, so this electrical potential energy is conventionally designated by the Greek letter "phi" (Ø). Its value in the solution ($Ø_s$), with its surplus of positive charges, will be different from its value in the metal ($Ø_m$) with its accumulation of surplus electrons. There is, therefore, a potential energy difference, referred to as the Galvani potential energy difference ($\Delta Ø$), across the metal-solution interface:

$$\Delta Ø = Ø_m - Ø_s \qquad (2.6)$$

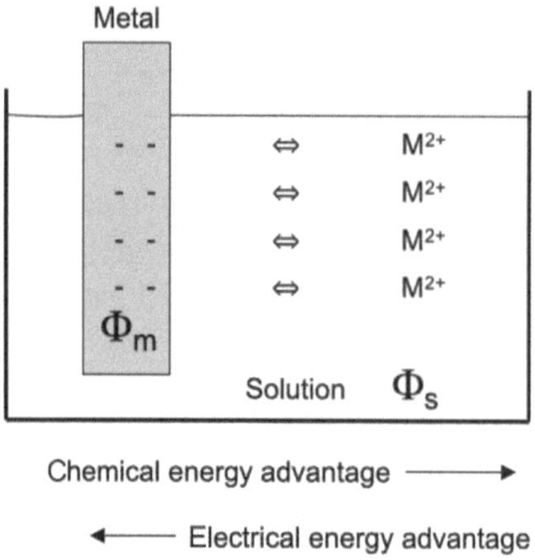

Figure 2.4 An equilibrium "potential".

For convenience, the world of CP, and electrochemistry more generally, shortens the term "potential energy difference" to "potential".

2.4.2 How do we measure the potential?

2.4.2.1 The problem

There is a simple answer to this question. We can't.

Take a look at any voltmeter. You will see that it has two terminals. This reminds us that we cannot measure electrical potential energy; but we can, in practice, measure the difference between two electrical potential energies. For convenience, we contract this expression to the simple term potential difference; a parameter that we measure in units of volts. Sometimes, we also use the expression electromotive force (EMF) to describe this potential difference.

Now consider a piece of steel corroding naturally in seawater. We would like to measure its potential, by which we mean the potential difference between it and the seawater. Our first thought might be to grab our voltmeter and connect one terminal to the steel, the other to the seawater, and then simply read the dial. Connecting the steel to one terminal of the meter is not a problem. The difficulty lies in making the connection between the other terminal and the seawater. We might try the simple solution of using a copper test lead, plug one end into the meter's terminal and drop the other end of the lead into the seawater. Unfortunately, this will not work.

The potential difference registering on the meter will be the potential difference across the electrochemical cell:

steel|seawater|copper.

It will comprise the difference between two potential differences: the potential difference between steel and the seawater (the steel half-cell) minus the potential difference between the seawater and the copper (the copper half-cell).

We will see a reading on the meter, but we will not know what it means. We cannot distinguish between these two individual potential differences with a single measurement.

No matter what variations we might play on this exercise, for example using a different metal conductor to make contact with the seawater, we end up in the same situation. We cannot actually measure the electrical potential in the steel, or even the difference in electrical potential between the steel and the seawater.

2.4.2.2 *The solution*

2.4.2.2.1 The standard hydrogen electrode

You might have thought that our inability to measure potential would have put an end to electrochemical science just as it got started. Not so.

Practical electrochemistry is mainly interested in how *changes* in the potential difference between the metal (steel in this instance) and the electrolyte (seawater) bring about *changes* to the rates of the anodic and cathodic processes taking place on the surface of that metal. This means that we can get by without knowing what the potential difference is between the metal and the electrolyte. Providing we know that if we shift it by a measurable amount (say 100 mV) then we can observe the effect that shift has on the rate of the reactions taking place on the surface.

With this understanding, electrochemists agreed early in the last century on a standard reference half-cell. This is the standard hydrogen electrode (SHE),[1] which is illustrated schematically in Figure 2.5.

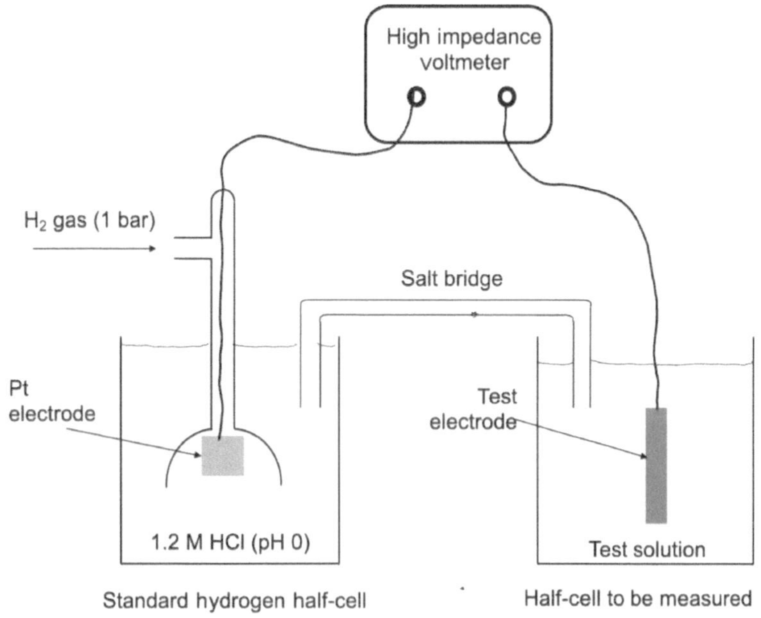

Figure 2.5 The standard hydrogen electrode.

As can be seen, the electrode in the standard hydrogen electrode is a piece of platinum immersed in 1.2 molar hydrochloric acid with hydrogen gas at 1 bar (100 kPa) bubbled over its surface. The platinum does not corrode, or react chemically in any way, with the acid or

[1] Strictly speaking, the SHE is mis-named. It is not an *electrode*. It is a *half-cell*. The difference between the two terms does not matter to us here.

the hydrogen. It "senses" the equilibrium established between hydrogen ions in the aqueous solution and hydrogen gas (equation 2.7).

$$2H^{+}_{(aqueous)} + 2e^{-} \Leftrightarrow H_{2(gas)} \tag{2.7}$$

The SHE half-cell has been assigned a potential of 0 V at all temperatures.

If this seems a rather arbitrary underpinning for electrochemical science, then consider that I am writing this in the year 2021. But: what exactly does that mean? It is not 2021 years from the creation of time in the Big Bang. It is simply 2021 years since an arbitrary reference date. For the most part, the world has opted to count dates from the assumed year of birth of an individual known as Jesus of Nazareth. He was an historical figure of whom most people have heard, and who is revered to some extent by about 31% of the world's population. It does not matter at all that alternative reference points are available. For example, the Hebrew year is 5781, and the Islamic is 1442. Similarly, it is not even relevant that the individual in question was probably born somewhere around 3 BCE. All that matters is that humanity can agree on a reference date for the start of the current era (CE).

The same holds for the measurement of potential. Once electrochemists had agreed on the SHE as the standard reference potential, everyone could get on with their experiments secure in the knowledge that their results would make sense to, and could be replicated by, others.

Although its arbitrary nature is not an issue, the construction of the SHE presents some practical difficulties. In addition to platinum, which is pricey, it requires strong acid and potentially explosive hydrogen gas. In practice, therefore, it is a rather expensive and unwieldy piece of apparatus to use in a laboratory. It would be wholly impractical for field use. Again, this is a problem for which electrochemists have found workarounds.

2.4.2.2.2 Secondary references

Once we have designated the SHE as our primary reference half-cell, there is nothing to stop us conjuring up other secondary reference half-cells for laboratory or field use. All that is required is that the secondary reference exhibits a stable potential in the environment in which we are using it. We can then agree to use the secondary reference for our work. Alternatively, we can measure the difference in potential between our secondary reference and the primary SHE, and then apply this difference as an offset to our recorded measurements.

As we mentioned in Chapter 1, it has become standard practice to use the Ag|AgCl|seawater half-cell as our reference. At its simplest, this consists of a length of silver wire onto which is fused a layer of silver chloride salt. This forms an electrode which, when immersed in seawater, becomes a half-cell. The Ag|AgCl is an example of an ion-selective electrode. Its potential depends on the concentration of chloride ions in the electrolyte. In typical seawater, with a chloride concentration of ~18980 mg/L, the potential of the Ag|AgCl|seawater half-cell has been measured against the SHE and found to be (in round figures) 250 mV more positive.

2.4.3 Potential measurement

As we saw in Chapter 1, armed with a high-input impedance voltmeter and an acceptable secondary reference electrode, we can now set about the process of measuring the potential of our steel specimen immersed in seawater. If we do this in practice, we will find that the reading on the meter changes quite rapidly during the first day or so. Eventually, however, it settles down to a value that will drift about; but might typically be somewhere in the range −600 to −700 mV. This illustrative figure is a typical value for the corrosion potential (E_{corr}) of the steel in seawater. We will return to the significance of corrosion potential when we discuss electrode kinetics in Chapter 5.

Two things need to be stressed: the polarity and the reference.

In Figure 1.3 in Chapter 1, we see the steel connected to the positive terminal and the reference to the negative. We also see a minus sign on the display. This tells us that the potential of the steel is *minus* 620 mV. It is vitally important always to include the sign when recording any electrode potential measurement. Furthermore, the measurement is made with respect to the Ag|AgCl|seawater reference. It is, likewise, essential that we always inform the reader what reference we are using. Indeed, if we want to be electrochemistry purists, we could convert our reading to the primary SHE scale. Since Ag|AgCl|seawater is ~250 mV more positive than the SHE, we can see that the potential difference between our piece of steel and the SHE would be −620 mV − (+250 mV) = −370 mV. The situation is illustrated in Figure 2.6. This also introduces zinc as a secondary reference.

Strictly speaking, reference half-cells should be systems that are at equilibrium, as is Ag|AgCl|seawater. Such references exhibit a stable potential which can be precisely measured against SHE. Zinc is not a "true" reference because it corrodes in seawater. So, it cannot be at equilibrium. Nevertheless, the potential it exhibits whilst corroding is reasonably stable. More important, unlike reversible reference half-cells, a block of slowly corroding zinc is very robust. For that reason, zinc is widely employed as a CP reference in the harsh marine environment. Its durability more than compensates for its lack of accuracy. We will have more to say about the use of zinc as a reference in Chapters 15 and 18.

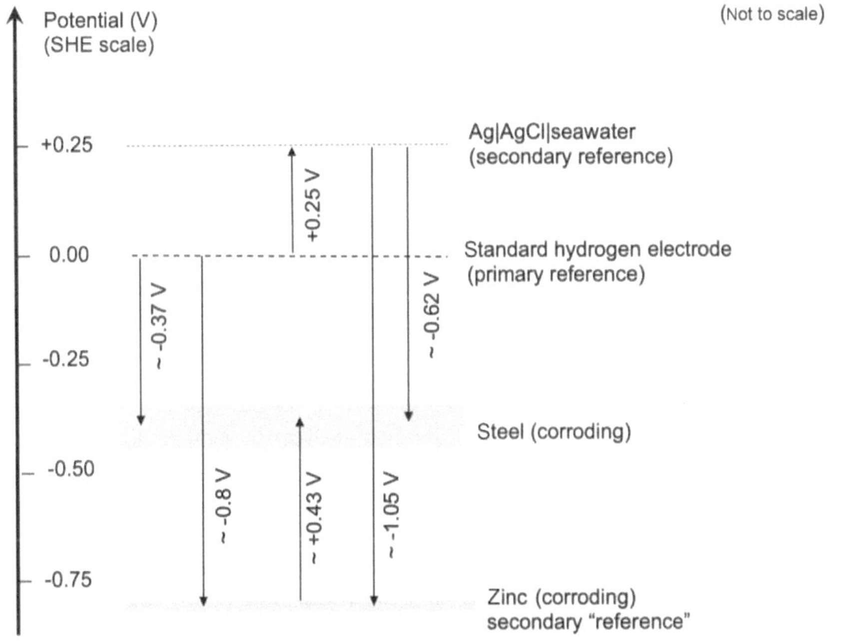

Figure 2.6 Relationship between primary and secondary references.

Figure 2.6 indicates that a measured potential of (say) −620 mV (vs Ag|AgCl|seawater) for a piece of corroding steel would be measured at around +430 mV against a piece of zinc. This emphasises why it is vital always to state the reference (and polarity) when recording potentials.

We will have more to say about reference half-cells and the measurement of potential throughout this book. In particular, we will look at the errors associated with potential measurement in Chapter 18. For now, we just need to bear in mind that what we are seeking to measure is the difference in electrical potential energy across a metal-solution interface with respect to an arbitrarily selected reference electrical potential energy difference. It is little surprise that CP engineers reduce this cumbersome phrase to the single word: "potential".

2.4.4 What is the potential needed for protection?

We have seen that in order to cathodically protect steel, or any other metal, we need to shift its potential to a more negative value. In the jargon of the electrochemist, we apply cathodic polarisation. The concept of "polarisation" is pivotal in CP. Indeed, we dedicate Chapter 7 to it. For now, we can accept that the noun polarisation simply means a change of the potential. For CP, our objective is simply to polarise the steel from its natural corrosion potential (typically somewhere between −600 and −700 mV), to a potential at which it is protected. This means that we need to know how much polarisation to apply. In other words: what is the protection potential?

This is a question that invites both a short answer and a long one. The long answer is explored in Chapter 6. Fortunately, for the purposes of this introduction, we can settle for the short answer. The codes tell us that steel is protected from corrosion in seawater at a potential of −800 mV versus Ag|AgCl|seawater.

We need to make a further very important point here. Although Ag|AgCl|seawater is not the only reference employed in marine CP, it is by far the most widely used. For that reason, unless stated to the contrary, ***all potentials*** quoted in this book ***are referenced to this half-cell.***

It will emerge that, although we can set a target potential (−800 mV), we cannot practically achieve a potential of exactly −800 mV over the entire surface of a real structure. Our practical target, therefore, has to be modified to −800 mV *or more negative*. Inevitably, to ensure that we achieve a potential at least as negative as −800 mV, some parts of the surface will be markedly more negative than this figure. There may also be some areas that just fail to reach −800 mV. We return to these issues and their implications later in this book.

2.5 CURRENT

2.5.1 Bare steel

Achieving cathodic protection by polarising steel to −800 mV, or more negative, requires the application of current. The obvious question is: how much current do we need? As with the subject of potential, this is a topic that merits a considerable discussion (see Chapter 7). At this stage, however, it will be sufficient to make some basic comments.

The first point is that the bigger the surface area of the structure we are trying to protect the more current we need. For this reason, CP practitioners refer to current per unit area, a parameter named *current density*. It is expressed in units of A/m^2 or mA/m^2.

The second point is that, unlike protection potential which depends only on the metal under consideration, the current density needed to protect a metal depends intimately, and in a complex manner, on environmental factors. For a metal in seawater, or seabed sediments, these factors include chemical parameters (dissolved oxygen content, pH and salinity) and physical influences (temperature and flow rate).

Furthermore, unlike the protection potential, current densities vary with time. This is because the electrochemical reactions taking place cause changes on the surface which, in turn, modify the current demand.

In fact, it is all but impossible for a CP designer to predict what actual cathodic current density will be required on the surface. However, as we shall see when we carry out some simple CP design exercises in the next chapter, we do not need to make that prediction providing we are content to rely on codes. These embody the experience and best guesses of the contributors to the codes. Thus, instead of risking our own guess as to the current density we need, we can take comfort in adopting the consolidated opinion of a committee.

2.5.2 Coated steel

Some structures, for example most space-frame platforms used for petroleum production and most harbour piling, are uncoated below the waterline. These structures, therefore, rely solely on CP for corrosion protection in the immersed and buried zones.

Other structures, such as ships' hulls, subsea petroleum production manifolds and pipelines, are coated. In these cases, the coating is often considered as the primary barrier against corrosion. The role of CP is then regarded as providing protection to any steel that becomes exposed to the seawater as a result of imperfections in, damage to, or natural degradation of, the coating.

This means that an engineer designing a CP system for a coated installation needs to predict the future performance of the coating. This is more problematic than predicting the required cathodic current densities. Again, the only real option is to guess. Fortunately, as for the selection of the design cathodic current densities, we can resort to the codes and make use of the consolidated guesses of the drafting committees.

We discuss coatings in Chapter 9. For now, it is sufficient to make the point that the effect of coatings in reducing the cathodic current demand, and therefore, the CP system cost is very dramatic. This is illustrated by a hypothetical example in Table 2.2. This illustrates the total cathodic current demand (I_{CP}) per kilometre of an 8" pipeline laid in seabed mud. The figures have been calculated by applying the current advised by DNV-RP-F103 for a 20-year design life. This calculation indicates that the average current demand for the coated pipeline is only about 0.2% of the requirement for the uncoated condition. On that basis, it is unsurprising that subsea pipelines are coated.

Table 2.2 Current requirement for 8" flowline in seabed mud

Coating	*Mean ICP (A/km)*
None	13.8
3-layer Polypropylene	0.026

2.6 POWER SOURCES FOR CP

2.6.1 Davy's work

We can start at the beginning. Sir Humphry Davy was obliged to consider this question when he carried out his investigations during 1823. His first thoughts [1] were... *of using a Voltaic battery*. However, he considered that this... *could hardly be applicable in practice*.

Accordingly, in his first experiment, he employed seawater which he had... *rendered slightly acidulous by sulphuric acid*. In this, he immersed a piece of polished copper... *to which a piece of tin was soldered equal to about one-twentieth of the surface of the copper*. He reported that... *after three days the copper remained perfectly clean, whilst the tin was rapidly corroded*. He obtained the same result when small pieces of zinc were soldered to the copper.

This result encouraged further experiments in which he discovered that both zinc and iron could protect copper. Crucially, he also discovered that zinc could protect iron. Thus, Sir Humphry was the first to cathodically protect iron using a zinc sacrificial anode; although as noted above, neither the term *cathodically* nor *anode* would have been words he would have recognised.

2.6.2 Sacrificial anodes

Sacrificial anodes are also widely referred to as *galvanic* anodes. This is after Luigi Galvani (1737–1798) who, as we have already mentioned, famously caused dead frogs' legs to twitch by prodding them with different metals. I am not much of an admirer of Galvani. He went through his entire life without grasping the concept that electricity could exist outside living organisms. He rejected the contrary rational scientific evidence discovered by his fellow Italian and near-contemporary Alessandro Volta (1745–1827). In my view, associating Galvani with cathodic protection gives him more credit than he deserves. Nevertheless, the name has stuck.

The terms *galvanic* and *sacrificial* are used interchangeably in CP. I use *sacrificial* because it immediately conveys the fundamental reality of causing one metal to corrode in order to protect another.

The first marine applications of CP used zinc blocks ("Davy's protectors") as anodes. If a steel structure is placed in seawater, it will corrode as discussed in Chapter 1. However, if we now connect a lump of zinc to the steel, the situation changes. This is illustrated in Figure 2.7. As Davy was aware, zinc is a... *more oxidable metal than iron*. He was also aware that zinc was slightly negative to iron when immersed in seawater. This means that, in seawater, there is a potential difference between the two.

Of course, Davy was not aware of the existence of electrons, which remained unknown until J.J. Thomson's publication in 1897. However, he would have been aware of the concept of an electric current. He corresponded with Alessandro Volta who had been producing electric current from batteries since about 1800. Even so, Davy's paper makes no mention of the word current, and there is nothing in it to suggest that his thinking involved any flow of electric charge.

We now know that, if we make an electrical connection between iron (or steel) and zinc, as Davy did when he soldered the two together, this potential difference causes electrons to flow from the zinc to the steel. At its most basic, this is a manifestation of the universal physical principle that opposite charges attract. Negative electrons flow to the (relatively) positive steel. Nowadays, we recognise a movement of electrons as an electric current.

As we inferred from our theoretical experiments, this supply of electrons causes the potential of the steel to become more cathodic (negative), and corrosion is reduced (Figure 2.2) or ceases altogether (Figure 2.3).

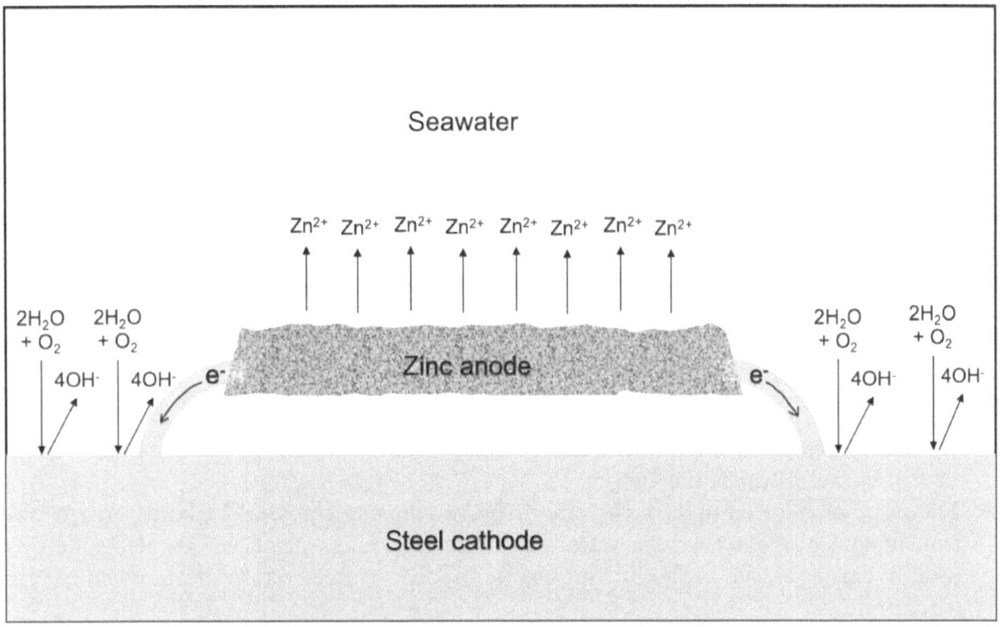

Figure 2.7 Cathodic protection using a zinc anode.

The converse situation applies to the zinc. Removing electrons from a metal causes the rate of its anodic dissolution to increase. Thus, the zinc is intentionally sacrificed to protect the steel structure as indicated in Figure 2.7.

For simplicity, the diagram in Figure 2.7 only shows the flow of electrons in the circuit. If nothing else happened, then the steel would charge up electrically and the flow of current would cease. This does not happen. This is because there is a return path for the current through the electrolyte (seawater). It is because of the need for this return path to complete the circuit that CP can only work in the case of buried or immersed structures, or inside equipment filled with conducting electrolyte.

Whereas the current in the electrical connection between the zinc and the steel is carried by electrons, the flow of current through the seawater is carried by ions. These include anions (Cl^- and OH^-) that migrate towards the anode (in this case the zinc), and cations (Na^+ and Zn^{2+}) that migrate towards the steel cathode (the steel).

This flow of current between the anode and cathode through the seawater gets us over one of the problems in our thought experiment. The idea of injecting electrons into a metal to protect it might have seemed to contravene the Law of Electroneutrality. However, now that we have introduced the completion of the electrical circuit by the flow of ions through the electrolyte, overall electrical neutrality is re-established.

We will have more to say about sacrificial anodes in Chapters 3 and, more particularly, in Chapter 10.

Two views of CP

We have already mentioned that the direction of flow of conventional (positive to negative) current in a metal is opposite to the direction of flow of the charge-carrying electrons. In addition, we have indicated that this means that we can view CP as either supplying electrons into the metal or draining current from it.

Figure 2.7 allows us to appreciate both of these views of CP.

- Starting at the positive (or least negative) metal (the steel), the conventional current flows through the metallic connection to the more negative zinc.
- From the zinc, the current enters the seawater.
- The seawater is an ionic, not an electronic, conductor. This means that there must be a change of charge-carrying species from electrons (in the zinc) to ions (in the seawater). This change in charge carrier takes place at the zinc-seawater interface. Any change of current carrier between electrons and ions is termed an electrode process. In this case, the electrode process at the zinc surface is anodic. Zinc atoms are ionised to zinc ions (equation 2.8).

$$Zn \rightarrow Zn^{2+} + 2e^- \tag{2.8}$$

- The ionic current, which is carried by all of the charged ions in the seawater, then moves through the environment towards the steel cathode. The positive cations (mainly and Na^+ and H^+) migrate towards the cathode, and negative anions (predominantly Cl^- and OH^-) migrate towards the anode.
- The ionic current completes the circuit by re-entering the steel cathode. Again, this transfer of charge across the water-steel interface takes place via an electrode process. In this case, it is cathodic and mostly involves the reduction of dissolved oxygen (equation 2.9).

$$O_2 + 2H_2O + 4e^- \rightarrow 4OH^- \tag{2.9}$$

This completes the circuit.

Thus, an electrochemist might care to think of electrons being pumped into a metal as giving rise to protection, and electrons being removed from a metal as stimulating corrosion. Equally correctly, an electrical engineer might view conventional current flowing from the environment *onto the surface* of a metal as conferring protection, and current exiting a metal's surface to flow *into the environment* as damaging. Both perspectives are correct, and both can be useful in interpreting CP.

2.6.3 Impressed current

The concept of impressed current cathodic protection (ICCP) is straightforward, as illustrated in Figure 2.8. Instead of relying on a sacrificial anode such as zinc to dissolve and provide electrons for our CP circuit, we simply connect our structure to the negative terminal of a DC power supply. Electrons are now available at the flick of a switch.

The fundamental laws of physics still hold. Electricity can only flow in a complete circuit.

Because we are talking about electrical equipment, it makes sense to adopt the electrical engineer's perspective on current. On this basis, we can interpret Figure 2.8 as follows:

- Current flows from the positive terminal of the DC power supply, via a conducting cable, to an electrode immersed in the seawater.
- It then enters the seawater by means of that electrode. Because current leaves this electrode to enter the seawater, we refer to it as an anode. We will return to the nature of this electrode, termed an impressed current anode, and to the anodic processes it sustains, in Chapter 11.
- The current then passes ionically through the seawater and is collected on the surface of the protected structure (the cathode). As noted above, current flowing from the environment onto the surface of a metal confers protection.

- The current then returns from the structure, via a conducting cable, to the negative terminal of the DC power supply.

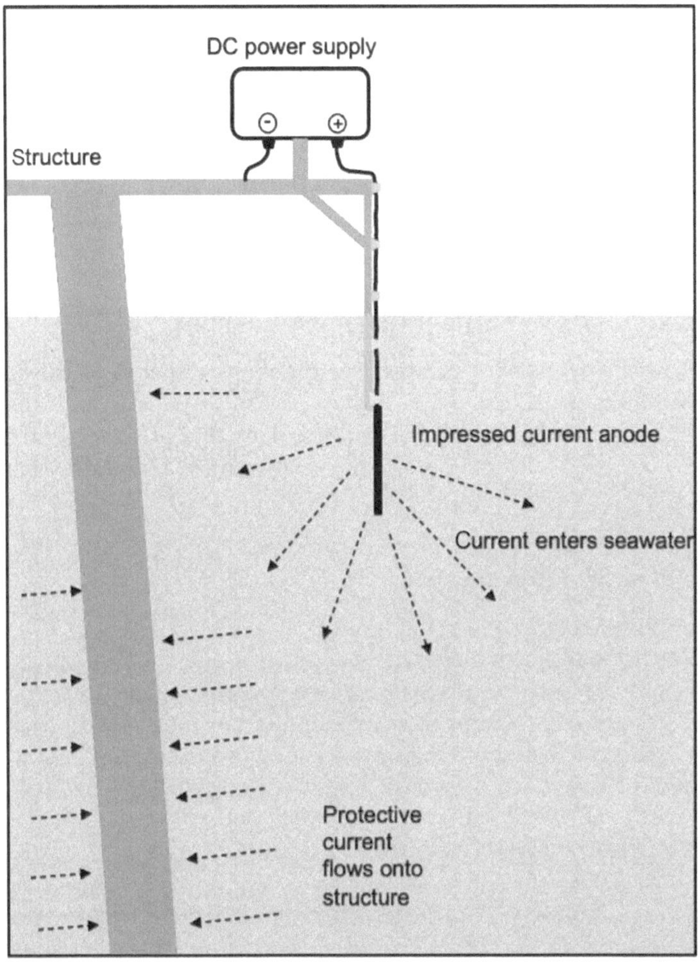

Figure 2.8 Impressed current CP.

As mentioned above, Davy originally envisaged what we would now call an impressed current system. He considered using a battery (which he termed a "Voltaic pile"), but could not see how to implement it in practice. Hence, he opted for what we now call sacrificial anodes.

The key to developing successful marine ICCP systems was to find materials for the anode that were both good electrical conductors, and that did not corrode when passing current anodically into the seawater. Copper would fulfil the first requirement of being a good conductor, but hooking it up to the positive terminal of a DC power supply causes rapid anodic dissolution (equation 2.10).

$$Cu \rightarrow Cu^{2+} + 2e^- \qquad (2.10)$$

For that reason, marine impressed current anodes are made from highly corrosion resistant, but electrochemically active, materials usually supported on an inert conducting substrate such as titanium. We will discuss these impressed current anodes in Chapter 11.

For the moment, we just need to consider the anodic reactions that take place on their surfaces to enable the current to pass into the electrolyte. There are two possibilities. The first is the anodic electrolysis of seawater (equation 2.11) to produce oxygen. The second is the anodic oxidation of chloride anions to form chlorine gas (equation 2.12). Since both of these processes liberate electrons, by definition, they are anodic. We will also need to consider the implications of these reactions in Chapter 11.

$$2H_2O \rightarrow O_2 + 4H^+ + 4e^- \quad (2.11)$$

$$2Cl^- \rightarrow Cl_2 + 2e^- \quad (2.12)$$

2.6.4 Sacrificial anodes versus impressed current

Almost every didactic work on CP summarises the relative merits of the sacrificial anode and impressed current approaches to CP. Perhaps the most widely replicated example was provided in British Standards CP 1021. We will not buck that literary trend here, so the following very brief comments set out some of the relative merits, and drawbacks, of the two approaches.

2.6.4.1 Sacrificial anodes: advantages

1. No external power required

   CP is an exercise in driving electric current, through the resistance of the environment, and onto the surface of the structure we wish to protect. As everyone who paid attention to physics lessons in school should be able to tell us: driving an electrical current against a resistance requires a voltage source. They might also recall the relationship:

   $$\text{Voltage}(V) \times \text{Current}(I) = \text{Power}(P) \quad (2.13)$$

   Thus, at its very simplest, CP is an exercise in delivering electrical power to a structure in a reliable, controlled manner. Obviously, in many marine situations, it is impractical to obtain electricity from an external source. This drawback often rules out the impressed current option. Sacrificial anodes then become the only alternative. They generate electrical power by releasing part of the energy that was used to win the metal from its ore in the first place.
2. Easy to install

   As we see in Chapter 10, the statement that sacrificial anodes are easy to install can often be a gross oversimplification. Nevertheless, it is correct to say that there is nothing philosophically complicated about the overall installation of a sacrificial anode system. Once the anodes are attached, directly or via cables, to the structure then they will operate spontaneously. That means that there is no need to provide any supplementary electrical control equipment.
3. Interference unlikely

   The electromotive force (EMF), or driving voltage, involved in a sacrificial anode CP system is the difference in potential between the protected structure (usually steel) and the sacrificial anode (either zinc or, more usually, an alloy of aluminium). This EMF is typically no more than about 0.25 V.

The received wisdom is that this EMF is too low to cause any electrochemical interaction with structures other than that to which the anode is intentionally connected to provide protection.

This view is almost always correct. However, there have been some exceptions. This is why this subsection is captioned "interference *unlikely*" rather than "Interference *impossible*".

4. All connections protected

   The stylised image of sacrificial anode CP shown in Figure 2.7 indicates a metallic, electronically conducting, connection between the sacrificial anode and the structure. In most instances, the anode is cast onto a steel core, and this core is directly welded to the structure. In other instances, the anodes may be installed slightly remote from the structure, and their steel cores are connected to the structure via copper cables.

   Either way, the sacrificial anode "sees" the connecting metalwork, steel or copper cable, as a continuous part of the cathode structure. This means that the connection between the sacrificial anode and the structure is likewise protected against corrosion.

5. "Fit-and-forget"

   This aspect of sacrificial anodes is included under the heading of "advantages". It is certainly an advantage that, once we have installed sacrificial anodes on the structure, there is nothing more that we need to do. The laws of electrochemistry take over. Providing we have designed the system properly (see next chapter), all ought to be well. The anodes should protect the structure for the duration of their design lives.

   However, this "fit-and-forget" attribute is something of a double-edged sword. The converse argument is that, once we have installed our anodes, there is nothing more that we can do. If the anodes work as intended, then all is well. But if they don't then we are probably looking at a CP retrofit to be carried out offshore. This might prove expensive.

   Accordingly, "fit-and-forget" is not really good advice. There would be some degree of hubris in assuming that, just because we believed our CP design to be correct, all would be well for the duration of the design life. History confirms that this has not always been the case.

   CP systems, sacrificial or impressed current, should never be installed and then simply forgotten. Some inspection is needed. We return to this in Chapter 18.

2.6.4.2 *Sacrificial anodes: disadvantages*

1. Limited power output

   Many of the disadvantages of sacrificial anode systems are just the obverse of the advantages. The low power output that renders sacrificial anodes unlikely to interfere with unconnected structures also presents a disadvantage when it comes to driving protective current against the resistance of the environment.

   Fortunately, this proves not to be too much of a problem when it comes to marine CP because the resistivity of seawater is conveniently low. However, it does become an issue when applying CP in situations where current demands are high, or where the environment resistivity is higher than for seawater itself; such as in river estuaries or the landward sections of offshore pipelines.

2. Multiple installations required
   An inevitable consequence of the low power output of sacrificial anodes is that, to protect sizeable structures, a relatively large number of anodes will be required. A corollary is that the larger the structure we seek to protect, the greater is the impetus to switch from sacrificial anodes to impressed current, providing the latter is feasible.
3. Adds weight to structure
   Sacrificial anodes inevitably add more weight to a structure than an impressed current system of the same current output capacity. Large offshore jackets might require many hundreds of tonnes of anodes to provide the quantity of CP current that could be delivered by impressed current anode systems with all-up weights of less than a tonne. For some structures, this weight difference could have profound implications. The attachment of sacrificial anodes might, for example, take the weight of a subsea installation above the capacity of a lift barge; forcing a different, more expensive installation method. On the other hand, the weight contributed by sacrificial anodes is inconsequential for a subsea pipeline.
4. Hydrodynamic effects
   A designer of an offshore structure has to account for the effect that the wave drag on the anodes has on the fatigue loading of the structure.
   In addition, more subtle hydrodynamic effects would be relevant to hull-mounted anodes on warships. As anodes dissolve away, their surfaces become roughened. The movement of these anodes through the water then generates acoustic signals that would be detectable by an enemy. For this reason, stealth-type submarines, and other warships, are invariably protected by impressed current systems rather than sacrificial anodes.
5. Finite anode life
   A sacrificial anode does just what its name implies. It sacrifices itself by dissolving electrochemically to provide the protective current. Its operating life is, therefore, finite. In principle, a sacrificial anode system is designed with just sufficient anode mass to last for the anticipated service life of the installation.
   For the most part, CP designers have been very proficient in ensuring that they provide sufficient anode to last for the design life of the offshore structure or pipeline. Unfortunately, the owners of the structures themselves have not always been particularly prescient in nominating what that design life should be. The result is that many installations have subsequently been required to operate considerably beyond the original design lives of the anodes. This has necessitated costly offshore CP retrofits. We return to this in Chapters 13 and 14.

2.6.4.3 Impressed current: advantages

The advantages of impressed current systems are generally the converse of the disadvantages of the sacrificial anode systems.

1. High current output
   Whereas a sacrificial anode is limited by electrochemistry to delivering, at most, about 10 A of anodic current per square metre of its surface area, modern marine impressed current anodes can easily deliver a hundred times that anodic current density. This means that far fewer, and smaller, impressed current anodes would be required for any given application.
2. High voltage output
   In principle, the output voltage on the DC power supply in Figure 2.8 could be stepped up to just about any required value. In practice, electrical engineering and

safety considerations limit power supply outputs: 24 V DC would be a typical value. In round figures, this is a hundred times the voltage available from a zinc or aluminium alloy marine sacrificial anode. This increase in voltage lends itself to delivering either large currents, counteracting resistive environments, or a combination of the two.

3. Long anode life

   In principle, the so-called non-consumable impressed current anodes do not undergo anodic dissolution reactions when delivering CP current. As we have seen in equations 2.11 and 2.12, the anodic electrode process involves either the electrolysis of the seawater, the liberation of chlorine, or a contribution from both reactions. On this basis, impressed current anodes might be expected to last indefinitely. In practice, they do not. This reflects the fact that chemical conditions that are generated at the surface of a working anode are very aggressive.

   Referring back to equations 2.11 and 2.12, we see that the electrolysis of the seawater produces hydrogen ions, chlorine gas or both. This combination of acidity (H^+), with a strong oxidising agent (Cl_2), renders the chemical environment at the surface of an anode very aggressive. Even the highly dissolution-resistant materials used for impressed current anodes suffer some attrition under these conditions. Nevertheless, the rates of dissolution are slow enough to permit lengthy anode design lives, as much as 60 years in some instances.

   We discuss impressed current anodes in more detail in Chapter 11.

4. Low weight

   We mentioned high weight as one of the disadvantages of sacrificial anode systems. It is certainly the case that the current output available from many tonnes of sacrificial anodes can be obtained from impressed current anodes weighing just a few kilogrammes.

   However, this weight advantage reduces considerably when it is recognised that impressed current anodes, their cabling and control systems, are inherently vulnerable in the harsh environments encountered offshore. This means that robust designs have to be developed that provide adequate mechanical protection to the anodes, cables and ancillary equipment. This mechanical protection inevitably adds weight, and cost, to the overall ICCP system.

5. Low profile

   The adverse hydrodynamic effects, associated with sacrificial anodes, are obviated by the impressed current alternatives. This is why many seagoing vessels, particularly large or fast ships (and submarines) are protected by ICCP systems.

2.6.4.4 Impressed current: disadvantages

1. Risk of incorrect connection

   Many years ago, I was told a story about an Eastern Mediterranean state that purchased a fleet of fast patrol boats for its navy. As might reasonably be expected for vessels of this type, they were protected by ICCP systems. During the commissioning, so the story goes, the ICCP systems were installed with the positive terminals of the onboard DC power supplies connected to the hull; and the negative terminals routed to the impressed current anodes. The "CP" was switched on and the vessels went to sea with the systems driving anodic current from the steel hulls, through the water and onto the anodes. Unsurprisingly, the hulls started to develop leaks in quite a short time: a situation not welcome in ships (military or otherwise).

   It is a lovely story, but I have never been able to find out whether or not it is actually true. Given the military and political sensitivity, not to mention the particularly humourless reputation of that state's security forces, it is probably best that I do not draw attention to myself by investigating too deeply.

All that I can say is that it is *possible* to connect an ICCP system the wrong way round. However, in over 40 years involvement with CP, offshore and onshore, I have never personally encountered such an event.

On the other hand, it is physically impossible to get the polarity wrong when installing a sacrificial anode

2. Risk of maloperation

   Whereas I have never encountered an ICCP system that has been connected up the wrong way around, I have come across all too many that have been operated with less than due diligence. We will discuss some of these instances, and their implications, elsewhere in this book.

   Again, because sacrificial anode systems are not adjustable this problem cannot arise with them.

3. Stray current interference

   The only performance attributes of an ICCP system that CP engineers can control are the currents and voltages in the metallic circuits. Neither the engineer nor the operator can dictate the pathways taken by the electrolytic current as it flows through the environment from anode to cathode.

   One thing we know for sure is that the current will not flow in straight lines. It is better to think of it as emanating radially from the anode, and flowing into the planet, which serves as an infinite conductor. It then returns from the planet to be collected on the cathode surface. No one can instruct the current to restrict itself to a prescribed pathway through the environment.

   Third-party metalwork, not connected to the system under CP, might intersect the domain of influence of this current. In some circumstances, it could pick up some of the currents at one location (possibly close to the anode) and discharge it at another point (closer to the protected structure). As we have already discussed, collecting current from the seawater confers cathodic protection. However, the current discharge from the third-party structure, back to the protected structure, can only take place by means of an anodic reaction on the third-party steelwork.

   Thus, it is possible for a CP system, whilst protecting its intended structure or pipeline, to cause damage to an unconnected installation. This is known as stray current interference. We will return to this in Chapters 11 and 13.

4. Anode-side connections

   Without exception, the wiring in ICCP circuits consists of insulated copper conductors. This includes the cabling that conducts the CP current from the positive terminal of the DC power supply to the anode.

   Connecting copper to the positive terminal of a power supply, and exposing it to seawater, would lead to accelerated corrosion of the copper (see equation 2.10). The cable would fail very quickly. For this reason, all cables and connections on the positive side of the circuit must be protected from the seawater. Even a minor nick in the insulation might create a path for water ingress. When that happens, the cable is sure to fail in short order.

   There is no point investing in a high-end ICCP system, with a design life of several decades, if the cables fail within months. Regrettably, such instances are not unknown.

5. Engineering, monitoring and management.

   The vulnerability of cables to seawater ingress is just one example that illustrates the need to apply rigorous engineering practices to ICCP systems. The need to provide robust mechanical protection to delicate components is another.

To this must be added the realisation that, unlike sacrificial systems, ICCP systems cannot be "fit-and-forget". They have to be monitored and adjusted on the basis of the monitoring data. The use of automatic systems which self-adjust on the basis of the monitoring information reduces the need for human involvement, but does not eliminate it altogether.

The implication of this is that any economic decision between a sacrificial and an impressed current CP option must take full account of the additional, sometimes hidden, engineering and management costs of the latter.

2.6.4.5 Selecting between sacrificial anodes and ICCP

Interesting as this comparison between sacrificial anodes and impressed current may be, the marine CP designer is rarely required to make a decision between the two. In many circumstances, the approach is a foregone conclusion dictated either by the unacceptability of one approach (for example, the impracticality of impressed current for a subsea pipeline where there is no source of electrical power), or else by the client's intransigent views on how CP should be provided.

On the occasions where there is a valid decision to be made between impressed current and sacrificial anodes, the situation is likely to merit a more profound consideration than can be gleaned from the above brief notes. In Chapter 13, we consider examples of how such decisions have been made in the past.

2.6.5 Hybrid systems

It might be inferred from the above that marine CP systems will be either sacrificial anode or ICCP. In many instances, this turns out to be the case. However, there are installations that are described as hybrid because they make use of both systems. We will also consider some of these instances in Chapter 13.

2.7 WHAT DOES CP ACHIEVE?

a. General Corrosion

   Suitably managed CP can reduce the rate of general corrosion of steel or any other metal, such as copper, to a value that is so low that it is of no practical interest. Indeed, it is theoretically possible to reduce the corrosion rate to zero as is indicated in Figure 2.3, although complete elimination of corrosion is likely to be both expensive and pointless. We will return to this aspect of CP when we discuss the protection potential in Chapter 6.

b. MIC and Pitting

   In this respect, it makes no difference if the steel is undergoing general corrosion in seawater, or its corrosion morphology is pitting due to MIC. Irrespective of the involvement of micro-organisms, the corrosion mechanism remains electrochemical. So, CP is effective.

c. Galvanic or Bimetallic Corrosion

   Similarly, CP can eliminate galvanic corrosion. In the case where CP is applied to a mixed-metal system, so long as both metals are polarised to the required protection potential of the more corrosion-prone metal (usually steel), the galvanic interaction is eliminated, and the corrosion is effectively controlled.

d. Corrosion Fatigue
We introduced fatigue and corrosion fatigue in Chapter 1. We discuss the involvement of CP with these processes in some detail in Chapter 12. However, for now, it is worth making the point that CP has the ability to increase the fatigue life of steel structures in seawater. In the extreme, it is capable of returning the fatigue behaviour, as depicted in an S-N curve, from the behaviour in seawater (corrosion fatigue) to the behaviour as if it were in dry air (fatigue), or even better. Indeed, for many structures its fatigue benefit outweighs its ability to control general corrosion.
e. Corrosion-Resistant Alloys
Although carbon steel is the dominant material of construction in the marine environment, occasional use is made of higher value corrosion-resistant alloys (CRAs) for some applications. We consider these alloys in Chapter 8, where we will find that they usually have good general corrosion resistance in seawater. They do not rust. However, they can be vulnerable to more pernicious localised forms of corrosion under some circumstances. Depending on the alloy type, examples of localised corrosion threats include pitting, crevice corrosion and stress corrosion cracking. For now, we need simply note that, subject to certain limitations, CP can also effectively control these localised forms of corrosion.

2.8 WHERE DO WE GO FROM HERE?

We now have enough basic knowledge to start considering how we might design CP systems for some straightforward marine installations. Indeed, we already have more grounding in the relevant theory than many "spreadsheet engineers" who find themselves assigned to carry out CP design work.

With this in mind, in the next chapter, we will go through the basic steps involved in designing CP systems, in accordance with the relevant codes, for two typical offshore structures pipelines not covered in chapter 3.

However, the fact that you have put your nose into this book suggests that you might be thinking about the subject a little more deeply than is required to slavishly implement a set of codified parameters in a spreadsheet. Furthermore, it also suggests that you may be considering some less than straightforward situations, or you might just simply be curious about what is actually going to happen when your CP design is put into practice.

For example, you may have questions such as:

- What is the basis of the designated protection potential, should it be different in different environments, at different temperatures, or for different metals?
- What happens if our CP system fails to polarise our structure to the intended (design) protection potential, or if we polarise to more negative potentials?
- What is the source of the published design current densities, and what current densities should we use for situations not envisaged in the codes?
- Where do the predictions of coating breakdown come from, how reliable are they, and how do they change for different coatings?
- How do sacrificial anodes work, and how do their performances change with changes in alloy composition?
- Having designed and installed a CP system, how do we know if it is working?
- What should we do if it is not working properly?

As we get deeper into this book, we will examine these and other questions.

REFERENCES

1. H. Davy, *On the corrosion of copper sheeting by sea water, and on methods of preventing this effect; and on their application to ships of war and other ships.* Philosophical Transactions of the Royal Society A 114, 151 1824.
2. K.R. Trethewey and J.R. Chamberlain, *Corrosion for Students and Science and Engineering.* Longmans (Harlow) 1988, p. 1.

Chapter 3

Designing according to the codes

Using a "Cookbook"

Although I have never studied cookery, I can follow a recipe and turn out an acceptable spaghetti carbonara. Anybody can. For much the same reason, any numerate engineer armed with a reliable CP design "recipe" can usually produce a fit-for-purpose CP design.

As with engineering in general, CP is a predominantly code-driven enterprise. A collection of codes exists to govern its design and operation across a wide spectrum of onshore and offshore applications. These codes include various standards, specifications, recommended practices and guidelines that have been developed by industrial, national and international committees of experts. This means that, with only the basic theory we covered in Chapter 2, and access to the appropriate code, we can set to work producing a CP design.

Before we do so, however, we should be aware of an underlying problem. My ability to churn out a bowl of pasta does not make me a culinary genius. For the same reason, an ability to read a design guide and to construct a computer spreadsheet does not transform an engineer into a CP expert. This then begs the questions: how far can engineers rely on codes; and how much of a grasp of the underlying principles of CP, including the electrochemistry, needs to be brought to bear? These are questions for later chapters. For now, we will content ourselves with "run-of-the-mill" CP designs.

3.1 DO WE NEED CP?

This is not a decision for the CP designer. It is a decision for the corrosion engineer, who may be a different individual. Even then, any views of the corrosion engineer will need to take account of local legislation or the views of certifying authorities.

The fact that steel will corrode in seawater does not necessarily mean that corrosion control is necessary, or even desirable. As we mentioned in Chapter 1, some corrosion is inevitable but that does not mean that it actually matters. It is a perfectly legitimate corrosion management strategy to allow corrosion to proceed, providing the corrosion damage that accrues over the lifetime does not compromise the fitness-for-purpose of the component. Sometimes, the relevant corrosion engineering option will be simply to incorporate a corrosion allowance into the thickness of the steelwork and to let corrosion take its course. Indeed, as we will see in Chapter 13, this approach has historically been adopted for harbour piling, although an emerging awareness of ALWC since the 1980s has produced an increased uptake of CP.

It follows that not every structure exposed to seawater actually requires CP, or if it does, it may not need the CP to operate for its entire life. Nevertheless, for the hydrocarbon

DOI: 10.1201/9781003216070-3

production structures that we use as examples in this chapter, dispensing with CP is unlikely to be an option. CP will usually be required, either for regulatory compliance or to satisfy the certifying authority.

3.2 WHO DOES THE CP DESIGN?

3.2.1 In the land of the blind

At its heart, providing a CP system for a structure is an exercise in applied electrochemistry. Nevertheless, from Sir Humphry Davy to the present day, successful CP designers have made no claims to be electrochemists. The lesson is simple: in many instances, a fit-for-purpose CP design can be produced without any deep appreciation of the underpinning science. That this is so bears testament to the general robustness of the technique itself, and to the usability of some CP guideline documents.

One result of this, albeit unintentional, is that some designs are produced using what we might call a "cookbook" approach, by engineers more familiar with Excel® or MathCad® software than with the fundamentals of CP. This is not necessarily a bad thing; any more than cooking according to a recipe book is a wrong. In many ways, CP design lends itself to the simple replication of previous examples.

Indeed, at least one major oil company dispenses altogether with the nuisance of producing CP design for its offshore installations. Pipeline contractors simply install stock bracelet anodes at prescribed intervals along the line. Platforms are protected by installing the appropriate number of standard impressed current anode sleds. (We will say more about impressed current in Chapter 11.)

The world's seas are filled with static and moving CP systems. The vast majority have been designed by individuals who would struggle to offer a coherent explanation for the basic concepts. Historically, CP design has often owed much to the adage: in the land of the blind, the one-eyed man is king.

Despite this, the corrosion of steel in the oceans is, for the most part, well-controlled. As a matter of experience, oil platforms are not toppling over, ships are not springing corrosion leaks, and aged subsea pipelines are being kept in service. That this has been so is due, in no small part, to the conservatism of the codes.

3.2.2 Who is the "Expert"?

Not everybody who claims to be a CP expert actually is. I encountered a spectacular example of this early in the millennium. The case involved a multi-metal effluent treatment tank which required internal CP to prevent corrosion penetration of the painted carbon steel tank wall. As we will see in Chapter 16, a complex multi-metal system provides a challenge for even the most seasoned CP practitioner. In the case of this tank, however, the owners saved the expense of a CP design specialist. The company accountant appointed himself as expert. Apparently, he had fitted an anode to his sailing boat, so he understood CP.

To be fair to our accountant/CP specialist, the outcome was reasonably successful. The tank lasted for almost 2 weeks before it started leaking! It brought to mind Einstein's remark... *The difference between stupidity and genius is that genius has its limits.*

Fortunately, such blunders in CP are rare. Nevertheless, mishaps do occur from time to time. Sometimes, these have arisen when CP experience has been transferred from one arena to another. For example, we will discover in Chapter 13, that some CP installations underperformed in the early days of the exploitation of the North Sea's petroleum reserves. This was partly due to the unfiltered import of the experience gained earlier in the more quiescent waters of the Gulf of Mexico.

Those issues were corrected soon enough, although usually not without expense. The lessons learned for such fixed petroleum production structures in harsher waters are now enshrined in standards such as DNV-RP-B401, to which we will return below. Nevertheless, in an example of history repeating itself, the late 20th century CP know-how gained for structures involved in hydrocarbon production in European waters was then reapplied, again unfiltered, in the early 21st century; this time to near-shore windfarm monopile foundations. Once again, the results were disappointing in some cases.

Major blunders are, of course, easy enough to spot; at least with the benefit of 20:20 hindsight. This is because it is a simple enough matter to measure the potential of the structure to see if it is protected. We will have more to say about potential and its measurement throughout this book, particularly in Chapter 18. On the other hand, cases where CP systems have been over-designed, even by the conservative measures customary in the industry, are much less obvious. If a structure is adequately protected throughout its design life, nobody is ever motivated to re-examine the original design to see if it could have been done at a lower cost.

We need to understand that CP is not a counsel of perfection. Its role is to provide sufficient protection for our installation to function safely for the duration of its design life. It is not at all necessary for the CP system to be designed, operated, maintained and managed by experts. By any yardstick that would be a frivolous over-indulgence. What is important is that the individuals responsible for carrying out a CP activity should be competent in that role. This brings us to the issue of certification.

3.2.3 Certification of competence

3.2.3.1 Is it needed?

This is a debate for another day. However, there can be little doubt that the engineering world is coming to be increasingly dominated by the requirement to be able to produce a certificate to show that you can do a job. There is a logic to this. It is simpler to check if potential employees or contractors have the appropriate certificates than it is to investigate their expertise yourself. For this reason alone, it is a fairly safe prediction that certification will become progressively more indispensable for those wishing to undertake CP work.

Nevertheless, we need to bear in mind that practically all of the advances in marine CP that we will come across in this book have been brought about by individuals who, at the time, possessed no CP certificate. Certification does not make you a more competent CP professional, but it does help you to convince others of your competence.

3.2.3.2 NACE

The professional body with the longest history in the certification of CP personnel is NACE, which re-named itself “AMPP” in 2021. It had developed four levels of CP competence: tester, technician, technologist and specialist. However, with its origins firmly in onshore CP, NACE’s certification schemes are aimed almost exclusively at land-based CP practice. Thus far, the organisation’s only foray into certification for offshore CP personnel has been a scheme for CP technicians working on ships.

3.2.3.3 ISO 15257

European standard EN 15257 was published in 2006. Although the bulk of the standard was directed at onshore CP, it extended the NACE range of certification into offshore CP, as well as to internal CP and reinforced concrete. It described three levels of competence. These broadly corresponded to NACE’s technician, technologist and specialist. The EN document was reissued as an ISO standard with the same number in 2017. It increased the EN's number of certification levels from three to five, introducing the tester (level 1) and the expert (level 5). These are summarised in Table 3.1 in so far as they are relevant to marine CP.

Table 3.1 ISO 15257 competence levels for marine CP

Level	*Indicative title*	*Knowledge areas*
1	Data collector (or tester)	Relevant safety issues, the collection and recording of CP operational data (potential and current) in accordance with instructions, sources of error in those measurements.
2	Technician	Level 1 plus: the fundamentals of electricity, corrosion, coatings, CP and measurement techniques, and relevant CP standards.
3	Senior Technician	Level 2 plus: the general principles of corrosion and CP, the principles of electricity, the significance of coatings and their influence on CP and a detailed knowledge of CP test procedures and safety issues.
4	Specialist	Level 3 plus: detailed knowledge of corrosion theory, principles of electricity, CP design, installation, commissioning, testing and performance evaluation, including systems affected by interfering conditions; establishing testing and performance criteria where none are available. General familiarity with CP in all application sectors.
5	Expert	Level 4 plus: knowledge derived from having made an original contribution to the science or practice of CP.

The ISO standard provides much more detail on the required levels of understanding of the various corrosion and CP-related disciplines for each level. We will return to the issue of competence at various points throughout this book.

In this chapter, we are concerned with basic CP designs and will be restricting ourselves to using only sacrificial anodes. In this respect, it is interesting to note what Table 4 in ISO 15257 says about the competence levels required to undertake offshore CP designs. According to the ISO, the minimum certification level to do even the most elementary design, such as for small boats or buoys, is a CP Technician (level 3). For the CP design of more substantial structures, such as coastal, offshore and submarine facilities, floating production and storage structures, and ships, a level 4 CP Specialist is required.

According to the ISO, an expertise in accountancy does not make the grade!

3.3 THE BASIS OF DESIGN

For large or prestigious CP projects, it is usual to prepare a basis of design (BoD) document. The scope of this document is to detail the information and input parameters needed to prepare the CP design. The following subsections indicate what is typically included in a BoD.

3.3.1 System life

The design life (T) for a CP system is invariably dictated by the design life of the structure it is to protect. The CP system life, however, is often longer than the planned operating life of the facility, since it might be required also to cover the fitting out and the decommissioning periods.

Ideally, the object of designing a sacrificial anode CP system is to ensure that exactly the right amount of anode alloy is installed to maintain polarisation of the structure until the last day of its design life. In practice, CP does not lend itself to such precision. The conservatism in the codes usually results in anode provisions that will outlast the intended design life. This conservatism is easily justified. It is relatively cheap to install anodes on a structure while it is being assembled on dry land, but very costly to retrofit anodes once it has been deployed offshore. Hence, some degree of erring towards over-design is both common and prudent.

That does not mean, however, that it is the CP designer's duty to second-guess the BoD. If the stated design life is (say) 20 years, then it is the CP designer's obligation to provide a 20-year system; not to anticipate any future extension of the service life.

3.3.2 Environmental parameters

The BoD should provide the seawater properties relevant to preparing the CP design. If we are using DNV-RP-B401, we need the mean water temperature, depth and resistivity. A recent standard for the CP design of wind farm foundations recommends that CP design should be based on the more detailed metocean[1] data for the site. We will return to this in Chapter 13.

The temperature and water depth are relevant to determining the design cathodic current densities. The resistivity is needed to calculate the anode current outputs. If the BoD does not specify a seawater resistivity value, it can be derived from CP codes, many of which provide a graph of its known relationship to salinity and temperature (e.g. DNV-RP-B401, ISO 15589-2, or EN 12495). Other codes present resistivity data with respect to seawater density and temperature (e.g. EN 16222, EN 17243 and ISO 13174).

A typical salinity for open seawater is about 3.5% but is higher in some areas (such as the Arabian Gulf) and lower in others (e.g. the Baltic). It also varies with depth. Accordingly, the resistivity[2] ranges from as low as ~0.15 Ωm (in warm waters) to about 0.30 Ωm (in cold waters). In the absence of site data, it is prudent to assume a conservative value of 0.30 Ωm for open seawater in a temperate zone.

3.3.3 Coating

As we saw in Chapter 2, providing we deliver enough current, CP alone can fully protect the underwater areas of a marine structure. Coating alone cannot provide 100% protection, because a coating is always subject to incidental damage and natural deterioration which will leave some steel exposed. However, a coating greatly reduces the current demand, and thence the cost of the CP system. It would be economically impracticable, for example, to apply CP to an uncoated subsea pipeline. In other cases, coatings are required for reasons other than simply protecting against corrosion. Examples include antifouling coatings on ships and coatings designed to aid visibility for divers.

The decision whether or not to apply a protective coating to the underwater areas, like the decision to install CP itself, is the responsibility of the corrosion engineer. It is not always straightforward. For example, in Chapter 13 we will examine the CP of two very similar real structures. The owner of one decided to paint the immersed steelwork. The owner of the other left it uncoated. The corrosion of each structure was controlled entirely adequately throughout their service lives. For this reason, we will look at designing for both coated and uncoated structures.

It is the role of the BoD to capture the corrosion engineer's decision. If the structure will be coated, the BoD should also set out the quality of that coating. This is frequently achieved by referring to it in terms of its Norsok classification. We return to coatings in Chapter 9.

[1] "Metocean" is a composite term derived from "meteorological" and "oceanographic".

[2] Many publications give seawater resistance values in units of Ωcm. The conversion factor to Ωm is 0.01. (i.e. 25 Ωcm = 0.25 Ωm).

3.3.4 Sacrificial versus impressed current?

We outlined the various advantages and disadvantages of sacrificial anode and impressed current systems in the previous chapter. We will assume that the BoD calls for the CP system to be based on sacrificial anodes, as is the case for the majority of fixed offshore installations.

3.3.5 Which codes?

There is no international law telling us how we must design a CP system. A BoD could start from scratch and develop its own design parameters and methodologies. However, it makes more sense to learn from the experience, both successful and otherwise, of those who have gone before. This experience is embodied in published codes and the in-house practices of major operators. The owner usually chooses the design code. It is not the CP designer's role to overrule that choice. If, for some reason, the design needs to depart from the code, then the owner's written authority is required.

There are offshore CP codes for: pipelines, harbours, ships, floating production systems, windfarm foundations and internal spaces. We will touch upon each of these in later chapters. Here, however, we will assume that the CP is needed for hydrocarbon production structures and that the BoD requires CP to be designed in accordance with DNV-RP-B401.

3.3.6 Cathode parameters

From the previous chapter, we saw that CP is achieved by applying current to shift the potential of the steel from the value it adopts when freely corroding to a more negative value at which corrosion becomes negligible. By referencing a code such as DNV-RP-B401, the BoD effectively conveys what this target potential should be, and how much current is required in order to achieve it.

3.3.6.1 Protection potential

3.3.6.1.1 Uncoated steel

Remember that, in this book, we adopt the marine CP convention of referring all potentials to Ag|AgCl|seawater, unless we state otherwise.

The selection of the appropriate protection potential to use in the design can be trivially straightforward, or it can be the cause of much debate. We will return to the debate in Chapter 6. Here, because the BoD has instructed us to follow DNV-RP-B401, the selection of the protection potential is straightforward. It is −0.8 V for all parts of the structure in any marine environment.

Before moving on, we should note that some codes, and some CP experts, call for a more negative protection potential (−0.9 V) in the case of steel buried in seabed mud. The reason for this is a perception that a more negative protection potential is required where there is a threat of MIC, which we met in Chapter 1. This is something we will examine in some detail in Chapter 6. However, since DNV-RP-B401 does not require this more negative protection potential, we can ignore it here.

Nevertheless, an important point needs to be made. DNV-RP-B401 treats −0.8 V as a design input. It does not actually state that −0.8 V is also the least negative potential needed to secure protection. Similarly, it tells us nothing how long it takes to achieve the design potential once the CP is applied. This is also a subject to which we will return later in this book.

3.3.6.1.2 *Coated steel*

A coating does not alter the design protection potential.

3.3.6.2 *The protection current density*

3.3.6.2.1 *Uncoated steel*

Although the protection potential is fixed, the current needed to achieve, and then maintain, the steel at that potential varies according to the circumstances of the environment. As we have already explained, the current is directly proportional to the area of bare steel, so we refer to a current per unit area (or current density). We will have more to say about current densities and the complex way in which they vary with circumstance and time, in Chapter 7.

DNV-RP-B401, however, makes any attempt to characterise this varying current demand unnecessary. Instead, it prescribes "blanket" values that are agreed upon by the drafting committee. To this end, it provides the initial, mean and final cathodic current densities for bare steel.

- The **initial** current density (i_i) is the value needed to achieve polarisation. This is important for uncoated and part-coated structures. In practice, the initial current density requirement can be ignored in design calculations for fully coated structures and pipelines.
- The **mean** current density (i_m) may be thought of the value needed to maintain polarisation once it is achieved. For this reason, it is also referred to as the maintenance current density. It is the current needed to keep the surface protected.
- The **final** current density (i_f) is the value needed to re-polarise the structure if, for some reason, conditions have been so aggressive, for example during a storm, that the structure has depolarised. The final current density will not necessarily be applicable in all designs. For example, structures in deep water are unlikely to suffer depolarisation due to surface storms.

Table 3.2 gives the DNV-RP-B401 current density recommendations for bare steel in seawater at various geographical locations and in seabed sediments.

3.3.6.2.2 *Coated steel*

The mechanisms by which water, certain ions and ionic CP current are transported through a marine coating to the steel-coating interface are complex. However, from the standpoint of designing a CP system, the relevant effect of a coating is greatly to reduce the cathodic current demand by effectively reducing the area of exposed steel. However, over time, the coating degrades and its protective capabilities diminish. This arises from both natural deterioration of the coating material and the cumulative effect of incidental damage. This exposes more steel substrate, increasing the cathodic current demand.

The CP design accommodates this by introducing a coating breakdown factor into the design calculations.

This factor is not a scientifically determined parameter. It is a guideline prediction of how a coating might perform in the future. This prediction is itself the best guess of the drafting committee. It is based on the perception of how generically similar coating systems have performed in the past. Experience has shown that historical predictions of coating breakdown have generally been pessimistic. In response, the codes have cautiously evolved towards less dramatic predictions. Nevertheless, they all remain conservative, in keeping with their overall ethos.

Table 3.2 Seawater design cathodic current densities (DNV-RP-B401)

		Cathodic current density A/m²		
Location	*Depth (m)*	*Initial*	*Mean*	*Final*
Tropical (>20°C)	<30	0.15	0.07	0.10
	30–100	0.12	0.06	0.08
	100–300	0.14	0.07	0.09
	>300	0.18	0.09	0.13
Sub-tropical (12–20°C)	<30	0.17	0.08	0.11
	30–100	0.14	0.07	0.09
	100–300	0.16	0.08	0.11
	>300	0.20	0.10	0.15
Temperate (7–11°C)	<30	0.20	0.10	0.13
	30–100	0.17	0.08	0.11
	100–300	0.19	0.09	0.14
	>300	0.22	0.11	0.17
Arctic (<7°C)	<30	0.25	0.12	0.17
	30–100	0.20	0.10	0.13
	100–300	0.22	0.11	0.17
	>300	0.22	0.11	0.17
Any	Seabed mud	0.02	0.02	0.02

Table 3.3 provides the breakdown predictions offered by DNV for a typical generic coating system selected for subsea service. Factor (a) designates the initial coating breakdown, which largely reflects installation damage. Factor (b) is the assumed annual rate of coating breakdown, which is taken to include limited further mechanical damage together with the natural ageing processes within the polymeric film.

Table 3.3 Rates of coating breakdown advised for CP designs (DNV-RP-B401)

	Coating category			
	I	*II*	*III*	*IV*
Depth (m)	*Shop primed only*	*e.g. epoxy DFT > 250 μm*	*e.g. epoxy DFT > 350 μm*	*Norok M-501 System 3B or 7*
a) Initial coating breakdown factor				
Any	0.10	0.05	0.02	0.02
b) Annual coating breakdown factor				
<30	0.10	0.025	0.012	0.008
>30	0.05	0.015	0.008	0.005

The mean coating breakdown factor is calculated from the design life of the system (T years) and the coating breakdown factors ((a) and (b)) in equation 3.1 as follows:

$$\text{Mean Coating Breakdown Factor} = a + \frac{bT}{2} \tag{3.1}$$

Similarly, the final coating breakdown factor is calculated using equation 3.2

$$\text{Final Coating Breakdown Factor} = a + bT \tag{3.2}$$

For long-life structures, with coatings of lesser quality than considered in Table 3.3, it is possible for equations 3.1 or 3.2 to yield a nonsensical result greater than 1. In such cases, the final coating breakdown factor needs to be set to 1; and an arithmetic adjustment applied to the calculation of the mean coating breakdown factor. This adjustment is given in equation 3.3.

$$(\text{Adjusted})\,\text{Mean Coating Breakdown Factor} = 1 - \frac{(1-a)^2}{2bT} \tag{3.3}$$

It follows that the mean and final current densities for a coated cathode are arrived at by simply multiplying the bare steel current densities in Table 3.2 with the mean or final coating breakdown factor as appropriate. Initial current density requirements for coating structures are inevitably low. So, they do not normally feature in CP designs.

3.3.7 Anode parameters

3.3.7.1 General

We will cover anodes in more detail in Chapter 10. At this stage, we simply need to be aware of the electrochemical characteristics of an anode alloy that are relevant to a CP design. These are:

- the potential it adopts when delivering the CP current. This is often referred to as its closed-circuit or operating potential; and
- the total amount of CP charge that a given weight of anode can deliver. This availability of charge is termed its current capacity or, less usually, current efficiency.

In addition to the electrochemical properties of the alloy, we also have to consider how the manufactured anode itself will perform. Particular attention needs to be paid to how the anode alloy is supported on its steel core, since this determines how much of the anode mass can be effectively utilised in providing current.

3.3.7.2 Operating potential

As we saw in Chapter 2, for a metal to function as an anode in a CP circuit, it has to pass (negative) electrons through its metallic contact to the cathode (steel in this case). This means that the anode's potential must be more negative than the potential at which the cathode is protected. Furthermore, there needs to be a potential difference between the steel, when it is polarised to −0.8 V, and the anode, when it is operating. This potential difference, or electromotive force (EMF), is needed to drive the CP current against the resistance of the environment. In practice, this means that the CP of steel in seawater requires anodes that operate at about −1.0 V. This limits the available choices to zinc or aluminium alloys.

The operating potential of an anode alloy is not a fixed property. It is a characteristic that is mainly determined by the alloy composition but is also influenced by its microstructure which, in turn, depends on its thermal history during the casting process. It is also dictated by the salinity and temperature of the seawater.

The potential of an anode also varies depending on the current it is providing. This is a topic that we will revisit later. However, because in this exercise we are designing to RP-B401, we do not need to consider this feature. DNV provides a (conservative) figure for the operating potential of generic zinc and aluminium alloy anodes when they are connected to the structures (the "closed-circuit" potential). These values are reproduced in Table 3.4.

Table 3.4 Anode alloy properties in ambient conditions (DNV-RP-B401)

Alloy type	*Environment*	*Capacity (C) (Ah/kg)*	*Potential (E_a)(V)*
Al alloy	Seawater	2000	−1.05
	Seabed	1500	−1.00
Zn alloy	Seawater	780	−1.03
	Seabed	750	−0.98

3.3.7.3 *Charge availability*

3.3.7.3.1 Current capacity

Current capacity is defined as the charge available per unit weight of anode alloy. The charge is expressed as the product of current multiplied by time (ampere · hours). The capacity is then usually expressed in units of Ah/kg.

The capacity will depend mainly on the detailed composition and microstructure of the alloy. However, it will also be influenced by the environment, for example open seawater or seabed sediments, the temperature and the current it is delivering. We explore these factors further in Chapter 10.

Fortunately, where the BoD invokes DNV-RP-B401, there is no need to consider these factors. It is customary to adopt intentionally conservative capacity figures offered by the guidelines (see Table 3.4). Nevertheless, higher design capacities are permitted if supported by relevant testing.

3.3.7.3.2 Current efficiency

The maximum theoretical capacity of anode, that is the amount of electric charge produced when a given mass of anode dissolves, can be calculated quite easily from Faraday's first Law of Electrochemical Equivalence. In practice, the capacity will always be lower than the theoretical maximum.

In some spheres, mainly anode development rather than system design, it has been customary to express anode charge availability in terms of current efficiency rather than capacity. The efficiency is defined as the fraction of the theoretical charge, expressed as a percentage, that can be delivered. For example, an Al-Zn-In anode containing 5% zinc would have a maximum theoretical capacity of ~2870 Ah/kg. Testing shows that some commercial alloys of this type can deliver in excess of 2600 Ah/kg, which equates to an efficiency in excess of 90%. It may be noted that the default maximum efficiency to be used in a DNV design (Table 3.4) is conservatively set at around 70%.

3.3.7.4 *Utilisation factor*

The anode utilisation factor (U) is the proportion of the anode alloy that it is assumed will have been dissolved by the end of its life. This means that our CP design conservatively assumes that the anode has reached the end of its life, even when it will still be capable of outputting current.

The selected utilisation factor depends on the anode geometry, and it will be influenced by its design details, in particular the design and positioning of the steel core. The following summarises the DNV recommendations, where the description "stand-off" implies at least 300 mm separation between the underside of the anode and the structure.

Long slender stand-off anodes (length ≥ 4 × radius)	U = 0.90
Short slender stand-off anodes (length≥4 × radius)	U = 0.85
Long flush-mounted anodes (length > 4 × width or thickness)	U = 0.85
Short flush-mounted or bracelet anodes	U = 0.80

It is important to understand that these U values are not intrinsic properties. They are DNV's expectation of what an anode, with a carefully designed core, might be expected to achieve. Experience suggests that the U = 0.90 figure, meaning that 90% of the alloy will dissolve to produce useful CP current, is readily achievable for long slender stand-off anodes. There are anodes still outputting current even though much less than 10% of the original anode alloy remains.

On the other hand, ensuring even the more modest utilisation factor of 0.80 in the case of flush-mounted anodes, which are consumed only from the outer face, is more challenging. It requires careful core design.

3.4 THE DESIGN PROCESS

3.4.1 Overview

We will now examine the design process based on DNV-RP-B401. This requires estimating:

- How much current the CP system needs to provide to the structure at various stages of the life (initial, mean and final).
- The minimum mass of anode needed to supply the mean current over the design life.
- This is followed by exploring possible anode shapes and sizes that will allow it to deliver the maximum instantaneous current (initial or final) against the resistance of the environment.
- Finally, the design is optimised to find a reasonable balance between the anode numbers and the anode mass that will satisfy all aspects of the cathode's current requirements.

The process requires proceeding through a series of calculations which lend themselves to implementation using spreadsheets. This is why I have used the "cookbook" analogy. However, for balance, I should point out that DNV personnel have asserted at various conferences that RP-B401 is not a "cookbook".

3.4.2 Calculating the cathodic current demand

3.4.2.1 Interfaces

Most structures for which CP is required will also be in electrical contact with at least one other metallic component. For example, a hydrocarbon platform is likely to be in contact with pipelines. Those other components may, or may not, also be provided with their own

CP. The possible effect of the connected structure on the performance of the CP system designed for the target structure, therefore, also needs to be reviewed as part of the design process. For example, if a CP system is designed for a coated pipeline, and this is connected to an uncoated structure, then the structure will usually drain current from the pipeline anodes.

All such interfaces need to be evaluated for a CP design. We will say more about this later. However, for the purpose of this exercise, we will assume that all the connected structures are fitted with their own CP systems, designed to achieve the same target potential as our jacket. In effect, this means that, in this example, we do not need to consider the connected steelwork.

3.4.2.2 Uncoated zones

Step 1 Define the cathode zones

We can define each structure, or cathode, as the electrically continuous entity in contact with the seawater or seabed mud. The overall CP design has to accommodate the entire structure.

The structure may include substructures where the electrical continuity is uncertain. For example, a subsea christmas tree may have a protection cover, but it may not be certain that the two are in electrical contact. In such cases, we should assume that:

- Where electrical contact between substructures would be beneficial to CP, for example in permitting us to place anodes conveniently on only one substructure, it would be prudent to assume that continuity is absent. If the design requires continuity, then positive steps must be taken to provide it. Alternatively, separate CP installations are required for each substructure.
- Conversely, where electrical contact would be disadvantageous, for example in requiring us to provide CP current to items that we do not wish to protect, we must assume that continuity is indeed present. This is generally preferable to trying to achieve, and maintain, isolation subsea.

These two assumptions appear contradictory and pessimistic. Nevertheless, they are in keeping with the general conservatism required for subsea CP designs.

The structure is then divided into zones aligned with the water depths and exposure conditions (immersed or buried) recommended by DNV-RP-B401 for allocating protection current densities. These zones are adopted solely for the purpose of arriving at the total overall current demand for the structure. They must not be interpreted as imposing any physical restraint on the flow of CP current. For example, although the mud and seawater exposed zones of a structure merit separate current requirement calculations, it will be best to place all of the anodes in the seawater zone.

Finally, an obvious point needs to be made here. Just because some components may not need to be protected does not mean that they can be ignored in the CP design. If these components are connected to the protected structure, then CP current will flow to them, and this needs to be allowed for in the design.

A classic example failing to allow for all connected steelwork occurred in the early 1970s on a gas platform in the southern North Sea. The CP designers correctly calculated the exposed bare steel surface area of the structural steelwork, but they evidently omitted to include the area of the uncoated well conductors located at one end of the platform. Post-installation surveys revealed that was that the as-installed CP system was failing to provide sufficient current to protect both the structure and the electrically continuous well conductors. This necessitated an expensive CP retrofit.

Step 2 Calculate surface areas

The CP current requirement depends on the surface area of the cathode. This means that the surface areas will need to be calculated for each zone. Nowadays, surface areas are produced conveniently by computer-aided design (CAD) software. It is important to ensure that all the surface areas that can collect CP current are accounted for. This means that the design also needs to allow for features such as:

- roughness (welds, threads on fasteners, etc.),
- appurtenances, such as mud-mats and pile guides,
- connected third-party items (permanent and temporary),
- components to be bonded-in (e.g. riser clamp components), etc.

Step 3 Estimate the current demand

The current demand (A) for each zone (step 1) is then simply found by multiplying its area (step 2) by the current density for that zone (Table 3.2). This exercise is carried out separately for the initial, mean and final current densities.

3.4.2.3 *Coated zones*

The cathodic current demand for coated zones is carried out using the same three steps described above for uncoated zones. The only difference is that the mean and final bare steel current estimates are multiplied by the mean and final breakdown factors described above. Since the initial current demand for a coated structure is lower than the mean and final values, it does not feature in the design of a coated structure.

3.4.2.4 *Additional current demands*

The review of interfaces (Section 3.4.2.1) should identify any instances where the protected structure might be in contact with other items that may draw additional current. For example, when designing a CP system for a christmas tree, DNV-RP-B401 requires an allowance of 5 A for a subsea well. This 5A will need to be added to the design calculations for the initial, mean and final currents. This is an estimated allowance to cater for that part of the CP provision, intended for the tree, which will be consumed as a result of drainage to the outer surface of the well casing. This figure does not necessarily reflect the current needed to protect the casing. That depends on the nature of the strata the well passes through and the quality of the cementing.

3.4.3 Minimum anode mass

The anodes will be consumed in protecting the structure (cathode). We, therefore, need to calculate the minimum mass of anode alloy required. This is found from the sum of the mean currents (I_m) required for all of the zones.

$$W = \frac{8760\ T\ I_m}{CU} \tag{3.4}$$

W = weight of anode alloy (kg)
T= design life (years)
I_m = mean cathodic current demand (A)
C = current capacity (Ah/kg)
U = anode utilisation factor (see above)

Knowledge of the minimum anode mass required, however, is not enough to define the CP system. By way of an extreme example, equation 3.4 might tell us that we need 5 tonnes of anode alloy. That does not mean that we could cast, for example, five 1-tonne anode blocks, connect them to the structure, and all would be well. Almost certainly, the instantaneous current output would be insufficient to satisfy the current demand, and the distribution of what little CP current was generated would be poor.

We need to design anodes that, taken together, satisfy both the minimum mass requirement and the instantaneous output requirements. For a given alloy, the instantaneous current output depends primarily on the geometry and dimensions of the individual anodes.

3.4.4 Anode output

3.4.4.1 General

Unless the CP design is for a very large project, it will not be economic to commission an anode manufacturer to fabricate new moulds for casting the anodes. Accordingly, the majority of designs are based on anode dimensions, and anode core configurations, offered on an "off-the-shelf" basis by anode suppliers.

Table 3.5 Some typical trapezoidal stand-off platform anodes

						End-of-life dimensions (m)	
Type	*Length (m)*	*Width (m)*	*Depth (m)*	*Core diameter (m)*	*Al alloy weight (kg)*	*Length*	*Radius*
A	1.515	0.133	0.127	0.025	68.0	1.36	0.024
B	2.365	0.114	0.114	0.025	80.0	2.13	0.021
C	1.600	0.160	0.162	0.051	100.0	1.44	0.029
D	2.500	0.190	0.190	0.102	175.0	2.25	0.030
E	2.500	0.190	0.190	0.076	202.0	2.25	0.033
F	2.622	0.210	0.210	0.102	240.0	2.36	0.035
G	2.372	0.252	0.244	0.102	330.0	2.13	0.043
H	2.910	0.254	0.257	0.102	435.0	2.62	0.044
I	2.973	0.283	0.283	0.102	565.0	2.68	0.050

Table 3.5 shows a selection of stand-off trapezoidal anode dimensions, suitable for offshore platforms, from one of the major manufacturers.[3] If required, existing moulds can be shortened by inserting partitions. In addition, within certain limits, it is also possible to

[3] In this case we have used the examples from the product range of Impalloy Limited (www.impalloy.com). Other manufacturers produce similar ranges of anode size.

increase the diameter of the anode's steel core. This has the effect of producing a higher end-of-life current output. It will also produce an anode with a lower alloy weight and, therefore, a lower mean current availability.

For each of the anode designs in Table 3.5, the minimum total number of anodes that would be required is simply the quotient of the minimum required mass (Section 3.4.3) divided by the individual anode mass.

The objective of the next stages of the design exercise is to select one of the available anode types that:

- satisfies the minimum nett anode weight requirement, determined from the mean current requirement,
- satisfies the initial and final instantaneous current output requirements, but with minimum anode wastage.

To do this, we need to be able to calculate the instantaneous anode output. We assume that output is governed by Ohm's law:

$$I_a = (E_c - E_a) / R_a \tag{3.5}$$

where:
I_a = anode current output (A)
E_c = protected potential of the steel cathode (−0.8 V)
E_a = operating potential of the anode (−1.05 V)
R_a = anode resistance (Ω)

The fact that electrochemical systems do not follow Ohm's law is a subtlety we can let pass for the present purposes.

3.4.4.2 Estimating anode resistance

3.4.4.2.1 Initial

The anode resistance (R_a) is a function of the anode size and geometry, and the resistivity of the seawater environment (ρ). There are numerous anode shapes, and there have been various formulae advanced to calculate their resistances to earth. This subject is covered further in Chapter 10. For the purpose of carrying out basic designs, however, we can adopt the formulae given in DNV-RP-B401

For a trapezoidal anode which is at least 300 mm from the structure, RP-B401 advises that the anode resistance can be calculated using the modified Dwight formula:

$$R_a = \frac{\rho}{2\pi L}\left[\ln\left(\frac{4L}{r}\right) - 1\right] \tag{3.6}$$

where:
ρ = seawater resistivity
L = anode length (m)
r =anode radius (m)

The radius (r) is estimated from the average width and depth of the trapezoidal casting using:

$$r = \sqrt{\frac{\text{width} \times \text{depth}}{\pi}} \tag{3.7}$$

R_a is the only resistance term that we will calculate in this type of design exercise. It is the resistance between the anode and the seawater environment or, strictly speaking, between the anode and remote earth. The resistance of the cathode to the environment and the electrical resistance of the structure itself is ignored. For most marine CP applications, where the structure to be protected has a much greater surface area than the anodes, this is a reasonable approximation.

3.4.4.2.2 Final or end-of-life

The final or end-of-life resistance of the anode will be higher than the initial value because, as it will have dissolved away, it will have a smaller surface area exposed to the seawater. The standard way of calculating this for a stand-off type anode is to assume that (1-U) of the original alloy remains uniformly distributed around a length of core equivalent to U x the original length (L).

Once the final dimensions are determined, the final resistance is calculated from the modified Dwight equation, and the final output is then calculated using equation 3.5.

3.4.5 Anode optimisation

This step will involve trial-and-error calculations, most conveniently carried out using a spreadsheet. The object of the optimisation exercise is to select anode dimensions such that all of the following requirements are satisfied:

$$N \times W_a \geq W \tag{3.8}$$

$$N \times I_a(\text{initial}) \geq I_i \tag{3.9}$$

$$N \times I_a(\text{final}) \geq I_f \tag{3.10}$$

where:

W_a = weight of individual anode (kg) (supplier's data)
I_a (initial) = initial anode output (A)
I_a (final) = final anode output (A)
I_i = initial cathodic current demand (A)
I_f = final cathodic current demand (A)
N = number of anodes (calculation)
W = minimum required anode weight (kg)

3.5 EXAMPLE CALCULATIONS

3.5.1 Case 1 – uncoated structure

3.5.1.1 Life

The design life is stated to be 25 years.

3.5.1.2 Structure and area

To illustrate our design process, we can consider the steel jacket supporting a small drilling platform. This is in relatively shallow water (about 34 m depth[4]) in the southern North Sea. The relevant surface areas are set out in Table 3.6.

Table 3.6 Case 1 steel jacket area calculations

	Steel area (m^2)		
Structure components	*LAT to −30 m*	*−30 m to seabed*	*In seabed*
Legs	399.7	48.5	
Leg piles			322.3
Horizontal members	227.7	178.1	
Diagonal members	375.6	83.9	
Conductor guide frames	18.1	9.1	
Well conductors (5 off)	249.0	29.0	(see below)
Total	1270.1	348.6	322.3

For the purpose of this example, we will ignore the intertidal zone between HAT and LAT. This area will be coated with a high durability protective coating system. For the convenience of this illustration, we are assuming that the current demand of this zone is so much lower than that of the main uncoated underwater structure that it can be ignored. This is, of course, a non-conservative assumption, and a real CP design exercise would need to consider this zone.

3.5.1.3 Current densities

The design protection current densities are given in Table 3.7.

In addition, as recommended by DNV-RP-B401, our design will allow for a current drain of 5 A for each of the five wells.

Table 3.7 Case 1 steel jacket current densities

	Design current densities (A/m^2)			
Description	*Initial*	*Mean*	*Final*	*Current (A)*
LAT to −30 m	0.2	0.1	0.13	
Below −30 m	0.17	0.08	0.11	
Seabed sediments	0.02	0.02	0.02	
Wells				5

3.5.1.4 Current demand

The total design protection current is calculated from the information in Tables 3.6 and 3.7 The results for the total initial, mean and final current demands are presented in Table 3.8.

[4] We will follow the convention adopted by British Admiralty charts here water depths are related to chart datum which, in turn, is usually close to LAT.

Table 3.8 Case 1 current demand calculations

Exposure	Area (m²)	Current density (A/m²)			Current demand (A)		
		Initial	Mean	Final	Initial	Mean	Final
Seawater > −30 m	1270.1	0.20	0.10	0.13	254.0	127.01	165.1
Seawater < −30 m	348.6	0.17	0.08	0.11	59.3	27.9	38.3
Buried leg piles	322.3	0.02	0.02	0.02	6.4	6.4	6.4
Wells		(5 off @ 5 A per well)			25.0	25.0	25.0
Total					344.7	186.3	234.9

3.5.1.5 Minimum anode weight

We see from Table 3.8 that the mean current demand is 186.3 A. That enables us to calculate the minimum nett weight of anode alloy needed by inserting the relevant values into equation 3.4.

$$W = \frac{8760T\,I_m}{CU}$$

where

$T = 25$ years
$I_m = 186.3$ A (Table 3.8)
$C = 2000$ Ah/kg (Table 3.4)
$U = 0.90$ (see step 9).

This gives the result that the CP design must provide a minimum of 22.67 tonnes nett weight of aluminium alloy sacrificial anode.

3.5.1.6 Anode selection

We now need to work out the most efficient way of distributing that anode mass such that we can satisfy both the initial and final instantaneous current demand. For the purpose of this exercise, we will work with the selection of proprietary stand-off anodes presented in Table 3.5

We can use the initial (as cast) dimensions to calculate the anode resistance R_a using the modified Dwight formula given in equation 3.6. If we accept the default seawater resistivity $\rho = 0.3\ \Omega$m recommended in DNV-RP-B401, and employ a simple spreadsheet, the values of the resistance of each anode are as shown in Table 3.9.

Table 3.9 Case 1 initial anode resistance and output calculations

Type	Initial length (m)	Initial radius (m)	Initial R_a (Ω)	Initial current per anode (A)	Minimum No. of anodes
A	1.52	0.065	0.1114	2.24	154
B	2.37	0.057	0.0830	3.01	115
C	1.60	0.081	0.1007	2.48	139
D	2.50	0.095	0.0698	3.58	97
E	2.50	0.095	0.0698	3.58	97
F	2.62	0.105	0.0656	3.81	91
G	2.37	0.124	0.0672	3.72	93
H	2.91	0.128	0.0576	4.34	80
I	2.97	0.142	0.0551	4.54	76

The table also shows the initial current available from each anode. This is calculated from Ohm's law relationship (equation 3.5). The anode operating potential (E_a) is set at −1.05 V, and the cathode target protection potential (E_c) is set at −0.8 V. This gives a driving voltage ($E_c - E_a$) of 0.25 V.

Finally, the table gives the minimum number of anodes of each type that would need to be installed to satisfy the initial current requirement of 344.7 A.

This calculation is repeated for the end-of-life, or final, anode dimensions. The latter are estimated by assuming that (1-U) of the original anode mass remains uniformly distributed along U times the original core length. The results are given in Table 3.10.

Table 3.10 Case I final anode resistance and output calculations

Type	*Final length (m)*	*Final radius (m)*	*Final R_a (Ω)*	*Final current per anode (A)*	*Minimum no. of anodes*
A	1.36	0.024	0.1553	1.61	146
B	2.13	0.021	0.1122	2.23	106
C	1.44	0.029	0.1423	1.76	134
D	2.25	0.03	0.0998	2.50	94
E	2.25	0.033	0.0978	2.56	92
F	2.36	0.035	0.0930	2.69	88
G	2.13	0.043	0.0961	2.60	91
H	2.62	0.044	0.0815	3.07	77
I	2.68	0.05	0.0778	3.21	74

We can now compare the minimum number of anodes of each type needed to satisfy the initial, mean and final current demand. The minimum number needed to satisfy the mean current demand is obtained simply by dividing the individual nett alloy weights into the requirement of 22.7 tonnes. The number needed to satisfy the initial and final current demands is taken from Tables 3.9 and 3.10. The outcome is summarised in Table 3.11.

Table 3.11 Case I – anode optimisation

	Minimum number of anodes required				
Type	*Initial*	*Mean*	*Final*	*Result*	*Weight (tonnes)*
A	154	333	146	333	22.7
B	115	284	106	284	22.7
C	139	227	134	227	22.7
D	97	130	94	130	22.8
E	97	113	92	113	22.8
F	91	95	88	95	22.8
G	93	69	91	93	30.7
H	80	53	77	80	34.8
I	76	41	74	76	42.9

All of the anodes listed could do the job, providing the number in the "Result" column of Table 3.11 is installed. For anode types A to F, the number is dictated by the quantity of anodes required to satisfy the minimum total nett weight of the anode alloy. As the anodes get bigger then, obviously, proportionally fewer are required.

The number of anodes required continues to get smaller in the case of the large anodes (G to I). However, this further reduction in anode number incurs the penalty of having to install increasing weights of anode.

In practice, the anode selection would need to be made on the basis of comparing the installation costs with the anode costs. It seems likely that the final design would opt either for 116 type F anodes or, possibly, 93 Type G's.

3.5.2 Case 2 – coated structure

3.5.2.1 *Scope of CP design*

This example is for a petroleum production manifold assembly installed in the tropical regions. The assembly consists of three distinct parts:

- a guide base mounted on a monopile that is cement grouted in place. This serves to ensure that the main manifold is maintained at least 1 m above the seabed;
- the main manifold including its pipework; and
- a dropped-object protection cover.

It is likely that all three of these items will be in metal-to-metal contact, but this must not be assumed in the CP design. This means that separate, and sufficient, sacrificial anode provision is required for each. This example is concerned only with the CP of the main manifold. Similar design processes will be required for the guide base and protection cover. We will also assume that all pipelines serving the manifold are cathodically protected in their own right.

3.5.2.2 *Design parameters*

3.5.2.2.1 Areas

In this example, the areas for the main manifold have been calculated from design drawings as follows:

- coated steelwork structure: 1500 m^2
- uncoated pipework (operating at up to 55°C): 50 m^2.

3.5.2.2.2 Environmental

The environmental design parameters are listed in Table 3.12

Table 3.12 Case 2 example – design parameters

Parameter	*Value*	*Units*	*Remarks*
Life	20	years	
Water depth	350	m	
Seawater resistivity	0.30	Ωm	DNV-RP-B401 default value

3.5.2.2.3 Cathodic parameters

The design protection potential is set at −0.8 V for all alloys. This ensures the protection of the most vulnerable material (carbon steel). The cathodic protection current densities and the coating breakdown factors are set out in Table 3.13. In this design, it is assumed that all of the piping is (uncoated) stainless steel with a design temperature for the internal fluid of 55°C.

Table 3.13 Case 2 example – cathodic design parameters

Parameter		*Value*	*Units*	*Remarks*
Current density for structure	Initial	180	mA/m^2	DNV advice for >300 m depth
	Mean	90		
	Final	130		
Current density for hot pipework	Initial	210		DNV requirement to add 1 mA/m^2 for every °C above 25°C.
	Mean	120		
	Final	160		
Coating breakdown	Initial	2	%	DNV-RP-B401 Category III over 20 year life (at >30 m depth).
	Mean	10		
	Final	18		

3.5.2.3 *Calculations*

3.5.2.3.1 *Current demand*

In general, for structures that are (almost) fully coated, the key design current demands are the mean and the final (or end-of-life current). The mean current dictates the minimum mass of sacrificial alloy needed for the design life. The final current covers the instantaneous current output requirement needed at any time during the life but particularly towards the end of life when the coating breakdown has reached its maximum. Since comparatively little current is needed during the early stages of immersion, there is no need to consider the initial current requirement.

Based on the above areas, current densities and coating breakdown factors, we can calculate the mean and final current demands as indicated below:

Mean Current (I_m) =	*Area*	×	*Mean Current Density*	×	*Mean Coating Breakdown*	
Structure	1500 m^2	×	0.090 A/m^2	×	0.10	13.5 A
Pipework	50 m^2	×	0.120 A/m^2	×	1.00	6.0 A
Total						19.5 A

Final Current (I_m) =	*Area*	×	*Final Current Density*	×	*Final Coating Breakdown*	
Structure	1500 m^2	×	0.130 A/m^2	×	0.18	35.1 A
Pipework	50 m^2	×	0.160 A/m^2	×	1.00	8.0 A
Total						41.1A

3.5.2.3.2 *Anode outputs*

For the purpose of this exercise, we will use Al-Zn-In anodes with the DNV default design capacity of 2000 Ah/kg operating at a potential of –1.05 V (see Table 3.4). Using equation 3.4, we find that the minimum nett alloy weight we need for a 20-year life at a mean current of 19.5 A is 1898 kg.

For convenience, we will also examine the same proprietary anodes described in Table 3.5. Since the anode and cathode potentials, and the seawater resistivity, are the same as in case 1, we can use the final (end-of-life) anode outputs listed in Table 3.10. The resulting options are set out in Table 3.14.

Table 3.14 Case 2 – anode optimisation

	Minimum number of anodes required			
Type	*Mean*	*Final*	*Result*	*Weight (tonnes)*
A	28	26	28	1.90
B	24	19	24	1.92
C	19	24	24	2.40
D	11	17	17	2.98
E	10	17	17	3.43
F	8	16	16	3.84
G	6	16	16	5.28
H	5	14	14	6.09
I	4	13	13	7.35

Initial anode requirement not applicable to fully coated structures.

On the basis of these results, the designer would probably opt for 28 Type A anodes or 24 Type B's.

3.6 ANODE LOCATIONS

Having calculated the type, size and number of anodes required, there still remains the need to determine where we should attach them on our structure. For many structures, this does not present too much of a problem. For example, on the uncoated space-frame structure that we considered in Section 3.5.1, it is easy enough to distribute anodes more-or-less evenly over the surface. A rule-of-thumb of one anode for every 30 m^2 of steel surface would usually give a satisfactory outcome.

In the case of the more congested manifold structure that we considered in Section 3.5.2, it will probably prove more of a challenge to find space for all of the required anodes; but the anodes are smaller, so it should be easier to find room.

3.7 ANODE MANUFACTURE AND INSTALLATION

The CP design task is not yet complete. We still have to ensure that the anode is manufactured properly and attached securely to the structure. Fortunately, as with the design itself, there are codes to assist us. We will return to these issues in Chapter 10.

3.8 LIMITATIONS OF THE CODES

3.8.1 Can we use other codes?

Yes. Providing the BoD does not impose a restriction, we are at liberty to select other codes. Different codes recommend different values for the various design parameters However, some codes do not necessarily provide all the details needed to produce a design. For example, NACE SP0176 provides no guidance on coating breakdown factors. More important, we should never "cherry-pick" parameters from different codes just to make life easy for ourselves.

3.8.2 What if the codes gets it wrong?

At best, a CP design can do no more than capture the code drafting committee's view of what constitutes best practice at the time. This process is not infallible. Codes are subject to a continuing revision process which reacts to changes in industry practice, and evolution and refinement of design parameters.

Years ago, I always used to believe that rigid adherence to the codes was sufficient to protect against any future litigation if things happened not to turn out as hoped. It seems that things are not now so simple, at least in so far as English Law is concerned. As we will discuss further in Chapter 13, the English Supreme Court has recently pronounced that, in some cases, a designer must bear the responsibility for errors in a code. The case in question did not involve CP, although its consequences did.

However, the tendency of courts to look for precedents set by other judgements means that there is now a real possibility that rigid adherence to a code will not be a sufficient defence if you are unlucky enough to find yourself being litigated against on account of a CP system that the owner believes to be inadequate.

So be careful!

3.8.3 Where the codes are silent

If the limit of your ambition is to design CP systems for simple fixed structures, then you can stop reading here.

On the other hand, if you are still reading, then I suspect that you have realised that there is still a lot of ground that we have not covered. We have not yet paid any attention to ships or other floating structures, pipelines or to the CP of internal spaces, nor have we considered impressed current systems.

If you have an inquisitive or, preferably, cynical mind, you may well be wondering about the CP design parameters that we have taken from the codes. Where do they come from, and how reliable are they?

For example:

- What is the origin of the suspiciously round figure of −800 mV for the design protection potential? How reliable is it? Does it vary in different environments? Do different alloys require different protection potentials? How do we measure the potential in real situations? What are the implications if the target protection potential is not reached?
- How do the codes arrive at the design current densities? How reliable are the figures quoted? How do these change in different circumstances, for example on a moving vessel?
- Where do the figures for coating breakdown come from? How reliable are they? How does CP affect the coating?
- With regard to anodes: why do they have the compositions that they do? What would be the effect of changing the composition? Does the model of the anodes delivering current to remote earth really apply when all they are required to do is deliver it to the

adjacent structure? What is the origin of the resistance equations such as the modified Dwight? Are other equations more relevant in some cases, and if so, why?
- And so on...

If these questions interest you then read on.

However, it does mean that we need to dig a little deeper into the electrochemical theory behind CP. In the next two chapters we will look at thermodynamics, which tells us whether or not a reaction such as corrosion can happen; and kinetics, which tells us how fast, or slowly, it happens.

Chapter 4

Thermodynamics

4.1 INTRODUCTION

4.1.1 In the chemistry laboratory

Most of us will have encountered a dramatic example of corrosion during our elementary chemistry lessons at school. It was the preparation of hydrogen which was liberated by dissolving zinc granules in hydrochloric acid. We will probably also have written down equation 4.1 in our laboratory exercise books.

$$Zn + 2HCl \rightarrow ZnCl_2 + H_2 \tag{4.1}$$

Although this experiment was performed to demonstrate the production of hydrogen gas, it is also an example of a metal (zinc) reacting with its environment (HCl) and passing into the chemically combined state as zinc chloride ($ZnCl_2$). It is, by our definition, a corrosion process and, if memory serves, a very rapid one.

Having set the scene by dissolving zinc in hydrochloric acid, a devious chemistry teacher might then ask the class what salt would be formed if we were to dissolve copper in the same acid? The response would invariably be *copper chloride.*

Good guess; but wrong! It was a trick question.

Copper does not dissolve in hydrochloric acid. This is illustrated by the fact that we can use hydrochloric acid to clean a copper coin. The acid dissolves the oxide tarnish leaving the metal shiny, but otherwise unattacked.

It is a simple enough experiment to immerse some metal samples in various solutions for a period of time, and then to see whether or not corrosion has taken place. For example, we might leave samples of iron, copper and gold in solutions of seawater, hydrochloric acid and nitric acid. The results we would observe are summarised in Table 4.1. At first sight, there appear to be some inconsistencies in this table. For example, most of us would guess that concentrated hydrochloric acid is more corrosive than seawater. Generally speaking, we would be correct. However, as a matter of experience, copper corrodes in seawater, but not in hydrochloric acid. It requires an oxidising acid such as nitric acid to corrode copper, whilst gold is not corroded by any of these three solutions.

Table 4.1 Some metal-environment combinations

		Corrosion		
Environment	*Example*	*Iron*	*Copper*	*Gold*
Natural waters	Seawater	Yes	Yes	No
Mineral acid	Hydrochloric acid	Yes	No	No
Oxidising acid	Nitric acid	Yes	Yes	No

DOI: 10.1201/9781003216070-4

To make some sense of these observations, it is helpful to turn our attention to the discipline of thermodynamics.

4.1.2 In the real world

In Chapter 1, we made the point that *most* metals occur in nature in their chemically combined state, but that some, for example gold, are found in the uncombined, or native, state. We also remarked that an energy input was required to convert the chemically combined ores into usable metals. We should also have made the point that the amount of energy input required for extraction varies from metal to metal.

Considering the three metals in Table 4.1, gold requires no energy for its extraction. It is found in the elemental state. On the other hand, copper requires energy for its extraction but less than iron. This explains why the Bronze Age preceded the Iron Age. It took time for mankind to master the hotter fires, produced by charcoal or coal rather than wood, needed for the extraction of iron. Other common engineering metals, notably aluminium, require even more energy than iron for their extraction. Hence aluminium was not isolated until Sir Humphrey Davy achieved it in 1808, and it remained more than 100 times the price of steel until 1886 when it was discovered that it could be extracted from alumina by the application of electrical power.

The fact that different metals require different energy inputs to win them from their ores implies that the energy liberated when they corrode is also different. Intuitively, this suggests that the tendencies to corrode will also vary in a way that is linked to the amount of energy released. To explore this notion further, we need to explore the science of thermodynamics and examine what it can tell us about corrosion.

4.2 THE SCIENCE OF THERMODYNAMICS

4.2.1 Background

Thermodynamics has a very broad scope. For example, it is one of the disciplines brought to bear in our quest for a deeper understanding of the cosmos. However, its origins in the 19th century were more down-to-earth. It came to prominence as a result of endeavours to produce more powerful and efficient steam engines. The word itself, originally hyphenated "thermo-dynamics", was coined by William Thompson (later Lord Kelvin) in 1851. For those interested in the history of its development, I recommend Saslow's open-access review [9].

At its simplest, it is the study of the role of heat in physical and chemical changes. In the more mundane arena of corrosion, it adds to our understanding of why some metals corrode but others do not. Chemical reactions, including corrosion processes, can only occur spontaneously if they liberate energy. Such reactions are termed exothermic. Conversely, a reaction cannot occur spontaneously if it would require a nett flow of energy from the surroundings into the system. Such reactions are called endothermic.

4.2.2 Heat and mechanical energy

4.2.2.1 Parameters

The steam engine had demonstrated that heat could be used to perform mechanical work. So, 19th-century scientists set out to explore this mechanical equivalent of heat. However, the science could only advance once-reliable methods had been agreed for the measurement of mass (or rather weight), volume, pressure and (lastly) temperature. It also required an alignment of scientific thought on related parameters such as heat and energy. Early thinking was that heat was a fluid that flowed through a system; and the word energy, in its modern sense, only appeared in scientific discussion in the early years of the 19th century. However, by the middle of the century, these fundamentals had been effectively tied down.

Some of the main parameters, other than pressure temperature and volume, are listed in Table 4.2. This list includes two terms that are not in everyday usage: enthalpy and entropy. Enthalpy can be considered as equivalent to the total heat content of a system. Entropy is a thermodynamic quantity representing that part of the system's enthalpy (heat content) that cannot be extracted to produce mechanical work. William Thompson regarded it as "wasted energy". Entropy can also be viewed as the measure of the disorder in a system.

Table 4.2 Some key parameters in thermodynamics

Symbol	*Name*	*Definition*	*Units*
U	Internal energy	The sum of all the internal energy in a system	J
S	Entropy	A measure of the order in a system. It reflects the capacity for change. A highly ordered system can easily be disordered, but not vice-versa	J/K
H	Enthalpy	$H = U + PV$	J
G	Free enthalpy	$G = H - TS$ The free enthalpy content of a system (formerly referred to as the *Gibbs Free Energy*)	J
μ	Chemical potential energy	The free enthalpy of a mole of substance	J/mol

4.2.2.2 The laws

There are four laws of thermodynamics, numbered, somewhat clumsily, from zero to three.

The **Zeroth Law** tells us that if one system is in thermal equilibrium with two other systems, then those two systems are also in equilibrium with each other.

The **First Law** tells us that energy cannot be created or destroyed, but it can be converted from one form into another.

The **Second Law** tells us that if we do convert energy from one form to another, some will go missing. More precisely, for a process to occur spontaneously, there must be an increase in the entropy of the system (that is an increase in the disorder). Entropy always increases with time.

The **Third Law** states that the entropy of a pure substance at 0 K is zero.

For our purposes, we can ignore the clumsily named Zeroth Law. It tells us that if "a" equals "b", and "b" equals "c", then "a" also equals "c". Apart from being a statement of the obvious, it serves as a basis for defining temperature. Other than that, we can forget it.

Similarly, we need not dwell on the Third Law. It tells us that disorder disappears at the absolute zero of temperature.[1] However, since we are not going to get into the detail of entropy calculations, and since there could be no corrosion at 0 K, we can take this statement as read, and move on.

By contrast, we are very much concerned with the first and second laws, which were first enunciated in their present form by William Thompson. It is said pessimistically, but not inaccurately, that the First Law tells us that we cannot win in life. If that were not bad enough, the Second Law tells us that we cannot even break even! These laws tell us that we will never be able to make an engine which converts 100% of the available heat energy into useful work.

[1] The thermodynamic, and SI, unit for temperature is the Kelvin (K). However, in this book we use the more familiar Celsius (°C). 0 K is –273°C.

4.2.3 Chemical thermodynamics

To use thermodynamics for the study of chemical reactions, we have to introduce the fifth parameter listed in Table 4.2: the chemical potential energy (μ). This is defined as the free enthalpy of a gramme-molecular weight ("mole" or "mol") of a pure compound.

Applying the same First and Second Laws of Thermodynamics to chemical reactions tells us that, for a reaction to occur spontaneously, the change in the free enthalpy of the system (ΔG)[2] must be negative. By the sign convention of thermodynamics, this means that there is a nett outflow of heat energy from the reacting species to the surrounding universe.

In the case of a chemical reaction, such as might arise in corrosion, the free enthalpy change (ΔG) may be written as the difference between the sum of the chemical potential energies of all the reactants ($\Sigma\mu_{reactants}$)[3] and of all the products ($\Sigma\mu_{products}$).

$$\Delta G = \Sigma\mu_{products} - \Sigma\mu_{reactants} \tag{4.2}$$

For a reaction to proceed spontaneously, ΔG must be negative, so $\Sigma\mu_{reactants}$ has to be greater than $\Sigma\mu_{products}$.

4.2.4 Application to corrosion

We can illustrate the application of chemical thermodynamic laws to corrosion by considering the reaction of the three metals in Table 4.1 (iron, copper and gold) with water and oxygen. To simplify matters, we can assume that all reactants and products are in their standard states. By definition, this means that the reactants and products are at 298 K (25°C), and the oxygen is at a partial pressure of 1 atmosphere. We apply the superscript (°) to μ and G to indicate the standard state.

The expressions for the free enthalpy change for one mole of metal reacting then become:

For gold: $$\mathrm{Au} + 1\tfrac{1}{2}\mathrm{H_2O} + \tfrac{3}{4}\mathrm{O_2} \rightarrow \mathrm{Au(OH)_3}$$

$$\mu^o \quad 0 \quad -355.8 \quad 0 \quad +289.9\ \mathrm{kJ} \tag{4.3}$$

$$\boxed{\Delta G^o\ +645.7}\ \mathrm{kJ}$$

For copper: $$\mathrm{Cu} + \mathrm{H_2O} + \tfrac{1}{2}\mathrm{O_2} \rightarrow \mathrm{Cu(OH)_2}$$

$$\mu^o \quad 0 \quad -237.2 \quad 0 \quad -356.9\ \mathrm{kJ} \tag{4.4}$$

$$\boxed{\Delta G^o\ -119.7}\ \mathrm{kJ}$$

For iron: $$\mathrm{Fe} + \mathrm{H_2O} + \tfrac{1}{2}\mathrm{O_2} \rightarrow \mathrm{Fe(OH)_2}$$

$$\mu^o \quad 0 \quad -237.2 \quad 0 \quad -483.5\ \mathrm{kJ} \tag{4.5}$$

$$\boxed{\Delta G^o\ -246.3}\ \mathrm{kJ}$$

The simple calculation process is illustrated in equations 4.3–4.5. The values for the standard chemical potentials (μ^o) of water and the metal hydroxides are taken from standard texts (e.g. [6]). They have been obtained from thermochemistry measurements. For example, μ^o

2 The symbol Δ (Greek "delta") indicates a change.

3 The symbol Σ (Greek "sigma") indicates a summation.

for water is found by measuring the heat liberated when hydrogen and oxygen react together. However, the details of this thermochemistry need not concern us here. We are only interested in the results.

For example, the standard free enthalpy change for converting gold into its hydroxide is + 645.7 kJ/mol. The positive sign indicates that this reaction cannot proceed spontaneously. To set this thermochemical quantity in context: it tells us that to convert a gold Krugerrand (33.9g) to its hydroxide would need an input of energy equivalent to running a 100 W light bulb for about 18 minutes.[4] This is in accord with our experience that gold is found in nature in the uncombined, metallic state. It does not spontaneously react with natural, oxygenated waters.

In contrast to this, the ΔG^{o} values for the reactions involving copper and iron are negative. This tells us that it is thermodynamically favourable for both copper and iron to corrode in the presence of oxygen and water. Of the two, iron liberates the greater amount of energy. This tells us that it has a greater thermodynamic *tendency* to corrode. This, in turn, *may (or may not)* mean that it exhibits a higher corrosion rate.

As we will see later, the thermodynamic tendency to corrode does not equate to a rate of corrosion. Thermodynamics tells us nothing about rates of reactions. By way of an example, titanium has a higher thermodynamic tendency to corrode in water than iron, but in practice it is vastly more corrosion resistant.

Equations 4.3–4.5 are chemical equations. We have looked at the initial reactants and the final products, and calculated the standard free enthalpy differences between the two. This is a legitimate thermodynamic procedure. Indeed, equation 4.5 is the same as equation 1.1. However, we have already made the point that, although corrosion is a chemical process, it proceeds by an electrochemical mechanism.

In Chapter 1, we broke that equation down into its component anodic and cathodic half-reactions (equations 1.2 and 1.3), involving the release and consumption of electrons. We must now see how we can apply thermodynamics to electrochemical reactions. Our starting point is to return to the topic of potential for which we developed a non-rigorous workaday description in Chapter 2.

4.3 ELECTRODE POTENTIAL

4.3.1 The reversible electrode

The reversible electrode provides a handy starting point for our discussion of *potential.*

Let us begin by considering a piece of pure copper metal introduced into a solution of its ions. For example, we can take this to be a copper sulphate solution, which we might remember as being a pleasing blue colour. To keep things simple, we shall exclude oxygen.

There will be some tendency for copper atoms to leave the metal surface and enter the solution as positively charged copper ions.

$$Cu \rightarrow Cu^{2+} + 2e^{-} \tag{4.6}$$

The forward arrow (→) in equation 4.6 indicates that the reaction is proceeding from left to right. We should also recall that this is an anodic process. This anodic dissolution is favoured by the fact that the transition of copper from atoms in the highly ordered crystalline metal lattice to ions in the solution is thermodynamically favourable. It releases energy.

[4] Take my word for it. Don't bother trying the experiment.

However, each atom that dissolves leaves behind two electrons in the metal. This means that, in opposition to the *chemical advantage* achieved through the dissolution of the copper, we incur an *electrical disadvantage*. This is because the process of dissolution involves overcoming the energy barrier of separating positive and negative charges. To counteract this electrical disadvantage, there is some cathodic reduction of copper ions back to copper metal on the surface (equation 4.7).

$$\mathrm{Cu} \leftarrow \mathrm{Cu}^{2+} + 2\mathrm{e}^{-} \tag{4.7}$$

In this case, the reverse arrow (←) indicates that the reaction is proceeding from right to left.

The two reactions rapidly approach a situation whereby the chemical advantage of anodic dissolution (equation 4.6) is balanced by the electrical advantage (equation 4.7) gained in the cathodic redeposition process. The rate of dissolution balances the rate of deposition. The system is at *equilibrium* (equation 4.8). When an electrode is at equilibrium with its environment, it is termed *reversible.*

$$\mathrm{Cu} \Leftrightarrow \mathrm{Cu}^{2+} + 2\mathrm{e}^{-} \tag{4.8}$$

Since there is no nett change when a system is in equilibrium, there is no free enthalpy change. ΔG must therefore also be zero.

Because we are considering a process that involves chemical species and free electrons, we need to take account of both chemical and electrical potentials. We can do this by introducing the *electrochemical potential* (μ_e), which we define as follows:

$$\mu_e = \mu + nF\phi \tag{4.9}$$

where

n = number of electrons involved (2 in this case)

F = Faraday's constant 96,487 C/g equivalent.

where a coulomb (C) is an ampere-second, and g equivalent is the atomic or molecular weight divided by n. For example, since the atomic weight of iron is ~56, its g equivalent weight is 28 g.

ϕ = electrical potential (V)

As explained in Chapter 2, we are not so much concerned with the absolute values of the electrochemical potential, but with changes (Δ):

$$\Delta\mu_e = \Delta\mu + nF\Delta\phi \tag{4.10}$$

where copper is in equilibrium with its ions at a given concentration the electrochemical potential of the copper in the ionic and metallic phases must be the same ($\Delta\mu_e = 0$). In other words, the differences in chemical potential ($\Delta\mu$) and electrical potential ($nF\Delta\phi$) across the metal electrolyte boundary are in balance. Thus, equation 4.10 can be re-stated as equation 4.11, where E is termed the *electrode potential*, but should be termed the electrode potential *difference* between the metal and the solution.

$$\Delta G = -nFE \tag{4.11}$$

It is worth a brief diversion here to reassure ourselves that equation 4.11 makes sense in terms of units. As we mentioned in Table 4.2, the free enthalpy is a measure of energy so, in the SI system, ΔG is expressed in units of joules (J) per mole of material. The unit of potential difference (E) is the volt (V). This can be defined as energy (J) per coulomb (C or As) shown in equation 4.12.

$$V = \frac{J}{As} \tag{4.12}$$

Looking at the right-hand side of equation 4.11, we can express the quantities involved in terms of their fundamental SI units. Since n is dimensionless, the product $F \times E$ (Faraday's electrochemical equivalent times potential) has units of As/mol × J/As. Cancelling out the "As" element, we see that the units simplify to J/mol. This is the same as for ΔG. Thus, equation 4.11 is dimensionally sound.

Furthermore, as stated, when the reactants and products are in their standard state, we use the superscript "o" and equation 4.11 can be re-written as equation 4.13.

$$\Delta G^o = -nFE^o \tag{4.13}$$

where E^o is the standard equilibrium electrode potential.

E^o values for very many electrochemical equilibria are available in the literature. A list of some that are of interest to us here is presented in Table 4.3.

Conventionally, E^o values are reported with respect to the standard hydrogen electrode (E_{SHE}). However, because we use the more practical Ag|AgCl|seawater reference half-cell so widely in marine CP, Table 4.3 also lists the values on that scale.

Table 4.3 Some standard equilibrium electrode potentials

	E^o (V)	
Electrode reaction	*SHE*	*Ag|AgCl|seawater*
$Au \Leftrightarrow Au^{3+} + 3e^-$	+1.50	+1.25
$O_2 + 4H^+ + 4e^- \Leftrightarrow 2H_2O$	+1.23	+0.98
$Cu \Leftrightarrow Cu^{2+} + 2e^-$	+0.34	+0.09
$2H^+ + 2e^- \Leftrightarrow H_2$	0.000	−0.25
$Fe \Leftrightarrow Fe^{2+} + 2e^-$	−0.47 (see note)	−0.75
$Zn \Leftrightarrow Zn^{2+} + 2e^-$	−0.76	−1.01
$Mg \Leftrightarrow Mg^{2+} + 2e^-$	−2.37	−2.62

Note: Many older texts quote the E° value for the Fe/Fe^{2+} equilibrium as −0.44V (SHE). This figure was revised in the 1990's.

4.3.2 The Nernst equation

So far, we have only considered *standard* equilibrium potentials, where metals are in equilibrium with species at unit activity and at arbitrarily standardised temperature and pressure conditions. The real world is not so orderly. This means that our next step is to examine equilibrium potentials under non-standard conditions. Here we are indebted to the German scientist Walter Nernst (1864–1941). His equation 4.14 tells us how the potential changes as we change the concentration of the species participating in the electrode reaction, or as we change the temperature.

$$E = E^o + \frac{RT \ln K}{nF} \tag{4.14}$$

where n = number of electrons involved in the equilibrium reaction
F = Faraday's constant 96,487 C/g.equiv.
R = the gas constant (8.3145 J/mol.K)
T = temperature (K)
E = electrode potential under non-standard conditions
E^o = electrode potential under standard conditions
K = equilibrium constant

As we see below, this equation makes the construction of potential-pH diagrams possible.

4.4 E – pH DIAGRAMS

Thermodynamics tells us that any chemical process will only occur spontaneously if the process results in a decrease in free enthalpy. Furthermore, we have seen that we can relate free enthalpy to electrode potential (equation 4.11). We can now build upon this to explore the conditions under which any metal may, or may not, corrode. This brings us to E-pH diagrams. These are frequently referred to as Pourbaix diagrams in recognition of their development by the Belgian electrochemist Marcel Pourbaix (1904–1998).

4.4.1 The hydrogen electrode

We have already encountered the *standard* hydrogen electrode. It is now useful to consider that electrode under non-standard conditions. As seen in Table 4.3, we can describe the equilibrium involved in the following equation. In this case, we refer to this as equation (a) because of its significance in E-pH diagrams (see below).

$$2H^+ + 2e^- \Leftrightarrow H_2 \tag{a}$$

Since this equilibrium involves electrons, it is electrochemical in nature. It will have a potential that can be written in terms of the Nernst equation (4.15).

$$E_{H^+/H_2} = E^o_{H^+/H_2} + \frac{RT \ln K_{H^+/H_2}}{nF} \tag{4.15}$$

where
E_{H^+/H_2} = electrode potential for the H^+/H_2 equilibrium under non-standard conditions
$E^o_{H^+/H_2}$ = electrode potential for the H^+/H_2 equilibrium under standard conditions. By definition, this is the standard hydrogen electrode (SHE), so $E^o_{H^+/H_2}$ is 0 V on the SHE scale

K_{H^+/H_2} = the equilibrium constant written as: $K_{H^+/H_2} = \frac{\alpha^n_{H^+}}{\alpha_{H_2}}$
α = activity of the species

The activity, which is similar to "concentration", of the gaseous hydrogen at one atmosphere partial pressure is taken to be unity; and by definition, $\log(\alpha_{H^+})$ is (minus) the pH value of the solution.[5] Accordingly, equation 4.15 can be simplified to equation 4.16:

[5] The term pH was first coined by the biochemist S.P.L Sørensen in 1909. He defined it as: $P_H = -\log [H^+]$. The expression P_H, literally meaning "power of hydrogen", was changed to pH for the convenience of typesetters in the 1920s. Later, pH was redefined as $pH = -\log \alpha_{H^+}$. More recently, it has been redefined in terms of electrochemical measurements made in specific buffer solutions.

$$E_{H^+/H_2} = 0 - 0.059\,pH \tag{4.16}$$

where the value of E_{H^+/H_2} is expressed in volts on the SHE scale.

From equation 4.16, we can see, for example, that if we insert a piece of platinum foil into oxygen-free pure water, which has a pH value of 7, we would expect to observe a potential of $7 \times -0.059 = -0.41$ V (SHE scale). This would be approximately −0.66 V on the $E_{Ag|AgCl|seawater}$ scale.

We use platinum because, being a noble metal, it does not corrode in water. It is also very electrochemically active, so it supports the H^+/H_2 equilibrium on its surface. Thus, although this arrangement should, strictly speaking, be called a platinum electrode, it is universally referred to as a hydrogen electrode.

4.4.2 The oxygen electrode

If we now immerse our platinum into an *aerated* aqueous solution, it will also sense the equilibrium between dissolved oxygen and hydroxyl ions. By the convention of E-pH diagrams, this arrangement would now be termed an oxygen electrode, and its equilibrium is designated (b):

$$O_2 + 2H_2O + 4e^- \Leftrightarrow 4OH^- \tag{b}$$

To appreciate the relationship between the potential and pH for this oxygen electrode, it is convenient to rewrite equation (b) as follows:

$$E_{O_2/H_2O} = E^o_{O_2/H_2O} + \frac{RT\log\left(\dfrac{\alpha_{O2}\alpha^4_{H^+}}{\alpha^2_{H_2O}}\right)}{4F} \tag{4.17}$$

where

E_{O_2/H_2O} = electrode potential for the equilibrium:

$O_2 + 4H^+ \Leftrightarrow 2H_2O$

under non-standard conditions

$E^o_{O_2/H_2O}$ = electrode potential for the equilibrium under standard conditions.

As seen in Table 4.3 $E^o_{O_2/H_2O} = +1.223$ V (SHE).

The activity of pure water (α_{H_2O}) is unity. On the basis that a dilute solution approximates to pure water, we take ($\alpha_{H_2O} = 1$). Similarly, if we assume that the partial pressure of oxygen is 1 atmoshere then the activity of oxygen α_{O_2} will also equal 1. Equation 4.17 then simplifies to equation 4.18:

$$E_{O_2/H_2O} = +1.223 - 0.059\ pH \tag{4.18}$$

The equilibria for the hydrogen electrode reaction (a) and the oxygen electrode reaction (b) are seen by plotting equations 4.16 and 4.18 on a potential (E) versus pH diagram (see Figure 4.1).

At potentials more negative than line (b), dissolved oxygen is reduced. Equation (b) becomes:

$$O_2 + 2H_2O + 4e^- \rightarrow 4OH^- \tag{4.19}$$

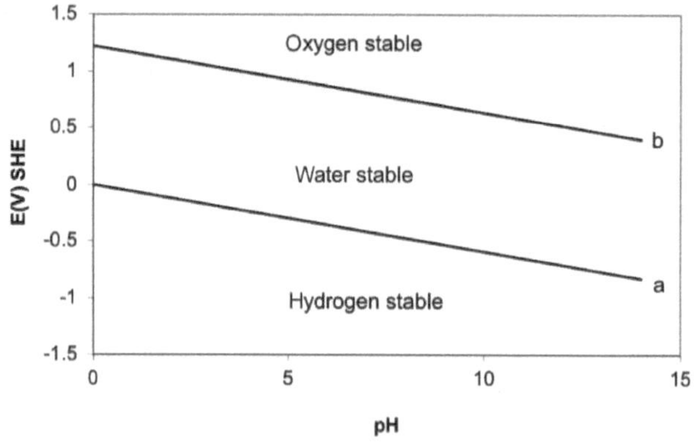

Figure 4.1 E-pH diagram for water.

where the change of symbol from ⇔ to → indicates a departure from a state of equilibrium to a situation where there is now a nett reaction taking place. It will also be recalled that, because this process consumes electrons, it is a cathodic process.

At potentials more negative than line (a), hydrogen gas is the stable species. In other words, the equilibrium (a) becomes the spontaneous cathodic process, and the water undergoes electrolysis to produce hydrogen gas:

$$2H^+ + 2e^- \rightarrow H_2 \tag{4.20}$$

Thus, lines (a) and (b) on an E-pH diagram indicate the domains where the cathodic processes may participate in corrosion reactions.

4.4.3 The metal and its corrosion products

Although much of our interest in CP lies in the corrosion of ferrous alloys and its control, it is convenient to develop our understanding of E-pH diagrams by using zinc. This is simply because the E-pH diagram for zinc is relatively uncomplicated. In water, zinc species can be involved in three electrochemical equilibria (equations 4.21–4.23).

$$Zn \Leftrightarrow Zn^{2+} + 2e^- \tag{4.21}$$

$$Zn + 2H_2O \Leftrightarrow Zn(OH)_2 + 2H^+ + 2e^- \tag{4.22}$$

$$Zn + 2H_2O \Leftrightarrow HZnO_2^- + 3H^+ + 2e^- \tag{4.23}$$

We need also to consider two chemical equilibria (i.e. not involving any change in valence state of the zinc). These are equations 4.24 and 4.25.

$$Zn^{2+} + 2H_2O \Leftrightarrow Zn(OH)_2 + 2H^+ \tag{4.24}$$

$$Zn(OH)_2 \Leftrightarrow HZnO_2^- + H^+ \tag{4.25}$$

Using the same processes as for the oxygen and hydrogen electrodes, these equilibria can be plotted on an E-pH diagram. The details need not concern us, but it will be readily apparent that 4.21 does not involve H^+ ions so is independent of pH. It appears as a horizontal line on the E-pH diagram. Its position on the E axis depends on the concentration of Zn^{2+} ions we choose for the equilibrium. Pourbaix's original work [2] considers various Zn^{2+} concentrations, producing a series of horizontal lines on the E-pH diagram. However, a convention has evolved whereby a metal ion concentration of 10^{-6}g ions/L is taken to set the equilibrium potential. The (rather woolly) thinking behind this is that, if a metal is in equilibrium with its ions at such a low concentration, it can be claimed not to be corroding. (We will return to this idea in the next chapter where we discuss the protection potential.)

Setting the Zn^{2+} concentration at 10^{-6}g ions/L means that the equilibrium constant (K) in the Nernst equation is also 10^{-6}. If we apply the Nernst equation (4.14) to the Zn/Zn^{2+} electrode equilibrium:

$$E = E^o + \frac{RT \ln K}{nF}$$

where

$n = 2$
$F = 96{,}487$ C/g equiv.
$R = 8.3145$ J/mol.K
$T = 298$ K (25°C is conventionally assumed)
$E^o = -0.763$ V (SHE) from Table 4.3
$K = 10^{-6}$ (arbitrary value)

we arrive at an equilibrium potential for the $Zn|Zn^{2+}$ electrode reaction of −0.955 V (SHE).

The chemical equilibria 4.24 and 4.25, which do not involve the transfer of electrons, are independent of potential. So, they appear as vertical lines on the diagram.

On the other hand, equilibria 4.22 and 4.23 are sensitive to both potential and pH, and therefore appear as sloping lines. The resulting diagram, showing the domains of stability of the various zinc species involved, is shown in Figure 4.2.

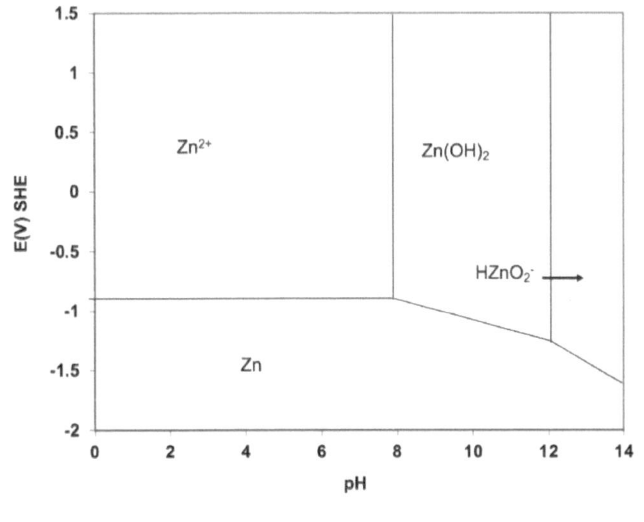

Figure 4.2 Simplified E-pH domains of stability for zinc.

4.4.4 The metal-water system

4.4.4.1 *Zinc*

If we now superimpose the E-pH diagram for water (Figure 4.1) on that for zinc (Figure 4.2), we obtain the E-pH diagram for the zinc water system (Figure 4.3).

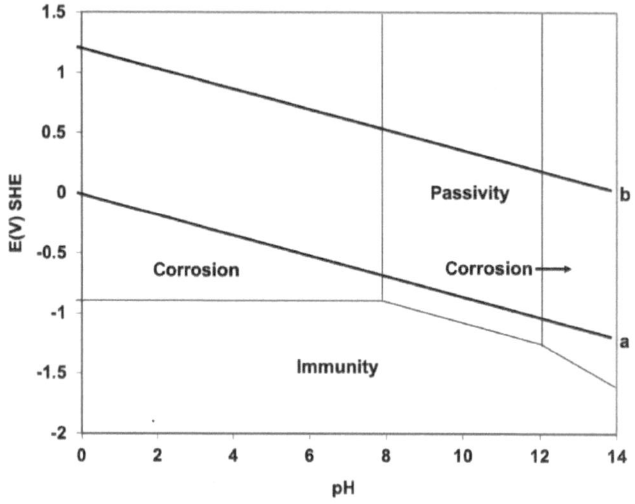

Figure 4.3 Simplified E-pH diagram for zinc in water.

This diagram provides us with some important insights into the corrosion of zinc. For example, if we consider a piece of zinc placed in a solution of hydrochloric acid (oxygen-free and at pH 0), then both the zinc equilibrium (4.21) and the hydrogen electrode equilibrium (a) seek to become established on the surface.

There is, however, a potential difference between these two equilibria. Under standard conditions, the potential of the hydrogen electrode (at pH 0) is, by definition, 0.0 V. The (theoretical) standard equilibrium potential of the zinc electrode is −0.76 V (on the SHE scale).

$$Zn \Leftrightarrow Zn^{2+} + 2e^- \qquad E^0 = -0.76\ V$$

$$2H^+ + 2e^- \Leftrightarrow H_2 \qquad E^0 = -0.00\ V$$

Nature is generally intolerant of differences, be they of pressure, temperature or, as in this case, electrode potential. It will always contrive to reduce such a difference. In the parlance of the electrochemist, the two electrodes will become polarised. That is, their potentials will shift towards a common value, intermediate between the two equilibrium values. The processes are no longer at equilibrium. Corrosion proceeds, with zinc going into solution via a process which liberates electrons. This is an anodic process. At the same time, hydrogen ions at the zinc surface consume these electrons to form hydrogen gas. This is a cathodic process.

$$Zn \rightarrow Zn^{2+} + 2e^-$$

$$2H^+ + 2e^- \rightarrow H_2$$

This is the reaction that we met at the start of this chapter. Although originally presented to us as a means of preparing hydrogen gas, it is also a dramatic example of a corrosion reaction.

In practice, no one would contemplate using zinc as a material to contain hydrochloric acid. So, as an example of a corrosion process, it is not particularly helpful. On the other hand, we are aware that zinc in the form of galvanising can give a reasonable service life in natural (near-neutral pH) waters. The E-pH diagram gives some clue as to why this is the case. As we move from left to right across Figure 4.3, we see that the equilibrium potential for the hydrogen evolution reaction (a) becomes more negative according to the Nernst equation. The result is that the potential difference between the zinc and hydrogen electrodes, which is the driving EMF for the corrosion cell, reduces. Thus, we are not surprised to find that the reaction between water at pH 7 (neutral) is less vigorous than it is at pH ~ 0 (acid).

If we continue to move further right across Figure 4.3, we find that, at slightly alkaline pH values (pH > 8), the stable corrosion product changes from soluble zinc ions (Zn^{2+}) to zinc hydroxide ($Zn(OH)_2$) which has a very low solubility. Under these conditions, the corrosion product may be expected to form a protective layer on the metal surface. This has the benefit of stifling subsequent corrosion.

At high pH values, on the far right of the E-pH diagram, formation of the soluble zincate ($HZnO_2^-$), rather than the insoluble hydroxide, is favoured. This illustrates the fact that zinc is soluble in both acids and alkalis; this behaviour is described as amphoteric.

4.4.4.2 *Copper*

We could construct an E-pH diagram for the copper-water system in the same way that we did for zinc. This would show us that the equilibrium potential for copper is more positive than the hydrogen electrode at all pH values. This explains why copper does not corrode in a mineral acid such as hydrochloric or sulfuric. However, it remains more negative than the equilibrium potential for the oxygen electrode. Thus, copper will corrode in natural aerated seawater. We could also infer from this that, although copper will not corrode in hydrochloric acid, it will dissolve in an oxidising acid such as nitric.

4.4.4.3 *Gold*

We could similarly also construct an E-pH diagram for the gold-water system in the same way that we did for zinc. This would show that the equilibrium potential for the dissolution of gold is more positive than the equilibrium potentials for both the hydrogen electrode and the oxygen electrode at all pH values. This explains why gold does not dissolve in natural waters. It is found in nature in its chemically uncombined elemental state. It also explains why gold remains unattacked by both mineral and mildly oxidising acids. If we wanted to dissolve gold in the laboratory, we would need to use a strongly oxidising acid such as the 3:1 mixture of hydrochloric and nitric acid known as *aqua regia*.

4.4.4.4 *Iron*

The same process used above for copper, zinc and gold may be used to construct an E-pH diagram for iron. However, the details of the process are more complex because more equilibria, between more species, are involved. Figure 4.4 provides one simplified version.

The diagram tells us that it is thermodynamically possible for iron to corrode due to the influence of dissolved oxygen or the hydrogen evolution reaction in aqueous solutions of any pH value.

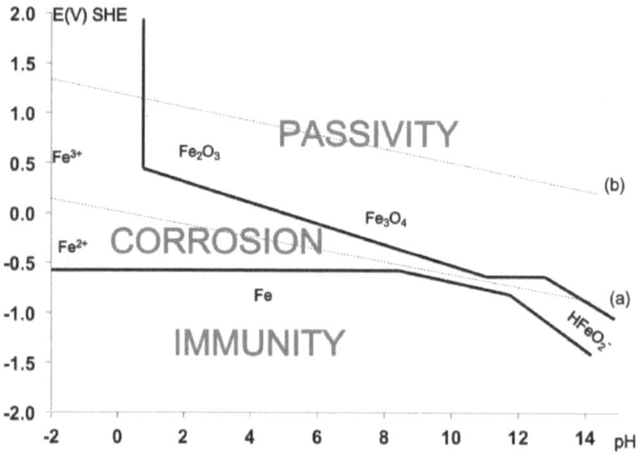

Figure 4.4 Simplified E- pH diagram for iron in water.

4.4.5 Limitations of E–pH diagrams

E-pH diagrams for metal-water systems play a useful role in helping us to appreciate which metal-solution combinations are likely to result in corrosion. Thus, it helps to explain why the metal-environment combinations shown in Table 4.1 behave as they do. However, E-pH diagrams have a number of fundamental limitations. We consider these below.

4.4.5.1 Pure metals

We have discussed Figure 4.4 which gives a simplified E-pH diagram for pure iron. However, we do not use iron for structures or equipment. We use steel, which is an alloy of iron, carbon, usually manganese and small levels of numerous other elements. Unfortunately, the thermodynamic data underpinning E-pH diagrams are only available for pure elements and compounds. It is not possible to construct an exact E-pH diagram for the more relevant steel-water system. We have to make a leap of faith that the E-pH domains for a pure metal are broadly similar to those for its alloys.

There is no justification for this leap of faith. By way of a trivial example, a typical steel contains about 1% manganese. A cursory examination of the E-pH diagram for this element [3] shows a more extensive range of stability for the soluble Mn^{2+} ion than the corresponding E-pH diagram for iron shows for Fe^{2+}.

4.4.5.2 Pure water

The construction of an E-pH diagram assumes that the only species present in the environment are those associated with pure water (H_2O, H^+ and OH^-). However, it is not possible to generate the range of pH, shown in the diagrams, without the addition of either acid-generating anions (such as chloride) or alkali-generating cations (such as sodium). Some of these ions will participate in the corrosion processes.

For example, the E-pH diagram for zinc in pure water (Figure 4.3) suggests that zinc will be passive at a pH above 8. However, as a matter of experience, zinc corrodes in seawater (pH ~ 8.2) at a rate in the region of 0.02 mm/year [5]. This is because the high chloride content of the seawater prevents the establishment of the protective zinc hydroxide passive film. Thus, a corrosion prediction based solely on the E-pH diagram for zinc would be incorrect.

4.4.5.3 *Thermodynamic basis*

E-pH diagrams derive from the laws of thermodynamics. Indeed, more percipient readers may be moved to point out that E-pH diagrams do not actually tell us anything that we do not already know. As we have seen, an E-pH diagram is constructed from thermochemical data. In other words, someone has taken the trouble to determine that reacting iron with oxygen liberates heat whereas converting gold to its oxide requires a heat input. A cynic might therefore aver that an E-pH diagram is just a particularly elegant re-statement of known experimental results. The cynic would be correct.

A more telling criticism is illustrated if we return to the E-pH diagram for zinc (Figure 4.3). We noted that in oxygen-free hydrochloric acid, there is a potential difference of ~0.76 V between the equilibrium potentials for the zinc and hydrogen electrodes. We also made the point that, if we place zinc in hydrochloric acid, corrosion takes place very rapidly indeed.

However, if we look at the situation for aerated, near-neutral water, we see that the equilibrium potential for the oxygen electrode (b) is very much more positive than the equilibrium potential for the hydrogen electrode (a), even where the former is in a neutral solution (pH 7) and the latter is in acid (pH 0). Thus, there is a greater driving voltage for the corrosion cell:

$$Zn \rightarrow Zn^{2+} + 2e^- \quad \text{anode}$$

$$\tfrac{1}{2}O_2 + H_2O + 2e^- \rightarrow 2OH^- \quad \text{cathode}$$

than for the cell where the cathodic reaction is hydrogen evolution. In other words, thermodynamics indicates that more energy is released when we corrode zinc in aerated natural waters than when we corrode it in acid. If all other things were equal, therefore, we would expect the zinc to corrode more rapidly in rainwater than in hydrochloric acid.

It doesn't.

Zinc is unusable in acid but has useful corrosion resistance to many natural waters. This is an observation that emphasises a fundamental limitation of E-pH diagrams. Being rooted in thermodynamics, it tells us whether or not corrosion is possible, but it cannot tell us how fast (or how slow) it will be.

We return to corrosion rates when we discuss kinetics in the next chapter.

4.5 CP AND THERMODYNAMICS

In Chapter 2, we introduced the concept of CP acting as a simple electrochemical switch that could, in principle, turn the corrosion off. Although this is by no means a bad image, the reality turns out to be more complex.

We have now developed outline theories of how the corrosion of steel can be explained in terms of thermodynamics, as represented in E-pH (Pourbaix) diagrams. This prompts us to consider whether thermodynamics can be relied upon to help us understand how CP works.

4.5.1 Immunity

From our simplified model, developed in Chapter 2, it would seem obvious to interpret CP as depressing the potential of the metal into the *immunity* region as indicated on an E-pH diagram. In fact, the immunity paradigm would work quite well for copper which, you will recall, was the first metal to be intentionally protected by CP.

Indeed, Pourbaix and co-workers stated [4]... *it should be possible to prevent copper from being corroded ... by judicious cathodic protection in such a way as to bring the potential of the metal to about +0.1 V (SHE) in acid solution and +0.1 to −0.6 V in neutral or alkaline solutions depending on pH.*

If we refer to the E-pH diagram for the copper-water system, we see that, at pH 8.2 (typical for seawater), we enter the zone of immunity at a potential of −0.1 V (SHE). Transcribing this to the Ag|AgCl|seawater reference gives us a value of about −0.36 V. It is also comforting to note that Pourbaix's mid-20th century view, that... *it should be possible*... to cathodically protect copper, successfully, albeit retrospectively, predicts Davy's early-19th century demonstration of that fact.

A final point that we should highlight here is that, according to Pourbaix, there would be no single target protection potential for copper. It varies according to the pH of the environment. More particularly, the protection potential becomes more negative as the pH increases. It will emerge that an electrode kinetic interpretation of CP would not agree with this view, but more of that later.

Nevertheless, the Pourbaix view might lead us to think that we can reconcile the successful history of the CP of steel in seawater with an E-pH diagram. All that we would need to do is shift the potential of the steel from its freely corroding value (E_{corr}) to a value within the immunity domain of the iron-water system. This would be equivalent to moving from the domain of "corrosion" to the domain of "immunity" in Figure 4.4.

Early theories of CP (see [1] for a review) considered that it would provide protection if the potential was lowered to the equilibrium potential of the iron electrode. At that point, the reaction:

$$Fe \rightarrow Fe^{2+} + 2e^{-}$$

could not proceed, so corrosion would cease. This would be immunity as envisaged on an E-pH diagram.

In principle, this view provides a straightforward way of determining the appropriate value for the protection potential. Unfortunately, a problem arises when we feed in the numbers. The standard potential for iron in equilibrium (E^{o}) with ferrous ions at a concentration of 1 g ion/L is −0.75 V. However, if the iron were in equilibrium with a solution containing 1 g ion/L of Fe^{2+} ions (i.e. 56000 mg/L), it could hardly be claimed not to be corroding.

Following the convention adopted by Pourbaix, it is often taken that corrosion can be discounted when the interfacial iron concentration is 10^{-6}g ion/L. This potential can be determined from the Nernst equation.

$$E = E^{o} + \frac{RT \ln K}{nF}$$

where

$n = 2$ for Fe/Fe^{2+}

$E^{o} = -0.75$ V (vs Ag|AgCl|seawater)

$K = 10^{-6}$

Because Fe is a solid (activity = 1), K equates to the activity of Fe^{2+} ions. Furthermore, at low concentrations: activity ≈ concentration.

The result, however, turns out to be (approximately) −930 mV.

The position becomes even more improbable if we challenge the, somewhat arbitrary, convention in E-pH diagrams that corrosion ceases if the concentration of metal ions in solution is below 10^{-6}g ions/L. There is a stronger philosophical case for assuming that the corrosion only ceases if we lower the potential to a value at which the iron is in equilibrium with Fe^{2+} ions at their natural concentration in seawater. The latter parameter is hard to tie down. We know that this figure is $<10^{-6}$g ions/L, but its exact value is uncertain. In total, iron is naturally present in seawater at levels of between 0.002 and 0.02 mg/L. However, much of this is in the form of suspended particulates, or else the iron is in complex form in biogenic matter. Based on the solubility product of $Fe(OH)_2$ and the pH of seawater (~8), the concentration of ferrous ions in solution is likely to be of the order of 10^{-10}g ions/L.

Using this concentration in the Nernst equation leads us to an estimate of a protection potential slightly more negative than –1060 mV.

However, as we saw in Chapter 3, the accepted protection potential for steel in seawater is –800 mV. This figure is supported by many decades of practical experience. It follows that we cannot explain CP simply as the depression of the potential of the steel into the region of "immunity" indicated on the E-pH (Pourbaix) diagram for the iron-water system.

In short, the immunity paradigm fails.

4.5.2 Passivity

The fact that the CP of steel can be effectively achieved without applying sufficient cathodic polarisation to depress the potential into the region of immunity has recently prompted a debate among onshore CP practitioners over the mechanism of CP. A view has emerged which may be summarised as follows. Because the action of CP is to increase the pH at the metal-environment interface, and because increased pH favours passivity, CP functions by rendering the steel passive [7,8]. The evidence advanced in support of the passivity paradigm has been challenged [10,11]. It has been concluded that there is no convincing evidence to support it in the onshore context.

Moreover, the passivity paradigm is a non-starter when it comes to CP in a chloride-rich medium such as open seawater. We will return to the issue of passivity in the next chapter, where we need to delve into a more rewarding branch of electrochemical theory: electrode kinetics. We will also need to re-visit the debate when we discuss the protection potential in Chapter 6. For now, however, we can park our discussion on thermodynamics.

REFERENCES

1. U.R. Evans, *The Corrosion and Oxidation of Metals.* Arnold (London) 1960.
2. N. de Zoubov and M. Pourbaix, Chapter 4 Section 15.1 Zinc, p 406, in *Atlas of Electrochemical Equilibria in Aqueous Solutions,* Eds. M. Pourbaix (English Translation). Pergammon Press (Oxford) 1966.
3. A. Moussard, et al. ibid. Chapter 4 Section 11.1 Manganese, p286.
4. N. de Zoubov, C. Vanleugenhaghe, and M. Pourbaix, ibid. Chapter 4 Section 14.1 Copper, p384.
5. Figure 1 *Typical lives of zinc coatings in selected environments*, in BS 5493 Protective coating of iron and steel structures against corrosion (1977) (superseded).
6. Table 21.5 *Standard chemical potentials*, in *Corrosion*, 3rd edition, Eds. L.L. Shreir, R.A. Jarman and G.T. Burstein. Butterworth-Heinemann (Oxford) 1994.
7. M. Büchler, *A new perspective on cathodic protection criteria promotes discussion.* Materials Performance **54**(1), 44 (2015).
8. R.A. Gummow, S. Segall and D. Fingas, *An alternative view of the cathodic protection mechanism on buried pipelines.* Materials Performance **56**(3), 32 (2017).

9. W.M. Saslow, *A history of thermodynamics: The missing manual.* Entropy **22**, 77 (2020).
10. J. Barthel and R. Deiss, *The limits of the Pourbaix diagram in the interpretation of the kinetics of corrosion and cathodic protection of underground pipelines.* Materials and Corrosion **72**(3), 434 (2021).
11. C. Googan, *The cathodic protection potential criteria: Evaluation of the evidence?* Materials and Corrosion **72** (3), 446 (2021).

Chapter 5

Electrode kinetics

The problem with thermodynamics

In Chapter 4, thermodynamics told us that iron dissolves in water at any pH value. This is useful information. However, thermodynamics also tells us that sugar dissolves in water. This may be unsurprising, but it is interesting to note that more energy is liberated when a given mass of iron corrodes in water than when the same mass of sugar dissolves. Intuitively, we might conclude that since more energy is released by the dissolution of iron, it would dissolve faster than sugar. Obviously, as a matter of everyday experience, this is not the case.

The flaw in my argument is that the two processes are chemically distinct: iron undergoes a chemical change when it dissolves, whereas sugar simply undergoes a physical dissolution process that is easily reversible. Nevertheless, the core point remains. Thermodynamics tells us whether or not a process can happen, but it is silent on the crucial matter of how fast or how slowly it will occur.

Corrosion engineers are as unsurprised by the fact that iron, or steel, corrodes, as they are by the fact that sugar dissolves in their coffee. What matters is not that corrosion can happen, but how fast it happens, or how slowly. We are interested only in the rate of corrosion, not the ultimate outcome.

This brings us to the topic of corrosion reaction rates. To explore this topic, we must turn to the science of electrode kinetics, where *kinetics* is derived from the Greek word for movement.

5.1 REVERSIBLE ELECTRODES

To get us started on electrode kinetics, it is convenient to re-visit the example of a piece of copper in equilibrium with a solution of copper ions:

$$Cu \Leftrightarrow Cu^{2+} + 2e^- \quad (5.1)$$

This is a reversible electrode. If we assume standard conditions 25°C and the concentration of Cu^{2+} ions as 1 M, which we could prepare by dissolving the molecular weight (249.7 g) of copper sulfate crystals ($CuSO_4 \bullet 5H_2O$) in 1 L of water, we would observe an electrode with a potential of +0.34 V measured on the SHE scale. This would be about +0.09 V if we measured it against the $E_{Ag|AgCl|seawater}$ half-cell used throughout this book.

From the thermodynamic perspective, this electrode is at equilibrium, $\Delta G = 0$ so nothing is happening. Left undisturbed, there will be no increase in the concentration of copper ions in the solution, nor any nett plating out of the copper ions back onto the surface of the metal. However, this does not mean that nothing is happening.

DOI: 10.1201/9781003216070-5

Electrode kinetics tells us that copper metal is indeed dissolving anodically to form cupric ions. However, it is doing so at a rate that is exactly matched by the rate at which cupric ions are plating back cathodically onto the metal surface. The anodic and cathodic rates have to balance, otherwise there would be a build-up of electrical charge on the copper and in the solution.

Thus, anodic and cathodic rates are in balance. There is no nett change, but the situation is not static. This prompts two questions:

1. What controls the rates of the forward (anodic) and backward (cathodic) reactions?
2. How can we measure these rates?

Historically, the methods used to measure electrochemical reaction rates were developed ahead of the relevant theory. In keeping with this history, we will touch briefly on the measurement, and will then proceed to the explanation.

5.2 ELECTROCHEMICAL EXPERIMENTS

In this part of our discussion, we will keep our focus on the relatively straightforward example of reversible electrodes; that is, electrodes that are in equilibrium with their environments, such as copper in a copper sulfate solution. These electrodes are not corroding.

5.2.1 Some terminology

All branches of science and engineering have evolved their own terminologies. Electrochemistry is no different. It has its own language. This has the advantage of speeding up communication between those already familiar with the science, but at the cost of making some concepts seem impenetrable to those wishing to get involved for the first time. With this in mind, we provide simple definitions of some basic concepts that we will be using as we go forward. We are also taking the opportunity to remind ourselves of some terms that we have already met.

5.2.1.1 Electrodes, electrolytes, anodes, cathodes and half-cells

By now we should already be familiar with these terms. For our purposes, an electrode is an electronic conductor (usually a metal such as steel), placed in contact with an ionic conductor (usually an electrolyte solution such as seawater). The means by which electric current passes between the electrode and the electrolyte solution is an electrode process. Electrode processes can be either anodic, which releases electrons into the electrode, or cathodic, which consumes electrons at the electrode's surface.

Anodic and cathodic processes usually occur together on the surface. However, it is possible for these processes to be spatially separated, in which case the different parts of the electrode surface are referred to as anodic and cathodic sites. In the extreme, the anodic and cathodic activity might become separated onto separate electrodes, termed anodes and cathodes. This is, of course, what we seek to achieve in CP, with the structure functioning as the cathode, and the CP current supplied by the anodes.

Finally, the combination of an electrode in an electrolyte solution is a "half-cell".

As with all language, familiarity lends itself to some laxity. Half-cells are now regularly, but imprecisely, referred to as electrodes, even in international standards. Similarly, the expression "electrolyte solution" is now almost universally truncated to "electrolyte". We will do the same in this book.

5.2.1.2 Potential

We introduced the term "potential" in Chapter 1 and added a little more detail in Chapter 2. It is the electrical potential difference between the electrode and the electrolyte solution with which it is in contact. More formally, because we are unable to measure this potential difference, the "potential" of an electrode in a solution is the measured difference in potential between it and a reference half-cell in contact with the same solution.

5.2.1.3 Polarisation

Polarisation is a major concept in electrochemistry and, therefore, in corrosion. It gets its own chapter later in this book. Ultimately, however, the word "polarisation" simply means a change in potential of the electrode (e.g. the potential of steel in seawater).

As we will see, this polarisation may be brought about electrically, chemically or a combination of the two. It may also be intentional, on the part of the experimenter or the CP engineer, or it may come about as a result of the natural circumstances of exposure.

Where polarisation is brought about electrically, this can be achieved in one of two ways:

- galvanostatically, in which case a fixed current is applied to the electrode, and the shift in potential (i.e. the polarisation) is observed; or
- potentiostatically, whereby the potential is fixed electronically, and the changes in current are observed.

The word "polarisation" is also sometimes linked with words indicating the cause of the change in potential, such as "activation polarisation", "resistance polarisation" and "concentration polarisation". We will discuss these as and when we come across them. For now, we just need to be aware that the word polarisation itself is still only referring to a change in potential.

5.2.1.4 Overpotential and overvoltage

The term "overpotential" is easy to define. As we mentioned above, we can polarise an electrode electrically using either a galvanostatic or potentiostatic technique. This intervention applies an overpotential, for which we use the Greek letter eta "η". Thus, the overpotential is easily defined. It is the difference between the natural potential of the electrode and the potential to which we have polarised it. It is not uncommon to say that we "apply" an overpotential.

The term "overvoltage" sounds similar. But it is distinct and less well defined. To understand the intent of the term, we can refer back to Figure 4.1 in the previous chapter. This shows the E-pH diagram for water. In this diagram, the line (a) represents the equilibrium between hydrogen ions (H^+) and hydrogen gas (H_2). At the left-hand side of the diagram (pH 0), the value of this equilibrium is 0.0 V (on the SHE scale). This means that, at potentials even slightly more negative than 0.0 V at pH 0, the water must decompose to form hydrogen gas. Indeed, this is what happens. However, the rate at which it happens turns out to be quite low. To get a meaningful rate of water decomposition, the potential has to be considerably more negative than 0.0 V (SHE). This difference between the theoretical potential at which a reaction takes place, and the practical potential at which it occurs at a noticeable rate, is termed the overvoltage. In the example that we have just considered, it is termed the "hydrogen overvoltage".

In principle, to define an overvoltage properly, we should also define the rate of the reaction, in this case hydrogen gas evolution, that we are trying to achieve. In practice, this precision of meaning is rarely expressed. Where the term is used, it is usually done so casually.

So, whereas overpotential is something we apply, and do so with precision; overvoltage is something ill-defined that we have to overcome. In this book, we use "overpotential" when needed, but we can forget about "overvoltage".

5.2.2 Galvanostatic polarisation

Early electrochemistry experiments were galvanostatic in nature. As indicated in Figure 5.1, a known current from a DC source, usually a battery (or "voltaic pile"), was passed through the specimen or working electrode ("WE"), and its resulting potential was measured and recorded. The process required that the experimenter, or usually the experimenter's lab assistant, had to attend the apparatus continually adjusting the resistor to keep the current constant. This tedious chore had to be persevered with for as long as it took the potential of the WE to stabilise. Usually, the current would then be increased to a higher value and the process repeated. In that way, a "galvanostatic polarisation" curve for that metal-electrolyte system could be compiled.

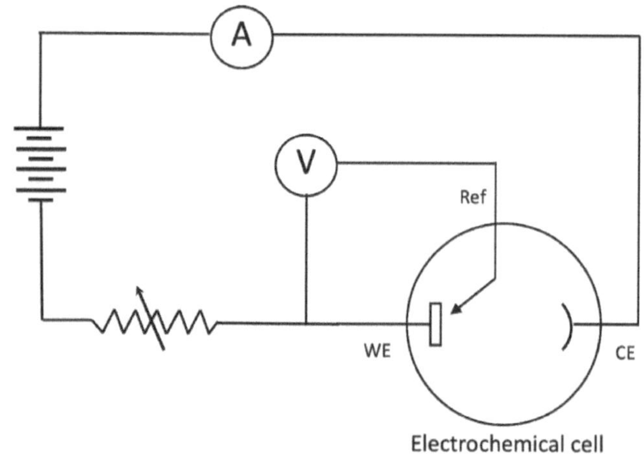

Figure 5.1 Controlling the current.

In later years, electrical equipment was developed that enabled the value of the current to be fixed and controlled automatically. Termed a "galvanostat", this later device removed the need for the continual, and tedious, adjusting of resistors. Galvanostats are used in various aspects of corrosion research. We will return to them when we discuss sacrificial anodes in Chapter 10.

5.2.3 Potentiostatic polarisation

5.2.3.1 The potentiostat

The copper electrode in equation 5.1 is an example of reversible metals M in equilibrium with their ions (equation 5.2).

$$M \Leftrightarrow M^{n+} + ne^- \quad (5.2)$$

If we polarise M by applying current, then we will upset the equilibrium between the anodic dissolution reaction and the cathodic reduction. For example, if we shift the potential in a positive direction (by removing electrons), we would expect the rate of the forward anodic dissolution to increase and that of the reverse cathodic reduction to decrease. The nett result would be that more M would go into solution as M^{n+} ions. Conversely, if we polarise in the negative direction (by adding electrons), we would get the opposite result. There would be a nett deposition of the metal from the solution onto the electrode.

Nowadays, this experiment can be carried out very simply, thanks to a piece of apparatus invented by Archie Hickling at the University of Leicester in 1942 [3]. The essential features of the device, which he originally named a "voltage clamp" before the coining of "potentiostat", are shown in Figure 5.2.

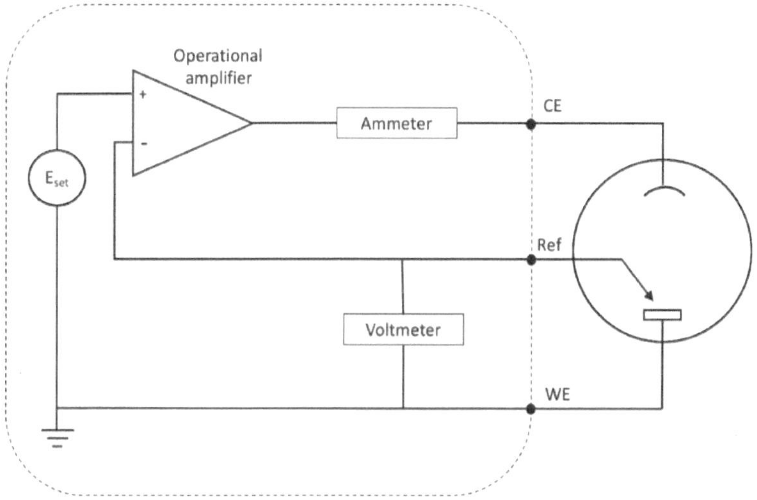

Figure 5.2 The potentiostat.

The key element is the operational amplifier. This is a piece of circuitry, the output of which is determined by the potential difference between its two inputs (marked "+" and "–" in the figure). This output is fed back to one of the inputs in a loop that incorporates the electrochemical cell. In this configuration, the output of the amplifier is always to try to eliminate the difference in potential between the two inputs. Thus, if a target potential difference (E_{set}) is imposed between the earth and the "+" input, the device will deliver current, via the counter electrode ("CE"),[1] to the working electrode. The feedback circuitry will ensure that the polarity and magnitude of this current will continually adjust in such a way as to (try to) make the potential of WE, with respect to the reference in the cell (Ref), the same as E_{set}.

Following Hickling's invention, electrochemists laboriously constructed their own potentiostats. This process became easier through the 1950s as modular operational amplifiers became commercially available [11]. However, it was not until the late 1960s that commercially available potentiostats started to be standard items of kit in corrosion laboratories.

In our polarisation experiment, the WE would be M and the solution in the electrochemical cell would contain a salt of M. For example, this could be a copper WE and a copper sulfate solution. The auxiliary electrode could be any metal, but the preferred practice is to use platinum since it will not dissolve, even if polarised anodically.

[1] The counter electrode is also commonly referred to as the *auxiliary* electrode.

5.2.3.2 *Plotting polarisation curves*

Many workers plot the results of a potentiostatic polarisation experiment with the dependent variable (the current or current density) on the *x*-axis. This contravenes the usual mathematical convention of plotting independent variable (potential in this case) on the *x*-axis, with the dependent variable (current) on the *y*-axis.

This habit probably had its origins in early polarisation experiments, which were carried out galvanostatically. The current was applied, and the resulting potential was measured. This made the potential the dependent variable. Hence, it was plotted on the *y*-axis. Generally speaking, this convention of plotting potential on the *y*-axis has persisted even though most polarisation experiments are now carried out potentiostatically rather than galvanostatically. The convention is not universal, however, and some workers, particularly in continental Europe, comply with the mathematical convention.

5.3 OBTAINING POLARISATION CURVES

5.3.1 The experiments

We can imagine a polarisation experiment in which we start at the equilibrium potential and gradually make E_{set} more positive. The potentiostat will react by making the WE more positive with respect to the reference. We can record the values of the current and potential as we do this. In the past, this involved frantically writing down the ever-shifting readings on the voltmeter and ammeter. This gave way first to the use of motorised chart recorders. Nowadays we enjoy the convenience of digital data loggers. We can also repeat the exercise, but this time by reversing the polarity of E_{set} and polarising in the negative direction.

A graph of potential (*E*) of our specimen M versus the current (*I*) flowing in the external measuring circuit is indicated schematically in Figure 5.3. In this case, since we can measure the surface area of the specimen, we report current per unit area, or current density (*i*).

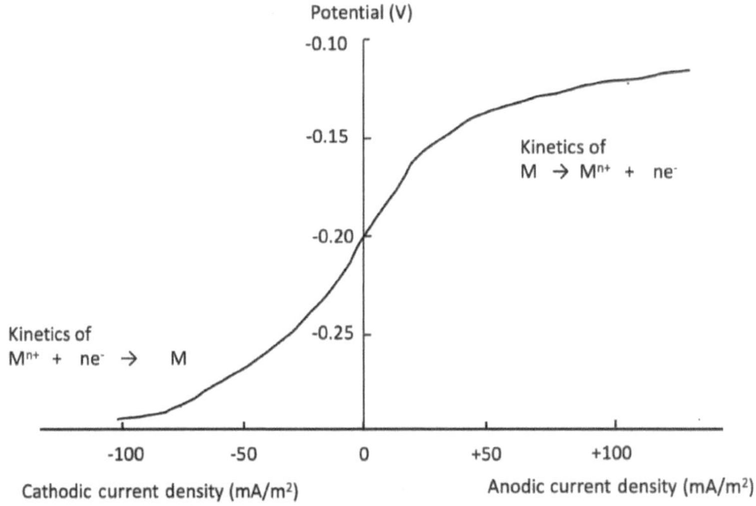

Figure 5.3 Polarisation of a reversible electrode.

In the upper right-hand part of the graph, we see the relationship between the positive (anodic) current and the potential. We refer to this as the anodic polarisation curve. The lower left-hand side of the graph shows the cathodic polarisation curve. This is an electrode kinetic graph. It depicts the rate of the reaction (represented by the current) as a function of the external energy input represented by the potential.

We can replot these polarisation data points making a couple of changes in the way that we represent them.

- First, for convenience, we can ignore the positive and negative signs that tell us whether the current is anodic or cathodic. We simply plot the absolute value of the current density "$|i|$" instead of positive and negative values of "i". This is equivalent to folding the graph along the vertical axis, so that we plot both anodic and cathodic currents in the same quadrant.
- Second, we can plot the current density values on a logarithmic rather than a linear scale. This gives us the form of presentation shown in Figure 5.4.

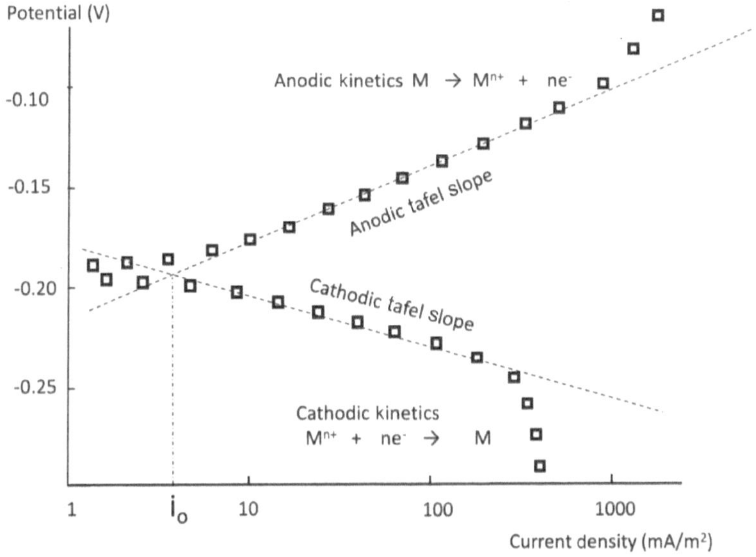

Figure 5.4 Tafel behaviour.

5.3.2 Tafel behaviour

In the case of the anodic part of the polarisation, we make an interesting observation. Over a significant part of the polarisation curve, the response is linear. That is the rate of the anodic dissolution of M increases logarithmically as the potential is made more positive. This linear relationship between potential and current behaviour was first published in 1905 by the Swiss electrochemist Julius Tafel (1862–1918). Hence, the linear portion of such polarisation curves are referred to as the Tafel region, and the gradient of the E–log $|i|$ part of the graph is termed the Tafel slope.

You may also note that there appears to be some Tafel-type behaviour in the cathodic part of the curve. However, this does not develop beyond a certain degree of cathodic polarisation. The reason for this is that the cathodic current depends on the rate at which M^{n+} ions in the solution diffuse to the WE surface under the influence of the applied electric field. However, because the concentration of M^{n+} ions in the solution is limited, this rate reaches a limiting value. Beyond this value of the cathodic current density, the Tafel behaviour can no longer be sustained.

5.4 ANALYSING POLARISATION CURVES

5.4.1 Fitting theory to data

There is one thing that a polarisation experiment is guaranteed to do. That is produce data. We can use graphs, which we call polarisation curves, to help us visualise the data; but the graphs do not interpret the data for us. We need to have a theory in mind as to what is happening on our specimen electrode. We then need to examine what curve would emerge from that theory and to compare it with the polarisation curve we actually obtain. With this in mind, we can start by considering the Tafel (linear E vs log $|i|$) behaviour we see in Figure 5.4.

5.4.2 The concept of activation control

5.4.2.1 *Activation energy*

In the same way that a rock will roll downhill, thermodynamics tells us that a chemical, or electrochemical, reaction will proceed if it releases energy. However, just as a rock might need a shove to get it going, a reaction needs an input of energy before it can happen. This generalised notion was formalised by Swedish chemist Svante Arrhenius (1859–1927). He introduced the hypothesis of a high-energy transition state – a concept that is now thoroughly established across the whole of chemistry.

This concept led to his producing his famous equation 5.3 linking the rate of a chemical process (k) to the absolute temperature (T).

$$k = Ae^{\frac{-E_a}{RT}} \tag{5.3}$$

where
k = the reaction rate (m^3/mol s)
A = a rate constant for the reaction (m^3/mol s)
E_a = the activation energy (J/mol)
R = the universal gas constant (J/mol k))
T = Temperature (k)

Put simply, as the temperature increases, more reactants gain sufficient thermal energy to exceed the required activation energy, so the faster the reaction goes.

We might visualise this activation energy as the energy needed to extract a metal atom, in this case copper, from its low-energy state in the metal lattice, whilst stripping away two of its electrons. Thus, we may interpret the "Reaction Coordinate" axis in Figure 5.5 as being either time-based or a measure of the distance away from the metal surface. In a roundabout way, this answers the question posed at the beginning of this chapter: why does sugar dissolve faster than iron? The answer is that the activation energy for the dissolution of sugar is very much lower than it is for the anodic dissolution of iron.

5.4.3 The Butler-Volmer equation

Whereas the Arrhenius equation deals with the effect of temperature on chemical reaction rates, the Butler-Volmer equation deals with the effect of overpotential on electrochemical reaction rates. The equation provides a more comprehensive electrode kinetic description

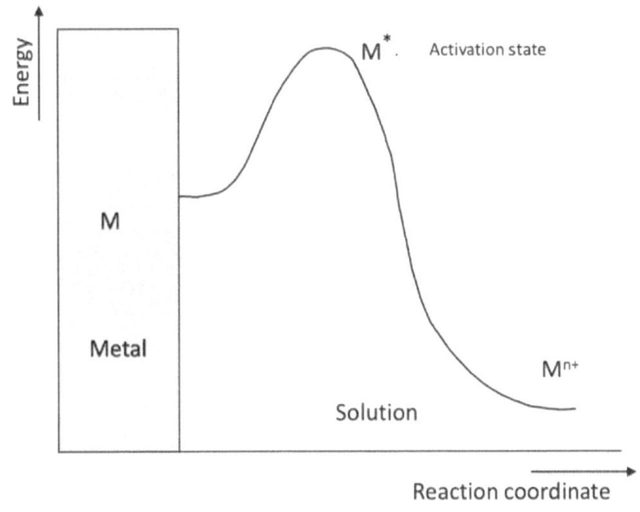

Figure 5.5 Activation energy – the transition state.

of polarisation curves of the type shown in Figure 5.3. It was developed by Max Volmer (1885–1965) on the basis of earlier work by John Alfred Valentine ("Jav") Butler (1899–1977). Hence, it is generally known as the Butler-Volmer equation 5.4.[2]

$$i = i_o\left(\exp\left\{\left[\frac{(1-\alpha)nF}{RT}\right]\eta\right\} - \exp\left\{-\left[\frac{\alpha nF}{RT}\right]\eta\right\}\right) \tag{5.4}$$

where

i=nett electrode current density (anodic or cathodic) (A/m^2)
n=number of electrons involved (n=2 for copper)
F=Faraday's constant (96485 C/mol)
i_o=exchange current density (see following) (Am^2)
η=overpotential (anodic or cathodic)
α=charge transfer coefficient (dimensionless quantity, usually$\approx$0.5)

As the overpotential increases, that is to say the electrode potential is made more positive, the left-hand exponential term increases and the right-hand one diminishes. At an η value in excess of about 40 mV, equation 5.4 approximates to equation 5.5.

$$i = i_o\left(\exp\left\{\left[\frac{(1-\alpha)nF}{RT}\right]\eta\right\}\right) \tag{5.5}$$

This can then be rearranged into the form of a Tafel equation (5.6.)

$$\eta = a - b\log(i) \tag{5.6}$$

[2] It is also less widely known as the Erdey-Grúz-Volmer equation after Volmer's co-author, the Hungarian chemist and politician Tibor Erdey-Grúz (1902–1976).

where a and b are the Tafel equation constants, and we have switched from $\log_e$ to $\log_{10}$ for convenience.

The value of b, the Tafel slope can be shown to be:

$$\frac{2.303 \ x \ RT}{(1-\alpha) \ nF}$$

This works out at 0.059 V at 25°C if $n=2$ and $\alpha=0.5$.

5.4.4 Tafel extrapolation

5.4.4.1 *Exchange current density*

You will see that the Butler-Volmer equation invokes the exchange current density (i_o). This is a measure of the rate at which the forward and backward (i.e. anodic and cathodic) reactions are taking place on the surface of a reversible electrode at the equilibrium potential.

How do we determine i_o?

In principle, this is not difficult. We can look more closely at the polarisation curve shown in Figure 5.4. The upper part of the graph describes the kinetics of the anodic oxidation half-reaction (M to M^{n+} ions), and the lower part represents the kinetics of the cathodic reduction of M^{n+} ions back to M.

At comparatively low values of overpotential, the nett current we measure in our circuit is the difference between the anodic and cathodic currents on the electrode. However, as the anodic overpotential gets greater, the cathodic reaction rate becomes so diminished that the ammeter only registers nett anodic current. At this point, the E–log $|i|$ relationship displays linear (Tafel) anodic behaviour. Conversely, at cathodic overpotentials in excess of about 40 mV, the contribution of the anodic half-reaction becomes negligible, so we only observe cathodic Tafel behaviour. If we now extrapolate the linear (Tafel) regions of anodic and cathodic plots backward (the dashed lines through the data points in Figure 5.4), they should intersect at the equilibrium potential. The value of the current at which this intersection takes place is the exchange current density (i_o).

5.4.5 Polarisation curves and polarisation diagrams

In passing, we should point out that there is an important difference between a polarisation ***curve*** and a polarisation ***diagram***. A polarisation curve, as shown for example in Figure 5.4, is the measured outcome of a polarisation experiment. A polarisation diagram is a sketch representing the assumed theoretical kinetic behaviour of the reactions giving rise to the observed polarisation curve. We will look at polarisation diagrams when we examine non-reversible electrodes below.

5.4.6 Departures from Tafel behaviour

In some instances, Tafel behaviour persists over several orders of magnitude of current. The copper electrode in copper sulfate solution described above would be one such example.

However, you will recall that Tafel (linear E–log $|i|$) behaviour indicates that an electrode process is under activation control. That is where the rate of reaction is determined solely by the likelihood of the reactants achieving the transition state. The greater the applied overpotential, the greater this likelihood is. However, this only holds providing there is an unlimited supply of reactants, and the reaction products themselves do not interfere with the reaction.

In practice, this means that Tafel behaviour does not persist indefinitely as we continue increasing the overpotential. At some point, other rate-controlling factors come into play. We can readily appreciate such a controlling factor. Copper refining, for example, involves the cathodic deposition of high purity copper from a solution of acidified copper sulfate. A refinery seeks to maximise the cathodic deposition rate. To this end, it might seem obvious to ramp up the cathodic overpotential.

This does not work.

There are two problems. The first of which is that the cathodic reaction requires Cu^{2+} to diffuse from the bulk solution to the metal surface in order to be cathodically reduced and plated out. At high enough overpotentials, it is this rate of mass transport of reactants, rather than the energy needed to achieve the transition state, that ultimately controls the corrosion rate. We will return to this mass transport control when we discuss the electrode kinetics of corrosion reactions below.

The second problem is that, as we polarise progressively to more negative potentials, we introduce the possibility of another cathodic reaction. These are the reduction of hydrogen ions and the electrolysis of water, both of which produce hydrogen gas at the cathode. The presence of these reactions on the surface would render the copper deposition very inefficient.

This is all we need to say about reversible electrodes such as copper. They provide a convenient example upon which to develop the basic electrochemical principles we need to study corrosion and CP. However, the electrodes of interest in corrosion and CP are never at equilibrium in water, so cannot be reversible electrodes.

5.5 NON-REVERSIBLE ELECTRODES

The concepts of polarisation developed above for reversible electrodes, such as copper in copper sulfate solution, are informative, but beg an obvious question. How is this relevant to corroding electrodes which, by definition, are not at equilibrium, and are therefore not reversible?

5.5.1 The mixed potential electrode

We can start by returning to our example of zinc corroding in hydrochloric acid. This is described by the following overall chemical equation.

$$Zn + 2HCl \rightarrow ZnCl_2 + H_2 \qquad (5.7)$$

Equation 5.7 can be broken down into its anodic and cathodic half-reactions 5.8 and 5.9. The chloride ions do not participate in the reaction so they can be cancelled from both sides of the half-reactions.

$$\text{Anodic} \qquad Zn \rightarrow Zn^{2+} + 2e^- \qquad (5.8)$$

$$\text{Cathodic} \qquad 2H^+ + 2e^- \rightarrow H_2 \qquad (5.9)$$

Now we can carry out another of our hypothetical experiments. We insert a piece of zinc into hydrochloric acid. Initially, two equilibria might seek to become established on the surface: the equilibrium between zinc and its Zn^{2+} ions, and the equilibrium between hydrogen ions (H^+) and hydrogen gas (H_2). However, as we have seen in Chapter 4, the equilibrium potential is more positive for the hydrogen electrode than for the zinc electrode.

This natural difference in potential between the two electrodes on a single surface is intolerable. Separate states of equilibrium for the two processes cannot be sustained. The potentials of the two processes move towards a common value. We can represent this polarisation behaviour schematically in Figure 5.6. In this case, we have simplified the diagram by assuming Tafel behaviour applies to the full polarisation curves for both reactions. That is, we have ignored the reverse equilibrium processes associated with each reaction.

$$Zn \leftarrow Zn^{2+} + 2e^- \tag{5.10}$$

$$2H^+ + 2e^- \leftarrow H_2 \tag{5.11}$$

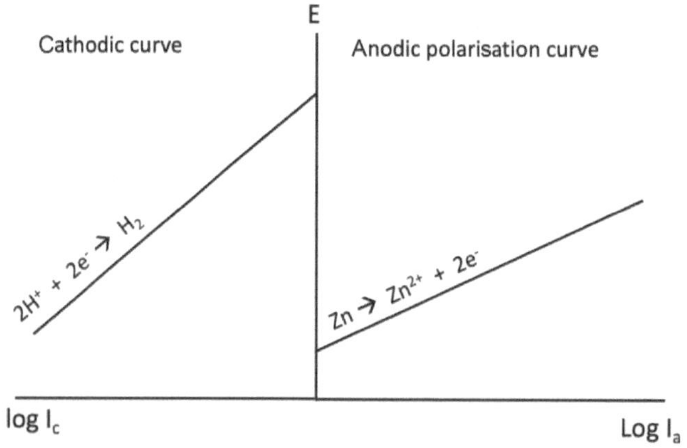

Figure 5.6 Schematic polarisation diagram for zinc in acid.

As for the reversible copper electrode, it is convenient to present an *E*–log *i* data in form *E*–log |*i*|, with the sign of the current direction ignored. This is the equivalent of folding the diagram on its *E*-axis. The result is shown in Figure 5.7.

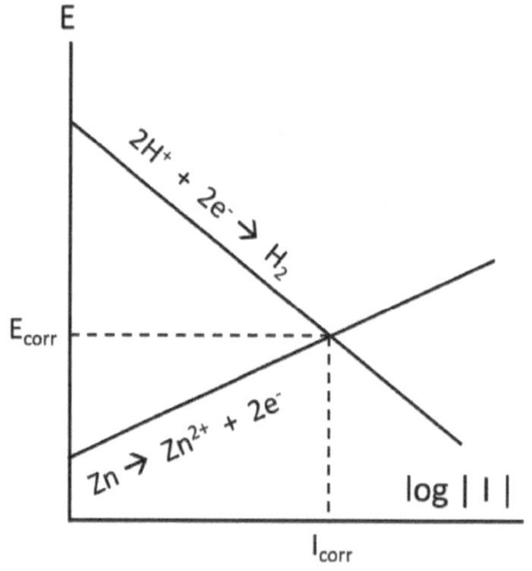

Figure 5.7 Schematic *E*–log |*i*| diagram for zinc in acid.

This figure illustrates an important point. There is a unique potential at which the nett anodic current due to the dissolution of the zinc is exactly balanced by the nett cathodic current due to the evolution of hydrogen gas. That potential is known as the corrosion potential (E_{corr}). The magnitude of the anodic or cathode current densities at E_{corr} is termed the corrosion current density (i_{corr}). As we explained in Chapter 2, the corrosion current can be converted to a rate of metal loss by applying Faraday's Law of Electrochemical Equivalence. More important, if we work in current density (mA/m^2) then this can be converted into a rate of metal penetration (mm/year).

Figure 5.7 illustrates the concept of the mixed potential theory of aqueous corrosion developed by U.R. Evans (1889–1980). For this reason, it is common to refer to diagrams of this type as Evans' diagrams.

It is worth pointing out here that the schematic E–log $|i|$ diagram for iron corroding in hydrochloric acid would be similar.

5.6 CORROSION IN SEAWATER

5.6.1 Oxygen-free seawater

This book is concerned with corrosion and CP in seawater. The fact that iron fizzes rapidly to destruction in hydrochloric acid is interesting but of no practical relevance. We should now consider the effect of changing the environment from a strong acid (pH$<$0) to seawater, which typically has a pH in the range of 8.0–8.2.

For convenience, we will first consider the artificial situation where there is no dissolved oxygen present. In principle, we can create a family of E–log $|i|$ diagrams for iron corroding in deaerated solutions of increasing pH (Figure 5.8). The kinetics of the anodic oxidation of iron to Fe^{2+} ions remain unchanged. However, the effect of increasing the pH is to make the potential for the H^+/H_2 equilibrium more negative (as described by the Nernst equation). The result is a family of curves, each with the same Tafel slope, but displaced downwards (towards more negative values) on the diagram.

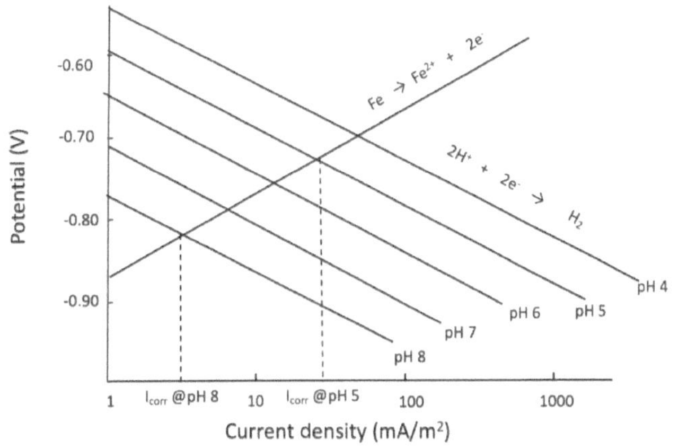

Figure 5.8 Schematic E–log $|i|$ diagram for iron corrosion – effect of pH.

The point of intersection of the H^+/H_2 Tafel line with the Fe/Fe^{2+} line represents the corrosion potential (E_{corr}) and corrosion current (i_{corr}) in the solution at the stated pH. At pH 8.0, which is close to the natural pH of seawater, the value of i_{corr} would be very low indeed. This is why, in many cases, the corrosion of steel in deaerated seawater can be ignored for all practical purposes.

5.6.2 Aerated seawater

One of the definitions of pH is *minus the logarithm to the base ten of the hydrogen ion activity*. Since the activity and concentration are almost identical in dilute solutions, this means that the concentration of the cathodic reactant (i.e. hydrogen ions) in seawater at pH 8.2 is $10^{-8.2}$g equivalents/L. On the other hand, the concentration of dissolved oxygen in open seawater is somewhere in the region of 8 mg/L, or eight parts per million. This may not sound like very much, but it amounts to about 10^{-3}g equivalents/L. This means that in aerated seawater the effective concentration of dissolved oxygen is five orders of magnitude greater than the concentration of hydrogen ions. It is not surprising, therefore, that the dominant cathodic process when steel corrodes in seawater is the oxygen reduction reaction.

You might think that we could redraw Figure 5.8 by simply replacing the Tafel lines for the reduction of hydrogen ions to hydrogen gas (the hydrogen evolution reaction) with a Tafel slope depicting the kinetics of the reduction of dissolved oxygen. However, if we were to do so, as suggested by the dotted line in Figure 5.9, we would conclude that the corrosion rate of steel in seawater would be phenomenal, and much higher even than in a strong acid. This conflicts with reality. The dotted line in Figure 5.9, therefore, cannot represent reality.

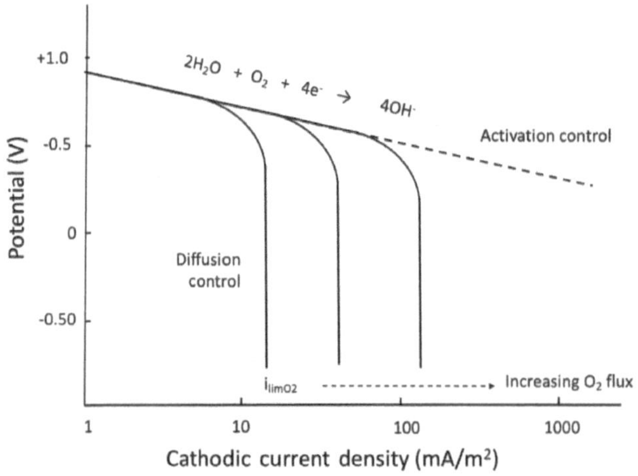

Figure 5.9 E–log |*i*| diagram for oxygen reduction.

The problem with Figure 5.9 lies in one of the key assumptions underpinning the Butler-Volmer equation. That assumption is that the rate of the reaction is controlled by the activation energy required to cause the current to flow across the electrode interface. This assumption holds as long as there is no restriction on the availability of reactants; as is the case for the availability of hydrogen ions in a strong, concentrated acid.

In the case of the oxygen electrode, shifting the potential in a negative (cathodic) direction initially gives rise to Tafel-type behaviour. The rate of the reaction is controlled by the activation energy, so the Butler-Volmer equation is obeyed. However, the rate of reduction of oxygen on the surface soon reaches a point at which the molecules are reduced as fast as they can diffuse to the metal surface from the bulk solution. Once that situation is reached, it makes no difference how much we shift the potential in the cathodic (negative) direction. We cannot increase the cathodic current. This polarisation behaviour, which is limited by the availability of the diffusing reactants, is termed “concentration polarisation”.

Figure 5.9 shows this by the kinetics of the oxygen reduction reaction becoming a vertical line. When this value of the current density is reached, further polarisation has no effect on increasing the reaction rate. The oxygen molecules are electrically neutral. This means that making the potential of the electrode progressively more negative has no effect on the rate at which it is transported through the solution.

This limiting rate at which dissolved oxygen can diffuse to the surface is called the limiting current density (i_{limO_2}). As indicated in Figure 5.10, in seawater, the corrosion current for steel is, for all practical purposes, identical to the limiting diffusion current for dissolved oxygen. This reminds us of the work of Ashworth that we discussed in Chapter 1.

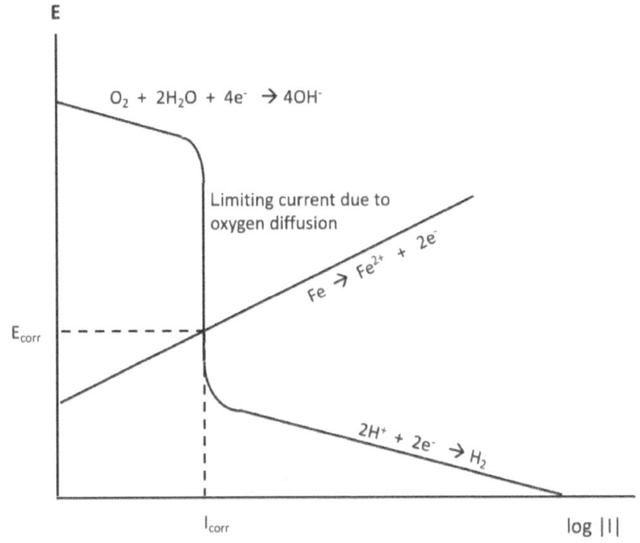

Figure 5.10 E–log |*i*| diagram for the corrosion of steel in seawater.

5.7 ELECTRODE KINETICS AND CP

5.7.1 General

In the absence of a convincing theoretical explanation of CP on the basis of the thermodynamic concepts of immunity or passivity (see Chapter 4), we need to turn our attention to electrode kinetics. This means that we abandon interest in the abstract concept of whether or not we can stop steel corroding. Instead, we turn our attention to the question: how much we can reduce the corrosion rate?

5.7.2 The theory

Most workers cite a 1938 paper by Mears and Brown [1] as being seminal in the interpretation of CP in terms of electrode kinetics. However, in the published discussion on that paper, T.P. Hoar refers to his own very similar work published a year earlier. Mears and Brown studied the galvanic corrosion between copper and zinc in 20% NaCl solution. They showed that when the copper was externally polarised to the same potential as the zinc then zero galvanic current flowed between the copper and the zinc. From this observation, they concluded that... *in order for cathodic protection to be entirely effective, the local cathodes on the corroding specimen must be polarised to the potential of the unpolarised local anodes.*

In the paper itself, the authors fell some way short of explaining how their proposed theory translated from a zinc-copper galvanic couple to the case of (say) steel in a corrosive medium such as seawater. However, the message unfolds in the appended discussion section between the authors and Hoar.

The electrode kinetic interpretation of CP has, at its heart, the mixed potential theory of corrosion, as developed by U.R Evans, which we outlined in Section 5.5.1. Evans himself [4] and others [6–8,10] have applied this theory to CP.

The theory is re-capped in the following set of polarisation diagrams. In this case, we are repeating the "theoretical experiment" we described in Chapter 2, but this time we are describing it in electrode kinetics rather than the ideograms of the corrosion cell. In Figure 5.7, we drew a stylised E–log $|i|$ diagram for zinc dissolving in acid. On the same basis, we can construct a generic E–log $|i|$ diagram for any notional metal undergoing anodic dissolution supported by a cathodic reduction process. Figure 5.11 shows the kinetics of the anodic dissolution of and the metal and the cathodic reduction of the environment.

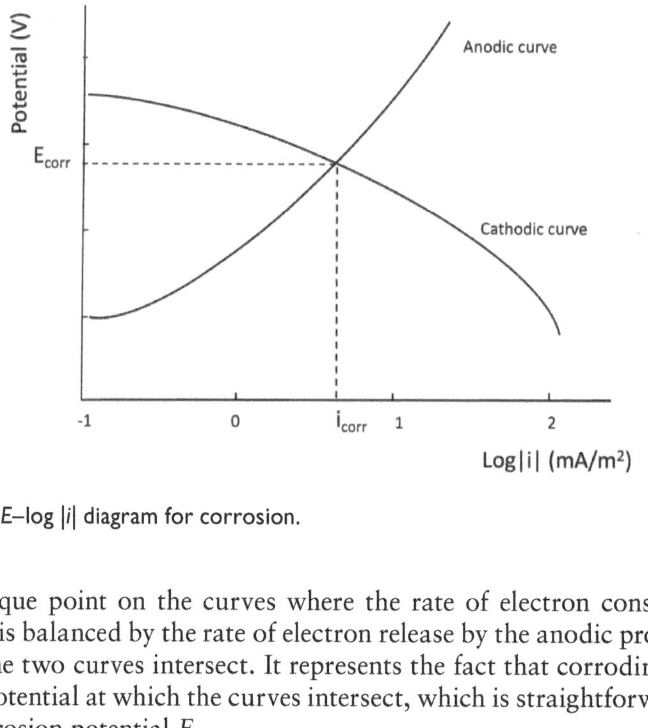

Figure 5.11 Generic E–log $|i|$ diagram for corrosion.

There is a unique point on the curves where the rate of electron consumption by the cathodic process is balanced by the rate of electron release by the anodic process. This is the point at which the two curves intersect. It represents the fact that corroding metals do not charge up. The potential at which the curves intersect, which is straightforward to measure, is the unique corrosion potential E_{corr}.

It is also possible, but less straightforward, to estimate the value of the current where the curves intersect. This is the corrosion current density i_{corr}. As we saw in Chapter 1, we can apply Faraday's First Law of Electrochemical Equivalence to convert this corrosion current density to a rate of metal weight loss, and, thence, to a corrosion rate measured in (say) mm/year.

If we now apply some CP, we cause (positive) current to flow onto the surface of the metal. As we have discussed, this is equivalent to pumping electrons into the metal. The potential of the metal shifts in a negative (cathodic) direction. As Hoar succinctly pointed out [2], at any potential: *the applied cathodic current is equivalent to the rate of the cathodic reaction on the corroding metal minus the rate of the anodic reaction on the corroding metal.* This is shown schematically in terms of electrode kinetics in Figure 5.12. The potential is shifted to *E*'. The anodic current density on the metal at that potential is given by i'_a, and the cathodic current density is given by i'_c. The CP current being applied is the difference between i'_a and i'_c.

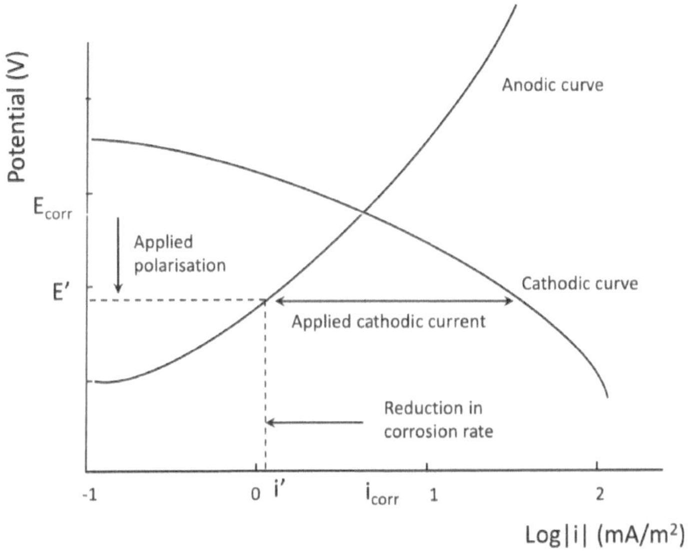

Figure 5.12 Generic *E*–log |*i*| diagram for partial CP.

In principle, we can continue the polarisation until the potential is shifted to E_a, the reversible potential for the anode reaction. Once we have reached this, we can be sure that we have attained full CP. Corrosion of the metal is now no longer possible. This is indicated in Figure 5.13.

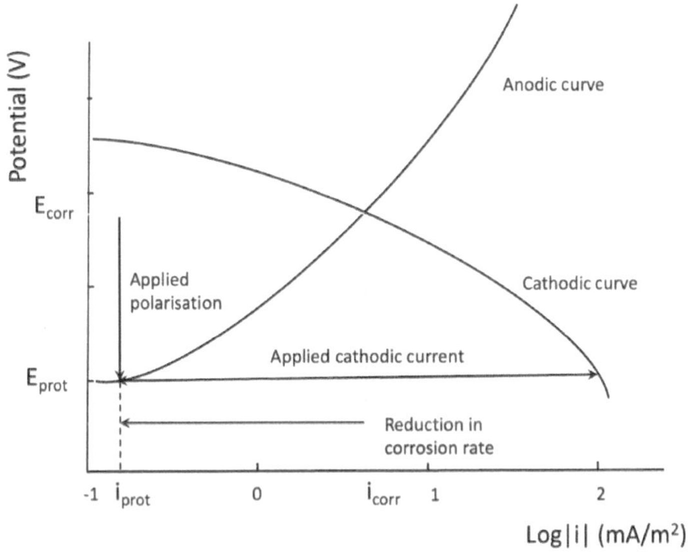

Figure 5.13 Generic *E*–log |*i*| diagram for full CP.

5.7.3 Implications for CP

At first sight, it might seem that there is no practical difference between describing CP in terms of achieving thermodynamic immunity, as discussed in Chapter 4, or in terms of reducing the nett anodic dissolution rate to zero, as we have just described. Both concepts would require us to polarise the metal to a potential at least as negative as its equilibrium potential if we are to achieve full cathodic protection.

The important difference between the two approaches is revealed when we realise that achieving full CP, which we might define as a corrosion rate of zero, is a pointless ambition. It implies an infinite life. In practice, a life of a few decades is sufficient for the metallic structures we expose to seawater. Rarely would we need to design for a life of more than a century. Even if we did, a century still leaves a lot of change when subtracted from eternity!

This points to the benefit of electrode kinetics over thermodynamics when dealing with CP. Our real-world objective is to reduce the rate of corrosion, not to stop corrosion altogether. Since thermodynamics gives no information on reaction rates, it is useless in this respect. The electrode kinetic interpretation, on the other hand, is just what we are looking for.

An underpinning principle of the electrode kinetic description of corrosion, and of CP, for steel is that the nett rate of anodic dissolution (corrosion) at any given value of potential depends solely on the kinetics of the anodic reaction:

$$Fe \rightarrow Fe^{2+} + 2e^-$$

It has been demonstrated experimentally that this reaction follows Tafel-like behaviour in many electrolytes, including seawater. In other words, there is a logarithmic relationship between the rate of the reaction and the applied overpotential (η). Furthermore, the slope of the Tafel plot for this reaction is known to be very insensitive to the actual nature of the electrolyte. Baboian [9] has published a figure of 0.060 V/decade for steel in 0.1 N Na_2SO_4. This is consistent with polarisation curves published by Mor and Beccaria [5], which show that the Tafel behaviour for steel in seawater is close to 60 mV/decade and do not change very much over the temperature range 5°C–20°C.

The current is related directly to the rate of metal loss by Faraday's Laws. This means the Tafel plot is telling us that: for every 0.06 V negative (cathodic) shift in potential, the corrosion rate of steel reduces by a factor of ten.

In Chapter 1, we found that a reasonable ball-park figure for the corrosion rate of steel in seawater, which we discussed in Chapter 2, was around 0.1 mm/year. Applying ~60 mV of cathodic polarisation reduces this to 0.01 mm/year, equivalent to losing 1 mm of metal thickness in a century. There are very few, if any, engineering applications where this degree of corrosion control would be inadequate. A further 60 mV of cathodic polarisation reduces the anodic dissolution rate by another order of magnitude, meaning that it now takes a thousand years to corrode a mm of steel.

The electrode kinetic interpretation of CP also forces two important conclusions about the potential and current needed for protection.

The protection potential depends only on the nature of the metal. The protection current density depends on the environment. We discuss these points in more detail in the next two chapters. For now, it is useful to return to Figure 5.13 where we see that the current density needed for protection is dictated by the rate of the cathodic reaction at the protection potential.

This tells us that if the cathodic reaction is exhibiting Tafel-type behaviour then the current that we would need to apply to provide CP would be very high indeed. In practice, this means that, although CP might be theoretically possible in a strong acid solution, it would be wholly impracticable because the current needed would be so high.

The reason CP is practicable in seawater is because, at pH ~ 8, the dominant cathodic reaction is the reduction of dissolved oxygen. As we have illustrated in Figure 5.10, the rate of this reaction is limited by the rate at which the dissolved oxygen molecules can diffuse through the seawater to the surface of the steel. This is shown as the limiting cathodic current density in the figure.

When we apply our electrode kinetic view of CP to the situation shown in Figure 5.10, we see that the current needed to protect the steel is, for all practical purposes, limited to the value of the oxygen diffusion current. This is illustrated in Figure 5.14. This is essentially an electrode kinetic interpretation of the corrosion cell description of CP that we advanced in Chapter 2.

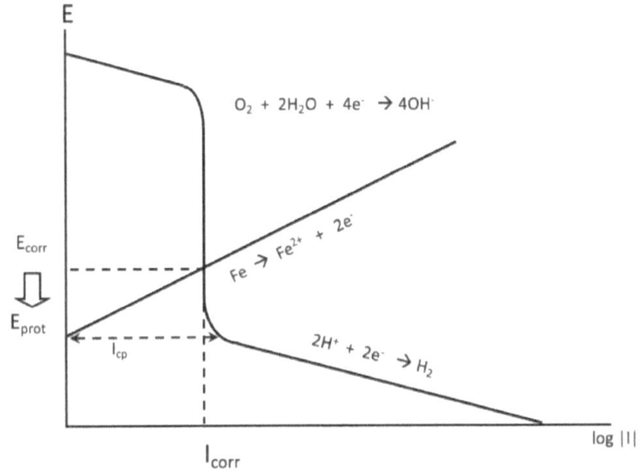

Figure 5.14 Schematic *E*–log |*i*| diagram for the CP of steel in seawater.

REFERENCES

1. R.B. Mears and R.H. Brown, *A theory of cathodic protection.* Transactions of the Electrochemical Society **74**, 519 (1938). Reprinted (with discussion) in *Cathodic Protection Criteria: A Literature Survey,* Eds. R.A. Gummow et al. NACE (Houston, TX) 1989.
2. T.P. Hoar, Published discussion in Ref [1].
3. A. Hickling, *Studies in electrode polarisation Part IV: The automatic control of the potential of a working electrode.* Transactions of the Faraday Society **38**, 27 (1942).
4. U.R. Evans, *The Corrosion and Oxidation of Metals.* Arnold (London) 1960.
5. E.D. Mor and A.M. Beccaria, *Influence de la Pression Hydrostatique sur la Corrosion de Fer dans l'Eau de Mer.* Proceedings of International Symposium on *Corrosion & Protection Offshore*, Paris Cefracor, 1979.
6. D.A. Jones, *Electrochemical fundamentals of cathodic protection*, Paper no. 317 CORROSION/87, NACE (Houston, TX) 1987.
7. V. Ashworth, *The theory of cathodic protection and its relation to the electrochemical theory of corrosion,* Chapter 1, in *Cathodic Protection: Theory and Practice,* Eds. V. Ashworth and C.J.L. Booker. Ellis Horwood (Chichester) 1986, pp. 13–30.

8. W. Schwenk, *Fundamentals and concepts of corrosion and electrochemical corrosion protection,* Chapter 2, in *Handbook of Cathodic Corrosion Protection*, 3rd edition, Eds. W. von Bæckmann, W. Schwenk and W. Prinz. Gulf Publisher (Houston, TX) 1989, 40–46.
9. R. Baboian, p. 61, in *NACE Corrosion Engineers Reference Book*, 2nd edition, Eds. R.S. Treseder et al. NACE International (Houston, TX) 1991.
10. V. Ashworth, *Principles of cathodic protection,* Section 10.1, in *Corrosion*, 3rd edition, Eds. L.L. Shreir, R.A. Jarman and G.T. Burstein. Butterworth-Heinemann (Oxford) 1994, pp. 10.3–10.9.
11. J.E. Harrar, The potentiostat and the voltage clamp. *The Electrochemical Society Interface*, p. 42 Winter (2013).

Chapter 6

Protection potential – carbon steel

6.1 INTRODUCTION

In Chapter 2, we considered a simple picture of CP in which we caused electrical current to flow from the environment onto the surface of our steel structure. This made the potential more negative and led to the anodic half-reactions on the surface being progressively stifled. This view of CP prompts two primary questions:

- What potential does the structure need to reach in order that we can be confident that CP is effective? In other words, what is the protection potential?
- How much current do we need to apply to the surface in order to achieve that potential?

In this chapter, we deal with the protection potential and what happens if we fail to achieve it. Chapter 7 then deals with the current that we need to apply to the steel surface to achieve this potential. It will also consider how quickly or slowly polarisation occurs. Then, in Chapter 8, we will consider the potential and current we need if we are to apply CP to metals other than carbon steel.

6.2 WHAT DOES CP NEED TO ACHIEVE?

The CP of steel can have two objectives. The first, and most obvious, is to reduce the rate of corrosion. On some structures, there is also a requirement to control corrosion fatigue. This chapter is concerned with corrosion. We consider fatigue in some detail in Chapter 12. However, it will emerge that the potential we need to effectively control corrosion is the same as that which we need to reduce corrosion fatigue to simple metallurgical fatigue.

The first thing that we need to recognise is that our earlier concept of CP of electrochemically switching off the anodic half-reactions (as indicated in Figures 2.2 and 2.3) is an over-simplification. In particular, achieving a "zero" corrosion rate turns out to be difficult. It also has no real practical relevance to the real world. We have no intention of making things last forever. Much of the literature on CP, and most codes, are directed towards achieving "full protection". This is not the same as zero corrosion. For example, Section 6.1 of ISO 15589-1: 2015 states... "*the metal-to-electrolyte potential at which the corrosion rate is less than 0.01 mm per year is the protection potential... This corrosion rate is sufficiently low so that corrosion will be acceptable for the design life*". Although ISO 15589-1 relates to the CP of land-based pipelines, this definition also holds for offshore CP systems.

DOI: 10.1201/9781003216070-6

The requirement of this ISO standard to target a corrosion rate of <0.01 mm/year is reasonable from an engineering perspective. It amounts to the loss of <1 mm of steel thickness per century. This is unlikely to be an issue for a pipeline, even one with a design life of many decades. However, 0.01 mm/year is not *zero* corrosion. This point is emphasised when one considers that even a very low corrosion rate of 1 mm per century means that ~3.7 trillion iron atoms are being lost from each square centimetre of the surface every second!

6.3 WHAT DO THE CODES SAY?

A CP code may be regarded as a digest of the opinion available at the time of its publication. With regards to the protection potential, the code developers have not been short of information to mull over. For example, Wanklyn [19] reviewed the published advice on the least negative protection potential for steel in seawater. He collated recommendations produced between 1950 and 1980 in the USA, Europe, Japan and the Soviet Union. These offerings ranged from −770 to −990 mV. As we will see below, code drafting committees have considerably narrowed this 220 mV span.

6.3.1 Aerated seawater

The code advice for the protection potential of steel in aerated seawater is summarised in Table 6.1. As previously, all potentials here are referenced to Ag|AgCl|seawater. Most codes specify a least negative and most negative potential limit. The least negative limit, which is sometimes (but casually and incorrectly) termed the *minimum* potential, is the potential at which the steel is fully cathodically protected according to the code. The most negative limit (likewise frequently misnamed the *maximum* potential) signifies the limit beyond which the cathodic polarisation might have damaging effects. These damaging effects can apply to a protective coating applied to the steel, which we discuss in Chapter 9, or to the mechanical properties of the metal, which we discuss in Chapter 12.

As we can see in Table 6.1, there is full agreement among the codes concerning the least negative protection potential for steel in aerated seawater. It is −800 mV. There is some variation between the codes as to the most negative potential that can be tolerated by some higher-strength steels (see Chapter 12).

6.3.2 Anaerobic environments

In addition to setting design protection potentials in aerated seawater, many of these codes also recommend design protection potentials for anaerobic environments. Typically, the codes also explain that these are anaerobic environments where MIC is considered to be a threat.

As can also be seen in listed in Table 6.1, the codes are less aligned when it comes to this requirement. The EN and ISO codes all call for the design protection potential to be made 100 mV more negative if there is a perceived threat of MIC. Other authorities are less clear, or even ambivalent. For example, neither DNV-RP-F103 nor RP-B401 requires polarisation to −900 mV, but RP-B101 does. Similarly, NACE SP0176 has no such requirement, but SP0607, which is a modified version of ISO 15589-2, requires −900 mV.

Moreover, in addition to some of the codes, some CP reference books [20,24] endorse the −900 mV criterion.

6.3.3 Elevated temperature

Most of the codes listed in Table 6.1 relate to structures that will be at near-ambient temperatures at all times. Generally speaking, therefore, these codes do not consider whether or not any adjustment is required for elevated temperatures.

On the other hand, ISO 20313, ISO 15589-2 also require the protection potential to be shifted from –800 to –900 mV if the steel surface is at a temperature above 60°C.

As with the requirement for an additional 100 mV of polarisation to mitigate the threat of MIC, this additional polarisation in the case of elevated temperature conflicts with the evidence. We return to this in Section 6.6.

Table 6.1 Protection potentials for carbon steel

			Protection potential (mV)			
			Least neg.			
Code	*Year*	*SMYS (MPa)*	*Aerobic*	*Anaerobic*	*Most neg.*	*Scope*
DNV-RP-0416	2016	<550	–800	–800	n.s.	Wind farm foundations
DNV-RP-B101	2019	n.s.		–900	–1100	Floating installations
DNV-RP-B401	2021	<500		–800	n.s.	Structures
DNV-RP-F103	2016	n.s.		–800	n.s.	Pipelines
EN 12473	2014	<550		–900	–1100	General
EN 12495	2000	<550				Offshore structures
EN 13173	2001			n.s.		Floating structures
EN 16222	2012			–900		Ships
ISO 13174	2012			–900		Harbours
ISO 15589-2	2024					Pipelines
ISO 20313	2018	<550				Ships
		>550			–0.83 to –0.95	Ships
ISO 24656	2022	<550			–1100	Wind farm foundations
NACE SP0176	2007			–900	–1100	Offshore structures
NACE SP0607	2007					Pipelines (modified version of ISO 15589-2)
Norsok M-503	2018			–800	–1100	Structures and pipelines

n.s. = not stated or no not applicable within the scope of the standard.

6.4 AEROBIC ENVIRONMENTS: THE –800 mV CRITERION

Table 6.1 shows the uninamity of the standards when it comes to defining the protection potential for steel in seawater at ambient temperatures. It is –800 mV. This figure has been in codes since the early 1970s, before which it had been well established in the CP industry for some decades. Of course, the fact that codes all agree does not necessarily mean that they are correct. If it were to be demonstrated that –800 mV was incorrect, then even such a unanimous consensus would have to be overturned.

The support for –800 mV arises from three sources. In order of increasing relevance, these are theoretical considerations, laboratory testing and, finally, practical experience. We consider these below.

6.4.1 Theoretical considerations

6.4.1.1 Thermodynamics: immunity

In Chapter 4, we estimated what the protection potential of steel, in any aqueous environment, would need to be on the basis that it would be necessary to polarise steel into the immunity domain of the E-pH diagram. For seawater, the result was a protection potential, more negative than –1060 mV.

Although this figure can be supported on the basis that it marks the potential at which the corrosion of steel becomes thermodynamically impossible, to adopt it as the protection criterion for steel in seawater would overturn almost a century of practical experience. Furthermore, it would mean that applying CP by means of zinc or aluminium alloy sacrificial anodes simply could not work. As we know, the evidence, dating back to the work of Sir Humphry Davy [1], is to the contrary. The concept of thermodynamic immunity, although it might work for (say) copper in seawater, does not provide a basis for the CP of steel.

6.4.1.2 Thermodynamics: passivity

We touched on passivity at the end of the previous chapter, and we now continue the discussion. We will also revisit the subject of passivity when we discuss corrosion-resistant alloys in Chapter 8.

The protection potential for steel in soils of any type has been accepted as –850 mV (with respect to Cu|$CuSO_4$(satd.))[1] for about a century. However, it is now being questioned. This challenge coincides with a fundamental debate about CP mechanisms that has developed in recent years among some CP practitioners working with onshore pipelines. The reasons for this challenge are broadly as follows. It seems that some pipeline operators claim evidence of corrosion on nominally protected coupons and pipelines. This has prompted re-examination of CP theory which, in turn, has given rise to a novel theory. This is that CP works because the increase in pH at the cathode surfaces causes the steel to passivate.

An implication of this model is that it is not possible to define a protection potential that applies to all soils. Instead, the criterion for protection will need to be determined, individually for different soil types, on the basis of the current density needed to generate passive conditions at the steel surface. This passivity model is supported in a review paper by Angst [32] but challenged by others [33].

The debate is relevant to us because, although the passivity model has been envisioned in the context of land-based CP, some of its proponents have also advanced it as applicable in seawater. Furthermore, even an offshore pipeline system will make landfall at some point. As we explore in Chapter 14, this means that even practitioners of offshore CP may need to enter the onshore debate.

For what it is worth, my own rather lengthy analysis [34] recognises the complexity of the issue but, ultimately, rejects the passivity paradigm. However, a book on marine CP is not the forum to continue the theoretical debate. It is more sensible to examine what the passivity model offers in the way of predictions for the protection potential in seawater, and what evidence there is to support it. We return to this in Section 6.5.

1 –850 mV (Cu|$CuSO_4$(satd.) is, as near as makes no difference, –800 mV Ag|AgCl|seawater. As with the rest of the book, unless stated otherwise, potentials are expressed with reference to –800 mV Ag|AgCl|seawater. Where necessary, these values have been transposed from the reference in the original publication.

6.4.1.3 Electrode kinetics

In Chapter 5, we discussed the mechanism of CP in terms of electrode kinetics. Figures 5.12–5.14 provide polarisation diagrams that offer a convincing explanation for the way CP operates. However, as we also mentioned in Chapter 5, a theoretical polarisation diagram is not the same as an experimental polarisation curve. In other words, the fact that we can conjure up a theory does not necessarily mean that it is correct. What matters is what the theory predicts, and whether or not we can test it.

To this end, we can draw on three pieces of evidence:

1. As a matter of experimental observation, steel corroding in aerated seawater adopts a steady-state potential somewhere in the region of −650 mV. This is its natural corrosion potential (Ecorr). It is not a fixed property, but experience suggests that, over the long term, it is very unlikely to depart from this figure by more than ±50 mV.
2. As we saw in Chapter 1, there is a very considerable body of test data that shows that the corrosion rate of steel in seawater, measured in testing of up to 5 years duration, is somewhere around 0.1 mm/year. Again, as we discussed, this figure is not fixed since corrosion is a process, not a property. Nevertheless, the consensus from the test data reviewed is that figure is unlikely to vary by more than ±0.05 mm/year.
3. There is a body of electrochemical test data [22] that shows that the slope of the anodic Tafel slope for the reaction:

   $$Fe \rightarrow Fe^{2+} + 2e^{-}$$

   is around 60 mV/decade. That is, for every 60 mV of applied polarisation, the anodic reaction rate changes by a factor of ten. If the applied polarisation (or "overpotential") is positive, then the anodic reaction speeds up. If it is negative, the reaction slows down.

This means that the electrode kinetic theory offers us a prediction. This is illustrated in Table 6.2 which indicates that if the natural corrosion potential in seawater is −650 mV, then polarisation by 60 mV to −710 mV renders the corrosion rate sufficiently low for most practical purposes. A further 60 mV of polarisation (to −770 mV) takes us to rates of corrosion that are far too low to be easily measured, let alone of any practical concern.

Table 6.2 Electrode kinetics – predicted effect of cathodic polarisation

Cathodic polarisation	*Corrosion rate (typical) mm/year*	*Life of a 1 mm corrosion allowance (years)*
Freely corroding	0.05–0.15	7–20
−60 mV	0.005–0.015	70–200
−120 mV	0.0005–0.0015	700–2000
−180 mV	0.00005–0.00015	7000–20,000

6.4.2 Laboratory testing

6.4.2.1 The predictions

As we have seen above, electrode kinetics offers a prediction: if we progressively apply a cathodic overpotential (η) to a steel specimen, then the corrosion rate will reduce on a logarithmic basis. The passivity model, on the other hand, does not offer a theory linking η

with a progressive reduction in corrosion. The passivity model only envisages two surface conditions: passive and not passive. There is no progressive change in corrosion rate when transiting from one state to the other.

6.4.2.2 The results

6.4.2.2.1 Moore and Knuckey

In 1978, Moore and Knuckey [14] published results of a series of experiments designed to determine the appropriate protection potential for steel hulls of Australian Navy ships. They examined the weight loss of steel specimens held at different set potentials for periods of between 6 and 25 days. Unlike earlier workers, these researchers enjoyed the benefit of the commercial availability of laboratory potentiostats. Furthermore, and crucially, they also took the trouble to carry out blank weight loss tests and to conduct triplicate experiments.

Their results are shown in Figure 6.1. In this, I have replotted the corrosion rate data, which were originally presented on a linear scale, to a logarithmic scale. As can be seen, cathodic polarisation led to a rapid reduction in the rate of corrosion down to a potential of −790 mV. They concluded that this was... *the least negative potential that will ensure complete cathodic protection.*

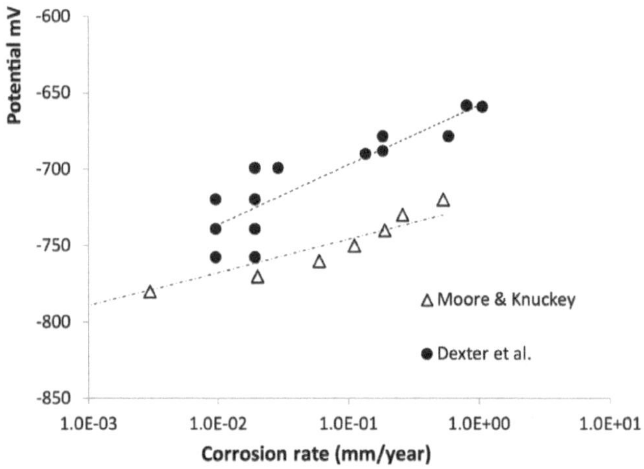

Figure 6.1 Corrosion rate versus potential. (Data from Refs. [14,18].)

6.4.2.2.2 Dexter, Moettus and Lucas

Dexter et al. at the University of Delaware carried out similar work [18]. Their main objective was to test a 1965 theory proposed by LaQue and May [10] that CP works by progressively reducing the area of the anodic sites on the steel. If this theory held then there might be serious practical implications. Marginal underprotection might introduce an undesirable change of corrosion morphology from general attack to pitting.

In the event, Dexter et al. were able to refute the LaQue and May description of CP. In doing so, they polarised steel coupons at different potentials and measured the weight losses to determine the corrosion. Their experiments were broadly similar to earlier results of Moore and Knuckey, of which they seemed to be unaware. They also used potentiostats and exercised the precaution of carrying out blank weight loss determinations.

Although Dexter et al. presented their corrosion data as the percentage weight loss of their specimens, they also gave the plate thickness (1.7 mm). Based on the test duration (120 days), and assuming corrosion from both faces, it is possible to manually translate their weight losses into corrosion rates (mm/year). Similarly, their reference potential scale is transposed from saturated calomel electrode (SCE) to Ag|AgCl|seawater. The transposed individual results are also plotted in Figure 6.1.

Their results show that the corrosion rate is reduced to low values (<0.02 mm/year) when the specimens are polarised cathodically to about −700 mV. This supports the view of the codes that the target value of −800 mV is sound, albeit conservative. In that respect, their results are similar to, albeit they do not directly align with, the results of Moore and Knuckey.

6.4.2.2.3 Other testing

As part of their investigation into hydrogen embrittlement problems, a topic we cover in Chapter 12, Batt and Robinson carried out weight loss versus potential experiments in natural seawater and sterile seawater on a high-strength low-alloy steel. They found that corrosion rates were reduced to 0.001 mm/year, which was the detection limit of their technique, at potentials in the range from −760 to −790 mV [23].

On the other hand, there are test results that conflict with the data shown in Figure 6.1 and the Batt and Robinson results. In my analysis [34], I point out that some of this work suffers because of experimental shortcomings. For example, early work of La Que and May [10] was conducted quasi-galvanostatically, so there was no real control of the potential. Other results lend themselves to challenge but cannot simply be dismissed. For example, the results published by Leeds and Cottis in 2006 [25] and 2009 [26] contradict the electrode kinetic model and challenge the conclusion that −800 mV is a safe, albeit conservative, protection potential. For example, they report a corrosion rate for steel as high as 0.25 mm/year at a code-advised protection potential of −800 mV in seawater. According to their results, even polarising to −1400 mV failed to achieve adequate CP.

6.4.2.3 Evaluation of the evidence

The 1978 work of Moore and Knuckey and the 1985 contribution of Dexter et al. provide support for the electrode kinetic explanation for CP since there is good evidence for a logarithmic (Tafel-type) relationship between potential and the anodic dissolution rate. There is no way that any model based on CP causing passivity can be reconciled with this type of potential versus corrosion rate characteristic.

Furthermore, these results, taken together with those of Batt and Robinson, endorse the selection of −800 mV as a suitable protection potential.

Nevertheless, the case is not closed. The two data sets shown in Figure 6.1, although similar, are by no means an exact fit. This may seem surprising since they are results from notionally the same experiment. Furthermore, the contrary evidence of Leeds and Cottis has to be considered.

It should be stressed that the object of the latter polarisation experiments was primarily to examine surface films on polarised electrodes. Their specimens were selected more with a view to carrying out surface analysis than for gravimetric work. Their surface-to-area weight ratio was ~0.88 cm^2/g, compared with Dexter et al. (~1.76 cm^2/g) and Moore and Knuckey (~4.59 cm^2/g). All things being equal, a higher specimen area, and area to weight ratio, might be expected to increase the accuracy of the work. In addition, Moore and Knuckey conducted triplicate experiments, and Dexter et al. carried out duplicates.

Leeds and Cottis report only single specimens polarised at each potential. Furthermore, unlike the earlier workers, Leeds and Cottis did not carry out blank determinations to compensate for any weight loss arising from cleaning the specimens after the test. This would have exaggerated the calculated corrosion rate.

On balance, because we are obliged to make a choice between conflicting sets of experimental results, my vote goes to Moore, Dexter, Batt and their colleagues. It would be nice if, one day, a more definitive set of experimental data could be produced.

6.4.3 Practical experience

We have already mentioned Sir Humphry Davy's lectures to the Royal Society in 1824 [1] reporting the effectiveness of small lumps of zinc in protecting both copper and iron in seawater. That evening can reasonably be regarded as witnessing the birth of cathodic protection. Even so, at no point did Davy use the word *potential*, let alone attempt to measure it.

As we have discussed, the best part of a century passed before Davy's invention returned to the fore. It re-emerged in the 1920s with the application of forced current drainage to protect onshore welded petroleum pipelines in the USA. In 1933, Kuhn [2], having worked in the field for a decade, proposed –850 mV (measured against the Cu|$CuSO_4$ satd. half-cell) as the optimum potential for the protection of onshore steel pipelines. Eighteen years later, Schwerdtfeger and Dorman [4] endorsed Kuhn's view. However, they pointed out that Kuhn's "optimum" potential had, in the interim, become adopted as the least negative potential for onshore pipeline CP.

In the early days of marine CP, attention focussed more on the current needed to control corrosion than the potential. By way of example, a 1949 NACE conference presentation by Humble [3] on CP in seawater is based mostly on current density.

It would appear that Kuhn's onshore value migrated, somewhat belatedly, into the sphere of offshore CP. A potential of –850 mV (v Cu|$CuSO_4$ satd.) is as near as makes no difference –800 mV when expressed on the Ag|AgCl|seawater reference scale. This may, or may not, be simply a coincidence. I have not been able to find any documentation supporting the view that the –800 mV offshore protection criterion is simply an adoption of Kuhn's onshore value. Nevertheless, this protection potential for steel has been in CP codes since 1969 (NACE RP-0169), and specifically advised for seawater since 1973 (BS CP 1021). No less an authority than DNV (in RP-0416) reported that it ... *is not aware of any documentation that a potential (IR free) in the range –0.80 to –0.90 V has ever led to any corrosion damage (including corrosion damage by bacteria).* Similar statements followed in RP-F103, and the 2021 edition of RP-B401.

6.4.4 Implications

The applicability of a theory of CP based on electrode kinetics, the majority of the laboratory testing, and the impressive track record, all support the adoption of –800 mV as the appropriate criterion for the CP of steel in aerated seawater. It is little surprise, therefore, that all the codes (Table 6.1) endorse this value.

However, when referring to codes there can only be two outcomes: compliance and non-compliance. If we measure the potential on a structure and get a reading of –801 mV then, self-evidently, we are in compliance. On the other hand, a reading of –799 mV would, by definition, indicate non-compliance. But does it matter?

Unfortunately, there is insufficient documented field experience to help us with this question. The problem is that there is no body of hard evidence regarding the corrosion that has occurred as a result of long-term immersion at (recorded) potentials in the region of (say) –700 to–800 mV. The DNV experience cited above is not accompanied by complementary evidence that immersion at potentials in the region from –700 to –800 mV has led to observable corrosion damage.

In this context, it is also worth mentioning that, up until the late 1960s, the British Royal Navy was content to set a target protection potential of –750 mV for its warships [7,12]. Even when the target protection potential was moved to –800 mV, hull potentials in the range from –750 to –850 mV were deemed acceptable up until the 1990s [21]. Similarly, Moore and Knuckey [14] referred to –770 mV as a value which... *does not necessarily correspond to complete protection but is stated to be a practical value giving an economically acceptable degree of protection.*

In practice, indications of potentials failing to reach the target of –800 mV have often prompted CP retrofits. In some instances, such retrofits have been commissioned even when a more broad-based corrosion or fatigue engineering assessment would have revealed that retrofitting CP was unnecessary. This means that there is no accumulation of in-service experience covering the corrosion rates of steel structures at potentials anywhere between the free corrosion potential and –800 mV.

6.5 ANAEROBIC ENVIRONMENTS: THE –900 mV CRITERION

6.5.1 The codes

Examination of Table 6.1 confirms that the majority of the codes relevant to offshore CP require a potential more negative than –900 mV to ensure protection against MIC. My view is that the majority is incorrect. Others obviously disagree. Accordingly, we will examine the evidence for –900 mV in anaerobic environments in the same way that we discussed, and endorsed, –800 mV for aerated seawater. That is, we will look at theoretical considerations, laboratory testing and, finally, practical experience. However, before we do this, it is worth a brief recap of how the –900 mV requirement became enshrined in some of the codes.

6.5.1.1 British Standards Institution

The first internationally recognised code to specifically require –900 mV for the protection of steel in anaerobic environments, where sulfate-reducing bacteria (SRB) might be active, was BS Code of Practice CP1021 (1973). The criterion remained in place when CP1021 was extensively redrafted and re-issued as BS 7631-1 in 1991.

6.5.1.2 European and ISO standards

As is clear from Table 6.1, the requirement for a –900 mV protection potential is now replicated in EN and ISO standards dealing with situations where CP might be applied in anaerobic marine environments. Indeed, it even crops up in ISO 20313 in relation to ships' hulls. Other than as a result of particularly inept navigation, a ship to end up in seabed mud during prolonged lay-up.

It would appear that the requirement for –900 mV in EN and ISO owes much to simple replication. Once in a standard such a fundamental criterion is unlikely to be challenged. Indeed, I have sat on various standard drafting committees and pressed for a revision, but have had it pointed out that such a change would be inconsistent with other standards.

6.5.1.3 NACE

The first internationally recognised standard on CP was the NACE Recommended Practice RP0169 published in 1969. Although its title includes "immersed" pipelines, it focuses on onshore CP. It originally gave five criteria for achieving full CP:

- two based on achieving potentials of –800 mV[2]
- two based on potential swings of –100 and –300 mV
- one based on the inflection in an *E*–log *i* polarisation curve.

Over the various revisions of the code, the –300 mV swing and the *E*–log *i* criteria were dropped, and the requirement to achieve –800 mV was clarified to emphasise that the potential had to be a true polarised value.

The original edition of RP 0169 did not include any modification of this requirement for anaerobic environments.

However, from 1996 the document, now issued as a Standard Practice (SP0169), stated that the –800 mV criterion might be inadequate... *in some situations, such as the presence of sulphides, bacteria, elevated temperatures...* However, it was not until the 2013 edition that a requirement for a further 100 mV of polarisation was advised *...where MIC has been identified or is probable.*

NACE, on behalf of the American National Standards Institute (ANSI), did not adopt ISO 15589-2 for offshore pipelines. Instead, in 2007 it issued a modified version of the ISO document as SP0607. However, like ISO, NACE recommended –900 mV for anaerobic environments.

6.5.1.4 DNV

The pre-2005 editions of RP-B401, and its 1981 pre-cursor TNA 703, covered both structures and pipelines. All editions clearly required a design protection potential in anaerobic seabed conditions of –900 mV.

From 2005, subsea pipelines were removed from the scope of RP-B401, and the design protection potential was reset to –800 mV. This was accompanied by the statement... *It has been argued that a design protective potential of –0.90 V should apply in anaerobic environments, including typical seawater sediments. However, in the design procedure advised in this RP, the protective potential is not a variable.*

This statement did not confirm that DNV had actually rejected the –900 mV criterion. As it also pointed out in the RP, where a CP system was designed to achieve a potential of –800 mV, the... *potential will for the main part of the design life be in the range –0.90 to –1.05 (V).* This wording has remained essentially unchanged up to and including the 2021 edition of RP-B401.

RP-F103, which applies to submarine pipelines, and was first issued in 2003, failed to give a clear steer on the –800 versus –900 mV question. Because it was aligned with ISO 15589-2, the design protection potential for calculating anode output was, by inference, –900 mV in anaerobic environments. However, its design protection potential for calculating potential attenuation[3] was –800 mV.

Fortunately, the position of the 2016 edition is clearer. It assumes that all submarine pipelines encounter the *possibility of SRB activity* and that a potential more negative than –0.80

[2] I have changed –850 mV vs Cu|$CuSO_4$ (satd.) to –800 mV to align with the rest of this book.
[3] We deal with pipeline potential attenuation in Chapter 14.

V... *is then considered to give adequate protection for such conditions for CMn steel line pipe material.* In a guidance note, it also adds... *DNV is not aware of any published evidence that a potential between −0.80 and −0.90 V... has resulted in corrosion damage to carbon steel exposed to marine sediments normally being anaerobic due to bacterial activity.*

6.5.2 Theoretical considerations

6.5.2.1 Thermodynamics

Thermodynamics only offers indirect support for selecting a more negative protection potential than −800 mV. A 1964 paper by Horváth and Novák [11] presented E-pH diagrams for the Fe-S-H_2O system. On the basis of these diagrams, the authors concluded... *that in nearly neutral media (where sulphate-reducing bacteria are most active) an even more negative potential...* (than −900 mV)... *is necessary for complete cathodic protection if the media are saturated with* H_2S. As Stott remarks [28] in his chapter on MIC in the 2010 edition of the textbook *Shreir's Corrosion*, ...*There is a long-standing precept, current in the industry, that structures need to be held at potentials more negative than ~−0.9 V (vs Ag|AgCl) in order to protect against anaerobic corrosion by SRB. Horvath and Novak presented the most reasoned argument for this view, which has appeared numerous times in print, though the evidence is not very convincing.*

Stott's opinion is telling, but he does not elaborate on why he considered the paper *not very convincing.* I understand that, from a microbiological perspective, there might be a difficulty in positing the activity of SRB in an environment that is *saturated with* H_2S. SRB, like all micro-organisms, cease activity when their metabolic waste products, in this case H_2S, accumulate.

From my own, non-microbiological perspective, the problem with the Horváth and Novák interpretation of their E-pH diagram is their assumption that the potential of the steel has to be depressed into the immunity domain in order to achieve protection. As we have already seen, achieving immunity is not necessary to achieve adequate levels of protection.

Elsewhere, some workers who favour the equilibrium thermodynamics model of CP are obliged to view the action of CP as raising the pH of the surface to a value high enough to sterilise it. This is a flawed perspective. It is predicated on a misunderstanding that MIC is a biological corrosion mechanism. This is not the case. The mechanism of corrosion remains electrochemical. The role of the organisms is to modify the environment. It is not necessary to kill the microbes in order to control corrosion. As a matter of experience, marine structures are inevitably covered in microbiologically active biofilms. However, they do not corrode providing adequate (−800 mV) levels of CP are applied.

6.5.2.2 Electrode kinetics

The electrode kinetic standpoint on MIC is very straightforward. It accepts that, in some cases, microbiological activity can exert a profound effect on the environment at the metal surface. This in turn stimulates the kinetics of the cathodic reaction, thereby increasing the corrosion rate. However, the shape and slope of the anodic polarisation curve are independent of the environment. This means that making the potential more negative forces a reduction in the rate of anodic dissolution. Furthermore, as described in the Butler-Volmer equation, this rate reduction is logarithmic in character, as illustrated in Table 6.2.

Thus, from an electrode kinetic perspective, the position is unequivocal. There is no requirement to polarise to a more negative potential just because MIC is a threat.

6.5.3 Laboratory investigations

A 1952 paper by Wormwell and Farrer [6] was pivotal in introducing the −900 mV criterion into BS CP1021, and thence into other national and international codes. It appears in the bibliography of the standard. Perhaps more relevantly, Dr Wormwell was easily the most highly qualified corrosion scientist on the drafting committee for CP1021. Because of its influence, I analysed this paper in some detail in a 2015 review of the −900 mV criterion [29]. Unfortunately, the experimental details were poorly reported by the authors. So, it was difficult to draw any firm conclusions. However, since they did not actually measure any corrosion rates for their polarised specimens, one has to be circumspect in accepting their conclusion that a more negative protection potential was required in the presence of SRB.

Significantly, Wormwell and Farrer cite an earlier paper by Ewing [5], in which they state that he ... *has shown that the minimum potential at which steel will be protected varies with the corrosive environment. This is reasonable on theoretical grounds and is confirmed by our own work.* I also reviewed the Ewing paper in 2015 and found that he had actually said no such thing. The important point here is that, intuitively at least, it might seem entirely reasonable that a more negative protection potential is required for a more corrosive environment. Many corrosion specialists have accepted this proposition.

For example, no less an authority than UR Evans, in his classic 1960 textbook [8] explains that the potential that must normally be achieved to ensure protection is −800 mV. However, he then adds that, in the presence of SRB, the potential... *must be kept at a still lower level; Wormwell and Farrer advise −0.90 V.* That Evans added this clause might be associated with the fact that Dr Wormwell acted as the scrutiniser for that chapter.

However, later in the same book (pp. 890–891), Evans presented the principles of cathodic protection based on electrode kinetics. Although he did not explicitly say so, his theory makes it clear that, although the current needed to achieve protection varies according to the environment, the actual potential at which protection is achieved does not. Thus, even the great can occasionally be inconsistent.

The flaw in the intuitive argument that the protection potential must vary with the corrosive environment has been demonstrated by Dexter et al. [18]. They reported that to avoid having to wait 120 days for the results of experiments in seawater, they conducted a few additional experiments in 0.1 M HNO_3. These tests were of only 36 hours duration. Nevertheless, on converting the as-reported data into a potential versus corrosion rate format, they provide an interesting picture, as is seen in Figure 6.2. Allowing for the inevitable experimental errors in this type of ad hoc experiment, which I discuss elsewhere [34], it is clear that −800 mV is an effective protection potential for steel in nitric acid, just as it is in seawater.

This demonstrates that the view held by many that a more negative protection potential is required for a more corrosive environment, whilst intuitively reasonable, is simply incorrect.

It is also worth pointing out that there is no way that enthusiasts for the passivity paradigm of CP can explain this result using equilibrium thermodynamics.

As mentioned, I reviewed the experimental evidence supporting the −900 mV criterion in 2015 [29]. That paper ended up being rather lengthy. I do not propose to reproduce it here. Nevertheless, it is worth commenting briefly on some of the papers included in the review, and some more recent publications.

A case in point is the 1968 paper Booth and Tiller published on SRB [13] which concluded with the following (impressively wordy) sentence: *The practical criterion for the cathodic protection of ferrous materials that, in the presence of sulphate-reducing bacteria, the protective potential should be depressed by 0.10 V below the protective potential in the absence of bacteria, appears to be a correct estimate in so far as it is possible to make a direct comparison between the system used in the present work and practical cases in the field.*

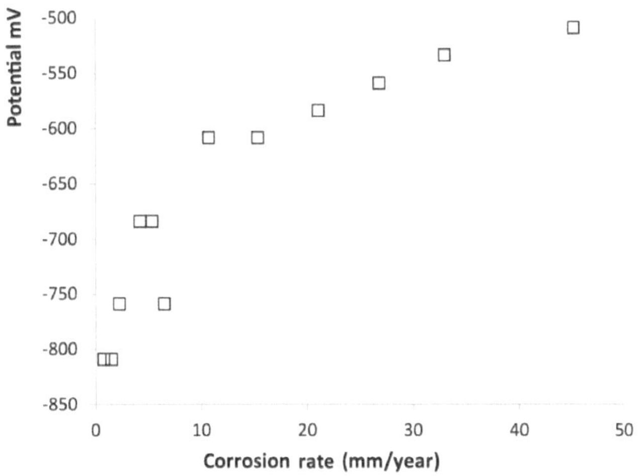

Figure 6.2 Corrosion rate versus potential in 0.1 M HNO_3. (From Ref. [18].)

Given that the work was carried out at the UK's prestigious National Physical Laboratory, it is unsurprising that it was reflected in CP1021. However, as explained in my review, their data do not actually support the proposition that a more negative protection potential is required in the presence of SRB.

Similarly, other reports of laboratory studies, when critically analysed, fail to demonstrate corrosion at potentials more negative than −800 mV. Very often the corrosion rates reported have not been measured under true potentiostatic conditions, or the experimenters failed to run replicates or, crucially, omitted blanks to serve as gravimetric controls. Some publications are based on experiments that suffered from all of these limitations [17]. The running of gravimetric blanks or employing some other technique (e.g. ASTM G1 [9]) to compensate for errors incurred when descaling specimens is of fundamental importance.

An additional problem we find when reviewing the relevant corrosion literature is that microbiologists have often studied the effects of CP on microbial activity, rather than making corrosion rate measurements (e.g. [27]). Alternatively, they have mis-interpreted corrosion data. For example, Liu and Cheng [31] immersed steel electrodes in solutions of soil extracts inoculated with SRB. These were maintained at either the free corrosion potential, −800 or −950 mV for a short duration (7 days). Then after a fixative treatment, the agglomerations of microbes on the surfaces were studied in a scanning electron microscope. More relevantly from our point of view, they removed the biomass and calcareous scales, which they referred to consistently but without evidence as "corrosion products", by pickling in inhibited hydrochloric acid. They then examined the surfaces using an atomic force microscopy (AFM). As a result of this, they reported that... *pitting corrosion would still occur on the steel under the biofilm even when the steel is at a CP potential of...* (−950 mV).

Obviously, if correct, the codes listed in Table 6.1 would all need to be revised. However, their result, and hence their conclusion, are not correct. They interpreted the pits detected by AFM as deriving from the 7 days of polarised exposure. Unfortunately, they ignored the fact that even inhibited hydrochloric acid (e.g. Clarke's solution) causes micro-pitting [30]. Crucially, like so many other corrosion investigators, Liu and Cheng did not run any control blanks through their acid cleaning process. Had they done so they would, presumably, have revised their conclusion.

In addition to experimental problems, further confusion is found where some authors have published reviews which do not align with what the reviewed paper actually said. One example of many is provided by de Romero et al. [27] who stated that, when the media contain SRB, the... *potential criterion changes to...* (−900 mV)... *to polarize the metallic structure adequately. This statement was proposed by Butlin and Vernon in 1949 and by Horvath and Novak in 1964, based on thermodynamic data of Fe-S-H_2O system, and was afterwards experimentally verified by Fisher in 1981 who studied the cathodic protection of pipelines in North Sea sediments and muds containing SRB.*

This is a misreading of Fisher's work. The question that he sought to address [15] was not: does the presence of SRB require polarisation to −900 mV? It was: is −900 mV adequate to protect in the presence of SRB? The difference is important because he did not investigate whether or not −800 mV would have been sufficient. Thus, he concluded that −900 mV was adequate; but he did not, and could not, conclude that −800 mV was inadequate.

6.5.4 Field test data

6.5.4.1 Onshore pipelines

All of the instances of claimed corrosion of MIC activity leading to corrosion on cathodically protected pipelines are from onshore. Unfortunately, this body of onshore information is unlikely to enable us to take this issue forward. This is because the reported instances of alleged MIC on onshore pipelines relate to corrosion under disbonded coatings on pipelines with an uncertain CP history.

Thus, we might be able to confirm the cumulative corrosion that has taken place: either by excavating the pipeline or by running an in-line inspection tool. We might also be able to conduct a root-cause analysis that points to MIC. However, we cannot know when the corrosion commenced, nor can we assume that its rate has been constant. More important, it is unlikely that we would be able to claim precise knowledge of the potential, or the range of potentials, that has been sustained on the corroded area at all times through the pipeline life.

6.5.4.2 Offshore pipelines

I am aware of no instances of reported external MIC, or any other external corrosion for that matter, on offshore pipelines that are under CP. This is encouraging but, like the experience with onshore pipelines, it does not help us resolve the −800 versus −900 mV question.

The reason for this is that we have yet to acquire sufficient relevant offshore pipeline history. Practically, all pipelines have been so well-coated when laid that they have spent most of their lives polarised to near anode potential (about −1050 mV). In the few cases where the pipeline CP has been compromised for some reason, retrofit cathodic protection systems have usually been installed with some alacrity. Thus, almost all of the cumulative surface of offshore pipelines have been at −900 mV or more negative. This means that we simply have very little experience of pipelines protected at less negative potentials than −900 mV in seabed mud.

The experience of DNV that it ...*is not aware of any documentation that a potential (IR free) in the range −0.80 to −0.90 V has ever led to any corrosion damage (including corrosion damage by bacteria...* is informative; but alone, it is not conclusive. There is not a lot of buried pipework in the potential range from –800 to –900 mV.

Advocating a relaxation of the –900 mV criterion to –800 mV on the basis of either field experience or laboratory work, therefore, runs into the philosophical conundrum of trying to prove a negative. Just because nobody has seen corrosion, does not prove beyond all doubt that it is not there.

The relaxation can only be supported by demonstrating irrefutably that electrode kinetics, and not equilibrium thermodynamics, is the appropriate explanation for CP. If that can be demonstrated, and universally accepted, then the sufficiency of –800 mV automatically becomes self-evident.

6.6 THE EFFECT OF TEMPERATURE

6.6.1 What the codes say

The codes have plenty to say about the effect of temperature on the current density needed for protection (Chapter 7) and its effect on the performance of sacrificial anodes (Chapter 10). By contrast, they have much less to say about any relationship between the target protection potential and the temperature.

Most marine CP is applied to unheated steelwork (offshore structures, ships' hulls, harbour piling, etc.) in ambient temperature seawater. Within the temperature range of the world's oceans, which is at most from 0°C to 35°C, it is evidently not considered necessary for any adjustment to the design protection potential. For example, DNV-RP-B401 only advises a single protection potential (–800 mV) for any seawater temperature.

Over the years, however, offshore pipelines have been required to handle progressively hotter crudes, such that the pipe-seawater interface can exhibit elevated temperatures. Whereas DNV-RP-F103 does not require any modification of the protection potential for elevated temperatures, Table 1 of ISO 15589-2 calls for a shift in the protection potential from –800 to –900 mV where the pipeline is buried in seabed sediments and subject to elevated temperature.

However, the view of the standard is somewhat obscured. It advises a potential of –900 mV to cover... *the possibility of SRB activity and/or high pipeline temperature ($T > 60$°C).*

As with our analysis of the –800 and –900 mV criteria, we can examine the code requirements by examining the theory, laboratory testing and field experience.

6.6.2 The theory

ISO 15589-2 does not disclose the basis of its recommendation, and I have not found any supporting test evidence. Nevertheless, we can indulge in a little theorising as to the effect that elevated temperature might have on the protection potential. For example, if we set up the Nernst equation for iron in equilibrium with its ions at a concentration of 10^{-8} g.ions/L (a figure which we discussed in Chapter 4), then we see that increasing the seawater temperature from (say) 15°C–65°C shifts the equilibrium potential about 40 mV more negative. Thus, if we were viewing CP in terms of equilibrium thermodynamics (which, in my view, we should not be), we would conclude that adding an extra 100 mV to cater for the possibility of a 40 mV requirement would be very conservative. However, given the historical approach of drafting committees, this conservatism would not be surprising.

In Chapter 5, we introduced the Butler-Volmer equation as the theoretical explanation underpinning Tafel-type electrode kinetic behaviour. This tells us that the value of the

Tafel slope on an E–log|i| graph is a function of temperature. Simple calculations show that, in theory, the Tafel slope increases by about 15 mV per decade, that is from about 60–75 mV/decade, if we raise the temperature from 15°C to 65°C. This behaviour provides a prima facie case for setting a temperature-dependent target protection potential. So, if −800 mV is acceptable for seawater at 15°C, and we assume that two orders of magnitude reduction in the corrosion rate are needed, then the revised target protection potential would be −830 mV. On this basis also, the ISO standard errs comfortably on the side of caution.

6.6.3 Laboratory testing

A UK Department of Energy study into the effect of temperature on marine CP parameters, including the protection potential was published in 1982 [16]. This concluded that... *adequate cathodic protection can be provided to either heated or unheated steel by polarisation to a potential of −0.8 V.* Figure 6.3 shows results from the study, converted from gravimetric corrosion rates to mm/year, which support this conclusion [34]. Indeed, if we accept that a corrosion rate of <0.01 mm/year represents adequate corrosion control, then, based on Figure 6.3, any potential more negative than –700 mV would appear to be sufficient; although nobody is suggesting that the codes be rewritten.

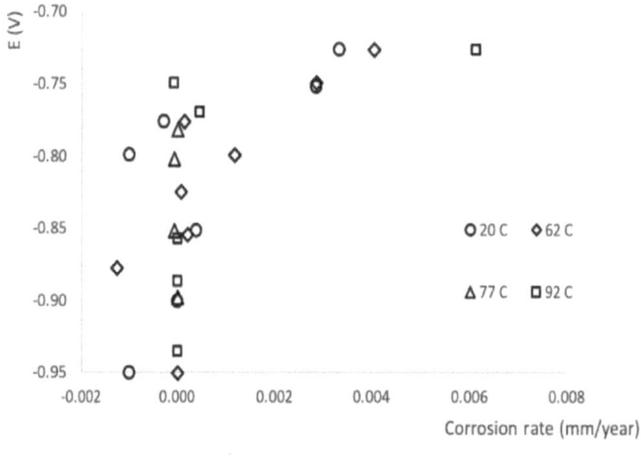

Figure 6.3 Effect of steel surface temperature on protection potential [16,34].

Before moving on, it is worth observing that Figure 6.3 also contains some apparently "negative" corrosion rates. This does not mean that CP reinstates lost metal! These results simply illustrate that, even in the most carefully performed experiments, random uncontrollable errors need to be accounted for.

6.6.4 Field experience

As noted in the context of the −800 versus −900 mV debate for buried pipelines, practically all submarine steel hydrocarbon pipelines are protected at potentials more negative than −900 mV. No operator would knowingly let the potential of an operational, or indeed a mothballed, pipeline drift to potentials less negative than −800 mV without instigating a CP retrofit. This means that the conclusion of the UK Department of Energy study that −800 mV is adequate for temperatures up to at least 92°C has not actually been tested in service.

6.7 EXCESSIVELY NEGATIVE POTENTIALS

It may be noted from Table 6.1 that, in addition to prescribing the least negative potential needed for the protection of steel, most codes also give a maximum tolerable level of cathodic polarisation. For low and medium-strength steels, where a figure is specified, it is –1100 mV. This potential limit is intended to minimise damage to protective coatings (see Chapter 9). For higher-strength steels, even less negative limits are recommended. This is intended to mitigate the threat of hydrogen embrittlement. We deal with this in Chapter 12.

6.8 OPTIMUM POTENTIALS

In this chapter, we have gone to some length establishing the "correct" potential for the protection of steel in seawater. There is no reason for challenging the unified code view that this figure should be –800 mV for aerated seawater. We are also now aware that this figure embodies some conservatism. We also advanced the case that –800 mV was fully sufficient for steel in anaerobic seabed sediments, even where there is a threat of MIC. However, that remains a view that is not universally accepted in the CP industry. Beyond that, there is a body of experimental evidence that –800 mV comfortably applies to steel in any environment, even dilute mineral acids, and at elevated temperatures (up to 92°C). Again, notwithstanding the evidence, this is a view that does not carry the judgement of all CP professionals.

The protection potential is the target least negative potential to be achieved on the structure. However, this is not necessarily the same as the optimum potential. The latter parameter might be more negative than the protection potential. It reflects the fact that CP may be achieved more efficiently in seawater, in terms of current consumption, at potentials of (say) –900 to –1000 mV. The reasons for this are explored when we consider current requirements for protection in the next chapter.

6.9 POTENTIAL DISTRIBUTION

Our above discussions have been phrased in terms of "a potential" that needs to be achieved on the surface. In reality, CP does not generate a single potential. It produces a distribution of potentials that broadly reflects the distribution of the current delivered by the CP system. We will return to issues of current and potential distribution when we discuss fixed and floating structures, pipelines and internal spaces (Chapters 13–16). In Chapter 17 (Modelling), we will look at how computational techniques can help us predict how the potential might be distributed on a structure. Then, in Chapter 18 (CP System Management), we examine how potential and current distribution can be measured.

REFERENCES

1. H. Davy, *On the corrosion of copper sheeting by sea water, and on methods of preventing this effect; and on their application to ships of war and other ships*. Philosophical Transactions of the Royal Society A **64**, 151–158 (1824).
2. R.J. Kuhn, *Cathodic protection of underground pipe lines from soil* corrosion. API Proceedings, vol. 14, pp. 153–157, section 4, November 1933. (Reprinted in *Cathodic Protection Criteria: A Literature Survey*. NACE International (Houston, TX) 1989.

3. H.A. Humble, *The cathodic protection of steel piling in sea water.* Corrosion **5** (9) 292 (1949).
4. W.J. Schwerdtfeger and O.N. Dorman, *Potential and current requirements for the cathodic protection of steel in soils.* Journal of Research of the National Bureau of Standards **47**(2), 104 (1951).
5. S.P. Ewing, *Potential measurements for determining cathodic protection requirements.* Corrosion 7 (12), 410–418 (1951).
6. F. Wormwell and T.W. Farrer, *Protection of steel in presence of sulphate-reducing bacteria.* Chemistry & Industry **30**, 973–974 (1952).
7. J.T. Crennel and W.C.G. Wheeler, *Zinc Anodes for use in sea water.* Journal of Applied Chemistry **6**, 415 (1956).
8. U.R. Evans, Chapter VIII *Buried and Immersed Metal-Work*, p. 285, in *The Corrosion and Oxidation of Metals: Scientific Principles and Practical Applications.* Edward Arnold Ltd (London) 1960.
9. ASTM G1, *Standard practice for preparing, cleaning, and evaluating corrosion test specimens*, 1967 (Updated to G1-03 in 2003, current edition 2017).
10. F.L. LaQue and T.P. May, *Experiments relating to the mechanism of cathodic protection of steel in sea water.* Proceedings of 2nd International Congress on *Metallic Corrosion,* 1965, p. 789. Re-printed: Materials Performance **21** (5), 18 (1982).
11. J. Horváth and M. Novák, *Potential/pH equilibrium diagrams of some Me-S-H_2O ternary systems and their interpretation from the point of view of metallic corrosion.* Corrosion Science **4**, 159–178 (1964).
12. T. Howard-Rogers, *Marine Corrosion* (Table 23, p. 192). Newnes (London) 1968.
13. G.H. Booth and A.K. Tiller, *Cathodic characteristics of mild steel in suspensions of sulphate-reducing bacteria.* Corrosion Science **8**, 583–600 (1968).
14. B.T. Moore and P.J. Knuckey, *Minimum protective potential of mild steel in sea water.* Corrosion Australasia **3** (3), 4 (1978).
15. K.P. Fischer, *Cathodic protection criteria for saline mud containing SRB at ambient and a higher temperature.* Paper 110 CORROSION/81.
16. G.P. Rothwell, P.E. Francis and K.F. Hale, *Society for underwater technology for UK department of energy.* Summary report OT-R-8292 (London) 1982.
17. T.J. Barlo and W.E. Berry, *An assessment of the current criteria for cathodic protection of buried steel pipelines.* Materials Performance **23**(9), 9 (1984).
18. S.C. Dexter, L.N. Moettus and K.E. Lucas, *On the mechanism of cathodic protection.* Corrosion **41** (10), 598 (1985).
19. J.N. Wanklyn, *Input data for modelling marine cathodic protection*, Chapter 4, in *Cathodic Protection: Theory and Practice,* Eds. V. Ashworth and C.J.L. Booker. Ellis Horwood (Chichester) 1986, pp. 69–71.
20. J.H. Morgan, p. 97, in *Cathodic Protection*, 2nd edition. NACE International (Houston, TX) 1987.
21. D.J. Tighe-Ford, R.A. Botten and R.D. Hughes, *Study of design criteria for ship impressed-current cathodic protection by stylised modelling.* Corrosion Prevention & Control **37** (2), 5 (1990).
22. R. Baboian, p. 61, in *NACE Corrosion Engineers Reference Book*, 2nd edition, Eds. R.S. Treseder et al. NACE International (Houston, TX) 1991.
23. C.L. Batt and M.J. Robinson, *Optimising cathodic protection requirements for high strength steels under marine biofilms.* Corrosion Management Issue 31 Sept./Oct., 13 (1999).
24. R.L. Bianchetti, p. 97, in *Peabody's Control of Pipeline Corrosion*, 2nd edition, NACE International (Houston, TX) 2001.
25. S.S. Leeds and R.A. Cottis, *An investigation into the influence of cathodically generated surface films on the mechanism of cathodic protection.* Paper 06084 CORROSION/2006.
26. S.S. Leeds and R.A. Cottis, *The influence of cathodically generated surface films on corrosion and the currently accepted criteria for cathodic protection.* Paper 09548 CORROSION/2009.
27. M. de Romero, O. de Rincónet and L. Ocando. *Cathodic protection efficiency in the presence of SRB: State of the art.* Paper 407 CORROSION/2009.

28. J.F.D. Stott, *Corrosion in microbial environments*, in *Shreir's Corrosion*, vol. 2, Eds. J.A. Richardson, et al., Elsevier (Amsterdam) 2010, pp. 1169–1190.
29. C. Googan, *Offshore Pipelines: Do we need −900 mV?* Paper 100 Eurocorr, 2015.
30. S.A. Wade and Y. Lizama, *Clarke's solution cleaning used for corrosion product removal: Effects on carbon steel substrate*. Proceedings Australasian Corrosion Association Conference Adelaide, 2015.
31. T. Liu and Y.F. Cheng, *The influence of cathodic protection on the biofilm formation and corrosion behaviour of an X70 steel pipeline in sulfate reducing bacteria media*. Journal Alloys and Compounds **729**, 180 (2017).
32. U.M. Angst, *A critical review of the science and engineering of cathodic protection of steel in soil and concrete*. Corrosion **75**(12), 1420 (2019).
33. J. Barthel and R. Deiss, *The limits of the Pourbaix diagram in the interpretation of the kinetics of corrosion and cathodic protection of underground pipelines*. Materials and Corrosion **72** (3), 434 (2021).
34. C. Googan, *The cathodic protection potential criteria: Evaluation of the evidence?* Materials and Corrosion **72** (3), 446 (2021).

Chapter 7

Current and polarisation

7.1 WHAT WE NEED TO KNOW

Although the protection potential for steel is difficult to define precisely, there is unanimity among the codes that −800 mV works. As discussed in Chapter 6, there is a lack of consensus about the figure for anaerobic sediments, but we need not revisit that debate here. Furthermore, the protection potential is independent of the ambient seawater conditions and does not vary through the life of the structure.

There is less certainty when it comes to the cathodic current. The values we select for a design will depend on the ambient seawater conditions and will alter during the operational life. The factors that determine the cathodic current demand are:

- the surface area of the steel structure,
- the presence and performance of any coating,
- the cathodic current density needed for bare steel,
- the manner in which that current density will change over time,
- current drain to connected metalwork.

Because the surface area should be straightforward to determine, we need not consider it further here. We will consider coatings in Chapter 9 and current drain to connected structures in Chapters 13 and 14. In this chapter, our focus is on the bare steel cathodic current density.

7.2 WHAT THE CODES ADVISE

7.2.1 Current densities for seawater (offshore)

Since we used DNV-RP-B401 in Chapter 3, we can continue with it here. As we saw, it provides three design current densities:

- the initial value is needed to achieve polarisation to the protection potential,
- the mean value is the current density needed to maintain that protection potential once it has been achieved,
- the final current density includes an allowance to recover full protection in the event that it is lost in a depolarising event such as a severe storm. Instead of "final", this could be termed the "repolarisation" figure.

The mean current demand dictates the minimum mass of anode alloy that has to be installed. The ability to satisfy the instantaneous initial and final current demand depends on the geometry and surface area of the anodes.

DOI: 10.1201/9781003216070-7

It is worth noting that not all codes listed at the start of this book call for an initial, mean and final current density. Indeed, it would be easy, although tedious, to compile a list of all of the design CP current densities advised by these codes. Instead, by way of an example, it is more illuminating to look at the changing advice offered by DNV over the years. Table 7.1 shows how its recommendations have changed for tropical waters (0–30 m depth).

Table 7.1 DNV cathodic current densities – tropical waters 0–30 m depth

		Cathodic current density (mA/m²)		
Year	*Code*	*Initial*	*Mean*	*Final*
1981	TNA-703	120	90	80
1986	RP-B401	130	70	90
1993		150		
2005				100
2010				
2017				
2021				

7.2.2 Current densities for seawater (near-shore)

The development of wind farms in near-shore waters, where tidal ranges and tidal currents are more exaggerated than offshore, has prompted codes such as DNV-RP-0416 and ISO 24656 to recommend higher design current densities. For example, DNV-RP-0416 recommends a 50% uplift in the design values that DNV-RP-B401 would advise for structures further offshore. We return to this topic in Chapter 13.

7.2.3 Current densities for seabed burial

Current densities for the seabed are lower than for aerated seawater. This reflects the restricted diffusion of dissolved oxygen through the sediments. Interestingly, the codes make no distinction between the nature of the seabed sediments or the depth of burial. This might seem surprising since both factors are likely to have a profound effect on the rate of oxygen diffusion. Nevertheless, by way of illustration, the figures given in some current (2021) and superseded codes for structural steelwork in seabed sediments are set out in Table 7.2.

Table 7.2 Cathodic current densities – structures in seabed sediments

	Cathodic current density (mA/m²)	
Code	*Initial*	*Mean/Final*
DNV-RP-B401	20	
Norsok M-503		
NACE SP0176	11–33	
EN 12495	25	20
ISO 13174	25	20
ISO 13174[a]	30–50	25–30

[a] ISO 13174 relates to harbour piling under... *conditions with established microbial corrosion.*

These codes adopt the reasonable view that, unlike pipelines, structures will not be internally heated. Similarly, buried surfaces will experience no depolarising events during the operating life. So, there is no need for an increased final, or repolarisation, current density. A search of the literature has unearthed little in the way of any rigorous justification for the figures in Table 7.2. Like so much of the information in the codes, they are best regarded as representing well-intentioned, albeit conservative, opinions of experts.

7.3 THE PROBLEM WITH THE CODES

For the most part, the CP design codes, such as DNV-RP-B401, envisage only a single potential for the steel cathode (−800 mV), a single operating potential for the anodes (−1050 mV for an Al-Zn-In alloy) and the three current density states: initial, mean and final. This design scenario is illustrated in Figure 7.1.

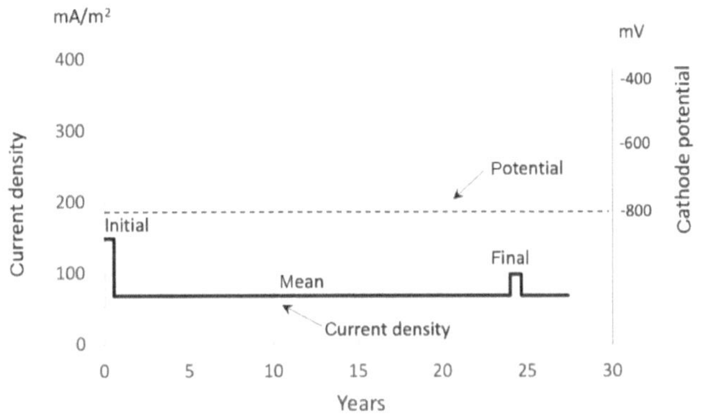

Figure 7.1 Polarisation prediction – code-based design.

In reality, CP systems are not simple electronic components. They do not "switch state" instantaneously as Figure 7.1 might imply. Both potential and current densities change over time. Figure 7.2 is a schematic illustration of a more realistic time-based profile for the current density and profile for cathodically protected steel.

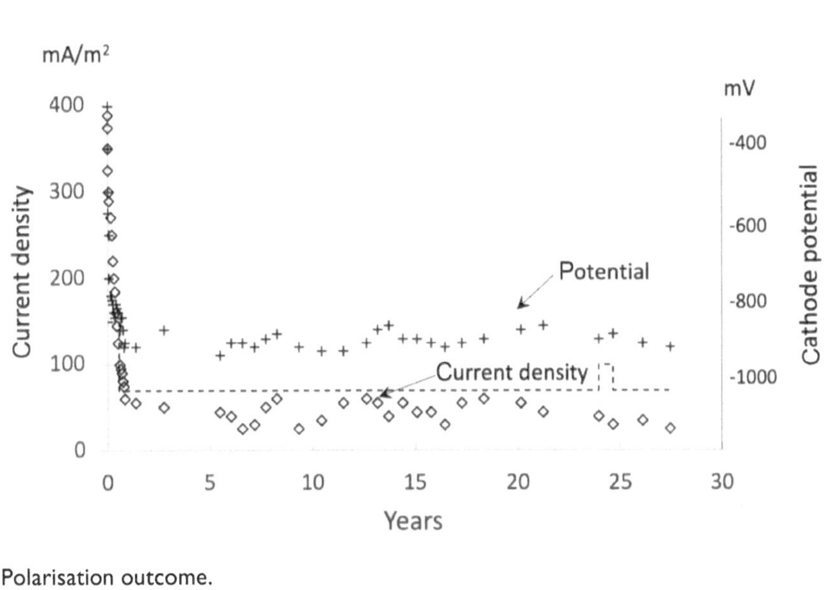

Figure 7.2 Polarisation outcome.

In viewing these two figures, it is important to understand that the *initial* design current density values given in the codes are based on the arbitrary, but incorrect proposition that the cathode has already reached −0.8 V. In fact, the potential of bare steel with atmospheric rust or mill scale will be very much more positive when it is first immersed in seawater. This means that sacrificial anode CP systems, designed according to DNV to give the initial current densities in Table 7.1, will actually deliver very much higher currents than the code values during the early stages of polarisation.

Importantly, just because a code such as RP-B401 uses an "incorrect" view of the initial current density does not mean that the code itself is "incorrect". We need to remember that most codes do not try to mimic the electrochemical processes taking place during polarisation. Their sole function is to present a methodology that will produce a CP system that does the job.

When it comes to cathodic current densities, there is not much theory that is useful to us. So, we will focus on laboratory testing and field experience. Here, we need to keep in mind that such testing usually provides short-term (<1 year) data of the type seen on the left-hand side of Figure 7.2. The information is interesting and obviously meaningful but cannot simply be substituted for the codes-based values indicated Figure 7.1. The substitution of such short-term current density results for the values recommended in the codes, without adjusting the potentials to match, will lead to CP over-design.

7.4 LABORATORY TESTING: CLEAN STEEL

Thermodynamics offers no insight into the current density needed to cathodically protect any metal. This is not surprising since current density is a rate parameter, and thermodynamics is silent on rates. Electrode kinetics, on the other hand, points us in the right direction. As we saw in Figure 5.14, the cathodic current demand on a protected surface is the arithmetic difference between the anodic and cathodic currents at the protection potential (E_{prot}). Since the anodic current is close to zero at E_{prot}, the CP current density is approximately equal to the cathodic current density at that potential.

We have already seen that, in seawater at a pH of ~8, the dominant cathodic reaction is the reduction of dissolved oxygen. In Chapter 5, we also saw that this cathodic current is under diffusion control. It is limited by the rate at which the oxygen molecules can diffuse from the bulk seawater to the steel surface. You may recall that in Figure 5.9 we termed it the limiting, or diffusion, current density $i_{\lim O_2}$.

For a clean steel surface, the value of $i_{\lim O_2}$ varies according to the dissolved oxygen concentration, the flowrate of the seawater across the surface and the temperature. The relationship between these parameters has been explored by Ashworth [22] who measured corrosion rates using a steel rotating cylinder electrode to produce controlled conditions of hydrodynamic flow. The corrosion rates were essentially measurements of the values of $i_{\lim O_2}$ at the different rotational speeds. He then demonstrated that, using mass transfer analogies developed for chemical engineering, the $i_{\lim O_2}$ data could be extrapolated to other flow rates and dissolved oxygen concentrations. Some of his results are reproduced in Table 7.3.

If we had no other information, we might use these data as a basis for selecting current densities for a CP design. Indeed, the figures are broadly similar to values in the codes. However, it needs to be borne in mind that the data in Table 7.3 were obtained in short-term experiments conducted on polished steel electrodes. If the experiments had been run for longer, the observed currents would have diminished. This brings us to the important topic of calcareous deposits.

Table 7.3 Effect of flow on cathodic current density

	O_2 *concentration (mg/L)*				
	6	*7*	*8*	*9*	*10*
Flow rate (m/s)			$i_{lim\,O_2}$ *(mA/m²)*		
0	68	80	91	102	114
0.3	78	91	105	118	131
0.4	82	95	105	118	131
0.6	89	103	118	133	148
1	102	119	136	153	170
2	136	159	182	205	227
4	205	239	273	307	341

From [22].

7.5 CALCAREOUS DEPOSITS

7.5.1 The chemistry

We are interested in seawater because of its salt content (~3.5% NaCl), which makes it conductive, and because of its dissolved oxygen content (up to ~12 mg/L), which makes it corrosive to steel. However, seawater also contains numerous other dissolved species, including carbonate (CO_3^{2-}) and bicarbonate (HCO_3^-) ions. These ions are present in concentrations governed by the equilibrium:

$$HCO_3^- \Leftrightarrow CO_3^{2-} + H^+ \quad (7.1)$$

Because hydrogen ions (H^+) appear on the right-hand side of the equilibrium, we know that more acid conditions (i.e. lower pH) favour the formation of bicarbonate, whilst more alkaline conditions, such as provided by seawater, favour formation of the less soluble carbonate.

As we saw in Chapter 2, when CP is applied in aerated seawater, the primary cathodic reaction is the reduction of dissolved oxygen:

$$O_2 + 2H_2O + 4e^- \rightarrow 4OH^- \quad (7.2)$$

This production of hydroxyl ions (OH^-) at the surface causes a local increase in the pH. This is readily demonstrated in the laboratory, for example adding a pH indicator to the solution in an electrochemical cell clearly shows the pH rise at a cathode surface. On the other hand, the magnitude of this pH rise is less easy to predict or measure. Various workers have tried. For example, Dexter and Lin [18] produced results that might be regarded as encouraging but not definitive.

If the cathodic polarisation is applied over-enthusiastically, another cathodic reduction reaction comes into play: hydrogen evolution. This cathodically reduces hydrogen ions to liberate hydrogen gas:

$$2H^+ + 2e^- \rightarrow H_2 \quad (7.3)$$

This removal of hydrogen ions from the solution adjacent to the metal likewise causes an increase in pH. The CO_3^{2-} concentration increases since equilibrium 7.1 is shifted to the right. The removal of H^+ ions forces the dissociation of bicarbonate (HCO_3^-) ions to produce more carbonate (CO_3^{2-}).

Seawater also contains magnesium (Mg^{2+}) and calcium (Ca^{2+}) ions. If we conduct experiments in synthetic seawater, we find that magnesium reacts with the cathodically produced OH^- to produce magnesium hydroxide ($Mg(OH)_2$), also known as brucite:

$$Mg^{2+} + 2OH^- \rightarrow Mg(OH)_2 \tag{7.4}$$

Similarly, Ca^{2+} reacts with carbonate ions to give calcium carbonate ($CaCO_3$), in a mineral form known as aragonite:

$$Ca^{2+} + CO_3^{2-} \rightarrow CaCO_3 \tag{7.5}$$

The solubilities of both $Mg(OH)_2$ and $CaCO_3$ are likely to be exceeded, leading them to precipitate as solid crystals on the surface. Under ambient temperature conditions (~20°C), seawater is naturally close to being fully saturated with $CaCO_3$. The action of CP is to increase the degree of super-saturation, thereby increasing the tendency for solid $CaCO_3$ material to precipitate onto the surface.

Taken together, these precipitated salts of magnesium and calcium are referred to as calcareous deposits. The formation of these deposits, which are porous in nature, is a natural consequence of the seawater composition, and the pH increase that occurs on any metal surface under CP.

7.5.2 Importance

7.5.2.1 Benefits

The development of calcareous deposits is almost uniformly beneficial for CP. The fact that the mean design current density is lower than the initial (see Table 7.1) is a direct result of these deposits. Simplistically, but not inaccurately, they can be viewed as hampering the diffusion of dissolved oxygen to the steel surface.

Indeed, it has been claimed that... *cathodic protection in seawater is a practical proposition only because the discharge of the cathodic current leads to the deposition of a protective layer of calcareous material on the surface being protected* [23]. Although that claim might overstate their importance somewhat, it is certainly true that the quality of the calcareous deposit formed during the initial polarisation has a profound effect on the subsequent current density needed for protection throughout the remainder of the life.

7.5.2.2 Possible drawbacks

The only area where I have encountered claims that calcareous deposits might present a problem is where they might change the profile of engineered surfaces. For example, calcareous deposition has been raised as issues of concern for underwater hydraulic equipment and subsea connectors.

Any concern would need to be evaluated on a case-by-case basis. Nevertheless, it is worth noting that there are tens of thousands of subsea connectors and hydraulic actuators on subsea equipment under CP. I have never actually seen a failure investigation report that implicates calcareous deposits. So, it would seem that the problem, if there is one at all, is not endemic.

7.5.3 Laboratory investigations

Over the years, calcareous deposits have been the subject of numerous research projects. They are relevant to the CP of high capital value offshore structures, making research funding readily obtainable. The topic offers research students the opportunities in both electrochemistry and surface analysis. The electrochemistry involves conducting polarisation experiments to create the deposits and to measure their effects on electrode behaviour. The surface analysis employs sophisticated techniques for characterising the composition and structure of the deposits. Any research students embarking on a study in this area were guaranteed plenty of results for a thesis.

As a result of this research effort, there is an extensive body of literature on calcareous deposits. Practically all of this is interesting, but not all of it is useful to an engineer looking to select current densities for a CP design. For this reason, I apologise for the omissions of worthy work which space prevents me from referencing in this chapter.

7.5.3.1 Deposit growth

There is not a clear relationship between deposit thickness and its ability to restrict the cathodic current demand of the surface. This should not be surprising. It takes current to grow the film. Thus, a thicker film suggests that more current has passed, suggesting that thicker films are less protective. Thus, although the literature contains plenty of current versus time curves from potentiostatic tests, we are not able to draw conclusions regarding film thickness. Even where the mechanism of film growth has been investigated, information on deposit thickness or growth rate has often been limited.

The potentiostat, which we met in Chapter 5, has been used very widely in the study of calcareous deposits. A typical experiment might involve holding a steel specimen at a fixed potential (say −800 or −900 mV) and monitoring the cathodic current over a period of time. Experiments of this type have shown four stages in the formation of a calcareous deposit on steel [17].

Stage 1	A rapid decrease in current density over a matter of minutes from an initial value that may well exceed 1 A/m^2
Stage 2	An upper plateau that extends to ~2000 minutes with a near constant current density often around 300 mA/m^2
Stage 3	A period of steady current density decay over a further 2000 minutes
Stage 4	A final lower plateau with a constant current density ~30 to 40 mA/m^2

Stage 4 persists for the life of the system unless the calcareous deposit is damaged in some way. If that happens, the deposit layer heals rapidly with a current time curve showing a current transient followed by a sharp decline.

These four stages in the current versus time curve reflect changes in chemistry. During the first stage, a thin magnesium-rich layer is deposited. This evidently restricts oxygen diffusion quite profoundly, as evidenced by the rapidity of the current decay. Even so, it is clearly not impermeable since the cathodic current remains relatively high (~300 mA/m^2).

Precipitates of $CaCO_3$ form during the second stage. These appear as small nodules on the surface of the Mg-rich film. During this stage, these deposits provide no protective benefit because the current density remains constant.

The $CaCO_3$ nodules continue to grow laterally in the third stage. As the nodules start to coalesce, the deposit becomes increasingly protective as shown by the steady fall of the current with time.

The fourth stage arises when a nearly uniform layer of $CaCO_3$ has developed on top of the initial Mg-rich layer. The current density is relatively low but does not reduce further. This tells us that the deposit is a good but imperfect barrier to the diffusion of dissolved oxygen. It also suggests that the thickness of the deposit is no longer changing.

Such films have been examined in scanning electron microscopes (SEMs) with analytical facilities. It is observed that the "layer" of $CaCO_3$ consists of a regular array of hemispherical growths, each growth appearing like the florets of a cauliflower. These adhere to a primary featureless thin layer of $Mg(OH)_2$.

7.5.3.2 Deposit thickness

The effectiveness of calcareous deposits in reducing CP current demand is not directly related to their thicknesses. Accordingly, there has been little incentive to characterise thickness in any detail. The published information lacks consistency. For example, Pathmanaban and Phull [3] refer to thicknesses in the range 50–500 µm but do not tell us the source of these figures. On the other hand, Luo et al. [10] report lower values: 40–169 µm. Hartt et al. [6] studied deposits formed at −900 mV, observing that they grew to 40 µm at 3°C and 80 µm at 24°C.

Finnegan and Fischer in their studies on the effects of flow on calcareous deposits [11,12] reported that the deposit thickness varied between 8 and 40 µm but that occasionally nodules up to 500 µm were observed. Hartt and Wolfson [2] measured deposit thickness as a function of time. At low flow (0.08 m/s) and a potential of −1.03 V, the film grew steadily to ~28 µm over 400 hours, after which there was limited further growth.

7.5.3.3 Factors affecting deposit growth

The morphology and rate of deposition of the calcareous film depend intimately on the chemical changes in the seawater brought about by the flow of the cathodic current. The current is, in turn, continually affected by the film it is causing to develop. This mutually interactive process is, therefore, remarkably sensitive to changes in the conditions. In this respect, Finnegan and Fischer [11] have remarked that the... *inherent sensitivity of calcareous deposit composition to minor changes in the test conditions (potential, current density, flow velocity, etc.) necessitates rigorous experimental control and statistical analysis of the data before a clear understanding of the results can be obtained.*

In other words, we might be able to draw general inferences about how individual parameters affect calcareous scale deposition, but we cannot predict how two or more of these parameters will interact.

7.5.3.3.1 Temperature

Luo and Hartt [17] investigated the effect of temperature on the adhesion of deposits. In experiments at 10°C, 22°C and 30°C, they found that deposit adhesion increased with temperature. They related the adhesion of the deposit to its thickness.

They pointed out that the solubility of $Mg(OH)_2$ increases with increasing temperature, whereas that of $CaCO_3$, unlike most salts, decreases. Moreover, under comparable cathodic current densities, lower temperature resulted in lower interfacial pH. This suggested that lower temperature deposits have a thicker Mg-rich inner layer and a thinner Ca-rich outer layer. The converse holds for the higher temperature (30°C) deposits. The fact that Mg-rich deposits do not increase in thickness after stage 1 means that low-temperature deposits are thinner. Anecdotal evidence from the North Sea supports this view. It is the received wisdom that calcareous deposits are extremely thin, if they form at all, in polar waters.

7.5.3.3.2 Current density and pH

As noted, it is easy to see that CP increases the interfacial pH but problematic to quantify. Estimates in the range of 10.9–11.5 appear in the literature. However, according to a model developed by Dexter and Lin [18] at potentials where oxygen reduction is the only available cathodic process, the maximum pH value that their model predicts is 9.9. However, higher pH values can arise if potentials are sufficiently negative to favour the hydrogen evolution reaction (equation 7.3).

In practice, the pH will depend on the applied current density. The higher this is, the faster OH^- is generated and, up to a limit, the higher the pH. $CaCO_3$ is expected to deposit when the pH rises to between 8.3 and 8.7 and $Mg(OH)_2$ between pH 9.3 and 9.7. As the natural pH of surface seawater is usually in the range of 8.0–8.2, it is clear that it is already close to being saturated with $CaCO_3$.

Viewed in isolation, the threshold pH values for saturation might suggest that, as we increase the interfacial pH by applying CP, $CaCO_3$ would precipitate first. That this does not happen is because the precipitation kinetics for $CaCO_3$ in seawater are slower than those for $Mg(OH)_2$. The reason for this is that Mg^{2+} ions inhibit aragonite ($CaCO_3$) crystal nucleation. By contrast, the kinetics of brucite ($Mg(OH)_2$) deposition are rapid. This means that, providing the pH quickly rises above about 9.3–9.7, brucite deposits before aragonite. This explains stage 1 formation of a Mg-rich film in potentiostatic tests. The stage 2 nucleation of the Ca-rich precipitate is slow; but once it has nucleated, the stage 3 $CaCO_3$ film growth is more rapid.

7.5.3.3.3 Flow rate

The rate of flow of seawater across the surface fundamentally affects the development of calcareous deposits. Finnegan and Fischer [11] showed that low flow rates favour thicker deposits. However, there is a tendency for thick deposits, formed under low flow conditions, to be damaged easily if the seawater velocity subsequently increases. Thin films, on the other hand, are more persistent and less easily damaged by the fluid flow.

The effect of very high flow rates on the behaviour of calcareous deposits was studied by the US Navy in the 1980s [13]. Hack and Guanti used a rotating cylinder electrode, spinning at up to 1000 rpm, to generate surface velocities of over 27 m/s (>53 knots) on a number of metals. Although they did not examine carbon steel, they used a high-strength low alloy material for which the cathodic behaviour and calcareous deposition characteristics would be similar. Their results showed that, as expected, once calcareous deposits had formed, they greatly reduced the cathodic current demand needed to polarise the specimens to −1000 mV. What was perhaps more surprising was the observation that the deposits were evidently not damaged by the hydrodynamic shear forces generated. This held even up to the maximum rotation speed.

It is worth remarking that, given this US Navy evidence, the requirement of some code to specify a final or repolarisation current density appears to be another example of the code drafting committees leaning towards conservatism.

7.5.3.3.4 Excessive current density

If the current density is too high, hydrogen evolution (equation 7.3) becomes the dominant cathodic reaction. The hydrogen gas emerging from the surface can physically disrupt, or completely destroy, calcareous deposits [4]. This is one of the reasons that codes specify a negative limit for the target protection potential for steel. Such excessive polarisation is unlikely in CP systems using zinc or aluminium alloy sacrificial anodes, but it has to be considered when using ICCP or, as occasionally still happens, magnesium anodes.

7.5.3.3.5 Surface finish

Mantel et al. [14] carried out experiments which showed that deposit composition, morphology and protectiveness were broadly independent of surface finish. Further work [17], carried out to examine the influence of surface finish on deposit adhesion, revealed that, in the short term (<2000 minutes), mechanical interlocking due to a rough surface increased adhesion. However, over the longer term (>6000 minutes), internal cohesion between the deposit particles was more important than mechanical interlocking to the surface.

7.6 SITE TESTING

7.6.1 The limitations of the laboratory

7.6.1.1 The microbiological dimension

When a metallic surface is immersed in natural seawater, it rapidly attains a surface film of adsorbed organic compounds and micro-organisms. In time, this film develops to form a complex, and occasionally thick, fouling layer. This microbial colonisation of surfaces has been the subject of much study in marine biology research facilities around the world. It has also attracted the attention, albeit to a lesser extent, of investigators into corrosion and CP.

There is comparatively little information on the interaction between cathodic polarisation, calcareous deposit formation and marine fouling. The literature is not particularly enlightening; the position is best summed up by Hernández et al. [27] who concluded that... *biofilms and calcareous deposits, formed simultaneously at the time of initial metal immersion, can either enhance or decrease the effectiveness of cathodic protection.*

That such an equivocal conclusion should arise is not surprising. The formation of marine biofilms and the development of calcareous deposits are complex and variable processes. Their mutual interaction inevitably makes things more complex. Moreover, most of the experimental work reported has involved situations that have been artificially tailored to facilitate laboratory work. For example, the majority of the microbiological studies involve stainless steel substrates, even though Edyvean et al. [20] have shown differences in the microbial colonisation of carbon and stainless steels polarised to the same potential.

The same workers [20] also pointed out that the role of organic compounds naturally occurring in seawater is crucial. In artificial seawater, the scale formed at the initial high current densities is magnesium hydroxide and magnesite ($MgCO_3$). As the current density decreases, and moving away from the steel surface, this is followed by the deposition of aragonite and dolomite ($CaMg(CO_3)_2$). Finally, at the lower current densities, calcite (a polymorph of $CaCO_3$) precipitates. On the other hand, in natural seawater with a high organic loading, no aragonite or dolomite is formed; only magnesite and calcite. These observations undermine a great deal of the laboratory studies described above.

Thus, the only conclusion that may be safely drawn from the body of work to date on the role of microbial films is that... *artificial seawater cannot replace naturally occurring seawater in laboratory experiments* [7].

7.6.1.2 Modes of polarisation

Most laboratory workers have studied calcareous deposit growth using one of two methods of polarising the cathode:

- potentiostatic tests in which a potentiostat impresses a constant potential on the specimen (see Figure 5.2), and much less frequently,

- galvanostatic tests, in which the potentiostat is reconfigured to apply a constant cathodic current to the specimen (see Figure 10.10).

In the potentiostatic mode, the current progressively reduces as the cathode surface accumulates the calcareous film. In the galvanostatic mode, the potential of the specimen progressively becomes more negative as an increasing EMF is needed to overcome the resistance presented by the calcareous film developing on the surface.

Neither mode truly reflects the behaviour of a working CP system. In reality, there is a complex interactive relationship between potential and current.

It is probably for this reason that Hartt et al. at Florida Atlantic University, in a programme of investigations from the early 1980s, endeavoured to mimic the actual potential-current relationships observed when steel is polarised in seawater by sacrificial anodes. We return to this work in Section 7.9.

7.6.2 In-situ measurements

7.6.2.1 Monitoring of existing structures

We will deal with monitoring in some detail in Chapter 18. For now, we need simply note that, in the 1970s, there were some instances of offshore hydrocarbon production structures installed with fixed equipment for monitoring local cathodic current densities. A typical installation involved a steel panel mounted around a structural member but electrically insulated from it. The panel was connected to a structure via a low-resistance shunt. The current collected on the panel was detected using a topside voltmeter wired across the shunt. Such early installations were few and generally unreliable and produced only limited useful data. More robust equipment has been developed subsequently (see Chapter 18). Ultimately, however, a current density measurement device can only collect data at one location. However, its benefit is that it can monitor that current density over time.

Arguably more useful in situ determinations of cathodic current density have subsequently been made using ROV-mounted equipment to measure the electrical field gradients in the seawater. This technique is also described in Chapter 18. It requires computer modelling (Chapter 17) of the raw field gradient data to predict the cathodic current densities on the steel. This process has the advantage that the entire structure can be surveyed. Published results universally demonstrate that the mean design current densities recommended by the codes have been conservative, at least as far as long-lived structures are concerned.

7.7 SITE EXPERIENCE

7.7.1 South China Sea

DNV-RP-B401, which we used for design exercises in Chapter 3, was developed largely on the basis of North Sea experience and opinion. It was also intentionally conservative. Since its 1993 edition, it has specifically offered operators the option of substituting their own, experienced-based parameters for the guideline figures. If operators accepted this option, it could lead to substantially reduced CP costs. This point was illustrated by a 1997 paper [31] in which Shell challenged the DNV current densities for a tropical location, based on its own historical data. These differences are summarised in Table 7.4. As can be seen, the Shell experience in the South China Sea permitted markedly lower cathodic current densities than the contemporary edition of RP-B401. The DNV 2021 figures are also included in Table 7.4. These show that the code remains conservative when viewed against the operator's experience. On the other hand, as seen in Table 7.10, the 2007 edition of NACE SP0176 is less conservative than DNV.

Table 7.4 DNV and Shell parameters (South China Sea)

			Cathodic current demand (mA/m²)	
Phase	*Water depth (m)*	*Year published*	*RP-B401*	*Shell*
Initial	0–30	1993 & 2021	150	
	0–50	1997		94
	>30	1993	130	
	30–100	2021	120	
	50–100	1997		100
Mean	0–30	1993 & 2021	70	
	0–50	1997		32
	>30	1993	60	
	30–100	2021		
	50–100	1997		35
Final	0–30	1993 & 2021	90 & 100	
	0–50	1997		32
	>30	1993	80	
	30–100	2021	80	
	50–100	1997		35

In addition to the lower current densities, Shell also expressed its view that 3 A was a sufficient allowance for current drain to wells, instead of the 5 A recommended by DNV.[1] These less conservative, but nonetheless realistic, modifications led to a CP installation for a new jacket that was calculated to be considerably cheaper than if the design had slavishly followed the DNV guidelines.

7.7.2 Middle East – operator 1

7.7.2.1 *The requirement*

Whereas the South China Sea experience has been published by Shell, this case is not in the public domain. I am, therefore, obliged to mask some of the details to preserve anonymity. In 2015, an operator commissioned two engineering firms independently to estimate the quantity of anodes needed to extend the service lives of its offshore platforms. Neither firm departed from the design mean current density offered by the 2010 edition of DNV-RP-B401 for seawater >20°C. This was despite the fact that they were now dealing with a population of existing platforms with many years of accumulated polarisation history.

I was commissioned to review the reports issued by these engineering firms. I rejected the notion that we should rely on DNV for what was to be a retrofit design. Instead, I made use of historical subsea inspection records to estimate the actual cathodic current densities drawn by the structures. This approach was entirely in keeping with DNV-RP-B401's statement that: *Owners of offshore structures may specify a less, or in certain cases a more conservative design data, based on their own experience or other special considerations.*

[1] We consider current drain to well casings in Chapter 13.

7.7.2.2 *Approach adopted*

Of the numerous platforms considered, there were three (structures A, B and C) for which:

- original CP design reports were available;
- diver surveys showed full protection (despite the CP systems having exceeded their original design lives);
- no supplementary anodes had ever been retrofitted.

A fourth (D) was a small simple structure for which I could not find the original CP design report. Despite this absence, I was able to estimate the original anode provision. A summary of the information available for the review is set out in Table 7.5.

All of the jackets are uncoated below the water line and are protected by aluminium alloy sacrificial anodes. The proprietary anode formulations were not available, but it is likely that they were all generic Al-Zn-In alloys.

Table 7.5 Summary of information available (as at August 2015)

Structure	*A*	*B*	*C*	*D*
Year installed	1990	1988	1982	1980
Water depth (m)	51	48	58	32
Number of legs	8	8	6	3
Number of wells/conductors	0	0	22	0
Number of pipelines/risers	4	7	4	0
Pipelines isolated	(Not known at the time)			No Pipelines
Original anode nett mass (kg)	273	273	330	Not recorded
Original anode number	102	110	178	18
Most recent CP survey	2009	2015	2014	2011

7.7.2.3 *Example of analysis – structure A*

7.7.2.3.1 *Original CP design*

The original design process had estimated a mean current requirement of 409 A for the immersed areas plus 107 A for buried components and well casings, giving a total demand of 516 A. The CP design report gave the anode dimensions and nett weight (273 kg) but did not explain the anode design. It is likely to have been a simple procurement decision to use an available proprietary anode. In total, 102 anodes were installed (27,846 kg) to satisfy the estimated mean current demand (516 A). It is informative to note that a CP designer using the 2017 DNV-RP-B401 guidelines would have come up with 408 A instead of 516 A.

We should note that the original CP design envisaged no current flowing to or from the pipelines. This is normal CP design practice. I later verified that the pipelines had their own bracelet anodes of adequate capacity.

7.7.2.3.2 *Protection status*

A 2009 CP survey had shown that the platform potentials fell in the range of –0.975 to –0.999 V, confirming that the jacket was well protected and had developed good quality calcareous deposits.

7.7.2.3.3 Anode consumption surveys

The extent of anode consumption had been estimated by divers in 1999 and 2009: 9 and 19 years, respectively, after installation. There had been no physical measurement of the remaining anode dimensions. The survey reports indicated that consumption was generally within a range (e.g. 25%–50%) for each anode. By using the median value for each of these assessment ranges, and then averaging the medians across all of the anodes, I estimated the total anode consumption as 24% consumed (1999) and 33% consumed (2009).

Attempting to estimate the degree of anode consumption based on a diver's visual assessment can never be definitive. Nevertheless, taking the readings at face value, it seems that about 6.5 tonnes of anode had been consumed by 1999, but that this only increased by a further 2.6 tonnes from 1999 to 2009. Whereas we would expect a somewhat higher rate of anode consumption in the first few months of immersion, the difference in the rate of consumption indicated by the two sets of survey data was probably a result of the imprecision inherent in the assessment methodology, rather than any real change in the rate of anode consumption.

7.7.2.3.4 Estimated mean current demand

The mean current was estimated using the equation we met in Chapter 3 for calculating the minimum required weight (W) of anode.

$$W = \frac{8760 T I_m}{CU} \tag{7.6}$$

where

$W = 6592$ kg, inferred from the 1999 survey
$T = 9$ years, from installation to survey
$C = 2420$ Ah/kg, the original design value
$U = 1$, since the anodes had not reached the end of their lives

Equation 7.6 was simply rearranged to solve it for I_m giving an estimated mean current value of 202 A. The exercise was then repeated on the basis of the 2009 survey results. The value of W changed to 9214 kg, and T changes to 19 years. The estimate of the mean current demand over the period reduced to 134 A.

Both the estimated original design current (516 A) and the current arrived at by application of the 2017 DNV code (408 A) are seen to be at least double the current demand implied by the *worst-case* diver assessment of the rate of anode consumption (202 A). The result endorses the view, widely held in the CP industry, that the design codes are intentionally conservative.

7.7.2.4 Results of analyses – structures B - D

The results of the analogous exercises for structures B, C and D are summarised in Table 7.6. Again, in each case, the current demand conservatively calculated from the estimated anode consumption was considerably lower than would have been predicted from application of the DNV guidelines.

7.7.2.5 Application to other structures

An interesting observation was made when the current demands estimated from field experience (Table 7.6) were plotted against the immersed bare steel surface areas of the structures. The result is shown in Figure 7.3.

Table 7.6 Mean current demands DNV versus field experience

	Current demand (A)	
Structure	*DNV*	*Field experience*
A	408	202
B	401	200
C	627	387
D	54	43

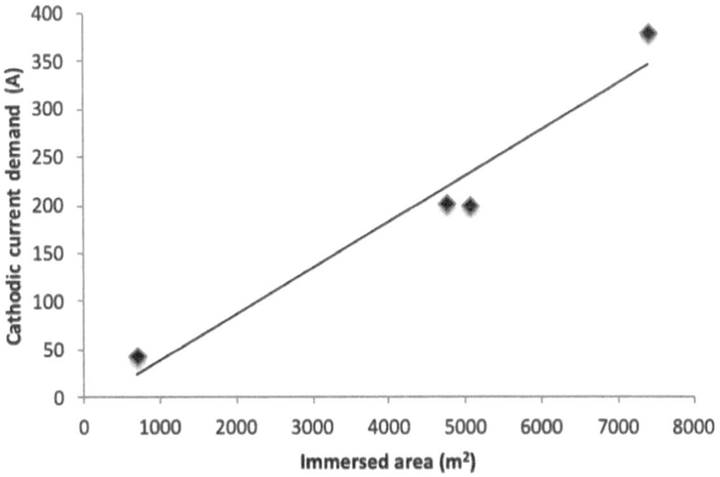

Figure 7.3 Total mean cathodic current demands versus immersed steelwork area.

For a group of structures in the same marine area, and at broadly similar depths, we can advance the following propositions:

1. although CP current flows to both immersed and buried steelwork, by far the greater part flows to the immersed areas,
2. the ratio of the current densities for open seawater and seabed sediments is approximately constant, with much higher values applying in seawater,
3. for any structure, there is a degree of proportionality between the area exposed to seawater, and the area of buried steel (e.g. piles and well casings) associated with it.

If these propositions hold, as seems reasonable, then it follows that a retrofit CP design for a structure in that field can be based solely on the area of the immersed steelwork. All that is required is to allocate an overall cathodic current density for the immersed conditions. The current drain to the buried parts of the structure will then be included within this overall figure.

The results in Figure 7.3 fall close to a straight line, which also passes through the origin. The slope of the line is about 46 mA/m^2.

Thus, where original CP design and historical retrofit data are unavailable for other structures in this field, these preliminary results indicate that life-extension CP calculations could be based on this field's sampled experience. The data suggest that retrofits could be sized solely by allocating (say) 50 mA/m^2 to the known immersed area of the structure. This approach would result in a very considerable saving compared with a conventional design approach using code criteria.

7.7.3 Middle East – operator 2

Other operators have also adopted less conservative approaches to CP design than advised by codes. For example, one operator, with considerable experience in the warm waters of the Arabian Gulf, uses a design current density of 50 mA/m². This figure applies to the initial, mean and final phases of life, and applies to all water depths (up to 90 m). This is a decidedly lower figure than indicated in Table 7.1 but, as we have shown, is compatible with the observations made elsewhere in the region.

7.8 DEEPER WATERS

7.8.1 Codes

The quest for hydrocarbons has required installing structures in ever deeper waters, particularly during the last two decades of the 20th century and the beginning of the 21st century. With steel remaining the principal material of construction, this has meant that CP has had to adapt to the new environments. Code recommendations for deep water were needed to support these projects, even before any long-term experience could be collated. The code drafting committees were inevitably cautious.

By way of example, the evolution of the DNV-RP-B401 guidelines for deep water is summarised in Table 7.7. For convenience, the recommendations for warmer waters (>20°C surface temperature) have been singled out because it is warmer regions (West Africa, Brazil, Gulf of Mexico) that have seen much of the deep-water application of CP. NACE SP0176, also included in Table 7.7, limits its recommendations to deep water in the Gulf of Mexico. However, we should remember that, irrespective of their surface temperatures, all deep waters are cold.

Table 7.7 Recommended design current densities – deep waters

				Cathodic current density (mA/m²)		
Authority	*Code*	*Year*	*Depth*	*Initial*	*Mean*	*Final*
DNV	TNA 103	1981	(Any)	100	80	70
	RP-B401	1986	>20 m	130	70	90
		1993	>30 m	130	60	80
		2005–2021	30–100 m	120	60	80
			100–300 m	140	70	90
			>300 m	180	90	130
NACE	SP0176	2007	"Deep"	194	75	86

The codes agree on a requirement for a final, or repolarisation, current density. It seems unlikely that calcareous deposit could suffer storm damage at such depths, particularly when the work of Hack and Guanti, discussed above, shows that these deposits are stable at velocities in excess of 50 knots. Again, this seems to be an example of code caution.

7.8.2 The theory

In 1992, a NACE Task Group reviewed the extant literature on deep-water CP. Its report No. 7L192 [19] emphasised the difficulty of predicting cathodic current requirements at depth and related this directly to the difficulty of predicting the rate of formation, and protective quality, of calcareous deposits. Table 7.8 presents the Task Group's summary of the probable interactions among depth-related parameters.

Table 7.8 Effect of depth on cathodic protection parameters

		Effect of depth variation on	
Parameter	*Variation with increasing depth*	*Calcareous film formation*	*Other parameters*
Temperature	Decreases (not uniformly)	Antagonistic	Increases: • resistivity • O_2 solubility • $CaCO_3$ solubility
Salinity	Increases (generally)	Negligible	Decreases: • resistivity • O_2 solubility
Dissolved O_2	Decreases (initially) may increase at greater depth	Antagonistic	
pH	Decreases (initially) may increase at greater depth	Antagonistic	Increases $CaCO_3$ solubility
Flow	Decreases (generally)	Uncertain	Reduces O_2 flux to surface
Pressure	Increases	Inconclusive	Complex
Fouling	Heavy fouling limited to <30 m biofilms found at any depth	Complex	Lowers: • O_2 diffusion • pH

Of particular concern is the general tendency for deeper waters to be colder, of lower pH and depleted in dissolved oxygen compared to the near-surface strata. Dexter and Culberson [1] have explained that the world's deep oceans are of a relatively constant temperature (<5°C) due to circulation emanating from polar regions. The lowering of the dissolved oxygen content with depth is due to the biochemical oxidation of organic matter. This produces CO_2 which lowers the pH. The extent to which these parameters alter with depth is dependent on the geographical location and the seasons. For a North Atlantic test location, the dissolved oxygen content reduces from ~5 to ~3.5 mg/L over the first km of depth, whilst for a North Pacific location, of comparable latitude, the reduction is from ~6.5 to ~1 mg/L over the same depth. Interestingly, at both sites, the oxygen levels then increase at depths beyond ~1 km. It was also found that the pH values dropped in a similar profile: at the Pacific location, the pH value fell from ~8.25 at the surface to 7.60 at ~1 km; whilst at the North Atlantic test location, it fell less dramatically from ~8.25 to ~7.96 over the same depth interval.

The implications of these changes are quite profound for calcareous scale formation, and hence for CP design. Whereas the trend towards lower dissolved oxygen contents suggests both lower natural corrosion rates and lower cathodic current requirements, it also means that, using conventional sacrificial anode designs, the diffusion limited cathodic current densities during the early stages of polarisation will also be low. This is not necessarily beneficial, however. Low initial current densities are unlikely to encourage the formation of effective calcareous deposits. It follows that CP systems for deep-water structures might need to be sized for greater mean-life current densities than their shallower counterparts. The predicted lowering of pH and temperature with depth, each of which increases the solubility of the calcareous compounds, exacerbates this situation.

One of the papers [9] cited in the NACE review makes the telling point that... *the geographical location may be of greater significance to the initial cathodic polarisation behaviour than depth.* Thus, other than emphasising the desirability of site-specific survey data, and reporting a tendency of CP designers to allow for increased mean current and final current requirements, the literature published up to the end of 1990 and reviewed by the NACE Task Group, did not cause any revision of the existing guidelines in respect of deep water.

Fischer et al. [28] also pointed out that the rate of dissolution of the calcium carbonate polymorphs, aragonite and calcite increases with increasing depth. They then presented a critique of the thermodynamic and kinetic stability of calcareous deposits in deep seawater. On this basis, they tentatively concluded that, in the Gulf of Mexico and the Norwegian Sea, calcareous deposits, once formed, would be stable down to 3000 m.

However, they also pointed out that this situation need not apply to all waters. A case in point is offshore Angola where the pH and oxygen content are low (<1 mg/L). The under-saturation of aragonite occurs at ~100 m, and of calcite at ~300 m. Thus, there is a thermodynamic tendency for deposits to dissolve in deeper waters in that region. However, dissolution is controlled by kinetic, rather than thermodynamic, factors. It requires a sufficient degree of under-saturation of the seawater in order for it to proceed at a meaningful rate. They anticipated that, at 500 m depth, aragonite would dissolve, but that calcite would not show an accelerated dissolution rate.

The implications of this work for practical CP systems in deep-water offshore Angola remain to be determined. On the one hand, there is a strong suggestion that benefits from calcareous deposition will be muted. On the other, the relatively low-oxygen content which is likely to be antagonistic to calcareous deposition will in itself give rise to lower mean-life current demands.

Parametric studies might be contemplated to complement the theoretical analysis. These involve developing an understanding of how each variable of seawater composition and circumstance influences the polarisation characteristics of the cathode. In principle, once the effects of these variables, and more crucially their interdependence, are established the CP design parameters could be predicted by reference to the local seawater conditions, and at any depth. In practice, this is a tall order!

The variables themselves are generally known from the body of oceanographic research. For example, Goolsby and Ruggles [16] assembled a database for the northern area of the Gulf of Mexico which contains more than 7000 profiles detailing: temperature, salinity, dissolved oxygen content, density, pH, water chemistry, etc.

It is the correlation of these properties, with the polarisation behaviour, that proves more problematic. For example, a model was produced by Yan et al. [21,26] and subsequently enhanced by Blackburne and Griffin [29]. In my view, the effort has been commendable, but the results unconvincing. There are a number of reasons for this, including the fact that the polarisation tests underpinning the model have been carried out potentiostatically and in synthetic seawater. Neither applies to the real world.

7.8.3 Laboratory testing

Laboratory testing for deep-water CP is restricted to organisations with hyperbaric facilities. The few investigations that have been carried out have focussed on the effect of CP on the mechanical properties of steels. A particular concern has been the effects of pressure on the hydrogen evolution reaction (equation 7.3) and its implications for hydrogen embrittlement of the steel (we discuss these matters in Chapter 12). Festy reported [35] no specific effects of hydrostatic pressure (up to 300 bar) on the hydrogen embrittlement of steels (<1000 MPa). Since it is impracticable to replicate, and sustain, natural oceanic seawater conditions inside sealed hyperbaric chambers, the data on cathodic polarisation it produces have to be viewed with some caution. Nevertheless, he observed higher cathodic current densities at 300 bar than at atmospheric pressure. He suggested this was due to the higher solubility of carbonate compounds at higher pressure, causing a reduction in the protective quality of the calcareous deposit.

7.8.4 Site testing

In 1987, Fischer et al. [9] commented that... *quantitative current density values applicable to CP design can be determined only from practical experience or field testing.* With this in mind, they assembled test cells comprising an aluminium alloy anode (area 40 cm^2) coupled to a grit-blasted steel disk (600 cm²). In these "free-running" tests, the cells were placed at depths of up to 400 m off the Norwegian coast, and the steel potential and CP current monitored on data loggers. The results from this work influenced the subsequent revisions of the RP-B401. This paper was also significant because it influenced the subsequent work of Hartt and co-workers on what is now termed the slope parameter. We return to this in Section 7.10.

There have been a limited number of publications relating to deep-water testing: Brazil [24,32] Norway [28] and the Gulf of Mexico [25,33]. Baptista and da Costa [32] reported on-site testing at 102, 290 and 975 m in the Campos Basin offshore Brazil. Their multiple test specimens comprised monitored carbon steel cathode discs galvanically coupled to zinc anodes with a cathode:anode area ratio of 10:1. A resistor was added in series to control the initial current density. For the tests at 975 m, for example, a dozen different initial current densities were used in the range of 52–470 mA/m².

The tests ran in the free-running mode for periods of 8–13 months. The galvanic current in each test was measured by recording the voltage drop across a 1 Ω resistor in series with that employed to control the current. In presenting their results, they reported the minimum initial current density that was found to be sufficient to achieve cathode polarisation to <–0.8 V in 20 days or less. Their results, which are summarised in Table 7.9, endorsed the NACE Task Group's view that greater initial and mean current densities are required to achieve and maintain protection in deep water.

Table 7.9 Cathodic current densities for carbon steel – offshore Brazil

		Current density (mA/m²)	
Depth (m)	*Temperature mean (°C)*	*Initial (min.)*	*Mean*
102	18	208	55
290	12	291	120
975	4	363	130

From [32].

Also working in the Campos Basin, Vianna and Pimenta [24] studied the nature of the calcareous scale formed at different depths and at five initial current densities: 50, 100, 200, 300 and 400 mA/m². Their tests lasted for 14 months, and their results confirmed the received wisdom that high initial current densities favoured the formation of a denser and, therefore, more protective, calcareous layer. They also observed that the calcareous deposits, formed at any given initial current density, increased in calcium content as the depth increased. This is, at first sight, surprising given that Ca^{2+} solubility in seawater increases with depth. According to the authors, it emphasises the need for in situ testing in deep waters.

A paper by Goolsby and McGuire [33] on the Gulf of Mexico Cognac platform has some relevance. By way of background, the authors explained that the base section of the platform (255–320 m) was fitted with monitored reference electrodes. When it was installed in 1977, it polarised satisfactorily. It reached –835 mV within 24 days and was still shifting to more negative values. This means that, at one relatively deep location at least, a CP system designed according to the rules in force in 1977 yielded an acceptable result.

Fischer et al. [28] carried out studies in the Barents Sea (>73°N) that revealed that cathodic current density requirements *decrease* with depth. For a given cathodic current density (300 mA/m² in this case), the subsequent protective quality of the calcareous deposit increases with depths down to ~470 m. Although this is in agreement with earlier work for the Northern North and Norwegian Seas [9], it runs counter to the experience noted above for the warmer waters of the Gulf of Mexico or offshore Brazil. Nevertheless, the target *initial design* current densities implied by the work in the Barents Sea (~300 mA/m² at 471 m) compare reasonably with the *minimum adequate* initial current densities in the Campos Basin (>291 mA/m² at 290 m, and >363 mA/m² at 975 m).

Festy et al. subsequently modified the testing equipment that they had employed in hyperbaric chambers so that it could be deployed offshore. A diagram of the core component of the test equipment, called the CPC sensor, is reproduced in Figure 7.4. It comprises a steel cylinder (~0.27 m long and ~0.04 m diameter). This is connected via a resistor (R) to an anode. The CP current flowing (I) is recorded as the voltage drop across R, and the potential of the steel (V) is measured against a zinc reference electrode. To conduct an in situ investigation, an array of CPC sensors is mounted on a frame and deployed on the seabed at the test location for an extended period (at least a year). Each of the CPC sensors is fitted with a different combination of resistor and anode alloys in order to produce a range of polarising currents. The result is that, taken as a whole, the array produces in-situ polarisation experiments. In a field deployment, results were successfully obtained at a water depth of 1350 m in the Gulf of Guinea [44].

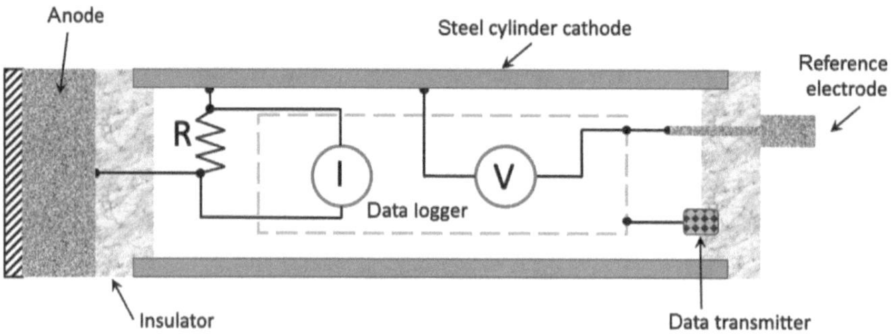

Figure 7.4 CPC sensor. (From Festy et al. [35,44].)

Rather than reproduce their West Africa data, however, Figure 7.5 provides a schematic representation of the type of data that are obtained using this approach. The individual readings, represented by the markers on the plots, represent time-averaged current density and potential data from each sensor. At any time, higher currents are produced by combinations of lower resistances and more active anode alloys. Following the data over a period of months reveals that there is a trend towards lower current densities and more negative potentials. This confirms the beneficial effect of the developing calcareous deposit.

A further remark is relevant. The polarisation process was observed to be continuing between the 6- and 12-month marks. There is no reason to suppose that it will have stopped there.

In Figure 7.5, a dashed line, which in practice is reasonably linear, is drawn through the data sets associated with each resistor-anode alloy combination. We return to the implications of this when we discuss the slope parameter in Section 7.10.

7.8.5 The future

In 2009, NACE issued an update [43] of its 1992 Task Group report 7L192 [19]. The new edition was intended as a supplement to the 2007 edition of NACE SP0176. This update included a review of information gained in the 17 years since the original publication, including revised codes and 30+ new papers. Much of this material has been discussed above, or else we will cover it in Section 7.10.

It seems likely that the military will have an interest in CP systems operating at great depth. However, if this is the case then, unsurprisingly, the relevant information is not in the public domain. Thus far, the publications on CP in deep water have all been in the context of producing hydrocarbons. In this respect, there is an important remark in the Task Group Chairman's overview of the updated report [42]... *all of the production systems deeper than 500 meters involve floating systems...Compared to the traditional bare steel jacket structures for which the industry has long experience, these floating systems often involve complex geometries, coated steel and long cables or tubulars with CP by remote anodes...* Importantly, the prevalence of coatings on subsea production hardware, particularly at greater depths, does not mean that CP is no longer required. Rather, it means that the estimation of design CP currents will inevitably be dictated more by the long-term performance of the coating than the subtleties of calcareous deposit formation.

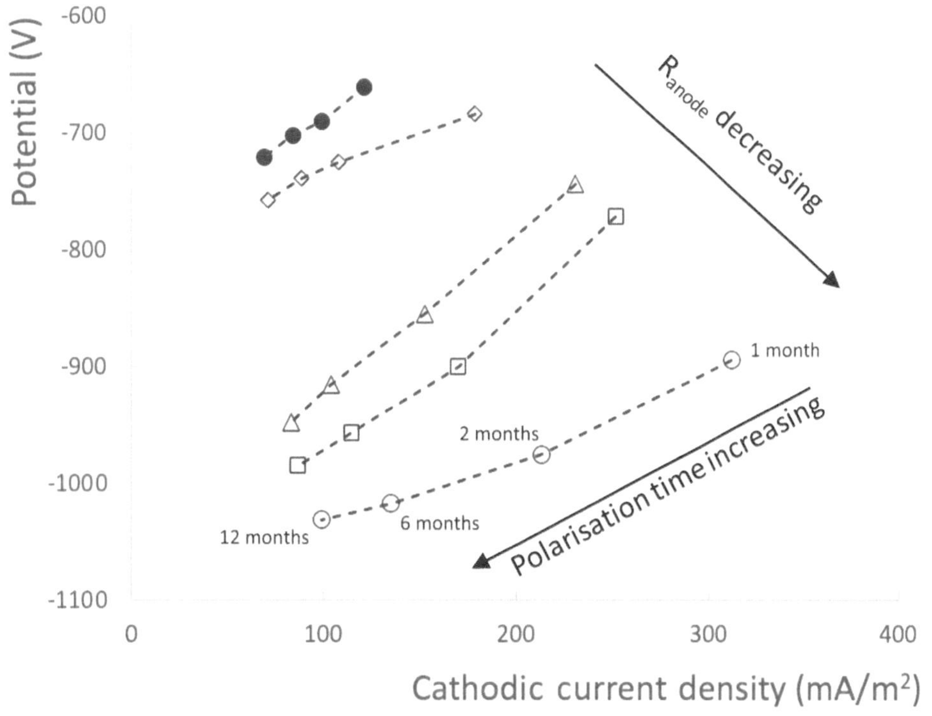

Figure 7.5 Schematic representation of in situ polarisation data.

7.9 S-CURVES

Festy et al. acknowledged [35,44] that the design of their CPC sensors owed much to the extensive body of work carried out by Hartt et al. Unlike researchers who used potentiostats to hold steel specimens to a pre-set potential, the Hartt approach was to use galvanic coupling tests. In these tests, the steel was connected to a sacrificial anode via a pre-set resistor, the various values of which were set to represent the overall resistances in actual CP circuits. So, instead of experiments where the steel potential stayed constant and the current decayed as calcareous deposits built up, in Hartt's experiments, both the potential and current changed over time in a manner closer to reality.

Each set of experiments produced families of polarisation curves similar in form to those shown in Figure 7.5. Hartt and Chen [36] have plotted the potential-current density values arrived at the end points of these tests. Their laboratory results, transposed to the Ag|AgCl|seawater reference scale, are reproduced (square markers) in Figure 7.6. As can be seen, there is much scatter in the data points. This reflects the vagaries of carrying out polarisation experiments of this type in seawater. Nevertheless, we see a tendency for the results to fall into a sigmoidal pattern or "S-curve". The implication is that, although full protection is achieved at –800 mV, the long-term (mean) current density needed is higher than if the potential is held in the range –900 to –1000 mV.

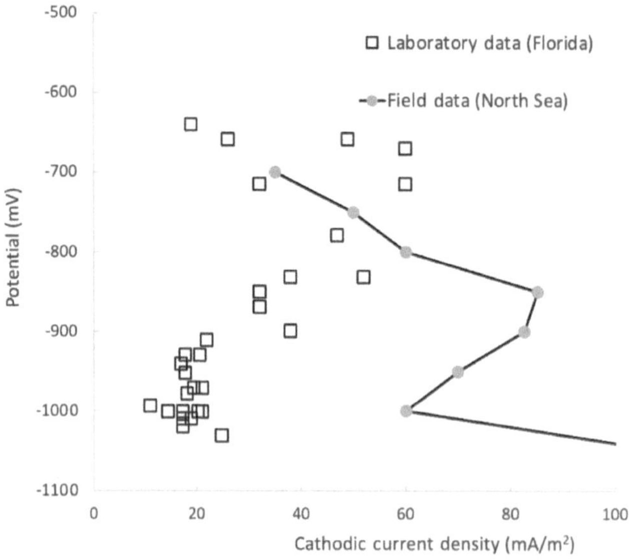

Figure 7.6 Laboratory and field polarisation data [8,36].

There is also evidence for S-curve type behaviour from field data. A set of mean current density versus potential survey data is included as a solid line in Figure 7.6 for comparison. These data were reported some years earlier by Strømmen et al. [8]. They were obtained by field gradient measurements (see Chapter 18) on an undisclosed number of structures in the cooler waters of the North Sea. The field data also indicate that, for well-polarised structures, lower cathodic current densities are required to maintain potentials at –900 to –1000 mV than at –800 mV.

The S-curve behaviour in Figures 7.5 and 7.6 tells us that, to produce a CP design requiring the minimum mean current provision, and therefore the minimum anode mass, it would be beneficial to force the system over the "knee" of the S-curve. This would ensure that through-life potentials would be maintained, for the most part, in the optimum –900 to–1000 mV potential range.

Unfortunately, there is no obvious way that we can implement this in a CP design methodology of the type we worked through in Chapter 3. For example, we cannot simply pick a single current density value out of Figure 7.6 and insert it into Table 7.7, and then use it in a "cookbook" design approach.

This problem is recognised in the standards and elsewhere. For example, DNV-RP-B401 tells us that the design protection potential for steel in seawater should be –800 mV. However, recognising the relevance of a sound calcareous deposit, it adds that an... *appropriate final design current density (and hence CP polarising capacity) will further ensure that the protection object remains polarised to a potential of –0.95 to –1.05 V throughout the design life. In this potential range, the current density demand for maintenance of CP is lowest.*

In other words, the code adopts something of a "work-around" that ends up with the desired result. Unfortunately, the wording in the code has given rise to misunderstandings. Some workers have interpreted it as either meaning that the design potential should be –900 mV, or that potentials more negative than –900 mV become a specified performance requirement.

Neither view is correct. It just so happens that using the DNV-RP-B401 methodology, as we did in Chapter 3, usually results in space-frame structures, for which the code is intended, spending most of their lives in the optimum potential range.

Other workers, likewise aware of the long-term benefits of ensuring that the initial polarisation of the structure is carried out at high current densities, have suggested departing from the codes. For example, Schrieber and Reding [15] considered the addition of short-lived magnesium anodes to CP systems to bring about rapid polarisation. We will return to this topic when we discuss sacrificial anodes in Chapter 10.

An alternative approach, developed by Hartt's group, is the so-called "slope parameter". We need to focus on this here because, unlike rapid polarisation techniques, it is written into one of the codes.

7.10 THE SLOPE PARAMETER

7.10.1 What is it?

The slope parameter has emerged from a desire to find a single "Unified Design Equation" for CP systems [30]. The core of the approach appeared in a 1987 paper by Fischer et al. [9]. It emerges from the sigmoidal behaviour of the time-dependent polarisation behaviour described above. The original idea was subsequently subject to a great deal of further investigation, using both laboratory and in situ data by Hartt et al. Over a period of more than 20 years, this group published numerous papers at conferences and, more relevantly, in peer-reviewed journals [2,5,34,37,38,39–42]. At some point early in this period, the term "slope parameter" emerged.

We can use the schematic data in Figure 7.7 to gain an insight into how the slope parameter approach to a CP design is intended to work. The upper curve, labelled "initial state", represents the cathodic polarisation behaviour of a steel surface soon after it is immersed. As can be seen, it is similar to the upper curve in Figure 7.5.

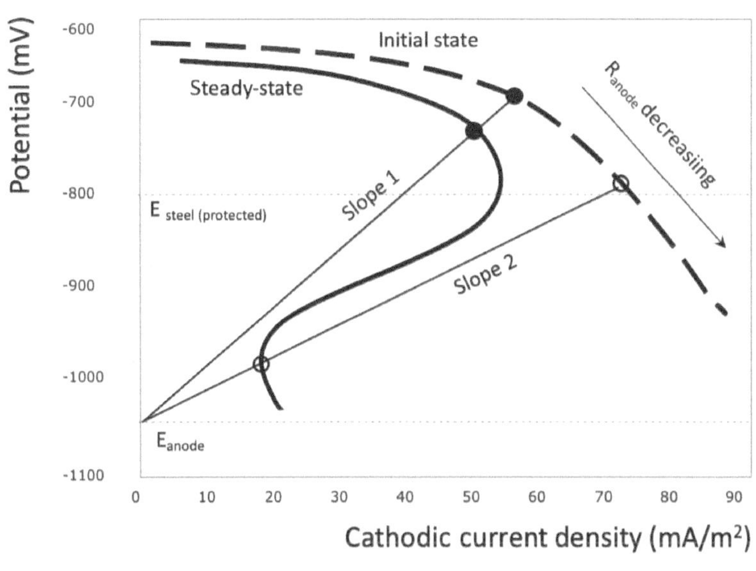

Figure 7.7 The slope parameter concept.

As we discovered in Chapter 3, most of the resistance in a CP circuit is at the anode. The CP designer has the option to change the circuit resistance by modifying either the number of anodes or the individual anode resistance. This means that the CP designer has the ability to select where on the "initial state" curve the long-term polarisation process effectively begins.

Figure 7.7 shows two starting options; represented by the closed and open circles. Designing a CP circuit with a lower resistance, that is with either more anodes or with anodes designed to produce a lower individual resistance, means that the initial current provided to the steel will be higher. This is indicated by the open circle.

As we also see from the dotted lines added to Figure 7.5, once polarisation has started it progresses with time. Allowing for the vagaries of real-life testing, this progress is approximately linear. Figure 7.7 shows two such idealised linear slopes.

Both curves start at the selected initial design current density on the initial state curve. In addition, both lines notionally track back to an origin represented by the anode potential and zero current. This origin is, of course, entirely notional. It may be regarded as representing the situation where the steel cathode has completely polarised to the potential of the anode. Because the potentials of the two metals are the same, no current can flow. Physically, this may be interpreted as reaching the unattainable goal of a calcareous deposit that is 100% effective.

Of more relevance than the fictitious zero-current point are the points at which slopes 1 and 2 intersect the steady-state curve. In Figure 7.7, slope 1 crosses at a moderately high current density but, crucially, does not result in the structure being polarised to the required protection potential (−800 mV). Slope 2, on the other hand, starts at a higher current density and, therefore, bypasses the "knee" in the sigmoidal curve. It then intersects the steady-state curve at a much more negative potential, confirming full protection. Also, it intersects at a much lower cathodic current density.

The physical interpretation of this is that calcareous deposits formed at high initial current densities are more protective. Thus, designing for a high initial current density results in a lower steady-state, or mean, current demand, and a lower mass of anodes.

7.10.2 Slope parameter versus "cookbook"

7.10.2.1 Two perspectives

In the above description of the slope parameter concept, we started at the initial polarisation current density and then proceeded down the slope to discover the steady-state mean cathodic current density (i_m). In principle, the converse also applies. We could set a value for i_m, which, for example, might be constrained by the maximum mass of anodes the structure could support. From this value of I_m, we could project up the slope to discover the magnitude of the initial polarising current we would need to supply.

Irrespective of which way we were going about things, we would need to know the value of the gradient of the line (the *slope parameter*).

7.10.2.2 SP0176

The only code guidance for a CP designer employing the slope parameter approach is found in the non-mandatory Annex E of NACE SP0176. Unfortunately, I find it difficult to see how the design process, as described in the annex, constitutes a worthwhile advance on the methodology based on the DNV-RP-B401 code that we illustrated in Chapter 3.

The SP0176 approach is based on equation E1 (reproduced as equation 7.7). For simplicity, this equation has been normalised to deal with the area of steelwork protected by a single anode.

$$R_aW = I_mTkS \tag{7.7}$$

where

R_a = Anode resistance (Ω)
W = Anode mass (kg)
I_m = Cathodic current (supplied by anode) (A)
T = System life (years)
k = Anode consumption (kg/year) obtained by rearranging equation 7.6.
S = Slope parameter (Ω)

The first problem relates to the value of the mean cathodic current density (i_m) used for CP design. As noted, Slope 2 in Figure 7.7 implies that the slope parameter approach offers the benefit of justifying a lower design value for i_m. However, there is no information in the code as to what constitutes a reasonable level of i_m to try for. Ultimately, most designers will select from the values advised in Annex A of the standard (see Table 7.10). At that point, the design methodology is likely to revert to the "cookbook" approach described in Chapter 3.

The second problem is that the value of the left-hand term in equation 7.7 (R_aW) can be arrived at via an infinite variety of anode designs. The CP designer is, therefore, left in the position of carrying out trial-and-error iterations, not unlike the processes described in Chapter 3.

Finally, there is only limited information in SP0176 regarding the appropriate value for the slope parameter S. This is reproduced in Table 7.10. The code advises that, like the value for i_m, the value for S should be based... *on prior experience of experimental data at the site of interest.*

To date, therefore, the slope parameter approach has failed to live up to the ambition of its developers who were looking for a unified design equation. It is hard to see that it adds much, if anything, of value to the "cookbook" approach described in Chapter 3. As far as I am aware, very few offshore CP designs have actually been carried out in accordance with Annex E of SP0176.

Table 7.10 A sample of SP0176 CP design parameters

	Typical design current density (mA/m²)			
Production area	*Initial*	*Mean*	*Final*	*Slope parameter (Ωm²)*
Gulf of Mexico	110	54	75	4.1
Deep-water Gulf of Mexico	194	75	86	No data
Cook Inlet	430	380	380	1.0
Northern North Sea	180	86	120	2.5
Arabian Gulf	130	65	86	3.5
West Africa	130	65	86	3.5
South China Sea	100	32	32	No data

7.11 THE RATE OF POLARISATION

The silence of the codes

Our discussion on the time-dependent formation of calcareous deposits tells us that polarisation is not instantaneous. However, there is nothing in any of the CP design codes to indicate how long a CP system, designed according to that code, should take to polarise a structure to its target protection potential.

In the case of a well-coated structure or pipeline, we can safely say that polarisation takes a negligible time. The new coating effectively insulates almost the entirety of the steel surface, so the resulting anode to bare steel surface area ratio is huge. This means that any steel exposed to seawater, or seabed mud, will polarise to a value very close to the anode potential (around –1050 mV) very quickly indeed.

For uncoated structures, the anode-to-cathode surface area ratio is far from huge. It is typically in the order of 1:100. Even so, providing the anodes are well distributed over the surface, and in temperate waters, we might reasonably expect a structure to polarise within a matter of weeks or, at the most, months. As mentioned above, for example, the Cognac platform, which was protected by sacrificial anodes, but which also had fixed potential monitoring, polarised to –835 mV within 24 days [33].

As we have established, the rate of corrosion of steel at a potential of –800 mV is too low to be of any practical concern. Indeed, the same claim can probably be made for potentials more negative than about –700 mV. However, there is a benefit in polarising the structure to potentials more negative than is needed for protection, preferably into the range of –900 to –1000 mV, to establish a calcareous deposit that is more protective, thereby reducing the mean current density requirement.

All of this begs the obvious question: how fast should a structure polarise to its protection potential? This is a topic on which the codes are uninformative. In some recent cases, involving offshore wind farms, this silence of the codes has led to disputes, and even litigation, between CP designers and owners.

Most offshore wind turbine generators are supported on pre-installed monopile foundations. We will have more to say about the CP of these structures in Chapter 13. For now, we need simply to appreciate that the anodes cannot be installed directly on the monopile because they would shear off as a result of the shock loadings encountered during piling. Hence, the anodes are installed on the transition piece that links the monopile to the tower and are then electrically bonded to the pile.

This means that there are two important differences between a monopile foundation and a typical offshore hydrocarbon platform support jacket. The first is that the monopile anodes are installed after the structure has spent some time immersed in the seawater. The second, and more important, is that whereas jacket anodes are distributed more-or-less uniformly over the structure, the monopile anodes are clustered at the top end of the monopile. This arrangement means that monopiles polarise more slowly. Over recent years, this effect has become more apparent as wind farms have been installed in deeper water.

A case in point relates to a wind farm of over 80 monopiles offshore UK. A few days after each pile was connected to its anodes, the potentials were measured at the seabed level, the part of the immersed structure furthest away from the anodes. It is little surprise, therefore, that these early readings showed negligible polarisation. This prompted the owner to carry out periodic potential surveys down the full length of the monopiles.

It was observed that polarisation occurred most readily at the tops of the piles and gradually progressed down the pile over a period of time. This can be explained as being due to the steel near the anodes receiving the highest current density, thereby being in the most favourable position to develop a good quality calcareous deposit. As that deposit became established it restricted the current flowing to that zone, permitting more to flow to lower levels. In this way, the calcareous deposits and the polarisation progressed down the pile.

This is illustrated in Figure 7.8 which shows the worst-case results from surveys carried out in February and August 2012. That is the least negative potential measured on any of the monopiles as a function of depth. This illustrates the progress of polarisation down a pile. It also demonstrates that, on this wind farm, polarisation was progressive with time; albeit that it took until August for the most slowly polarising pile to be fully protected along its full immersed length. The total polarisation time, measured from the connection of the anodes, was almost 2 years in some cases.

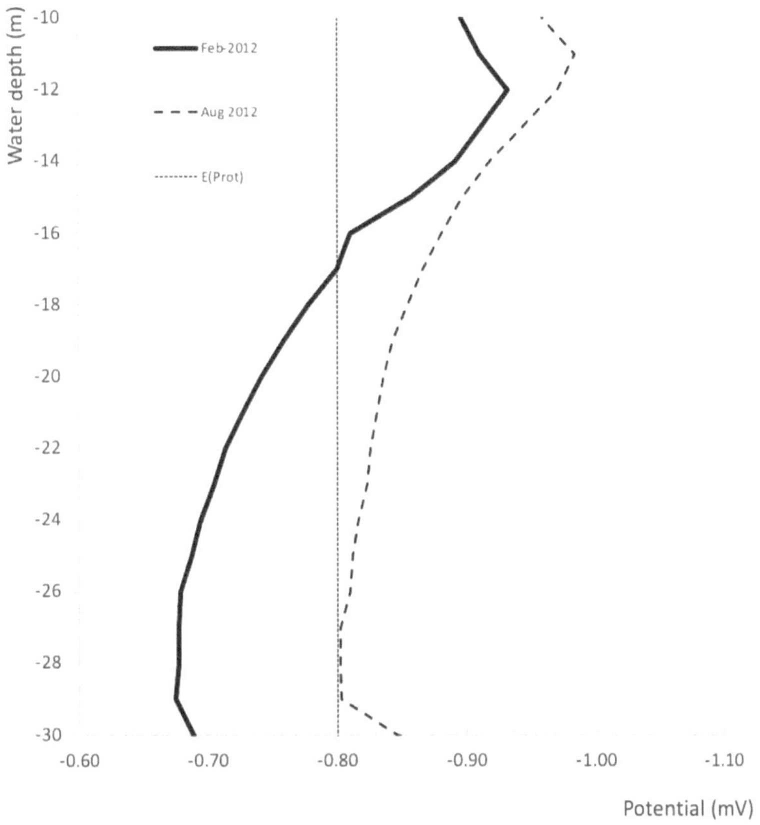

Figure 7.8 Rate of polarisation – wind turbine monopoles.

In the case of this wind farm, although it achieved polarisation to a fully protected level by the summer of 2012, and remained fully protected thereafter, the owner elected to retrofit supplementary anodes in 2014. Since the structure was fully protected, and the anode provision was demonstrably sufficient for the remaining life, retrofit was a very expensive solution to a non-problem.

REFERENCES

1. S.C. Dexter and C. Culberson, *Global variability of natural sea water.* Materials Performance **19**(9), 16 (1980).
2. W.H. Hartt and S.L. Wolfson, *An initial investigation of calcareous deposits on cathodic steel surfaces in sea water.* Corrosion 37 (2), 70 (1981).
3. S. Pathmanaban and B.S. Phull, *Calcareous deposits on cathodically protected structures in sea water.* Proceedings of UK Corrosion Conference, Hammersmith, Institute of Corroison, p. 165, 1982.
4. O. Yague-Murillo, *An aspect of calcareous film evolution.* PhD thesis, Victoria University of Manchester, 1983.
5. W.H. Hartt, C.H. Culberson and S.W. Smith, *Calcareous deposits on metal surfaces in seawater: a critical review.* Corrosion **40** (11), 609 (1984).
6. W.H. Hartt, M.N. Kunjapur and S.W. Smith, *The influence of temperature and exposure time on calcareous deposits,* Paper no. 291 CORROSION/86.

7. R. Johnsen and E. Bardal, *The effect of microbiological slime layer on stainless steel in natural seawater*, Paper no. 227 CORROSION/86.
8. R.D. Strømmen, H. Osvoll and W. Keim, *Computer modelling and in-situ current density measurements prove a need for revision of offshore CP design criteria*, Paper no. 297 CORROSION/86.
9. K.P. Fischer, T. Sydberger and R. Lye, *Field testing of deep water cathodic protection on the Norwegian continental shelf.* Materials Performance **27** (1), 49 (1988).
10. J.S. Luo et al., *Formation of calcareous deposits under different methods of cathodic polarization,* Paper no. 36 CORROSION/88.
11. J.E. Finnegan and K.P. Fischer, *Calcareous deposits: calcium and magnesium ion concentrations,* Paper no. 581 CORROSION/89.
12. K.P. Fischer and J.E. Finnegan, *Cathodic polarization behaviour of steel in seawater and the protective properties of calcareous deposits,* Paper no. 582 CORROSION/89.
13. H.P. Hack and R.J. Guanti, *Effect of high flow on calcareous deposits and cathodic protection current density.* Materials Performance **28** (3), 29 (1989).
14. K.E. Mantel, W.H. Hartt and T.-Y. Chen, *Substrate, surface finish and flow rate influences upon calcareous deposit structure and properties*, Paper no. 374 CORROSION/90.
15. C.F. Schrieber and J.T. Reding, *Application methods for rapid polarization of offshore structures*, Paper no. 381 CORROSION/90.
16. A.D. Goolsby and B.M. Ruggles, *Development of a water property database for deep water cathodic protection design.* Corrosion **47** (5), 387 (1991).
17. J.-S. Luo and W.H. Hartt, *Temperature, surface finish and electrolyte type influences upon adhesion of calcareous deposits*, Paper no. 423 CORROSION/92.
18. S.C. Dexter and S.-H. Lin, *Calculation of seawater pH at polarized metal surfaces in the presence of surface films. Corrosion* **48** (1), 50 (1992).
19. NACE. Publication 7L192, *Cathodic protection design considerations for deep water structures*, AMPP (Houston) 1992.
20. R.G.J. Edyvean et al., *Interactions between cathodic protection and bacterial settlement on steel in seawater.* International Biodeterioration & Biodegradation **29** (3–4), 251 (1992).
21. J.-F. Yan, R.B. Griffin and R.E. White, *Parametric studies of the formation of calcareous deposits on cathodically protected steel in seawater.* Journal of the Electrochemical Society **140** (5), 1275 (1993).
22. V. Ashworth, *Some basic design and operating parameters for cathodic protection*, Chapter 1 p2–7, in *Cathodic Protection Theory and Practice,* Eds. V. Ashworth and C. Googan. Ellis Horwood (Chichester) 1993.
23. T.E. Evans, *Mechanism of cathodic protection in seawater*, ibid Chapter 7 p125-132,
24. R. de Oliveira Vianna and G. de Souza Pimenta, *An initial investigation of calcareous deposits upon cathodically polarised steel in Brazilian deep water.* Proceedings of 12th SPE/NACE International Corrosion Congress, Houston, TX, p. 2278, September 1993.
25. J.G. Bomba, F.D. McCasland and P.J. Agosta, *Cathodic protection for deep water pipelines,* Paper no. 7204, Proceedings of 25th Offshore Technology Conference, (Houston) 1993.
26. J.-F. Yan, T.V. Ngugen, R.E. White and R.B. Griffin, *Mathematical modeling of the formation of calcareous deposits on cathodically protected steel in seawater.* Journal of the Electrochemical Society **140** (3), 733 (1993).
27. G. Hernández, W.H. Hartt and H.A. Videla, *Marine biofilms and their influence on cathodic protection: a literature survey.* Corrosion Reviews **12** (1–2), 29 (1994).
28. K.P. Fischer, W.H. Thomason and S. Eliassen, *CP in deep water: The importance of calcareous deposits and the environmental conditions*, Paper no. 548 CORROSION/96.
29. P.N. Blackburne and R.B. Griffin, *Mathematical model for predicting calcareous film formation,* Paper no. 562 CORROSION/96.
30. D.W. Townley, *Unified design equation for offshore cathodic protection,* Paper no. 473 CORROSION/97.
31. I. Rippon, *Platform cathodic protection design in the South China Sea,* Paper no. 475 CORROSION/97.

32. W. Baptista and J.C.M. da Costa, *In situ acquisition of cathodic protection parameters.* Materials Performance **36** (1), 9 (1997).
33. A.D. Goolsby and D.P. McGuire, *Cathodic protection upgrade of the 1050′water depth cognac platform,* Paper no. 472 CORROSION/97.
34. W.H. Hartt and E. Lemieux, *A principal determinant in cathodic protection design of offshore structures: The mean current density.* Corrosion **56** (10), 988 (2000).
35. D. Festy, *Cathodic protection of steel in deep sea: Hydrogen embrittlement risk and cathodic protection criteria,* Paper no. 01011 CORROSION/2001.
36. W.H. Hartt and S. Chen, *Deepwater cathodic protection laboratory simulation experiments,* Paper no. 01501 CORROSION/2001.
37. B. Bethune and W.H. Hartt, *Applicability of the slope parameter method to the design of cathodic protection systems for marine pipelines.* Corrosion **57** (1), 78 (2001).
38. S. Chen and W.H. Hartt, *Deepwater cathodic protection part 1: Laboratory simulation experiments.* Corrosion **58** (1), 38 (2002).
39. S. Chen, W.H. Hartt and S. Wolfson, *Deepwater cathodic protection part 2: Field deployment results.* Corrosion **59** (8), 721 (2002).
40. W.H. Hartt, X. Zhang and W. Chu, Issues associated with expiration of galvanic anodes on marine structures. Corrosion **61** (11), 1035 (2005).
41. E. Lemieux and W.H. Hartt, *Galvanic anode current and structure current demand determination for offshore structures.* Corrosion **62**(2), 162 (2006).
42. W.H. Thomason, *Overview of NACE state-of-the art report T7L192 "cathodic protection design considerations for deep water projects,* Paper no. 09064 CORROSION/2009.
43. NACE, *State-of-the art report T7L192 cathodic protection design considerations for deep water projects,* AMPP (Houston, TX) 2009.
44. D. Festy et al., *ICP-DATA: In situ data collection for cathodic protection design,* Paper no. 11050 CORROSION/2011.

Chapter 8

Corrosion resistant alloys

8.1 WHY CONSIDER CRAs?

Although CP was invented to tackle the corrosion of copper on ships' hulls, that problem is no longer relevant. Nowadays, the vast majority of the money invested in CP is directed at protecting carbon steel. Nevertheless, carbon steel is by no means the only alloy to which CP is applied. The shipping industry, for example, immerses some copper-based alloys, such as brass and bronze. These materials have some inherent resistance to corrosion in seawater, but they are not immune. In addition, modern shipbuilding also makes use of the high strength to weight ratio of aluminium alloys for the construction of specialist vessels. These alloys, likewise, can be vulnerable to corrosion in seawater.

Elsewhere, since the 1970s, the offshore petroleum industry has exploited progressively deeper, hotter and higher-pressure hydrocarbon reservoirs. This has given rise to the installation of subsea pipework and pipelines for transporting potentially more corrosive fluids. In many cases, carbon steel has had to give way to more corrosion resistant alloys (CRAs).

These CRAs are selected for their resistance to the particular corrosion threats posed by the reservoir fluids. A pipe alloy may be selected for its intrinsic resistance to internal corrosion, but the material selection exercise will not normally consider the external corrosion threat from the seawater or the marine sediments. This means that the selected CRA may, or may not, be adequately resistant to seawater corrosion in its own right. However, from the perspective of the CP engineer, the inherent corrosion resistance of these materials is largely irrelevant. Almost inevitably, these alloys will be connected to carbon steel components that are in need of CP. Our concern is, therefore, with how these alloys will influence the CP, and with how the CP will affect them. We deal with these issues at various points in this book.

Elsewhere, seawater is piped around offshore and nearshore facilities for process cooling or firefighting. For the most part, the piping and heat exchanger materials are selected to be resistant to seawater under the anticipated operating conditions. Sometimes, however, the degree of corrosion resistance errs towards the marginal, in which case internal CP might offer an option. We return to this in Chapter 16.

In this chapter, we cover enough information to help us with decisions about the cathodic protection of metals other than carbon steel. There is a very extensive literature on marine corrosion of these metals, and I have made no attempt to review any of it in detail. For readers seeking more information, I recommend as a good starting point the basic guide published by the EFC [19]. In addition, it is also worth thumbing through the EFC's report on examples of marine corrosion failures [10].

DOI: 10.1201/9781003216070-8

8.2 PASSIVITY

8.2.1 What do we mean by passivity?

The alloys mentioned above are not immune to corrosion in seawater. For example, the standard potential for aluminium in equilibrium with its ions is considerably more negative than that of the iron electrode (see Table 8.1). This tells us that aluminium has a greater thermodynamic tendency to corrode than iron. As we will see in Chapter 10, we make use of this fact when we exploit aluminium as a sacrificial anode.

Table 8.1 Standard equilibrium electrode potential for aluminium

	E^o	
Electrode reaction	*SHE*	*Ag|AgCl|seawater*
$2H^+ + 2e^- \Leftrightarrow H_2^-$	0.00	−0.27
$Fe \Leftrightarrow Fe^{2+} + 2e^-$	−0.47	−0.74
$Al \Leftrightarrow Al^{3+} + 3e^-$	−1.71	−2.03

Nevertheless, despite its thermodynamic tendency to corrode, aluminium is actually a very widely used material. It is used in aviation, automotive and structural engineering, as well as for many domestic appliances. That it proves to be so serviceable is due to the phenomenon of *passivity.*

The word *passive* first appeared in corrosion in the 1830s [1], and *passivity* has since been the subject of a number of definitions. All of these are along the lines that a metal is passive if it resists corrosion even where there would be a large decrease in free energy if it were to corrode. Burstein restated this in electrochemical terms when he defined passivity as ... *a state of low corrosion rate brought about under a high anodic driving force, or potential, by the presence of an interfacial solid film, usually an oxide* [11].

In the case of aluminium, its pronounced thermodynamic tendency to corrode is stifled by the formation of a highly insoluble corrosion product (Al_2O_3) on its surface.

8.2.2 Thermodynamics

Figure 8.1 offers an interpretation of passivity using the E-pH diagram for the aluminium-water system. This shows that in the pH range 4–9 the stable corrosion product is Al_2O_3. This is identified as a domain of passivity on the diagram, and this accounts for the corrosion resistance of aluminium in natural waters. It also highlights the fact that aluminium is amphoteric. It corrodes in acids (pH<4) to produce soluble Al^{3+} cations and in alkalis (pH>8.5) to form the soluble aluminate $\left(AlO_2^-\right)$ anion.

8.2.3 Electrode kinetics

In terms of electrode kinetics, passivation may be considered as a very pronounced departure from Tafel behaviour. The linear relationship between applied anodic potential and the logarithm of the anodic current breaks down. This is illustrated schematically for the anodic oxidation of notional metal M in the E-log|i| (Evans') diagram in Figure 8.2. At low anodic overpotentials, the Tafel relationship holds. This, in turn, tells us that the metal is undergoing active dissolution. The anodically produced M^{n+} ions leave the surface and enter the bulk solution.

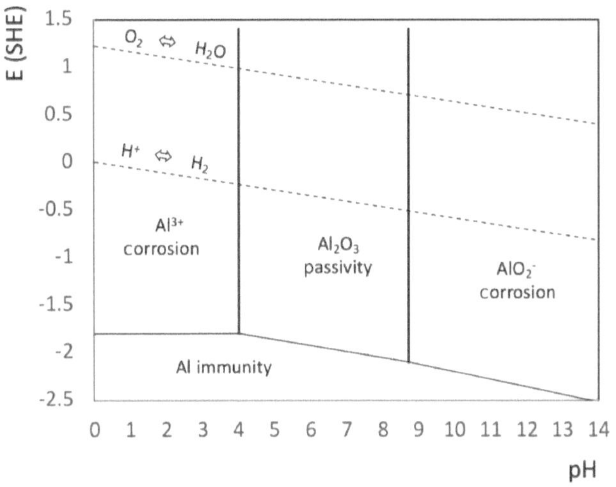

Figure 8.1 E-pH diagram for aluminium [2].

At potentials more positive than a certain value, the current density reduces dramatically. This is known as an active-passive transition. At potentials more positive than this transition, metal ions are being produced at such a high rate that their surface concentration exceeds their solubility limit. This leads to the precipitation of an oxide, or hydroxide, onto the surface. This effectively insulates the metal surface from the environment, drastically reducing the anodic dissolution rate. This is then equal to the very low anodic current that now flows through the low-conductivity oxide. This current is referred to as the passive current (I_p). We can now combine the anodic kinetics with the kinetics of the cathodic reactions available in natural waters. This is also illustrated in Figure 8.2. The corrosion potential, which falls in the passive region of the anodic curve, is very much more positive than we would have expected if anodic Tafel behaviour had persisted to higher current densities. On the other hand, the rate of corrosion (I_{corr}) is very much lower, being the same as I_p.

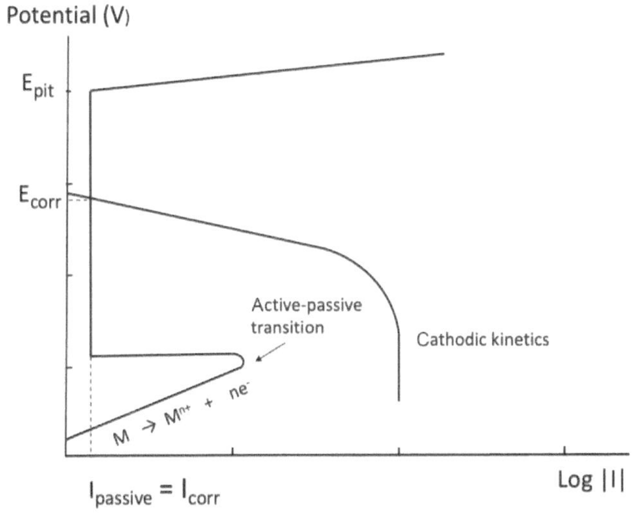

Figure 8.2 Anodic E-Log |i| diagram for a passive metal.

8.2.4 Passivity breakdown

8.2.4.1 Nature of the passive film

The passive films have extremely low solubilities in water. In some cases, such as stainless steels, the passive oxide films are very thin indeed: typically, no more than a dozen atomic layers. In other cases, such as aluminium, the passive films can be artificially thickened by anodic polarisation in specially formulated solutions. Passive films have been the subject of many thousands of erudite publications involving highly sophisticated electrochemical and surface analysis investigations.

This is all interesting stuff. But this is a book about marine CP, so we need to focus on the fact that passive films are not indestructible. Any passive film can be vulnerable in certain circumstances. Therefore, we need to know what these circumstances are, and what the implications of this breakdown might be for corrosion and CP.

8.2.4.2 Pitting

Pitting corrosion is a phenomenon that applies to passive metals. Like the nature of the passive film itself, the mechanisms and morphologies of pitting have been widely studied by corrosion scientists.

It is likely that different mechanisms apply to different metals, or to the same metal in different environments. However, because this book is about the marine environment, a factor common to all of the alloys that concern us here is the abundance of chloride in seawater (~19800 mg/L). This anion has the ability to initiate, and then sustain, pitting processes. Again, we need not be concerned here with the details of the chemical interaction between the chloride and the oxide film. We can be content to visualise the process in the context of an E-log|i| diagram.

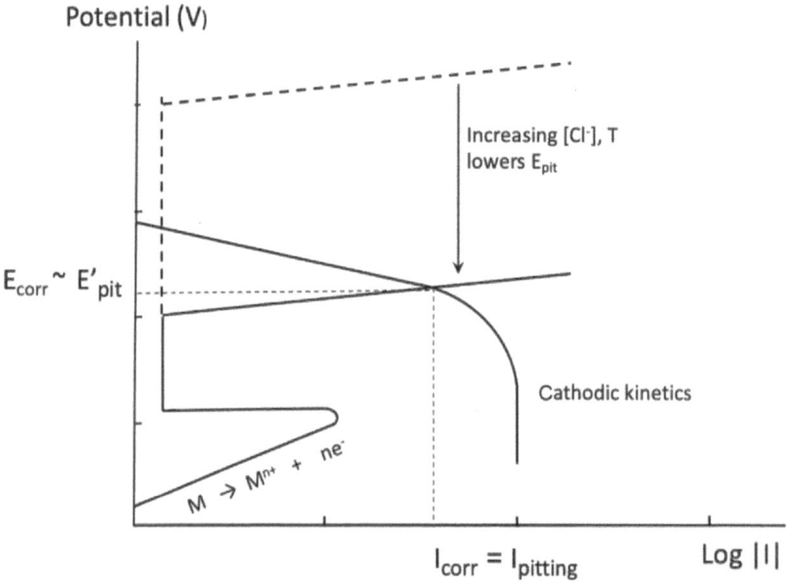

Figure 8.3 E-Log |i| diagram for pitting of a passive metal.

This is illustrated in Figure 8.3. It is based on the passive behaviour we saw in Figure 8.2, but we now see that the presence of increasing the chloride concentration lowers the potential at which the oxide film breaks down and pitting initiates (E_{pit}). If sufficient chloride is present, or the temperature is high enough, E_{pit} becomes more negative than the cathodic kinetic curve for the environment. At this point, pitting will initiate and propagate.

8.2.4.3 Crevice corrosion

Crevice corrosion bears many phenomenological similarities to pitting corrosion. At its simplest, and with some justification for our purposes, a crevice may be regarded as a two-dimensional pit that has already initiated. As a generalisation, most passive alloys will undergo crevice corrosion at slightly lower temperatures in seawater than they undergo pitting corrosion. Since it is very difficult to avoid crevices in the construction of offshore equipment using CRAs, crevice corrosion is, in practice, a more dominant threat than pitting.

8.2.4.4 Stress corrosion cracking

Load bearing offshore structures are not normally made from CRA, so there is no need for us to consider fatigue and corrosion fatigue as we do in Chapter 12 for carbon steel. However, certain CRAs employed in subsea engineering could be prone to a different form of cracking. This is stress corrosion cracking (SCC). Like so many of the localised phenomena that afflict passive alloys, there is a very extensive literature on SCC. It can only arise as the result of the interaction between a specific material, in specific environmental circumstances, and under certain conditions of stress.

The stress involved is always tensile but, depending on circumstances, can be either residual or applied. The literature catalogues numerous susceptible material-environment combinations. Their common feature is that SCC can only arise in an alloy exposed to an environment where it has a tendency to be passive. For this reason, carbon steel never undergoes SCC in seawater, because it cannot passivate. Certain stainless steels, on the other hand, can be susceptible to SCC in seawater, but the environmental circumstances usually also require elevated temperature (typically >50°C).

Fortunately, for the purposes of this book, we can simplify things by referring to the Venn diagram in Figure 8.4, which emphasises the requirement for the interaction of a specific material, a specific corrosive environment and stress. The message is simple: providing we can use CP to eliminate the "corrosion": then stress corrosion cracking does not occur.

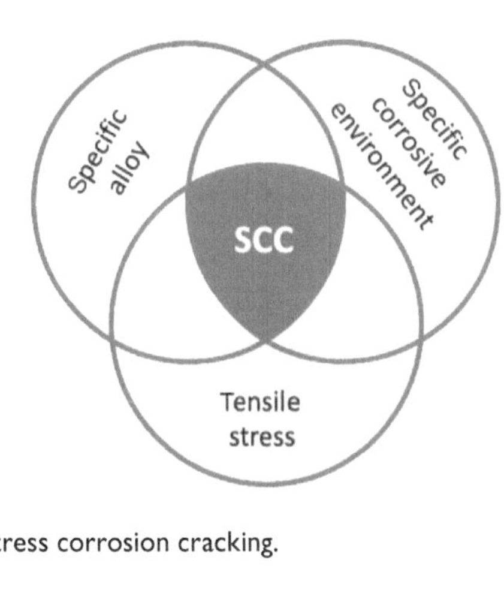

Figure 8.4 Venn diagram for stress corrosion cracking.

8.3 STAINLESS STEELS

8.3.1 Corrosion resistance

8.3.1.1 Passivity

Although the corrosion resistance of iron-chromium alloys was first recognised in the 1820s, the early materials were rich in carbon and too brittle for practical engineering applications. Nearly a century elapsed before it was possible to develop alloys with sufficiently high levels of chromium, and low levels of carbon, to be mechanically serviceable. There are now many hundreds of iron-chromium based alloys; collectively known as stainless steels. Their common feature is that they all contain sufficient chromium (>11% by weight) to result in the formation of a passive oxide film when exposed to air or moisture.

The passive film is responsible for the shiny, "stainless" appearance of the alloy. It has a very low, but not zero, solubility in water. Moreover, it is surprisingly thin, usually extending to no more than about ten atomic layers. Stainless steels may appear to be non-corrodible, but that is just the perception of viewing on a macroscopic scale. At the atomic scale, the passive film is constantly suffering attrition, either by dissolution or mechanical contact. All being well, it rapidly reforms itself. This process is termed "repassivation". In other words, despite their stainless appearance, the alloys are actually corroding, albeit very slowly.

This background corrosion rate of stainless steel is usually only of academic interest. It can be determined in the laboratory. For example, electrochemical polarisation techniques can be used to measure the passive current. This is an electronic measure of the rate at which iron and chromium atoms in the metal surface are oxidised to maintain the passive film. For example, Oldfield et al. [7] have measured passive currents on type 316 stainless steel in seawater and found values in the region of 1 mA/m^2. Applying Faraday's law, this gives us a ballpark corrosion rate of 1 µm/year, or 1 mm per millennium.

8.3.1.2 Passivity breakdown

If the natural attrition of the passive film were the only factor that affected stainless steel, then its corrosion rate would be of little interest to mankind. However, the uniform attrition of the passive film is not the only damage mechanism to which stainless steels are susceptible. In some situations, the passive film can locally fail to repassivate. When this occurs, the result can be pitting corrosion or the mechanistically related processes of crevice or under-deposit corrosion. If these localised forms of attack initiate on stainless steels, the resulting rates of corrosion penetration are much higher than encountered by the non-stainless carbon steels.

These localised mechanisms of corrosion of stainless steel are less straightforward to study, either in the laboratory or in natural exposure testing, than the general corrosion of carbon steel. Both crevice and, more particularly, pitting corrosion are characterised by two stages in their development. The first stage is initiation. This is the period of indeterminate length during which there are no readily observable changes occurring on the passive metal surface. Once initiated, however, the propagation rates are invariably rapid.

The various mechanisms believed to operate have been reviewed by Oldfield [3,7]. Initiation is governed by stochastic processes. This means that, in any laboratory test, all that can safely be concluded from a lack of observed pitting or crevice damage is that the corrosion did not initiate on that specimen during the period of exposure. Averring that it could not initiate at some point in the future involves the philosophical conundrum of trying to prove a negative.

For this reason, a great deal of the work on characterising the crevice and pitting resistance of CRAs has involved some form of accelerated testing to rank the relative resistance of the materials to corrosion propagation. This has involved employing elevated temperatures or unusually aggressive solutions such as ferric (Fe^{3+}) ions in the much-employed ASTM G48 test [24].

It would be too ambitious to attempt to summarise the immense body of corrosion research on the behaviour of stainless steels in seawater. However, what is clear is that the critical species responsible for interfering with the re-passivation process is chloride. Unsurprisingly, there has been a considerable body of work on developing stainless steels for seawater service. From this, it has emerged that a critical parameter is the pitting resistance equivalent number (PRE_n), which is defined for austenitic stainless steels (see Section 8.3.2 below) as:

$$PRE_n = \%Cr + 3.3\%Mo$$

and for duplex stainless steels (DSS) as:

$$PRE_n = \%Cr + 3.3\%Mo + 16\%N$$

As a matter of testing, and experience, a minimum PRE_n value of 40 is required for service in seawater (up to 30°C) where there is the possibility of crevices or the accumulation of deposits. It is worth noting that many stainless steels used in offshore engineering have PRE_n values well below this value. This is why the stainless steels in common use, even if they are judged suitable for transporting a crude reservoir fluid, cannot be exposed to the marine environment without supplemental protection in the form of coatings and CP.

Interestingly, as the seawater temperature reduces, the required PRE_n value also decreases. Laboratory tests have indicated that Type 316L (a widely used stainless steel) might be resistant to pitting or crevice corrosion in seawater at temperatures below about 15°C. Thus, since this temperature will not be exceeded in Northern European waters at the depths relevant to hydrocarbon production, it might be concluded that 316L stainless steel components will persist for at least centuries, or probably millennia, when left in-situ on the seabed. Unfortunately, however, experience in the North Sea shows that this is not the case. There has been at least one instance of type 316L connection flanges which have been placed on the seabed without the benefit of coating or CP. These were recovered after a few months and were found to be heavily pitted, to a depth of several mm.

This seemingly anomalous occurrence provides another illustration of the problem of trying to predict the performance of materials in natural, biologically active, seawater on the basis of test work carried out in abiotic laboratory solutions. This was emphasised by Johnsen and Bardal [4,5] who found that microbial slime layers formed on various stainless steels after about a week of immersion in natural seawater. This layer acted as a cathodic depolariser, meaning that it stimulated the rate of the cathodic reaction. This manifested itself as a positive shift in the corrosion potential for specimens not under CP, and an increase in the current density for specimens polarised potentiostatically at −0.4 V. These observations provide an electrochemical confirmation of the observed propensity for a stainless steel to pit in natural seawater, whereas it might not do so in a laboratory solution of the same chemical composition.

The mechanism of the change of electrochemical potential and the interaction of the biological films with the pitting or crevice initiation mechanisms have subsequently been the subject of extensive discussion in the literature. For example, these biofilms have been shown to stimulate crevice corrosion [12] and galvanic corrosion [15]. They have been also observed to stimulate the cathodic kinetics on passive alloys other than stainless steels

[9,17], and to be temperature dependent [21,22]. Thus far, however, a definitive mechanistic explanation has not been forthcoming, although some works point to the involvement of peroxides produced whilst micro-organisms are growing aerobically [16,17].

This work is of profound academic interest. However, from the point of view of this book, the practical conclusion is that many CRAs will require CP.

8.3.2 Designations

The UNS designation for most stainless steels is SXXXXX, where "S" stands, not unreasonably, for stainless steel. The numeric component is usually derived from the American Iron and Steel Institute (AISI) identification number. For example, a much-used stainless steel known as AISI Type 316L, mentioned above, becomes UNS S31603.

That much is easy enough to grasp. Unfortunately, the system is less obvious when it comes to some of the more highly alloyed stainless steels. For example, there is a group of materials sometimes referred to as superaustenitic. Some examples of which are classified in the UNS system SXXXXX series alloys, whereas other similar alloys are classified as nickel alloys (UNS NXXXXX). This is confusing, not least because the nickel content of these latter materials may be as little as 25%.

8.3.3 Grades used offshore

8.3.3.1 Allotropes

Some elements occur in nature in more than one crystal form, a phenomenon known as allotropy. The classic example is carbon which can exist as both graphite and diamond. The two materials are chemically identical. For example, at a very high temperature, both will burn in air to form carbon dioxide. If you are both curious and rich, give it a try.

Iron and its steel alloys also occur in allotropic forms, although the differences between them are not as obvious as they are between graphite and diamond. These allotropes arise because of the subtle differences in the way successive layers of atoms arrange themselves in the metal crystals.

8.3.3.2 Ferritic stainless steels

The allotrope of iron that is most familiar to us, because it is the form that is stable at room temperature, is called ferrite. Its crystalline form is termed body-centred-cubic (bcc). Crystallography, like so many other interesting scientific disciplines, is too far removed from the scope of this book for us to consider it further. The only learning that we need to take away from this is that ferrite or ferritic steels are magnetic.

Thus, the first stainless steels to be produced were ferritic. They comprised alloys of iron and chromium, which also has a bcc structure. Ferritic stainless steels are used in a wide variety of industrial applications. However, they are rarely, if ever, used subsea. If they are, we should never try to apply CP to them because they are particularly susceptible to hydrogen embrittlement (see Chapter 12).

8.3.3.3 Austenitic stainless steels

At elevated temperatures (typically >800°C depending on alloy composition), the layers of atoms in iron rearrange very slightly, changing the crystal structure from bcc to face-centred-cubic (fcc). This crystal form of iron is called austenite.

Although iron and low-alloy steels can only exist as austenite at high temperature, if the iron is alloyed with other elements, which also have a fcc crystal structure, then the austenite can become stable at normal temperatures. For example, this arises when sufficient nickel, itself naturally austenitic, is alloyed with iron or steel. Austenite is easily distinguishable from ferrite because it is non-magnetic. Other things being equal (which they never are) austenite is more naturally corrosion resistant than ferrite. Thus, austenitic stainless steels, of which Type 316L is perhaps the most widely used example in offshore engineering, will contain as a minimum chromium, to confer passivity, and nickel, to ensure that the austenitic structure is stable at room temperature. Its typical composition is shown, together with that of some other offshore steel types, in Table 8.2.

Table 8.2 Some stainless steels used subsea

		Typical composition (wt%)					
Generic type	*UNS No. (Example)*	*Cr*	*Ni*	*Mo*	*C*	*Others*	*Fe*
Austenitic	S31603	18	10	2.5	0.03		Balance
Supermartensitic	S41426	12.5		2	<0.03		
Duplex	S31803	22	5.5	3	0.03	N 0.15	
Superduplex	S32550	25	5.5	3	0.04	N 0.2, Cu 2	

8.3.3.4 Duplex stainless steels

As its name implies, duplex stainless steel (DSS) comprises crystals of both austenite and ferrite. It may be simplistically thought of an attempt to get the best of both worlds. Crude hydrocarbon systems might contain, for example, high levels of both H_2S and chloride. Ferritic stainless steels are susceptible to cracking in H_2S but not in chloride. Austenitics, on the other hand, are at risk of SCC in chlorides; but not in H_2S.

Creating a dual-phase structure, comprising a near equal mix of ferrite and austenite crystals, offers the possibility of resistance to fluids carrying both aggressive species. This dual-phase structure can be produced by judicious alloy formulation and, importantly, careful thermal control during the manufacturing process.

During the late 1970s, the Dutch operator NAM pioneered the use of DSS flow lines. This material offers an attractive combination of mechanical properties and resistance to corrosion by crude reservoir fluids. Although DSS steel offers better corrosion resistance to crude hydrocarbons than carbon steel, it is not sufficiently resistant to corrosion by seawater or seabed mud to be employed without supplemental protection. As for carbon steel pipelines, this protection takes the form of protective coating systems supplemented by CP.

Some high-profile cracking failures of DSS flowlines, and flowline manifold components, occurred in the early 1990s. These failures were ascribed, at least in part, to hydrogen embrittlement caused by the action of the CP system. This generated a body of work on controlling the operating voltage of the CP system to reduce the propensity for hydrogen generation on the surface of the duplex. However, it also emerged that duplex stainless steel is compatible with normal levels of CP providing adequate care is taken over its metallurgical composition, and the mechanical design does not introduce unacceptable levels of strain. In 2008, these requirements were enshrined in a DNV recommended practice [23].

8.3.3.5 *Superduplex*

The year 1990 saw the first application of 25% Cr, so-called "superduplex", stainless steels (SDSS), which offered even greater strength and improved resistance to crudes containing H_2S. In principle, these alloys are sufficiently resistant to seawater at ambient temperatures to be deployed without external corrosion protection.

Unlike DSS, there is a considerable body of experience of the behaviour of SDSS in seawater. This is because it is an attractive material for seawater piping systems. Previously, cupro-nickel alloys were regarded as the material of first choice for seawater handling systems. However, from the late 1980s, this material has been progressively superseded by 6 Mo austenitic stainless steel and SDSS.

SDSS and 6 Mo are both regarded as fully resistant to corrosion in oxygenated (and chlorinated) seawater at ambient temperatures. However, of the two, only SDSS is used for subsea flowline construction. In this service, it may be prone to localised forms of attack under certain conditions, such as when heated by internal fluids. These forms might include chloride-induced SCC, but pitting and crevice corrosion are more likely. As we discussed above, this means that we need to apply CP to protect SDSS. More usually, the SDSS will be connected to carbon steel equipment or structures, so it will receive CP anyway.

8.3.3.6 *Martensitic and supermartensitic*

In crystallographic terms, there is a "half-way house" between ferrite and austenite. If iron is heated and held above the austenite transformation temperature, all of the ferrite transforms to austenite. If it is then cooled rapidly, the reversion from austenite back to ferrite does not have time to progress to completion. The resulting allotrope of iron, which has a structure which crystallographers refer to as body-centred-tetragonal (bct), is known as martensite. The stand-out property of martensite is that it is very strong. However, this high strength is accompanied by a loss of ductility. This means that martensitic steels usually have to undergo further heat treatment, known as tempering, to make them less brittle and more useable. Another feature of martensitic steels is that, compared to the ferritic and austenitic allotropes, they are generally less easy to weld.

Like the austenitic and ferritic alloy systems, martensitic steels can also be alloyed with chromium to produce martensitic stainless steels.

The so-called supermartensitic stainless steels (SMSS) are a recent development of the martensitics. These are weldable versions of the high strength alloys widely employed for screw-jointed downhole tubulars. Their resistance to corrosion by produced fluids, whilst inferior to the more expensive duplex and superduplex materials, renders them suitable for constructing flowlines for moderately corrosive production fluids that would be too aggressive for carbon steel.

SMSS are formulated with lower PRE_n values than 316L. They are not expected to endure seawater exposure without being coated and provided with a CP system. Their known susceptibility to seawater means that there has been no great impetus to characterise their corrosion performance. It is reasonable to assume that they will exhibit better general corrosion resistance than carbon steel, but they will be particularly prone to pitting attack. The use of SMSS subsea brings with it a need to apply CP.

8.3.3.7 *Some stainless steels used subsea*

Table 8.2 sets out the compositions of some typical examples of stainless steels used subsea. Within the main groups, there are numerous individual alloy types in service.

8.4 HIGH NICKEL ALLOYS

The high price of nickel renders it unattractive for subsea engineering. Nevertheless, it finds use in some niche roles. Table 8.3 sets out the compositions of selected high nickel alloys used subsea. Of these, Alloy 625 (UNS N06625) merits comment as it is sometimes selected for small-bore chemical and production tubing on subsea equipment.

Table 8.3 High nickel alloys

		Typical composition (wt%)					
Common name	*UNS No. (Example)*	*Cr*	*Ni*	*Mo*	*C*	*Others*	*Fe*
Alloy 825	N08825	21	42	3		Cu 2.5	30
Alloy 625	N06625	21.5	61	9			4

In these instances, it is more likely to be selected for its resistance to the transported fluid than its seawater corrosion resistance which is excellent. For example, it is used in areas of seawater desalination plants where the elevated temperatures preclude the use of less resistant alloys such as 6 Mo or SDSS. Even so, it is not immune to all forms of corrosion in seawater. The effect of biofilms in ennobling the potential, and prompting the initiation of crevice corrosion, cannot be discounted. For example, Martin et al. [18] have stated that the material is... *susceptible to crevice corrosion in ambient temperature natural seawater and that crevice corrosion will initiate in seven days for a sample* ... the potential of which has been ennobled as a result of the activity of a biofilm. Thus, it seems that crevice corrosion of Alloy 625 is at least a credible process in natural seawater. That said, however, it needs to be understood that the *ambient* seawater used for the tests was from the Florida coast, so the temperatures were above those to be expected in deeper waters elsewhere. Moreover, their test panels involved artificial crevices, the geometry of which might be more severe than would occur naturally under deposits.

There has been little long-term exposure testing of Alloy 625 in deep, and therefore cool, seawater. However, Reinhart [1] reported that the material showed no crevice corrosion, and its general corrosion rate was less than 0.0025 mm/year.[1]

Even so, as with other CRA's, it may find itself involved in CP systems because it is connected to carbon steel.

8.5 COPPER ALLOYS

Copper alloys feature in marine construction, particularly in the case of ships which, for example, make considerable use of nickel aluminium bronze (NAB). These long-established materials exhibit passive behaviour and good corrosion resistance in seawater. Nevertheless, they are not immune from corrosion. Furthermore, where they are used, such as in ships' propellers or hull penetrations, they only have limited tolerance to corrosion damage.

The point is moot, however, because copper alloys earthed to the hull will be protected by the hull CP system. We consider the CP of propellers in Chapter 15.

1 Reported as <0.1 mpy (mils per year) in the original work. It is assumed that this result meant that the weight loss, recorded after removing the fouling, was indistinguishable from the weight loss of a blank specimen subjected to the same cleaning process.

8.6 ALUMINIUM ALLOYS

It is well known that, in American English, this metal is spelled and pronounced “aluminum” whilst those who adopt British English add the second “i” to make “aluminium”. Apparently, the blame for this difference lands at the feet of Sir Humphry Davy, whom we have already met in these pages. He first isolated the element from its double sulfate salt “alum”. As the privilege of naming a new element fell to its discoverer, he named it aluminum (without the "i"). The news of his discovery, and the new name, was published in America. It was only subsequently, and confusingly, that Sir Humphry tried to change the name to align it with other metallic elements that were being given names ending in “ium”. Britain changed, but America did not.

8.6.1 Alloy types

We have already encountered aluminium as an example of an element that has a strong tendency to form a passive film. Pure aluminium generally possesses too low a strength for structural purposes. Higher strength aluminium alloys are required. The most common group used in direct contact with seawater are the UNS A95XXX materials, where the “A” designates an aluminium-based alloy and the “9” identifies it as a wrought material. The designation “5” is incorporated from the Aluminum[2] Association's code for an Al-Mg alloy. These materials are commonly used for the construction of small vessels such as work boats and fast patrol boats [20]. They are also occasionally employed in the construction of high-speed commercial catamarans.

8.6.2 Corrosion threats and mitigation

These alloys have generally good resistance to general corrosion in seawater. However, as with passive metal more generally, the risks of pitting and crevice corrosion cannot be discounted. Generally speaking, it is found that the alloying additions found to increase aluminium strength, such as the addition of copper to make age-hardenable alloys, also increases the susceptibility to pitting. This means that the corrosion resistance afforded by the passive film is insufficient to protect aluminium alloys in seawater.

Most aluminium vessels will have a protective coating on the external hull surface as a first line of defence. This is often supplemented by a CP system.

8.7 CP OF CORROSION RESISTANT ALLOYS

8.7.1 Protection potential

8.7.1.1 Theory and practice

Both thermodynamic and electrode kinetic analyses lead to the conclusion that different metals have different protection potentials to that of steel. The value of that potential depends on whether the object of the cathodic polarisation is to maintain or restore a passive condition, or it is to reduce active anodic dissolution. In most practical situations, however, the point is of academic interest. Almost invariably, these other alloys will be in multi-metal installations. The target protection potential, therefore, has to be the most negative value

[2] The Aluminum Association is an American organization: hence, the spelling.

required for any of the materials. Almost invariably, a multi-metal subsea installation will contain carbons steel. So, the target protection potential needs to be −800 mV, or more negative.

8.7.1.2 The codes

The codes that prescribe protection potentials for alloys other than carbon steel do not differentiate between aerobic and anaerobic environments. The recommended values are given in Table 8.4. As noted above, the least negative potential to protect the alloy may not be particularly relevant. With some alloy systems, there is a perceived threat of hydrogen embrittlement (HE) as a result of cathodic polarisation. This is not a problem for austenitic materials which are highly resistant to HE. For duplex and martensitic materials, the susceptibility to HE increases with strength or hardness levels. In general, the codes urge caution over ensuring that the metallurgical condition of the alloy is not susceptible to HE. (We will have more to say about this in Chapters 12 and 14.)

Table 8.4 Some protection potentials – CRAs

Alloy	*Metallurgical condition*	*Protection potential (mV)*		*Code*
		Least negative	*Most negative*	
Stainless steel – piping or pipeline grade materials				
Martensitic and super martensitic		−600	Assess metallurgical condition	DNV-RP-F103
		−500		ISO 15589-2
Austenitic	$PRE_n \geq 40$	−300	−1100	ISO 15589-2
		−300	No limit	EN 12473
	$PRE_n < 40$	−500	−1100	ISO 15589-2
		−500	No limit	EN 12473
Duplex and superduplex		−500	Assess metallurgical condition	ISO 15589-2
				DNV-RP-F103
				EN 12473
Other alloys (used as fasteners, etc.)				
Copper alloys	Without Al	−450 to −600	No limit	EN 12473
	With Al		−1100	
Nickel-based alloys		−200	Assess metallurgical condition	

8.7.2 Protection current densities

DNV-RP-B401 tells us to apply the same design current densities for... *any stainless steel or non-ferrous components of a CP system.* This guidance is consistent with the work of Foster and Moores [6] who studied the CP current demand, at fixed potentials, of ten alloys in seawater. These included carbon steel together with brasses, cupronickels and an austenitic stainless steel. The experiments continued for up to 800 days. Although the current versus time traces were not identical, there was no discernible difference in the ultimate current demands of the different alloys.

This result is not surprising. As we have seen, the cathodic current demand depends initially on the rate of transport of dissolved oxygen to the metal surface, and then increasingly on the nature of the calcareous deposit. The rate of oxygen diffusion is dependent only on the seawater conditions. Furthermore, as we saw in Chapter 7, calcareous deposits are only slightly influenced by the initial profile of the surface. Thus, all other things being equal, we would expect the same steady-state cathodic current demand ultimately to be arrived at for (initially) rusty carbon steel and (initially) smooth stainless steel.

Lye [8] studied a variety of stainless steels, also in natural seawater, and produced conclusions that were consistent with Foster and Moores. He found no practical difference between different types of stainless steel, and his comparisons with carbon steel allowed him to conclude that, given the conservatism of the guidelines, the same current densities may be used.

Nevertheless, the detail of Lye's paper does reveal some interesting differences between carbon and stainless steel. For example, in free-running experiments, in which specimens were galvanically coupled to an aluminium alloy anode, the carbon steel drew an initial current density of ~800 mA/m^2 and was taking a steady-state current of 25 mA/m^2 after 120 days. A stainless steel cathode, on the other hand, took ~300 mA/m^2 initially, and the steady-state current had only dropped to 50 mA/m^2 after a similar period. Apart from reinforcing the maxim that calcareous deposits formed at higher current densities are more protective, this indicates that there are differences between the initial polarisation characteristics of carbon steel and that of the passive filmed stainless steel. The magnitude of the difference might not be of practical concern in near-surface waters. However, if the differences between carbon and stainless steel become accentuated as the depth increases, then this would have to be accommodated in the CP design.

In that respect, it may be noted that the availability of deep-water polarisation data for stainless steels is even more circumscribed than for carbon steel. In a deep-water testing programme, Baptista and da Costa [14] also polarised some panels of type UNS S31603 stainless steel. Their results indicate that, for some reason, a higher initial current density was required in the case of stainless steel.

Fischer, Thomason and Eliassen [13] polarised two stainless steel (UNS S31254) specimens in the Barents Sea. The limited number of tests precludes detailed analysis. However, at 100 m depth and 300 mA/m^2 initial (design) current density, the mean and final currents on the stainless steel specimen were lower than on carbon steel. Although this appears to run counter to the observations of Baptista and da Costa, it reflects, in all probability, the fact that different results can be expected from site testing in separate geographical locations.

8.8 SUMMARY

Marine and offshore engineering make use of very many CRAs. Whilst all of these alloys resist corrosion, to a greater or lesser degree, non is immune in seawater. Furthermore, the corrosion of CRAs is more likely to be localised, in the form of pitting or crevice attack, than general corrosion of the type exhibited by carbon steel.

All CRAs can be protected by the application of CP. Whilst the design protection potentials might be less negative than the −800 mV employed for carbon steel, this often has little relevance. Unless the CRA is isolated from carbon steel, a CP system will need to be designed to polarise the structure to −800 mV, or more negative. The issue that might then arises for some CRAs is to avoid potentials that are too negative.

To a good approximation, the design current densities to protect CRAs will also be the same as those required for carbon steel.

REFERENCES

1. F.M. Reinhart *Corrosion of Materials in Hydrospace Part II Nickel and Nickel Alloys.* Naval Civil Engineering Laboratory Technical Note N-915, 1967.
2. G. Wranglén, *An Introduction to Corrosion and Protection of Metals.* Inst. För Metallskydd (Stockholm) 1972.
3. J.W. Oldfield, *Corrosion initiation and propagation of nickel base alloys in severe sea water applications.* Paper no. 266 CORROSION/95.
4. R. Johnsen and E. Bardal, *The Effect of Microbiological Slime Layer on Stainless Steel in Natural Seawater,* Paper no. 227 CORROSION/86.
5. E. Bardal and R. Johnsen, *Cathodic Properties of stainless Steels in Sea Water and Some Practical Consequences of These Properties* Proc UK Corrosion, 1986.
6. T. Foster and J.G. Moores, *Cathodic Protection Current Demand of Various Alloys in Seawater,* Paper no. 295 CORROSION/86.
7. J.W. Oldfield, *Test techniques for pitting and crevice corrosion resistance of stainless steels and nickel-base alloys in chloride containing environments.* NIDI Technical series No. 10 016, 1987.
8. R.E. Lye, *Current drain to cathodically protected stainless steels in seawater.* Materials Performance **27** (10), 24 (1988).
9. R. Holthe, E. Bardal and P.O. Gartland, *The time dependence of cathodic materials in seawater stainless steels, titanium, platinum and 90/10 Cu/Ni.* Materials Performance **28** (6), 16 (1989).
10. EFC Publication No. 5 Working Party Report – *Illustrated Case Histories of Marine Corrosion* (1990).
11. G.T. Burstein, in *Corrosion,* 3rd edition, vol. 1, Eds. L.L. Shreir, R.A. Jarman, and G.T. Burstein. Butterworth-Heinemann (Oxford) 1994, p. 1:118.
12. H.-J. Zhang and S.C. Dexter, *Effect of biofilms on crevice corrosion of stainless steel in coastal seawater.* Corrosion **51** (1), 56 (1995).
13. K.P. Fischer, W.H. Thomason and S. Eliassen, *CP in Deep Water: The Importance of Calcareous Deposits and the Environmental Conditions,* Paper no. 548 CORROSION/96.
14. W. Baptista and J.C.M. da Costa, *In-situ acquisition of cathodic protection parameters.* Materials Performance **36** (1), 9 (1997)
15. S.C. Dexter and J.P. LaFontaine, *Effect of natural marine biofilms on galvanic corrosion.* Corrosion **54** (11), 851 (1998).
16. H. Amaya and H. Miyuki, *Laboratory Reproduction of Potential Ennoblement of Stainless Steels in Natural Seawater* Paper no. 168 CORROSION/99.
17. K. Ito, et al., *Potential Ennoblement of Stainless Steels by Marine Biofilm and Microbial Consortia Analysis* Paper no. 02452 CORROSION/2002.
18. F.L. Martin, et al., *Crevice corrosion of alloy 625 in natural seawater.* Corrosion **59** (6), 498 (2003).
19. *The corrosion performance of metals in the marine environment: a basic guide* ed. C. Powell and R. Francis EFC/NACE Publication No. 63. Maney Publishing (London) 2012.
20. C. Tuck, Chapter 6. *The Corrosion Performance of Metals in the Marine Environment: A Basic Guide,* Eds. C. Powell and R. Francis, EFC/NACE Publication No. 63. Maney Publishing (London) 2012.
21. D. Thierry, N. Larché and G. Rannou, *Monitoring of Seawater Biofilms on stainless steel for corrosion risk assessment.* Paper no. 1329 Eurocorr, 2012.
22. D. Thierry, N. Larché and C. Leballeur, *Corrosion Potential and Cathodic Reduction Efficiency of Stainless Steel in Natural Seawater.* Paper no. 1320 Eurocorr 2013.
23. DNV-RP-F112 *Design of Duplex Stainless Steel Subsea Equipment Exposed to Cathodic Protection* 2008; re-issued as *Duplex stainless steel – design against hydrogen induced stress cracking,* 2019.
24. ASTM G48 *Standard Test Methods for Pitting and Crevice Corrosion Resistance of Stainless Steels and Related Alloys by Use of Ferric Chloride Solution,* 2020.

Chapter 9

Underwater coatings

9.1 INTRODUCTION

In considering underwater coatings in this chapter, we are tackling an enormous subject. Its scale can be appreciated when we consider that, because of the variety of ingredients available to the formulators of protective coating systems, there is effectively an infinite number of possible anticorrosion coatings. The best we can do here is attempt to penetrate some of the coatings jargon and to gain an insight into how coatings protect the surface, and how coating systems are selected.

Broadly, there are three classes of material that can be used as a protective coating for steel: metallic, inorganic and organic. In some instances, steelwork subsea is provided with a metallic coating, usually zinc (in the form of galvanising) or aluminium (usually applied by thermal spray). However, this chapter is concerned only with organic or polymeric coatings.

9.2 SOME POLYMER BASICS

9.2.1 Polymerisation

Since most CP engineers were off school on the day of the polymer science lesson, it is worth running through some basics. This will help us understand some of the characteristics of organic coatings. The word polymer is a combination of the Greek words *poly* (meaning *many*) and *mer* (meaning *part*). A well-known example of polymerisation is the application of heat and pressure to ethylene gas[1] to produce polyethylene. This is described in chemical terms in Figure 9.1. Whereas the low-molecular weight ethylene is a flammable gas, its polymer is a relatively stable solid which finds very wide use in the protective coating industry.

9.2.2 Linear polymers

9.2.2.1 Polymers and copolymers

Ethylene is termed bifunctional, which is chemistry-speak for saying that each ethylene monomer molecule can only combine with two other ethylene molecules. This means that when we polymerise ethylene, we can only grow the polymer chain linearly. It cannot branch.

Linear polymers can be formed from a single monomer, as is the case with polyethylene; or they can be formed from two or more suitable bifunctional monomers. This results in the formation of copolymers as indicated schematically in Figure 9.2. For example, monomer A

[1] We should really use "*ethene*", which is the official IUPAC name for "*ethylene*".

DOI: 10.1201/9781003216070-9

might be ethylene and monomer B might be propylene. The resulting copolymer would be expected to be intermediate in character between polyethylene and its more rigid homologue polypropylene.

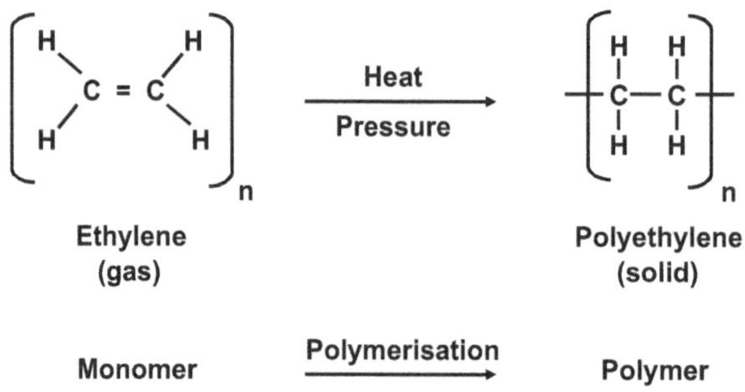

Figure 9.1 Example of polymerisation.

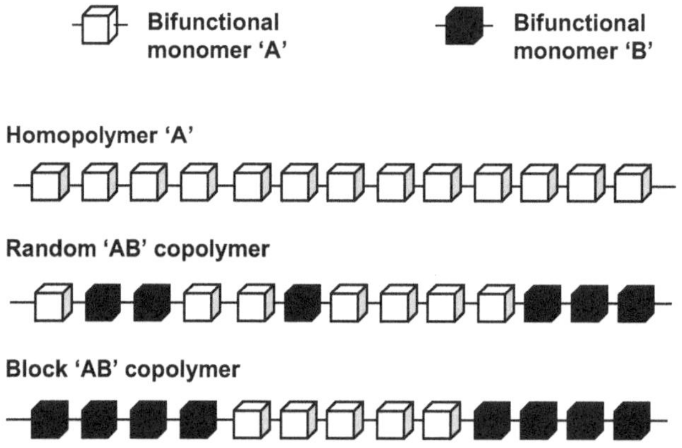

Figure 9.2 Linear copolymers.

9.2.2.2 *Flexibility*

The structure of a linear polymer is based on a chain of carbon atoms bonded together by single carbon-to-carbon chemical bonds. Two types of movement are possible for this type of bond. It can rotate or it can vibrate, as is shown schematically in Figure 9.3. Furthermore, the tendency to vibrate and rotate at each bond increases as the temperature increases. If the temperature is high enough, the effect of this movement at each carbon-carbon bond is to reduce the viscosity of the polymer. In other words, it melts.

This provides the practical advantage that linear polymers such as polyethylene can be shaped at elevated temperatures. This has important implications for their use as coatings. They can be applied as deformable solids and, when cooled, will form a solid film. This characteristic of being deformable at elevated temperature leads to these linear polymers being termed thermoplastics, from the Greek words for "heat" (*thermos*) and "of moulding" (*plastikós*).

Some linear (thermoplastic) polymers used in protective coatings are polyethylene (PE), polypropylene (PP), polyvinyl chloride (PVC), polyvinylidene fluoride (PVDF) and polytetrafluoroethylene (PTFE). The first two of these are very widely used for the coating of pipelines both subsea and onshore.

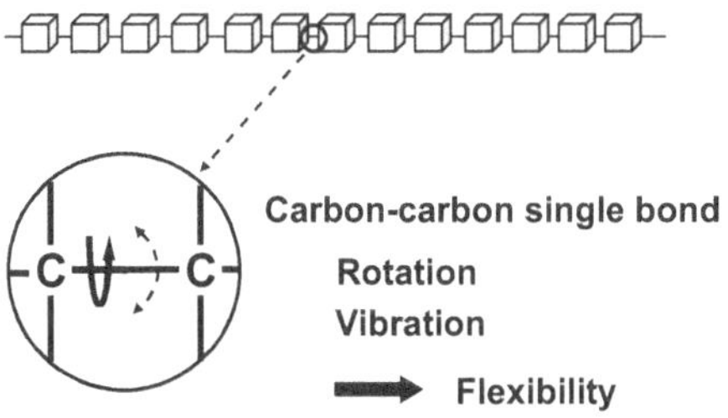

Figure 9.3 Carbon-carbon single bond.

9.2.3 3-Dimensional polymers

9.2.3.1 Cross-linking

Instead of polymerising a bifunctional monomer, we can polymerise a mixture of monomers which includes a tri-functional monomer. This produces a 3-dimensional structure as indicated schematically in Figure 9.4. Wherever a tri-functional monomer appears in the polymer chain, a new branch of the chain is developed.

This chain branching, which is also referred to as cross-linking, profoundly alters the nature of the polymer. In particular, it is no longer possible for there to be any meaningful rotation of the carbon-carbon bond because this would involve not just rotating a linear chain but disrupting the entire 3-D structure of the polymer.

Polymers which form in this 3-dimensional manner are termed thermosets to distinguish them from thermoplastics. As a general rule, a protective coating based on a thermoset material will be tougher and less permeable to water and dissolved oxygen than a thermoplastic material. However, these advantages may be offset by the difficulties arising from having to carry out the polymerisation (cross-linking) process in-situ on the surface of the structure rather than in the controlled environment of the polymer manufacturing facility.

There are numerous thermoset polymers used in protective coatings. Examples include epoxies, polyurethanes, polyesters, phenolic resins and furan resins. Since this is a book about CP, not coating technology, we will not be exploring all of these. Instead, we will focus on epoxy coatings since these are very widely used in subsea and in industry generally.

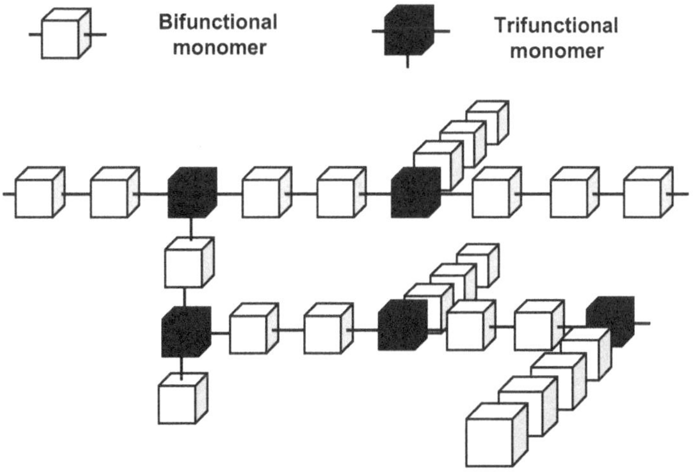

Figure 9.4 Cross-linked polymer.

9.2.3.2 *Epoxies*

At the risk of condensing many millions of man-years of research into a few words, we can say that the desirable features of a protective coating are that it should possess a good balance between its cohesive and adhesive strength. It must also be chemically stable. In other words, the polymer which forms the backbone of the coating must have enough internal strength that it does not fail when subject to sliding or impacting forces, it must stick firmly to the substrate, and it must not degrade when exposed to the environment.

The thermoplastic materials we discussed above possess adequate chemical stability but only modest cohesive strength. However, because they are relatively inexpensive, they can be applied at substantial thicknesses, thereby mitigating this disadvantage. Unfortunately, they also have the general disadvantage of exhibiting poor adhesion to steel. This is a more difficult shortcoming to work around.

Epoxies, on the other hand, offer the advantage of high cohesive strength, meaning that protection can be afforded with thinner films. More importantly, they possess very good adhesive strength; a fact which is unsurprising since they were originally developed as adhesives in the aircraft industry. To see why, as a class of polymers, epoxies exhibit both adhesive and cohesive strength, we need to take a brief look into their chemistry.

9.2.3.2.1 The resin

The first component to consider is the resin. This is sometimes referred to as the "base" in protective coating technology. Epoxy resins, of which there are many variations, are small molecular weight polymer chains with an epoxy (also called "epoxide") functional group at each end. Most epoxy resins are manufactured by reacting epichlorohydrin with bisphenol-A as shown in Figure 9.5.

You may not have studied much organic chemistry, or if you did, it may have been a long time ago. No matter. What we need to grasp here is that the result of this reaction is that the molecules combine by forming ether linkages (-CH_2-O-CH_2-) in which one oxygen atom is attached to two carbon atoms. This ether linkage is chemically very stable. The resulting resin molecule has a reactive epoxy group at each end.

This resin, which is a liquid, can further polymerise through the reaction of the -OH groups on the resin with the epoxy groups as indicated in Figure 9.6.

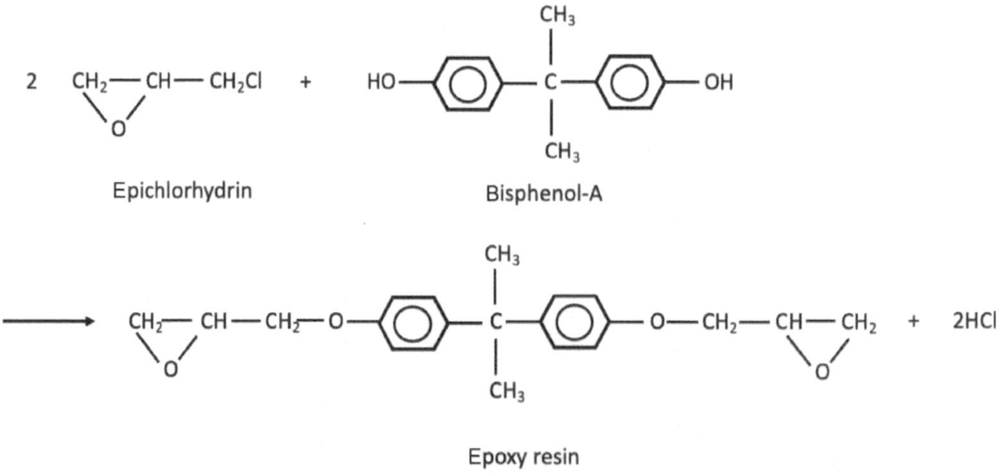

Figure 9.5 Epoxy resin formation.

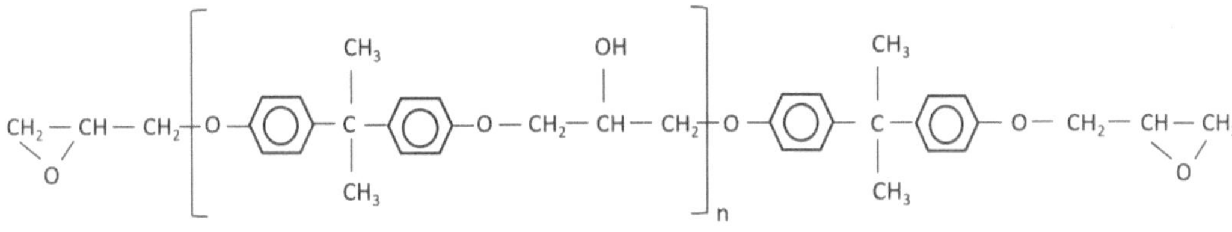

Figure 9.6 Epoxy resin – further polymerisation.

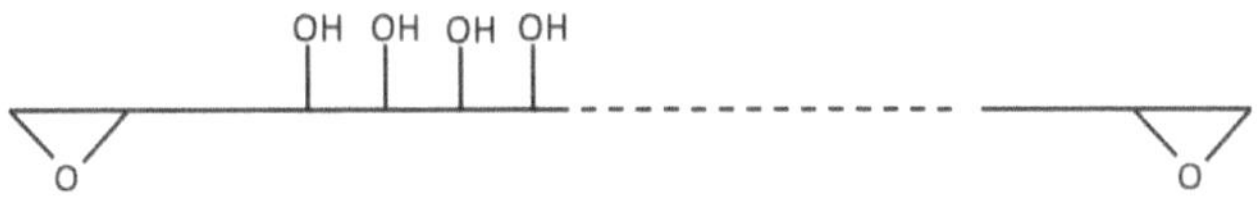

Figure 9.7 Epoxy resin – skeletal representation.

The parts of the formula in the square bracket will be repeated n times, where the value of "n" will typically be less than 20. For our purposes, it is convenient to simplify the above formula to a skeletal representation (Figure 9.7).

The important features of this resin molecule are:

- both the ether linkages and the phenyl groups are chemically stable,
- the hydroxyl groups (-OH) add polarity to the molecule,
- the epoxy groups provide reactive sites at each end of the molecule.

The polarity conferred by the -OH groups along the chain is important because this enhances the ability of the molecule to stick electrostatically to steel. Once more, I will take the risk of condensing the entire oeuvre of research on the surface science of adhesion to a few words. In a nutshell: polar substances, comprising molecules where there are some ionisable functional groups, are more likely to stick, through electrostatic adhesion, to a polar surface such as a metal (or more correctly the oxide that will be present on that metal). PTFE, which we have encountered above, is a completely covalent polymer, and, as such, it has very low adhesion. Thus, PTFE is an excellent material for a non-stick frying pan (once you have solved the problem of sticking it to the pan itself).

The presence of the epoxy groups is important because it provides the sites for chemical cross-linking (or "curing") of the resin.

9.2.3.2.2 The curing agent

In the parlance of coating technology, the curing agent which brings about the chemical cross-linking of the polymer is also known as the "hardener". Early epoxies were cured with organic amines as indicated in Figure 9.8. In the formulae, the symbol "R" indicates a small hydrocarbon chain (referred to by chemists as an alkyl group).

As each epoxy resin molecule will contain two epoxy groups, the polymerisation leads to a three-dimensional structure.

Subsequent developments have moved away from amine curing because amines are both toxic and carcinogenic. Most commercial epoxies are now cured by either less toxic amine adducts or polyamide resins. The details need not concern us here. All that we need to be aware of is that the end result is a chemically stable cross-linked polymer that is resistant to many chemicals (including water) and sticks well to the steel.

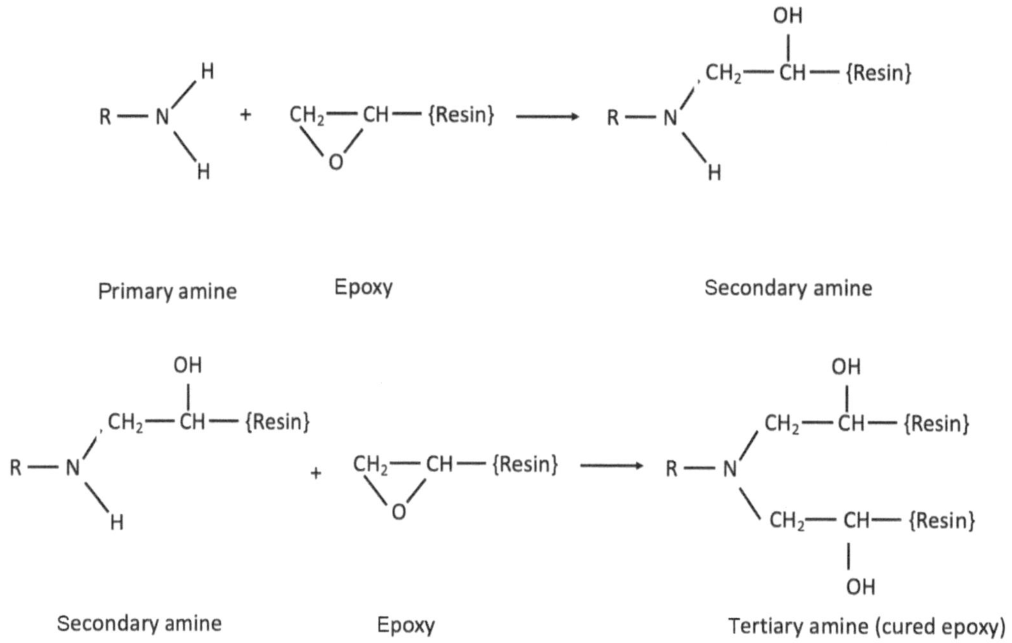

Figure 9.8 Epoxy curing – chemistry.

9.2.3.2.3 Epoxy paints

Due to the variations possible in the base epoxy resin itself, combined with the multitude of curing agents, there is a vast array of epoxy paints available. If we combine this variety with the permutations of other paint ingredients: solvents, diluents, pigments, fillers, wetting agents, levellers, stabilising agents and so on, then the number of possible paint products becomes essentially infinite. We should never make the mistake of assuming that all epoxy paints are the same or even similar.

9.2.3.2.4 Fusion bonded epoxy (FBE)

A liquid epoxy paint will come in two parts: resin (or base) and curing agent (or hardener). The applicator has to mix the two parts thoroughly. They must then apply the mixed paint to the correctly prepared surface before the curing reaction has proceeded too far, and the paint starts to gel and becomes too viscous for application.

The curing of the epoxy paint does not proceed linearly with time. This is illustrated schematically in Figure 9.9. The "β-stage" cure is reached when about 30% of the epoxy groups have reacted. At this point, the film will be reasonably tough but will still be well short of the properties anticipated for the fully cured condition. After this stage is reached, the rate of curing slows down. This is because the resin molecules become less mobile as they cross-link and grow in size. Chemists refer to this effect as steric hinderance.

In the manufacture of FBE powder, the curing reaction is stopped when the β-stage is reached. This is done by rapidly cooling the mixture, which is then dried and powdered. The product remains stable enough to be stored until it is required for application. It can then be sprayed, usually electrostatically onto an electrically charged heated surface such as line pipe. The electrostatic charging causes the powder particles to stick on the surface, and the elevated temperature (typically ~175°C) causes the part-cured resin to remelt and flow across the surface. The elevated temperature also promotes rapid completion of the curing process. FBE is widely used in the pipeline coating industry, either as a coating in its own right or as a primer for a 3-layer polyolefin coating.

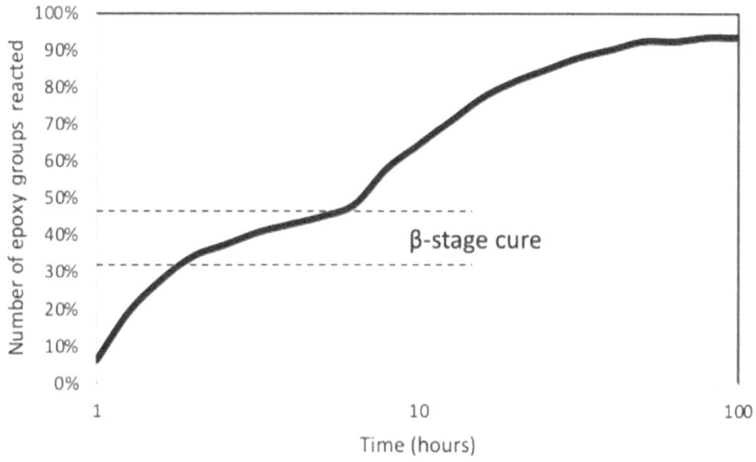

Figure 9.9 Epoxy curing – progress.

9.2.4 Elastomers

The elastomers are synthetic rubbers. Their properties may be thought of as intermediate between the pliable thermoplastics and the rigid thermosets. Certain types of linear polymer naturally tend to form into a spiral-type shape. The force responsible for this shaping is known as hydrogen bonding. This form of bonding is weaker than the carbon-to-carbon bonds responsible for the overall polymer structure. Hydrogen bonds stretch or contract in response to applied physical compressive or tensile stresses, giving the spiral coils a "spring" like character. This principle is illustrated schematically in Figure 9.10.

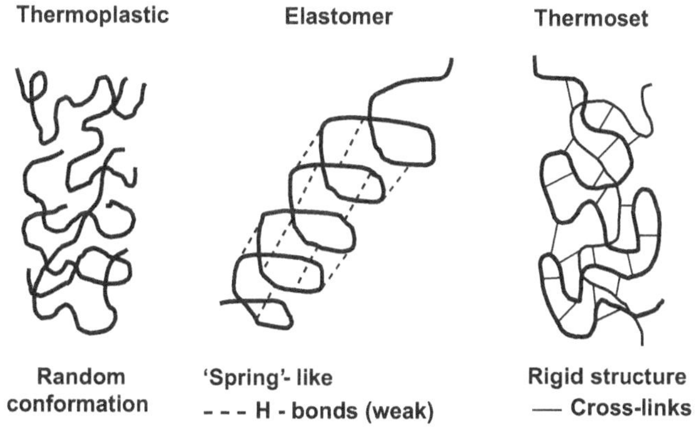

Figure 9.10 Polymer structures: thermoplastic, elastomeric and thermoset.

In addition to natural rubber, commercial elastomers include synthetic materials: polyisochloroprene, styrene-butadiene rubber, butyl rubber and nitrile rubber. Of these, polyisochloroprene, which was developed by DuPont in the late 1930s under the name *neoprene*, is used as a coating system where very high protective performance is required. For example, neoprene is used to clad hot oil production risers transiting through the particularly aggressive splash zone on offshore installations.

9.3 COATINGS AND CP

9.3.1 Do coatings benefit CP?

Yes.

As we saw in Chapter 3, the design of a marine CP system begins with an estimate of the current demand. This is obtained by multiplying the surface area by the cathodic current densities (Chapter 7). Coatings reduce the area of exposed steelwork, and thereby the overall current demand. This point is illustrated in Table 9.1, which presents an idealised characterisation of the effect of the coating in terms of the electrical resistance it interposes between the steel and the environment. This resistance is translated into a percentage of the pipeline area that remains exposed to the seawater. As can be seen, applying a "good" quality coating has a very dramatic effect in reducing the notional cathodic current demand.

Table 9.1 Benefits of a coating in reducing current demand

Coating resistance (Ωm²)	*Coating quality*	*Bare steel area %*	*Cathodic current density (mA/m²)*
0	Bare steel	100	100
10^3	Poor	3	3
10^6	Good	0.003	0.003
10^9	Excellent	(not worth measuring)	

Organic coatings degrade and lose their protective capacity over time, causing the cathodic current demand gradually to increase. This leaves the CP designer with the problem of predicting the future rate of coating degradation and its associated increase in current demand. In Chapter 3, we also saw how we could do this by using the coating prediction guidelines given in the codes. We return to this point in Section 9.7.

9.3.2 Does CP benefit coatings?

No.

Although coatings universally benefit CP, the converse does not hold. CP never improves the performance of a coating. It is always damaging, at least to some extent. We return to this point in Section 9.6 after we have introduced some coating systems for submerged steel.

9.3.3 Coating systems

Protective coatings do not last indefinitely. They are subject to various chemical and physical degradation mechanisms which, over time, lead to a diminishing of their protective capacity. Since this is a CP book, we will make no attempt to review the very substantial body of literature on protective coatings and their degradation mechanisms.

Nevertheless, we do need briefly to consider some of the coating systems used to protect steel, and some other alloys, exposed to seawater or seabed sediments. However, before we get into some of the detail of coating systems, we must say a few words about preparing the steel surface prior to applying a protective coating.

9.4 SURFACE PREPARATION

Most alleged coating "failures"[2] have arisen as a consequence of inadequate preparation of the steel surface. Deficiencies in the application of the coating are the second biggest cause, and problems with the coating material itself come in a distant third.

Although early paints, such as those based on natural oils, exhibited modest performance characteristics, they had the advantage that their protective capabilities were fairly insensitive to the condition of the steel surface onto which they were applied. Up to the middle of the 20th century, (partial) rust removal by wire brushing was generally regarded as sufficient preparation. The subsequent introduction of coatings based on synthetic polymers, in particular the two-component materials based on epoxy and polyurethane chemistry, required higher standards of surface preparation. Abrasive blast cleaning is now the normal mode of surface preparation for coatings to be exposed to seawater.

At the risk of once again reducing some millions of person-hours of paint research and development into a single paragraph, the situation might be summarised as follows. A protective polymeric coating has to achieve a balance between its internal cohesive forces and the adhesive force sticking it to the substrate. The development of tougher coatings meant the application of polymers with greater cohesive strength. Such materials can only function effectively if the adhesive forces sticking them to the steel surface are adequate. A tough, highly durable paint that peels off the surface is a disappointment.

That there was a relationship between the quality of the surface preparation and the subsequent performance of the coating was quickly appreciated. From the point of view of the painting industry, however, the problem was to quantify the quality of surface preparation in such a way that it could be included in coating specifications, and measured by inspectors.

Over the last four decades of the 20th century, various national and industry codes dealing with surface preparation were consolidated into a suite of ISO standards. These have all sought to quantify the rather abstract quality of surface preparation. There are now ISO standards with the primary title *Preparation of steel substrates before application of paints and related products.*

- ISO 8501 Parts 1–4 deal with rust removal,
- ISO 8502 Parts 1–15 deal with the measurement of surface cleanliness,
- ISO 8503 Parts 1–5 deal with the measurement of surface roughness.

These standards are now relied upon by the coating manufacturers in their product data sheets and are enshrined in project coating specifications, or equipment manufacturers' quality assurance procedures and inspection and test plans. Of course, this does not mean that the standard of coating preparation is universally perfect. However, it does mean that, with the occasional exception, it is at least fit-for-purpose.

9.5 COATING SYSTEM SELECTION

9.5.1 Fixed steel structures

9.5.1.1 Early paints

The need to apply protective coatings to iron or steel for seawater immersion service arose with the introduction of the first iron-clad and iron-hulled ships in the middle of the 19th

[2] By "failure", we are referring to a failure to meet a realistic durability expectation, not the natural ageing process of a polymeric coating.

century. Most marine paints employed up to the middle of the twentieth century were modified oil-based systems derived from natural ingredients such as linseed oil. When applied to the steel surface, these oil-based paints polymerised by reacting with oxygen in the air to form ester bonds. This bond formation resulted in the conversion from a liquid paint to a solid film.

However, ester bonds are susceptible to hydrolysis and bond breakage, particularly in an alkaline environment generated by CP. This bond breakage process is known as saponification. The chemical process is essentially the same as the traditional way of manufacturing soaps by reacting natural oils and fats with alkali.

With the benefit of chemistry hindsight, therefore, it was inevitable that these traditional paint materials would turn out to be wholly unsuitable for ships' hulls to which CP was applied. The world of coatings has moved on. These paint systems are no longer used on ships' hulls. So, we do not consider them further here.

After the Second World War, the development of synthetic organic polymers led to new classes of more durable and higher performance paints based on synthetic polymers of the type we have discussed above. These materials were more expensive than the earlier traditional products. So, it was traditional to blend them with less expensive, but beneficial, components. Thus, through the second half of the 20th century, most commercial and military ships' hulls were coated with proprietary formulations comprising vinyl or chlorinated rubber synthetic polymers blended with coal tar. A requirement for higher coating durability was invariably satisfied with formulations based on epoxy resins blended with coal tar pitch. These were commonly, albeit incorrectly, referred to as "coal tar epoxies" or "CTE"[3] rather than the more accurate "pitch epoxies".

Latterly, the use of coal tar products in underwater paints has been phased out due mainly to their carcinogenic nature. It is not thought that the cured paints are damaging to the environment. However, they pose a threat to the paint applicators, particularly during spraying applications where the carcinogens are atomised into the air.

9.5.1.2 Current systems

9.5.1.2.1 Operators' experience

Table 9.2 shows the generic underwater coating systems specified by four major offshore petroleum producers. All of these companies require the actual coating system to be compatible with both seawater immersion and with CP.

9.5.1.2.2 Norsok

System No. 7, as described in Norsok M501 [17], is often selected for coating immersed steelwork offshore. M501 only gives a generic description: two-coat epoxy to 350 µm MDFT. Rather than suggesting composition details, it sets the performance standard for proprietary coatings based on ISO 20340 [23]. This in turn sets minimum performance requirements in 6-month seawater immersion tests, and in seawater cathodic disbondment tests according to ISO 15711 [12] (see Section 9.6). All major coating manufacturers offer coating systems that have been independently tested for compliance with Norsok M-501.

[3] The abbreviation "CTE" causes confusion since it was used for two entirely different, but common, protective coating systems. It has been used for both coal tar (i.e. pitch) epoxy paints and hot-applied "coal tar enamel" pipeline coatings. Although both coatings have now fallen out of fashion for environmental reasons, there are still remain plenty of pipelines and structures with these coating systems. Many of these are presently candidates for CP life-extension exercises.

Table 9.2 Generic coating systems for fixed offshore structures (2016)

Coating system	*Norsok M-501 (note 1)*	*Offshore petroleum producer*			
		A	*B*	*C*	*D*
Surface preparation (note 2)	Sa 2½	Sa 2½	Sa 2½	Sa 2½	Sa 3
Primer	None	(note 3)	Zinc rich	None	None
1st build coat	Epoxy 175 μm MDFT	Glass flake epoxy or polyester 500 μm (NDFT)	Epoxy 250 μm (NDFT)	Glass flake epoxy 200 μm (NDFT)	Glass flake vinyl ester or polyester 600–1000 μm
2nd build coat	Epoxy 175 μm MDFT	None	Epoxy 250 μm (NDFT)	Glass flake epoxy 200 μm (NDFT)	(note 4)

Note

1 See Table A7 in NORSOK M-501.
2 See Section 9.5.1.2.2.
3 Operator A permits the use of manufacturer's holding primer (50 μm).
4 Operator D permits the full film thickness to be built up in either one or two coats.
MDFT: Minimum dry film thickness.
NDFT: Nominal dry film thickness.

9.5.2 Ships and floating installations

9.5.2.1 External hulls

The painting of a ship's external immersed hull steelwork has a twin objective. Corrosion protection is an important factor; but the performance of the antifouling coating, applied on top of the anticorrosion system, is also a vital consideration.

Over the years, ships' anticorrosion coating systems have improved such that nowadays a two-coat epoxy system, generically similar to those used on some fixed installations, is typical. The ship also has the advantage that it will be dry-docked several times during its life, during which time the opportunity can be taken to repair or replace the hull anticorrosion coating.

In principle, the same holds for floating installations employed offshore such as production storage and off-loading vessels (FPSOs), or semi-submersibles. However, irrespective of any originally intended dry-docking schedule, most floating oilfield installations remain on station for much longer periods than the 5-year dry-docking interval that is typical for a trading ship. It follows that CP designers would be prudent to assume a greater degree of coating breakdown for these installations.

9.5.2.2 Ballast spaces

Unlike the external hull, it would be very unusual for a ship's ballast tanks to be recoated during the lifetime of the vessel. The same applies to ballast tanks on floating installations.

Ships' ballast tanks are generally coated to a markedly lower standard than the external hull. Indeed, it has not been unknown for ship's ballast tanks to enter service coated in nothing better than the shipyard's holding primer[4]. More recently, however, the shipping industry has improved its act. Ballast tank coatings are now covered by IMO rules [15]

4 The ballast tank plating showing MIC damage in Chapter 1 is an example of such an uncoated ballast tank.

which require "hard coatings". This reflects the modern shipbuilding practice of lining ballast tanks with at least two coats of epoxy paint. It also implicitly recognises that CP cannot be fully effective in ballast tanks which spend up to 50% of their service lives empty.

9.5.3 Submarine pipelines

9.5.3.1 Factory applied coating systems

Up until the 1970s, pipeline coatings were predominantly based either on coal tar bitumen enamel or petroleum asphalt enamel. These relatively inexpensive materials were obtained as by-products of the conversion of coal into coke (bitumen), or from the distillation of crude oil (asphalt). The word "enamel" in this sense is used to indicate that the materials were applied as liquids at elevated temperature, and then cooled to form solid coatings.

Neither coating material offered attractive adhesive or cohesive properties. The shortfall in the natural adhesion to steel was overcome by applying the enamel onto an adhesive primer. Reinforcing the enamel with a glass fibre cloth embedded during application compensated for the inherent lack of cohesive strength.

Although relatively unsophisticated, these enamel-coating systems proved to be very durable, particularly when overcoated with concrete. They continued to be used on major pipeline offshore projects through the 1980s. Thousands of kilometres of enamel-coated pipelines remain in offshore service.

The 1980s saw the development of more durable pipeline coatings. Polyethylene (PE), which could be applied by either sintering or extrusion, showed promise. It possesses adequate cohesive strength and was sufficiently inexpensive to permit application at high film thicknesses (typically 3 mm). However, PE is non-polar in nature, which means that it has a very limited ability to stick to steel, or anything else. This proved a problem with early PE systems which were found to be susceptible to delaminating from the surface.

Elsewhere, development effort focussed on the fusion bonded epoxy (FBE) materials. In these systems, partly cured epoxy resin powder is sprayed, usually electrostatically, onto the heated pipe surface. The elevated temperature of the pipe causes the epoxy powder to melt (fuse) and flow out over the surface. At the same time, the thermal input restarts the curing process. The result is a protective coating exhibiting both high adhesion (due to the polar nature of the epoxy) and high cohesive strength (due to the cross-linking of the polymer chains). Unfortunately, epoxy resins are expensive. The high cost is largely responsible for limiting the thickness of FBE coatings to about 10% of the thickness of PE. They are covered by ISO 21809-2 [19].

This lack of thickness of epoxy coatings means that, despite their inherent strength, they are vulnerable to mechanical damage in offshore service. This damage can arise in the coating yard if a concrete weight coat is being applied, during transport and installation, and on the seabed.

Perhaps inevitably, the benefits of both coating technologies were incorporated into the 3-layer polyethylene (3LPE) coating system (Figure 9.11). 3LPE systems comprise a layer of FBE (typically ~50 µm), a layer of copolymer adhesive and a layer of extruded polyethylene (typically ~3 mm). Viewed simplistically, but not unrealistically, the FBE provides corrosion protection and resistance to cathodic disbondment (see Section 9.6), and the PE provides mechanical protection to the FBE.

3LPE exhibits good performance up to the temperature at which the PE polymer starts to soften (~70°C). Slightly higher temperatures, up to (~100°C), can be accommodated by polypropylene (PP). Nowadays, 3LPE and 3LPP coating systems dominate the offshore pipe coating market. Their application is covered by DNV-RP-F106 and ISO 21809-1 [24].

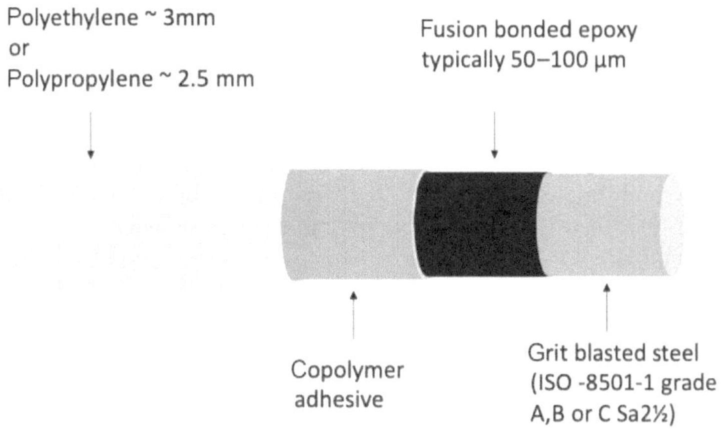

Figure 9.11 Three-layer polyolefin pipeline coating.

9.5.3.2 Field joint coatings

9.5.3.2.1 Do we need field joint coatings?

Probably Not.

In the mid-1980s, I presented a conference paper [6] that advanced the proposition that, since the CP design codes in force at the time were so pessimistic about the performance of field joint coatings, there was actually no point applying them in the first place. If they were omitted, the resulting additional cathodic current demand could easily be accommodated by a slight uplift in the anode provision. Moreover, there were positive advantages in the anticipated performance of the anodes if they were made to deliver a sizeable current from day one. Needless to say, this idea did not receive whole-hearted support from the purveyors of field joint coatings. More recently, the proposition has also been put forward by Surekein et al. [18]. I suspect that an opinion emanating from a paper jointly authored by ExxonMobil, Total and Statoil[5] carries much more weight than my own. Nevertheless, I would not expect to see pipelines being laid without field joint coatings. There is a great rush to be second when it comes to changing pipeline coating practice.

9.5.3.2.2 Systems

DNV-RP-F102 [21] provides a useful summary of pipeline field joint coating (FJC) systems. This includes comprehensive data sheets for the various options. These are referenced in the corresponding ISO 21809-3 standard [20].

9.6 CATHODIC DISBONDMENT

9.6.1 Characteristics

The characteristics of cathodic disbonding are now reasonably well catalogued. However, despite over 30 years of industrial and academic research, we do not really know a great deal about the mechanisms. A UK Marine Technology Directorate publication [9] described the

[5] Now known as Equinor.

state of knowledge up to 1990. Increments in our mechanistic understanding since then have been interesting, but marginal. Nevertheless, we are able to make the following broad-brush statements.

All polymeric coating systems undergo cathodic disbondment, but some are much more resistant than others. The more modern coatings have higher resistance which, in turn, makes them less easy to study experimentally. For example, it is reported [10] that polyester and vinyl ester systems applied at 750 μm DFT did not exhibit any cathodic disbonding in a BS 3900 F10 test [3], albeit that test is not particularly aggressive.

The phenomenon develops progressively over time. For example, Steinsmo and Drugli [10] carried out long-term testing on offshore coating systems under typical North Sea CP service conditions (–1.05 V, 7–10°C). They found that test durations of up to ~250 days were required to rank the systems. On the other hand, no workers have produced disbonding rate data that could feed directly into a CP design.

In the potential ranges associated with offshore CP, the more negative the potential, the more rapid is the rate of disbonding [1,4,7]. The effect seems to be progressive. There is no evidence of a step-change in the disbonding rate at any particular potential.

As we discussed above, the quality of surface preparation is crucial to coating performance, and it has long been recognised that painting over contaminated surfaces will exacerbate cathodic disbondment [4,8]. Increasing the surface roughness, thereby maximising the area of contact between the steel and the coating, reduces the rate of disbonding [5,7].

Unsurprisingly, the rate of disbonding increases with temperature [9]. There is also evidence that it is more rapid where there is a thermal gradient, such as with pipelines transporting hot fluids.

When it comes to the effect of film thickness, the situation is less straightforward. The intuitive assumption that increasing coating thickness reduces rates of disbonding does not hold for all coating types. Unfortunately, much of the work has been carried out on coating systems not used in offshore construction. This illustrates one of the problems when carrying out mechanistic studies on such slow processes. Researchers have only a limited time in which to carry out their investigations. To ensure results are obtained, test conditions are made more aggressive, or the study focuses on more susceptible coatings.

For example, Coulson and Temple [5] showed a thickness dependence for fusion bonded epoxy in the range of 100–700 μm. Similar results were reported by Zhou and Jeffers [14], who observed that increasing the thickness of FBE coatings from ~100 to ~200 μm produced a slight reduction in disbonding radius in short-term (48 hour) tests at 65°C. However, both sets of workers employed a test potential of –1.5 V. This would only be relevant to ICCP systems which are not normally encountered on offshore pipelines.

9.6.2 Corrosion threats under disbonded coatings

9.6.2.1 Onshore pipelines

It is important to distinguish between onshore and offshore pipelines undergoing cathodic disbondment. There are plenty of horror stories, some of which are true, relating to cathodic disbondment on onshore pipelines. The problems stem from the attenuation of protection under the disbonded coating. There are documented examples of onshore pipeline explosions resulting from a corrosion morphology termed carbonate/bicarbonate stress corrosion cracking occurring under disbonded coatings. Elsewhere onshore, there have been reports of MIC becoming established underneath disbonded tapes and bituminous wraps.

Although this book is about submarine pipelines, we also need to bear in mind that where "offshore" pipelines make landfall they can, technically, be subject to onshore corrosion threats. (We return to this in Chapter 14.)

9.6.2.2 Submarine pipelines

9.6.2.2.1 Field experience

Happily, the above onshore problems have not been encountered on submarine pipelines. Nevertheless, we cannot simply dismiss all corrosion threats under disbonded coatings. A 2007 report, prepared for the Norwegian Safety Authority (NSA) [16], contains the following rather oblique statement... *Field joints are the weakest part of any pipeline coating system, since if these are improperly applied or unsuitable, or incompatible materials selected, the (coating) can fail and lead to pipeline corrosion. The authors have been advised in a private communication that this has already happened in the North Sea.*

Conversely, Roche, then Head of Corrosion at Total, in a presentation on problems under disbonded pipeline coatings to an EFC working party in 2004 [13], stated: *No corrosion experienced on offshore pipelines exposed to seawater due to its high conductivity.*

At first sight, these two statements, both from authoritative sources, seem entirely at odds. In my view, the NSA comment needs to be regarded with some circumspection. Since we are not given an attributable source, we are not able to determine the cause of the corrosion alluded to. It may be conjectured that inadequate CP systems might also have been implicated. On the other hand, Roche referred to submarine pipelines that are provided with adequate levels of CP. The quotable experience of a major operator cannot be ignored.

9.6.2.2.2 Experimental evidence

A pivotal piece of work was carried out at NPL in the early 1980s [2]. Turnbull and May created artificial crevices and installed micro-reference half-cells at intervals down the crevice to determine the attenuation of potential from an anode located outside the crevice. It is worth repeating part of their conclusion verbatim... *Cathodic protection can be readily achieved in crevices in C-Mn steels in 3.5% NaCl and in seawater even for aspect ratios of 1:60,000. At an external applied potential of −1.0 V, the potential in the crevice was more negative than −0.950 V, i.e. a potential drop of ≤50 mV. Therefore, no additional measures are considered necessary to achieve protection of crevices of this type.*

Given that steel remains fully protected at these potentials, this is very encouraging. These observations are supported by the more extensive subsequent work of Benoit et al., who found no corrosion under disbonded 3LPE coatings at up to 65°C [22].

9.6.2.2.3 Discussion

The most relevant argument in respect of dismissing the threat of corrosion under a disbonded coating, on a cathodically protected pipeline in seawater, is based on *reductio ad adsurdum* reasoning.

- All subsea pipelines are protected by a combination of coating plus CP.
- Many of these coatings will have defects, often at the field joints, penetrating to the steel surface. After all, this is the reason CP is provided.
- There will be cathodic disbondment, to a greater or lesser extent, of the coating emanating from these local areas of coating damage. As discussed, although different coating systems vary in their resistance to cathodic disbondment, none is immune.

- If the benefits of CP could not penetrate along the crevice so formed then most, if not all, subsea pipelines would be suffering corrosion damage.
- However, as far as we are aware, there has been no documented case of a submarine pipeline, demonstrated to have been under CP for its full life, exhibiting external corrosion damage. This includes pipelines that have been in service for many decades.

9.6.3 Cathodic disbondment testing

The established test procedures are straightforward and similar. A coated panel is damaged to produce a defect penetrating through to the steel. It is then immersed in an electrolyte, and CP is applied. The extent of coating disbonding is assessed at the end of the test. This is usually achieved by no more sophisticated technique than using a penknife to remove the disbonded coating, and then measuring how far the disbondment front has spread from the edge of the artificial defect.

ASTM G8 [25], which applies to onshore pipelines, first appeared in the literature in 1968 and as a standard in 1972. Originally, a magnesium sacrificial anode was used to polarise an immersed specimen to a potential of –1.4 to –1.5 V. In 1996, an impressed current polarisation option was also specified. The electrolyte, intended to simulate ground water, is tap water containing 1% w/w each of NaCl, Na_2SO_4 and Na_2CO_3. Variations of the standard for testing at elevated temperature (ASTM G42) and a version specifically for pipeline specimens (ASTM G80) were subsequently issued and are revised periodically. ASTM G95 is a version in which the test cell is attached to a section of coated pipeline. Although the test electrolyte is 3% NaCl, so might be deemed more relevant to seawater, the test potential (–3 V vs Cu|$CuSO_4$) shows that its focus is on onshore ICCP systems.

BS3900 Part F10 was issued in 1985. It is based on an earlier COIPM test and was drafted with ships' hulls in mind. It uses either natural or synthetic seawater. It also specifies a more modest, but realistic, test potential of –1000±5 mV, applied by a potentiostat. This test potential is considerably less severe than those previously specified for example in the ASTM documents. This is not unassociated with the fact that the drafting committee included a paint company representative who was under instructions to ensure that the test should not be too demanding on his employer's products.[6]

Although BS 3900 F10 remains current (as at 2021), it has been effectively replaced by ISO 15711 which is similar. It requires testing at 23°C±2°C for 6 months. The only slight change is that the ISO calls for testing at a slightly more negative potential of –1050±5 mV. As we saw above, approval of a coating system per Norsok M501 requires 6-months disbondment testing in accordance with ISO 15711. To qualify, an initial 6 mm diameter artificial defect must not disbond to a diameter greater than 20 mm. In other words, the disbondment front is permitted to advance at a maximum rate of just over 1 mm/month.

9.7 COATING BREAKDOWN PREDICTIONS

Although the codes do not actually specifically identify cathodic disbonding as a component of the breakdown, it is implicit that the published guideline figures must accommodate some contribution from this deterioration mechanism. Thus, from the standpoint of conducting a CP design, the phenomenon of cathodic disbondment is not something that merits separate consideration. Providing the coating system has a demonstrated resistance to disbonding and is applied according to the manufacturer's instructions, the CP designer does not normally need to consider the phenomenon.

[6] That paint company representative is now sitting here writing this book.

9.7.1 Coatings for fixed structures

As seen in Chapter 3, when we carry out a CP design for a coated steel structure, a key variable is the rate of future coating breakdown, and thus how much steel will become exposed to seawater over the design life. Such information is enshrined in recommendations and codes of practice. Predicted values are expressed as an initial, mean and final percentage coating breakdown.

You should study these codes in detail if you are preparing or reviewing a CP design. We will not reproduce the code material here. However, it is informative to review how the predictions for coating breakdown given by various codes have changed over the years. This is illustrated for structures in Table 9.3 which collates the final (30 year) coating breakdown predictions for coatings given by codes published between 1981 and 2021. For convenience, the table only shows predictions in the 0–30 m depth range. Lower rates of breakdown are predicted for deeper waters. Circumspection is needed when viewing these figures since the coating types and thicknesses were not defined in the earlier documents. Nevertheless, it is clear that the assumed rates of coating breakdown have been progressively revised downwards from 50% to 60% to much lower values over this period.

This reflects the cumulative body of experience which points to the early predictions having been pessimistic. Indeed, it is likely that the existing figures remain conservative. However, the financial impact of the degree of CP over-design that this conservatism generates is marginal in the context of an overall offshore development. Moreover, the cost of adding surplus anodes in an onshore construction facility is often a reasonable insurance against the possibility of having to carry out an offshore retrofit at some point in the future.

Table 9.3 Selected coating breakdown guidelines for offshore structures

Year	*Code*	*Final breakdown (30 years)*	*Coating description*
1981	DNV-TNA-703	50%	High build epoxy (no other details)
1986	DNV-RP-B401	60%	"High quality thin film" <1 mm DFT
1993	DNV-RP-B401	38%	Multicoat system to 450 μm DFT
1997	Norsok M-503	40%	"Thin film"
2005 & 2010 2007 & 2016	DNV-RP-B401 Norsok M-503	38%	Marine epoxy, polyurethane or vinyl; 2 or more coats to 350° μm DFT (DNV Category III, Norsok M-501 system 7)
2021	DNV-RP-B401	38%	Marine epoxy; 2 or more coats to 350° μm DFT (DNV Category III)
		26%	Marine epoxy; 2 or more coats to 350 μm DFT per Norsok system 3B or 7 (DNV Category IV)

9.7.2 Ships' coatings

As we will discuss in Chapter 15, predicting coating breakdown on underwater hulls is not an activity upon which designers of CP systems for ships expend a great deal of time. Nevertheless, for what it is worth, ISO 20313 recommends using an initial coating breakdown factor of between 1% and 2%, and assumes an annual breakdown rate of 1.5% for a medium durability coating. The latter figure reduces to between 0.5% and 1% for a high durability material. In this context, it attempts to define durability in the accordance with ISO 12944-1 [11]. This is injudicious, since ISO 12944 does not apply to ships. Furthermore, the ill-defined durability of hull coatings is largely irrelevant because ships are dry-docked at intervals, rarely exceeding 5 years, for replacement of their antifouling topcoats. Repairs to the underlying anticorrosion, if required, are carried out at that time.

Nevertheless, a reasonable interpretation of ISO 20313 is that, for the hull of a vessel trading normally, mean and final design coating breakdown figures of 5% and 10% may be regarded as conservative. The standard adds that, for ships operating at >25 knots (>12.9 m/s) or ships in ice service, the annual coating breakdown rate may be higher. Unfortunately, it does not advise how much higher.

9.7.3 Pipeline coatings

We deal with pipeline coating breakdown predictions, and how they are used in pipeline CP designs in Chapter 14. Nevertheless, before leaving this chapter on coatings, it is illuminating to reflect upon how the advice of the codes has evolved since the early 1980s. Table 9.4 lists the figures advised by various authorities since that time.

As with the recommendations for coated steel structures, we have to bear in mind that these figures are not based on any realistic long-term performance testing nor are they based on any reliable measurements of pipelines in service. They simply represent a conservative consensus of the committees drafting the codes. It is clear that, over the decades, this conservatism has been trimmed back as the durability of modern pipeline systems has become more apparent. It seems likely that even the low breakdown figures given in current codes remain conservative.

Table 9.4 Generic breakdown guidelines for offshore pipeline coatings

Year	*Authority*	*Code*	*Coating description (non-buried; no concrete weight coating)*	*Final coating breakdown (30 years)*
1981	DNV	TNA 703	"Thick film" (no further details)	20%
1986	DNV	RP B401	"Thick film" (no further details)	30%
1993	DNV	RP B401	3LPE or 3LPP	11%
1997	NORSOK	M-503	3LPP	8.8%
2001	EN	12747	3LPE or 3LPP	5%–10%
2003	DNV	RP F103	3LPP with FBE+PP heat shrink sleeve field joint coating	0.27%
2004	ISO	15589-2	3LPE or 3LPP	1.1%
2007	NORSOK	M-503	(adopts ISO 15589-2)	
2016	DNV	RP F103	3LPP with FBE+PP heat shrink sleeve field joint coating	0.27%

REFERENCES

1. H. Leidheiser Jr., W. Wang, and L. Igetoft, *The mechanism for cathodic delamination of organic coating from a metal surface.* Progress in Organic Coatings **11**, 19 (1983).
2. A. Turnbull and A.T. May, *Cathodic protection in BS 4360 50D structural steel in 3.5% NaCl and in seawater.* Materials Performance **22** (10), 34 (1983).
3. BS 3900-F10:1985, Methods of test for paints. Determination of resistance to cathodic disbonding of coatings for use in marine environments, 1985.
4. D.W. Trotman, *Cathodic protection – overprotection and coating damage on ships and marine structures,* in *Cathodic Protection – Theory and Practice,* Eds. V. Ashworth and C.J. Booker. Ellis Horwood (Chichester) 1986.

5. K.E.W. Coulson, and D.G. Temple, *Impact, cathodic disbonding line – coating tests results vary with surface coating conditions.* Oil and Gas Journal 47 (1986).
6. C. Googan, *Field Joint Coatings*, in *Seminar on Internal and External Protection of Submarine Pipelines – Design and Practice.* Publ. Global Corrosion Consultants (Telford) 1986.
7. X.H. Jin, et al., *The adhesion and disbonding of chlorinated rubber on mild steel.* Proc. Symp. *On Corrosion Protection by Organic Coatings.* Electrochemical Society Publication 87-2, 1987.
8. J.H. Morgan, *Cathodic Protection.* NACE International (Houston) 1989.
9. Chapter 4 *Organic Coatings and Cathodic Protection* in MTD Publication 90/102 *Design and operational guidance on cathodic protection of offshore structures, subsea installations and pipelines* (ISBN 1870553-04-7) 1990.
10. U. Steinsmo and J.M. Drugli, *Assessment of coating quality in CP/coating systems.* Paper 561 CORROSION/96.
11. ISO 12944-1 *Paints and varnishes. Corrosion protection of steel structures by protective coating systems. Part 1 General Introduction*, 1998.
12. ISO 15711 *Paints and varnishes — Determination of resistance to cathodic disbonding of coatings exposed to sea water*, 2004.
13. M. Roche, *The Problem of Coatings and Corrosion with Buried Pipelines Cathodically Disbonded.* Presentation to EFC WP 10 meeting Nice, 2004.
14. W. Zhou, and T.E. Jeffers, *Application temperature, cure and film thickness affect cathodic disbonding of FBE coatings.* Materials Performance **45** (6), 24 (2006).
15. IMO Regulations Annex 1 Resolution MSC.215 (82) Chapter 11-1 Part A1 *Protective coatings of dedicated seawater ballast tanks in all types of ships and double-side skin spaces of bulk carriers*, 2006.
16. Petroleum Safety Authority (Norway) Report 17562/1/07 *International Experiences with cathodic protection of offshore pipelines and flowlines*, 2007.
17. Norsok M501 *Surface Preparation and Protective Coating*, 2012.
18. M. Surekein, et al., *Offshore Pipeline Coating for Field Joints – Why?* Paper 2258 CORROSION/2013.
19. ISO 21809-2 *Petroleum and natural gas pipelines, External coatings for buried or submerged pipelines used in pipeline transportation systems – Single layer fusion bonded epoxy coating*, 2014.
20. ISO 21809-3 *Petroleum and natural gas pipelines, External coatings for buried or submerged pipelines used in pipeline transportation systems Field joint coatings*, 2016.
21. DNV-RP-F102 *Pipeline field joint coating and field repair of linepipe coating*, 2017.
22. G. Benoit, et al., *Bare Field Joints for Subsea pipelines, a Possible Alternative?* Paper 10841 CORROSION/2018.
23. ISO 20340 *Paints and varnishes — Performance requirements for protective paint systems for offshore and related structures*, 2009. (Superseded by ISO 12944-9 *Paints and varnishes. Corrosion protection of steel structures by protective coating systems. Protective paint systems and laboratory performance test methods for offshore and related structures* 2018).
24. ISO 21809-1 *Petroleum and natural gas pipelines, External coatings for buried or submerged pipelines used in pipeline transportation systems – Polyolefin coatings (3-layer PE and 3-layer PP)* 2018.
25. ASTM G8-96 *Standard Test Method for Cathodic Disbondment of Pipeline Coatings*, 2019.

Chapter 10

Sacrificial anodes

What's in a name?

As we have already seen, marine CP is frequently achieved by attaching anodes. The intention is that the anode corrodes instead of the structure. We have referred to these anodes as *sacrificial*. The term is useful in that it conveys the concept of the anode sacrificing itself to provide protection. Other workers prefer the term *galvanic* anodes. This highlights the fact that the mode of action involves galvanic corrosion of the anode when coupled to the more noble structure.

Both terms are acceptable, but in this book we use "sacrificial".

10.1 WHAT PROPERTIES DO WE NEED?

At its simplest, this form of CP involves connecting the metal we wish to protect (the cathode) to a metal that has a more negative electrochemical potential in seawater (the anode). It is the method reported by Sir Humphrey Davy in 1824, although he could not use the words *cathode* or *anode* which only entered the English language after his death.

From this, we can deduce that there are two fundamental properties of an anode alloy that interest us: its potential and how much current it will deliver.

10.1.1 Potential

The potential of the anode when it is connected to the cathode and delivering current is variously described as its *operating* or its *closed-circuit potential*. This must be more negative than the target protection potential of the cathode. Furthermore, the potential difference between the operating anode and the protected structure must be sufficient to drive the CP current against the resistance of the environment. This means that, if we seek an anode material that will protect steel by polarising it to more negative than −0.8 V in seawater, then we need an alloy with an operating potential at least as negative of −0.9 V, and preferably more negative than that.

It is a simple enough matter to go through the periodic table of the elements to find metals which might be candidate anodes. However, we have to exclude from that list metals such as sodium or calcium that have potentials so negative that they react violently with water. We also have to exclude very expensive and radioactive metals. As we see from Table 10.1, when these unsuitable elements are excluded, our only options are aluminium, zinc and magnesium, or alloys of those metals. When viewing Table 10.1, however, we need to appreciate that the potential of an alloy is not an intrinsic property. It depends in a complex manner on the composition and metallurgical structure, the nature of the environment (composition, conductivity, temperature) and on the current being drawn from it.

DOI: 10.1201/9781003216070-10

Table 10.1 Anodes for the protection of steel in seawater

Metal	*Typical Potential*
Carbon steel (corroding)	–0.6 to –0.7V
Carbon steel (protected)	–0.8 to –0.9V
Aluminium alloys	–0.76 to –1.00V
Zinc	–0.98 to –1.03V
Magnesium	–1.60 to –1.63V

10.1.2 Current

10.1.2.1 Instantaneous output

There are two things we need to know about the current output of an anode. The first is: how high a current can it deliver to the cathode? The second is: for how long it can keep doing so.

As we saw in Chapter 3, and will explore further in Section 10.8, the instantaneous current that it can deliver depends on its potential and its resistance to the environment.

10.1.2.2 Capacity, consumption rate and efficiency

The length of time for which it can keep supplying that current is determined by its capacity. This is a measure of the total quantity of electric charge that can be obtained from a given mass of anode. It is usually expressed in units of ampere·hours per kilogramme (Ah/kg).

However, it is sometimes thought of as the consumption rate of the anode. This is usually expressed in units of kilogrammes per ampere·year (kg/Ay). For example, an aluminium anode alloy might have a capacity of 2500 Ah/kg. This could equally be expressed as 3.5 kg/Ay.

The theoretical capacity of a metal is an intrinsic property that can be calculated by the application of Faraday's Law of Electrochemical Equivalence. For example, this figure would be 820 Ah/kg for pure zinc, or 2980 Ah/kg for pure aluminium.

The capacity actually achieved in a CP circuit, however, is not a fixed property of the alloy. Like the potential, it depends on the metal composition and structure, the environment and the magnitude of the current being drawn. In practice, the theoretical capacity is never achieved. This is because not all of the alloy mass dissolves to provide the external CP current. Inevitably, some is consumed in what might be regarded as *self-corrosion.*

The ratio between the actual (measured) capacity of an anode and its theoretical capacity is termed its electrochemical efficiency. Thus, if an Al-Zn alloy with a theoretical capacity[1] of (say) 2850 Ah/kg exhibited a measured capacity of (say) 2350 Ah/kg, its efficiency could be expressed as 2350÷2850=82%.

10.2 ZINC ALLOYS

10.2.1 Background

As we have already seen, CP originated with Sir Humphry Davy's experiments on the protection of the copper sheathing on ships' hulls. Conducting experiments in Portsmouth Harbour [1], he observed that, even when the area of the *metallic protector* (zinc or cast iron) was only 1/40–1/150 of the copper, there was... *no corrosion or decay* of the latter

[1] The theoretical capacity of an alloy is the pro-rata sum of the capacities of the alloy components.

metal. In this early work, Davy also noted that the ability of zinc to protect diminished if attention was not paid to its purity.

In the event, the subsequent development of CP was slow. For more than a century, the fixing of zinc anodes, known as "protectors", to ships owed more to tradition than to any grasp of the principles of CP. However, in 1956 Crennel and Wheeler [6] observed that... *the ineffectiveness of zinc protectors as fitted to ships' hulls is generally recognized and has led many users to dispense with them.*

Despite this, the 1950s saw a considerable amount of zinc anode alloy development. All of this work was parametric in nature. Experimenters tested binary, ternary and even quaternary formulations to discover reliable anode compositions. There was little, if any, attempt to elucidate the mechanisms by which individual alloying ingredients interacted with the zinc, or with each other. By way of example, Teel and Anderson [4,5] related the progressive reduction in zinc output to the formation of a dense adherent corrosion product film. Further investigation linked the propensity of this film to develop to the iron content of the alloy, values above 0.0014% causing a marked fall-off in anode performance. They also noted the anode performance benefits of adding small amounts of cadmium (up to 0.07%) to the alloy. Subsequently, the US Navy incorporated these figures in its specification for zinc anodes (MIL-A-18001 – see Table 10.2). The details of this specification have changed little over the years.

The 1960s saw most of the patenting activity that there has been for zinc alloys. The patents applied for included binary, ternary and higher order alloys. The alloying elements included: aluminium, bismuth, cadmium, lead, mercury, silver, thallium and tin at various concentration ranges, either singly or in combinations. I have not listed the patents in this chapter. As is the nature of patents, they tended to be short on detail and long on claimed benefits. More to the point, all of these patents have long since lapsed. However, if you are interested in the patent history of zinc anodes you can find a summary in [20]. Frankly, however, there is little point.

Table 10.2 Zinc alloy anode materials

Code	*Designation*				
EN 12496	Z1		Z2	Z3	Z4
DNV-RP-B401		Table 8.5			
ISO-15889-2	Table 7				
ASTM B418	Type I		Type II		
US Military Spec.	A18001K				
UNS	Z32120		Z13000		
Element	*Composition (wt %)*				
Al	0.10–0.50		≤0.005	0.10–0.20	0.10–0.25
Cd	0.025–0.07	≤0.07 max.	≤0.003	0.04–0.06	≤0.001
Fe	≤0.005		≤0.0014	≤0.0014	≤0.002
Cu	≤0.005		≤0.002	≤0.005	≤0.001
Pb	≤0.006		≤0.003	≤0.006	≤0.006
Sn	-		-	≤0.01	-
Mg	-		-	≤0.05	0.05–0.15
Others (total.)	≤0.1	-	<0.005	≤0.1	≤0.1
Zinc	Balance				

10.2.2 Present day alloys

The CP industry still relies on pure zinc or the Zn-Al or Zn-Al-Cd alloy formulations that were in place by the late 1950s (see Table 10.2). Toxicity issues associated with cadmium mean that cadmium-free materials are generally preferred. In ambient seawater, these alloys are all expected to operate at potentials more negative than the –1.0 V design value advised in DNV-RP-B401. Their capacities are about 780 Ah/kg. Since this represents an electrochemical efficiency of 95%, it is unsurprising that there has been little interest in further development of zinc alloy anodes.

10.2.3 Limitations

10.2.3.1 Elevated temperature

All anode materials exhibit a drop-off in electrochemical efficiency as the temperature increases. In the case of zinc anode alloys containing more than 0.012% aluminium, however, the fall-off is dramatic because the anode undergoes a change in dissolution morphology above ~50°C from uniform dissolution, which is desirable, to intergranular dissolution, which is not. Figure 10.1 shows incipient intergranular dissolution on a zinc anode alloy after 2 weeks' coupling to steel at 70°C. As this intergranular dissolution continues, entire grains of alloy are detached from the surface without dissolving to deliver CP current.

The offshore CP industry became aware of this susceptibility of zinc anodes in 1977 when it was discovered on anode attached to a hot pipeline riser in the Ekofisk field in the North Sea [29]. It was studied by Ashworth et al. [24] who posited that the intergranular attack might be a result of grain boundary segregation of aluminium in the alloy. A similar view was also expressed by Jensen et al. [21], and subsequently by Haney [35]. However, subsequent work by Ahmed et al. [27] demonstrated that, although the presence of aluminium and cadmium in the alloy was necessary for the phenomenon to occur, grain boundary segregation was not involved. Subsequent work by the same group [40] demonstrated that it was a more subtle problem of hydrogen damage.

Figure 10.1 Intergranular corrosion of a zinc anode [20].

From the practical standpoint, although zinc alloys have been reported which resist intergranular dissolution, at up to 85°C in saline mud [36], the attitude of the marine CP industry, as encapsulated in DNV-RP-F103 is to set an upper temperature limit of 50°C for using zinc anodes.

10.2.3.2 Iron contamination

One would be forgiven for thinking that this problem of iron contamination was consigned to history through a combination of reliable standards and good foundry practice. This makes a 2016 paper [73] rather interesting. The authors reported on the performance of zinc anodes on a marine weather buoy. Some anodes had worked as expected, but some adjacent anodes remained unconsumed. Investigations revealed iron-rich intermetallic particles in the latter specimens. From this, the authors concluded this was an example of the known tendency of zinc anodes to passivate if contaminated with iron.

They may have been correct. Their conclusion aligns with the received wisdom regarding the effect of iron. However, they may have been wrong. Crucially, the authors did not carry out (or at least did not report) the following:

- any determination of the iron content of either the working or, apparently, non-working anodes,
- any examination of the working anode for iron-rich intermetallic particles of the type sought and found on the non-working anodes,
- any surface examination of the non-working anodes.

However, they did observe that the anodes were connected to the structure by bolting which, as they correctly point out, is less satisfactory than welding. They stated that after removal of the bolts a brown iron oxide was found between the anode core and the structure; adding that... *this oxide had increased the resistance between the nonsacrificed anode core...* and the structure.

In the absence of any data to demonstrate that the non-working anodes possessed higher iron contents than the working anodes, their proposed explanation must remain in doubt. Furthermore, it appears, but is not stated, that they believe the iron oxide (rust) between the core and the structure was formed because the anode passivated and did not throw current into this crevice space. However, it is no less possible that the anode cores were rusty as supplied, and that they were not properly de-rusted before the anodes were attached. In other words, the poor connections caused the anodes not to work, rather than the failure of the anodes causing the poor connections.

We shall probably never know the truth. The reason that I have raised this point here is to counsel circumspection when we come to examine the copious data accumulated in the literature for anode performance. As a matter of experience, experimental results in CP, and the conclusions drawn from those results, are occasionally not as dependable as we would like.

10.3 MAGNESIUM ALLOYS

Magnesium anodes operate at potentials in the range –1.5 to –1.7V. This is very much more active than zinc anodes. According to Humble [3], magnesium anodes were used to protect structures in seawater prior to 1949. Further background is provided by Doremus and Davis [9] who explained that, historically, a combination of magnesium ribbon, to achieve protection, and solid magnesium anodes, to maintain it, had been used in the Gulf of Mexico. They further explained that the position of magnesium... *was well established as the work-horse of marine galvanic anode installations during the 1950s and early 1960s.*

After the 1960s, the use of magnesium anodes to protect offshore structures faded. It is difficult to put a timeline on its demise. As late as 1972, Lennox [13] was referring to "dependable" magnesium anodes for seawater application. It is also included as an anode for seawater service in CP1021 issued the following year. However, that code of practice draws attention to the very negative potentials generated by magnesium and warns of the possibility of coating damage. It advised restricting the marine use of magnesium anodes to providing temporary protection for ships during fitting-out.

A precursor of what was destined to become DNV-RP-B401 was DNV-TNA-703 which was issued in 1981. This technical note related solely to fixed offshore structures. It excluded magnesium from consideration. This position has held in all subsequent national and international codes for offshore CP.

Nevertheless, this is not the last we will have to say about magnesium alloys. Even after their exclusion from DNV guidelines, they have occasionally been employed offshore. They are included in EN 12496 which provides compositional and performance information for two alloys (see Table 10.3). We will return to this when we discuss rapid polarisation in Section 10.5.2.

Table 10.3 Magnesium alloys for seawater use (EN 12496)

Alloy	*Type*	*Operating potential (V)*	*Capacity Ah/kg*
M1	Mg, 5-7 Al, 0.15-0.7 Mn, 2-4 Zn	−1.50	1200
M2	Mg, 0.5-1.5 Al	−1.70	1200

10.4 ALUMINIUM ALLOYS

10.4.1 The benefits

The benefit of aluminium compared to zinc is emphasised in Table 10.4. In principle, aluminium has more than three times the current capacity of zinc. Aluminium also delivers more current per unit volume. Its advantage is even more pronounced when we factor in the relatively low cost of aluminium compared with zinc. It is, therefore, not surprising that the CP industry committed considerable resources to the objective of developing aluminium-based sacrificial anodes.

Table 10.4 The theoretical benefit of aluminium

Property	*Zn*	*Al*	*Remarks*
Abundance in earth's crust	0.00014	1.38	relative to iron (=1)
Price per kg	$2.90	$1.98	May 2021
Atomic weight	65.39	26.98	
Valency	Zn^{2+}	Al^{3+}	
Maximum *theoretical* capacity	820	2980	Ah per kg
	5.82×10^6	8.95×10^6	Ah per m^3
	283	1505	Ah per $

10.4.2 Alloy research

The classic process of bringing any new product to the market involves research and development (R&D). The first stage is the fundamental research that discovers the concept and engenders understanding of how it works. This is followed by the development stage that produces the fine-tuning to make the concept marketable.

In the case of aluminium alloy anodes, the sequence of these two processes was reversed. It was a better commercial bet to get parametric tests underway, to produce data that could support patent applications, than it was to embark on mechanistic investigations that might turn out to be unrewarding. Where mechanistic studies have been reported, these have generally been on alloys that had already become commercially available.

A lot of this work is interesting or even fascinating. Ultimately, however, the fundamental research element of R&D has not produced any novel alloys. Indeed, there is an interesting debate to be had as to how far that body of research has advanced our understanding of how aluminium alloy anodes actually work. Unfortunately, that debate will need to be held elsewhere. This book is long enough as it is. So, we will skip the research and move on to the development part of the R&D.

10.4.3 Alloy development

It was not until the mid-1950s that enterprises embarked upon serious programmes of aluminium alloy anode development [15]. The problem to be addressed in exploiting the theoretical attractions of aluminium as an anode is that the metal and most of its alloys are normally protected by an air-formed passive oxide film. The exploitation of this element as an anode was, therefore, to find alloying additions that would diminish the protective capacity of this oxide.

As with the development of zinc alloys, aluminium was initially investigated entirely through parametric investigations. In essence, the process followed was as follows. Organisations prepared and tested binary, ternary and quaternary alloys seeking a result that would give a sufficiently negative operating potential and an acceptable capacity. This was followed by a headlong rush to patent offices to secure the rights to exploit their formulations. Once the patent rights were secured, the details of the testing started to emerge in the scientific literature.

Figure 10.2 Binary aluminium alloys (Reding and Newport [8]).

A classic of the genre is the 1966 paper by Reding and Newport [8] of the Dow Corporation. They reported seminal studies on the operating potentials of a series of binary aluminium alloys when anodically polarised in constant current tests at 10.8 A/m^2. Their results are summarised in Figure 10.2 using the layout of the Periodic Table. They observed that five metallic elements (gallium, mercury, indium, tin and bismuth) caused a negative shift in the operating potential of more than 0.3 V. Four further elements (zinc, cadmium, magnesium and barium) caused negative shifts of between 0.1 and 0.3 V.

This work formed the foundation of alloy development over the next decade or so. Workers patented numerous aluminium alloy formulations; nearly every one of which contained at least one of these nine elements. Figure 10.3 shows a list of alloying elements incorporated, either individually or in various combinations, in patented aluminium anode alloy formulations.

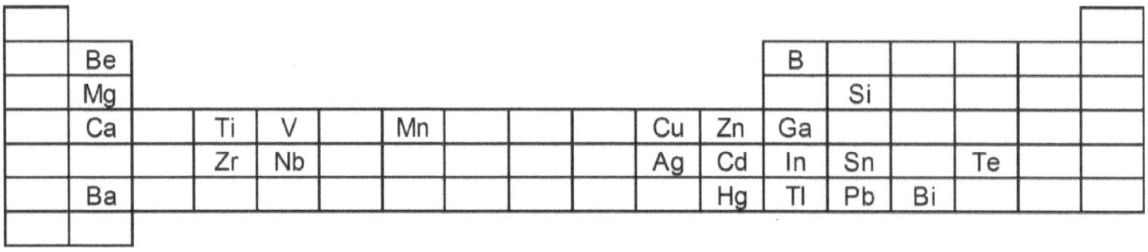

Figure 10.3 Elemental additions in patented aluminium anode alloys.

In general, the patent applications referred to alloying elements that had the most dramatic effect in making the anode's operating potential more negative as "activators" or "depassivators". Other, elements which were added in addition to the activators to further improve the performance, perhaps by improving the capacity or the dissolution pattern, were often referred to as "modifiers". However, it is cynical, but not inaccurate, to say that some of the activator or modifier elements could, more truthfully, be described as "patent makers" or "patent breakers".

When the dust had settled after this patenting frenzy, three main alloy systems had emerged as practical anode materials: Al-Zn-Sn, Al-Zn-Hg and Al-Zn-In. Some years later, the binary Al-Ga anode system was patented to compete in the niche market for anodes operating at more modest potentials than the ternary alloys.

10.4.4 Al-Zn-Sn and Al-Zn-Hg alloys

Numerous ternary Al-Zn-Sn alloys were patented. If you are interested in tracking down the individual patents, then details are summarised in an obscure, little-read document [23]. However, there is little point, since the patents have now all long since expired. Alloys containing tin usually required an expensive post-cast heat treatment in order to develop usable electrochemical properties.

Unlike the tin-containing formulations, mercury-containing anodes could be used in the as-cast condition. Through to the end of the 1970s, aluminium alloys containing zinc and mercury dominated the offshore anode market. Most major offshore structures installed during this boom decade were, and some still are, protected by Al-Zn-Hg anodes. Ultimately, however, the environmental notoriety of mercury weighed against these formulations.

Neither tin- nor mercury-containing alloys are being produced, and it is unlikely that they will return.

10.4.5 Indium-containing anodes

10.4.5.1 Al-Zn-In

Like the Al-Zn-Sn and Al-Zn-Hg systems, Al-Zn-In anode alloys have been developed entirely on the basis of parametric studies. Through the 1970s, innumerable alloy formulations were trialled. Any that showed promise, and some that did not, were patented. All of these patents have long since expired, leaving manufacturers unhindered in their choice of alloy compositions. The demise of Al-Zn-Hg on environmental grounds has left the Al-Zn-In alloy system dominant in the offshore CP market. Table 10.5 shows some currently used formulations.

Table 10.5 Currently used Al-Zn-In anodes

Element	*EN 12496* *Alloy A1*	*Alloy A2*	*Alloy A3*	*DNV-RP-B401 Table 8.5*	*ISO 15589-2 Table 6*	*US Military Spec. DTL-24779D (SH)*	*Al-Zn-Mg-In typical composition [70]*
Zn	2.0–6.0	3.0–3.5	4.75–5.75	2.5–5.75	2.5–5.75	4.0–6.5	3.2
In	0.01–0.030	0.016–0040	0.016–0.020	0.015–0.040	0.016–0.040	0.014–0.020	0.018
Si	≤0.12	≤0.10	0.08–0.12	≤0.12	≤0.12	0.08– 0.20	0.24
Ga						≤0.02	
Mg						≤0.01	1.2
Fe	≤0.12	≤0.09	≤0.06	≤0.09	≤0.09	≤0.08	0.073
Cu	≤0.006	≤0.005	≤ 0.003	≤0.003	≤0.003	≤0.005	≤0.001
Cd	≤0.002	≤0.002	≤0.002	≤0.002	≤0.002		
Others (each)	≤0.02	≤0.02	≤0.02		≤0.02		
Others (total)	≤0.1	≤0.1	≤0.1				
Al				Balance			

The existence of EN, ISO, DNV and US military formulations will inevitably lead to these alloys dominating the market. Furthermore, these specifications overlap to a considerable degree. As can be seen from Table 10.5, some formulations are very close, and it is possible to produce anodes with a composition that complies with more than one code.

We should also bear in mind that the compositions in Table 10.5 span relatively wide ranges. Individual anode manufacturers are likely to produce their products with tighter compositional tolerances.

EN 12496 gives *typical* properties for these alloys that are less conservative than the *design* values offered by DNV-RP-B401 and other codes. Operating potentials are expected to be –1.09 V in seawater (compared to –1.05 V in DNV) and –1.05 V in seabed mud (compared –0.95 V in DNV). Similarly, the typical capacity in either environment, at ambient temperature, is likely to be 500 Ah/kg higher than the DNV recommended design value. This emphasises the conservatism of the codes.

10.4.5.2 Al-Zn-Mg-In

You will also see an Al-Zn-Mg-In anode in Table 10.5, although magnesium-containing anodes do not appear under any of the codes. Indeed, magnesium is specifically proscribed by the US military specification. Nevertheless, these alloys have an acceptable track record in Japan.

The inclusion of magnesium in the alloy formulation was patented during the early alloy development when it was found that quaternary Al-Zn-Mg-In anodes exhibited high efficiencies. This, in turn, was ascribed to the more uniform dissolution morphology that resulted from the inclusion of the magnesium. In practice, however, cracking problems, associated with the age-hardening characteristics of these materials, were encountered on large anodes installed on North Sea structures. In some instances, apparently sound anodes were attached, only to develop cracks whilst the structures were still in the fabrication yards. Similar problems were also observed offshore Nigeria [56].

This cracking led to Al-Zn-Mg-In anodes ceasing to be used in Europe and North America. In Japan, on the other hand, the cracking problems were successfully addressed through suitable anode and core designs. The anode manufacturers have sufficient confidence in these anodes to use design values of −1050 mV and 2600 Ah/kg as design values [70].

10.4.6 Al-Zn-Ga and Al-Ga

Figure 10.2 reveals gallium as one of the five elements which causes the operating potential of aluminium to shift negatively by more than 300 mV. However, gallium-containing anodes did not impact the CP market. A likely reason for this is that mercury, tin and indium have very low solubilities in aluminium, and only need to be present at concentrations close to their solubility limits in order to produce effective anodes. Thus, a successful Al-Zn-In anode only contains around 0.02% indium. Gallium, on the other hand, which is soluble at up to around 20% in aluminium at room temperature, typically needs to be present at or above ~0.5% to produce an anode with a similar operating potential. So, apart from some academic laboratory-based study [53], there was little practical interest in Al-Zn-Ga.

Interestingly, however, the unimpressive lowering of the operating potential obtained in binary Al-Ga alloys using levels of gallium below 0.5% has engendered commercial interest in the niche market for anode operating at limited potentials. One of these alloys Al-0.1% Ga is described in EN 12496. It gives a typical operating potential of −0.83 V and a capacity of 1500 Ah/kg in seawater. These anodes have not been used in sediments. We return to the subject of anode with limited potentials in Section 10.5.1.

10.4.7 The future

10.4.7.1 The toxicity of indium?

As we have seen, mercury-activated anodes dominated the market from their introduction in the mid-1960s but were supplanted by indium-activated anodes in the mid-1970s because of the toxicity of mercury [18]. I cannot predict the future, but the possibility that indium will one day fall under the same environmental suspicion cannot be discounted. As pointed out by Talavera et al. [59], all indium compounds... *should be regarded as highly toxic. Indium compounds damage the heart, kidney, liver and may be teratogenic,*[2] This would leave the anode industry having to disregard the work carried out by Reding and Newport [8], amongst others. It would have to formulate aluminium-based anodes without low melting point toxic heavy metal activators. If this situation arises, it seems that the most likely avenue will be to re-examine Al-Zn and Al-Zn-Mg alloy systems.

10.4.7.2 Al-Zn and Al-Zn-Mg

In the Soviet Union, Lyublinskii was reporting on such materials in the early 1970s [14]. However, there has been little subsequent interest until recently when concerns emerged that indium might fall under an environmental cloud. This has prompted recent experimental work on aluminium alloy systems to which none of the classic activating elements (Sn, Hg, In and Ga) have been added [57,59,65,66,68].

[2] *Teratogenic:* capable of causing developmental problems.

Whilst not specifically researching indium-free anodes, Murray and Lenar observed that binary Al-Zn alloys could operate at −0.95 V, albeit with lower (but not unacceptable) efficiencies of ~75%. This is a more negative potential than observed by Reding and Newport in the 1960s [8]. It was suggested that the binary Al-Zn system might be showing promise as an anode because aluminium is now routinely produced at higher levels of purity. In the same work, Murray and Lenar observed that adding 0.1% silicon increased the efficiency to ~92%. They also observed that only 0.0075% indium was needed to shift the operating potential to between −1.05 V and −1.1 V. This latter observation suggests that, if indium does fall under environmental suspicion, a practical remedy might be to ensure that the indium content is kept to an even lower level than in current anode formulations.

Orzoco et al. have studied Al-Zn-Mg anodes with typically 5% Zn and 5%–9% Mg. They obtained electrochemical efficiencies of 63%–73%, and operating potentials in the range −1.05 V to −1.15 V. These results are promising.

On the other hand, further work by Quevedo and Genesca [66] is less so. They examined the performance of an Al-14%Zn-9%Mg alloy and reported efficiencies of only 50%–60% in tests carried out according to NACE TM-0190-98. Since this test is conducted at a high anodic current density (6.2 A/m^2) which favours high electrochemical efficiencies, these results are disappointing. It is not possible to extract information on the operating potential of this alloy from their published results because they focussed on potentiodynamic polarisation experiments and electrochemical impedance spectroscopy (EIS), and used a rotating cylinder electrode to simulate high water flow rates.

Thus, the work on indium-free Al-Zn and Al-Zn-Mg anodes has been interesting and has given some cause to believe that workable anodes could be produced using this alloy system. However, it seems fairly clear that the only way indium-free aluminium anodes could take the market share enjoyed by their indium-containing counterparts would be if indium, like mercury, finds itself banned on environmental grounds.

10.5 NON-STANDARD ANODES

10.5.1 Limiting the polarisation of the cathode

10.5.1.1 The need

Certain materials are susceptible to hydrogen embrittlement (HE) at less negative potentials than encountered in normal sacrificial anode CP systems. These materials include high-strength steels and, depending on their metallurgical condition, certain duplex and martensitic stainless steels. We will discuss HE, and its related phenomena, in Chapter 12 where we consider the mechanical effects of CP. For now, all we need to note is that it would be damaging to polarise materials which are susceptible to HE to potentials close to the operating potential of a zinc or Al-Zn-In anode.

For example, in the case of steels with a specified minimum yield stress >550 MPa,[3] ISO-15589-2 recommends that... *the most negative potential shall be ascertained.* EN 12473 offers the same advice but also notes that this negative potential limit will be in the range −0.83 to −0.95 V. The EN also contains advice to avoid excessively negative potentials for duplex and martensitic stainless steels.

In sacrificial anode CP systems, there are two ways in which we can limit the potential of the cathode. We can either change the operating potential of the anode, which means

[3] For old school readers: 550 MPa is approximately 80 ksi.

changing the anode alloy composition, or we can insert a passive current-controlling electronic device between the anode and cathode.

10.5.1.2 Alloy composition

We mentioned in Section 10.4.6 that, although gallium was identified as a prospective activator for aluminium anodes in the early Reding and Newport work [8], neither Al-Ga nor Al-Zn-Ga alloys have been developed for the mainstream anode market. However, Al-Ga alloys have found an application for applying CP at limited potentials.

Pautasso et al. [55] investigated the possibility of producing an anode that operated in the potential range −850 mV to −800 mV when delivering an anodic current density of 2–10 A/m^2. In laboratory trials, they observed that an Al-0.1Ga alloy came close. It operated at −825±15 mV at 0.05 A/m^2, but its potential shifted to 795±40 mV at 5 A/m^2. Lemieux et al. [62] tested the same alloy in seawater at Key West in Florida and found a slightly less negative average potential (−765 mV) than recorded in the laboratory testing. As noted above, EN 12496 advises −0.83 V (operating potential) and 1500 Ah/kg (capacity).

On this basis, Al-0.1Ga has been included in the US military specification (MIL-DTL-24779D(SH)) for aluminium anodes since 2013. Since then, Kidd and Druschitz [74] have shown that these alloys exhibit more consistent operating potentials when cast such that they cooled at a high rate.

10.5.1.3 Passive electronic components: resistors

It is intuitively obvious that installing a resistor between an anode and the structure it is protecting will reduce the current output of the anode and, as a result, could limit the polarisation of the cathode. Unfortunately, this does not provide us with a secure means of controlling the polarisation. To understand why this is so, it is useful to remind ourselves of the slope parameter in Chapter 7. Interposing a resistance in the circuit, and lowering the initial current, would certainly reduce the rate at which the structure polarised. However, assuming enough current was passed to ensure that protection was achieved, then polarisation would continue in the long term. Ultimately, a material susceptible to HE would polarise to a critically negative potential. The only benefit of the resistor would be that it would take longer to do so.

The only way that the goal of limited, but protected, potentials could be achieved would be if variable resistors were used, and their values adjusted throughout the life of the structure. This could hardly be practicable.

Although there is little or nothing to be gained from inserting resistors between anodes and carbons steel, resistors occasionally have a part to play in the protection of CRAs. We will return to this when we cover internal resistor-controlled CP in piping (Chapter 16).

10.5.1.4 Passive electronic components: diodes

Diodes are semiconductor devices which will only start to pass a significant current once the so-called threshold, or breakdown, voltage is exceeded. For any diode, operating at any given temperature, the value of this breakdown voltage is fixed. This diode property offers the prospect of interposing a fixed voltage drop between the anode and cathode.

For an anode operating at −1.1 V, and a target least negative protection potential of the cathode of −0.8 V, we would be looking for a diode with a breakdown voltage of ~0.3 V. When first connected, the corrosion potential of the susceptible cathode might be ~−0.6 V. Thus, the potential difference between anode and cathode would be (−0.6 V − (−1.1 V) = 0.5 V).

This would exceed the 0.3 V breakdown voltage. So, CP current would flow, and the structure would start to polarise cathodically. However, once its potential reached −0.8 V, the diode would switch off and polarisation would cease. From then on, potential of the structure would stay close to −0.8 V. Any tendency to depolarise would cause the diode to switch back on and pass some more CP current.

Most diodes have breakdown voltages in excess of 0.6 V, which makes them unsuitable for potential control. The only suitable solid-state devices are Schottky diodes, also known as Schottky barrier rectifiers (SBRs), invented by the German semiconductor physicist Walter H. Schottky. These have forward breakdown voltages in the region of 0.15–0.3 V. Although their primary function is very fast switching in power circuits, this feature renders them usable for potential limitation in CP.

An example of the voltage-current characteristic of the International Rectifier 55HQ030 device is shown in Figure 10.4. Its forward voltage-current characteristics, over the range relevant to anodes, are shown in Figure 10.5. The solid line shows manufacturer's data at 25°C. The dotted line shows the results for 10°C, derived from laboratory testing.

Figure 10.4 55HQ030 SBR diodes.

Figure 10.6 shows the arrangement for connecting an anode to a structure using a pair of SBRs. There are two reasons for using a pair of diodes when only the forward device is required. One is that using two diodes means that it is impossible to connect the unit the wrong way around. The second relates to stray current issues with welding carried out from a dockside or work barge when the structure is afloat. The reversed diode permits any stray currents, arising from the incorrect earthing of DC welding power sources, to exit into the seawater via the anode rather than the structure.

Some anodes for flush-mounting, fitted with potential limiting devices are shown in Figure 10.7. These were destined for installation on a high-strength steel jack-up drilling installation.

To date, I am only aware of the 55HQ030 SBR being used in CP. There is, however, an alternative device from the same manufacturer (95HQ015) which has a lower breakdown voltage. Depending on temperature and current, this would impose a voltage drop of between ~0.2 V and ~0.25 V between the anode and cathode. Such a voltage drop would be relevant where a cathodic potential somewhat more negative than −0.8 V were targeted.

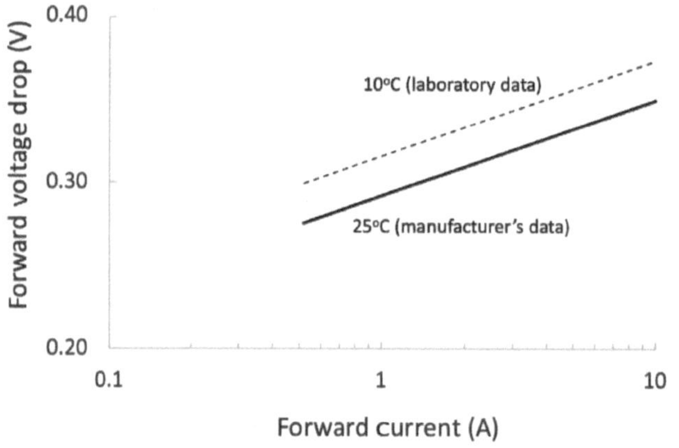

Figure 10.5 SBR 55HQ030 Characteristics.

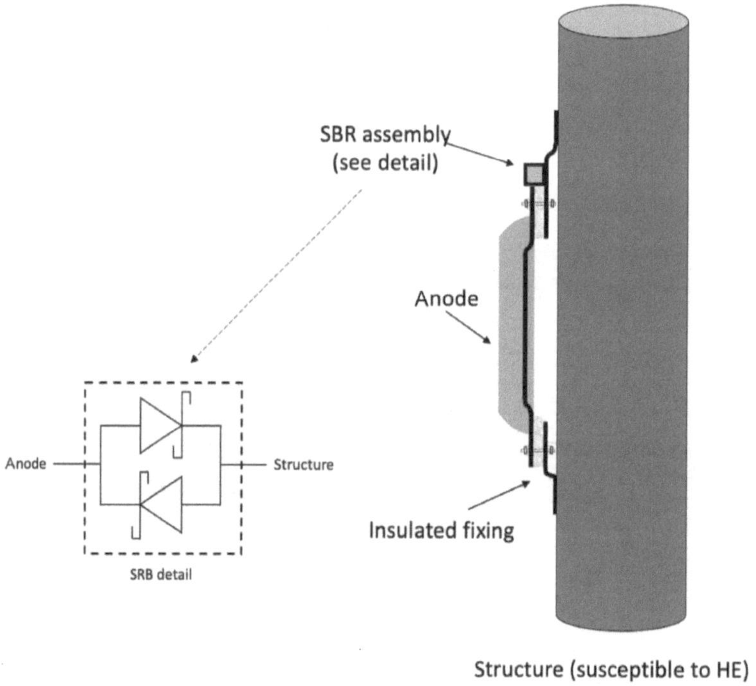

Figure 10.6 Schematic of potential limiting installation.

10.5.2 Anodes for rapid polarisation

10.5.2.1 Motivation

We discussed the importance of the calcareous deposit at some length in Chapter 7 and explained that the through-life (mean) current demand of a structure, and thence the weight of anodes required, was heavily dependent on the quality of the calcareous film formed on the steel. Furthermore, although a potential of −800 mV or more negative is sufficient to protect steel, calcareous deposits formed in the region −900 to −1000 mV are optimum for reducing the subsequent cathodic current demand.

Figure 10.7 Potential limiting flush-mounted anodes.

This knowledge has prompted a number of lines of anode R&D aimed at providing high initial current outputs in order to achieve rapid polarisation into the optimum potential zone for calcareous deposit formation.

10.5.2.2 Hybrid systems

In Chapter 13, it will emerge that only a minority of fixed offshore structures are protected by ICCP and that most of these are actually hybrid systems. They are also fitted with some sacrificial anodes. These were required to protect the structure initially whilst topside facilities, including the electrical power generation, were installed and commissioned.

Schrieber and Reding [41] suggested an alternative take on hybrid CP systems. In this application, the main CP system would use sacrificial anodes, but removable temporary ICCP anodes would be deployed to increase the initial current supply, thereby aiding calcareous film formation. Although this would have been an interesting approach, it never caught on.

10.5.2.3 Dual anodes

Another approach to achieving rapid polarisation is to make use of the very negative operating potential of magnesium anodes. Then, as the short-lived magnesium depletes, the continuing through-life protection is sustained by the longer-lived, more economic aluminium or zinc anodes.

As explained in Section 10.3, the concept of employing magnesium in offshore CP dates back to the 1950s. Moreover, from 1963 magnesium-jacketed zinc anodes (known as "Binodes") were available to CP designers seeking to achieve rapid polarisation [9]. In the event, there was little uptake of the dual anode concept until the mid-1980s. Then, for about a decade, there was interest in dual magnesium and aluminium anodes.

In principle, there are two ways of creating a dual anode. The first involves adding a magnesium rod anode to a stand-off aluminium anode as illustrated in Figure 10.8a. The second option is to cast the magnesium onto the surface of the aluminium anode. This can be achieved by casting the aluminium anode, and then topping up the mould with a layer of magnesium alloy as indicated in Figure 10.8b. Alternatively, the aluminium anode is cast in the conventional way. This anode is then positioned centrally in an oversized mould, and molten magnesium alloy is poured in to fill the intervening gap. The latter option results in total coverage of the aluminium anode with a layer of magnesium as illustrated in Figure 10.8c.

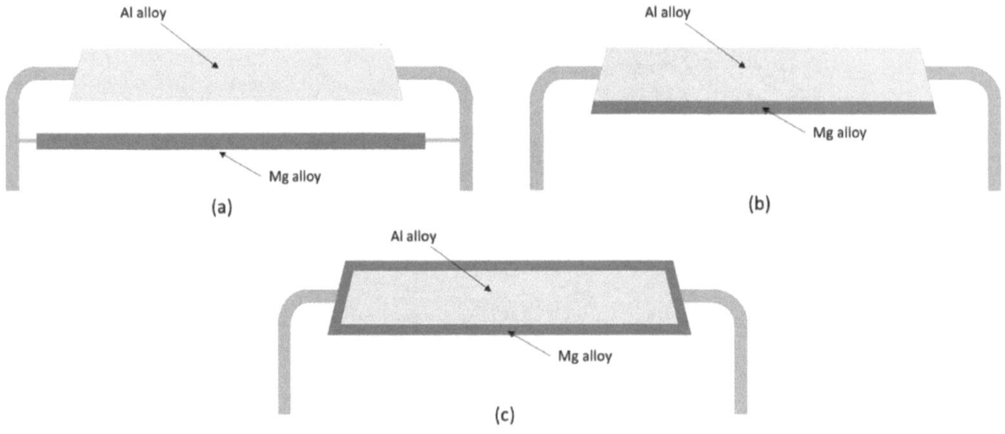

Figure 10.8 Options for dual anodes.

A version of the method illustrated in Figure 10.8a was installed on a small Amoco platform in the Rijn Field in the Dutch sector of the North Sea in 1985 [44]. The results were encouraging, with the structure polarising fully within 36 hours. Unfortunately, some of the benefits of the exercise were lost because of a failure of the current density measuring equipment installed as part of the exercise. Furthermore, although the objective of employing magnesium to achieve rapid polarisation was evidently achieved, there was no supporting evidence to confirm that the over-arching objective of reducing the through-life rate of aluminium anode consumption was also achieved.

A 1991 paper by Burk [45] published six years after the installation of the platform observed that the CP system was working satisfactorily but did not provide any further detail. Instead, it focussed on initial test data from an anode cast as illustrated in Figure 10.8b. There may have been some exaggeration in claiming that this was a "new development" since the concept had its origins 30 years earlier. More particularly, the paper averred that this "Dualanode" would lead to a 40%–60% weight saving compared with aluminium anodes alone. The paper notes that the magnesium layer was consumed in about 50 days. Unfortunately, however, it provides neither theoretical calculations nor experimental data to support this, possibly over-stated, claim. I have found no subsequent publication indicating that the dual anode was ever used on a real structure.

In 1993, Bonora et al. [46] described the anode type illustrated in Figure 10.8c and reported the intention to use them on two new Agip jackets the following year. Regrettably, due to pending patent applications, the 1993 paper is devoid of detail. Three years later, however, more detail was provided [52]. It seems that one of the platforms was the Garibaldi D. From the second paper, it is clear that the "composite" anode was installed as a test on a platform that was, otherwise, protected by conventional aluminium alloy anodes. The data, which are not very clearly presented, indicate that composite anode polarised the steel in its vicinity to about –1350 mV for about six months, after which time the magnesium layer was depleted. After about a year, the potential had settled back to about –1020 mV.

At about the same time, Kennelly and Mateer [47] described the installation of two anodes of the type shown in Figure 10.8b, on a deep-water jacket in the Gulf of Mexico. They referred to these as "bimetallic" anodes. Such measurements as were disclosed indicated that the bimetallic anodes delivered more current than the conventional anodes, and that the thin layer of magnesium was depleted relatively quickly. Neither observation was surprising. The information derived from the test was circumscribed, however, because only four of the 269 anodes installed on the structure were bimetallic. Thus, it was not possible to draw any practical conclusions about the putative benefits of the magnesium layer in reducing the through-life mean cathodic current demand on the structure.

Literature references to "dual", "composite" or "bimetallic" anodes dry up after the mid-1990s. As far as I am aware, no significant structures have subsequently been protected using this rapid polarisation technique.

Nevertheless, it seems that interest has not been lost entirely. A 2016 paper by Wang et al. [72] compared a sacrificial anode designs for an offshore jacket structure. Based on standard proprietary Al-alloy design, 234 anodes would be needed to satisfy the initial current demand, 146 for the mean and 160 for the final. However, reworking the design of the basis of dual (or "composite") Al-Mg anodes the maximum anode number reduced to 155.

10.5.2.4 Shaped anodes

As we saw when we went through some sacrificial anode designs in Chapter 3, the nub of the exercise is picking a combination of anode mass, shape and surface area that will satisfy the initial mean and final current demand of the cathode. For uncoated structures, it is usually the case that satisfying the initial current demand is the most onerous requirement. This frequently results in selecting anodes that are heavier than would otherwise be needed to meet the design life mean current demand. This has cost implications.

It is advantageous to optimise anode designs such that the initial, mean and final anode current outputs are in proportion to the initial, mean and final current demands of the structure. However, this objective needs to be set against the practical constraints of anode manufacture and installation. For example, Schrieber and Reding [41] reported 1986 work of Evans who demonstrated the benefits of using 3 m long anodes in an Arabian Gulf installation. On the other hand, de Waard [49] has pointed out that it was desirable to make stand-off anodes as short as possible to reduce wave loadings.

Recently, Mim and Siqi [71] reported laboratory tests in which high initial current densities were obtained using wing-section Al-Zn-In anodes. Their results are interesting, but the experiments were conducted on very small specimens machined from a commercial casting. The specimens had no core, the largest weighing only ~0.05 kg. It remains open to question whether the trial shapes could practicably be manufactured as full-size anodes with the requisite steel core support. It seems unlikely.

10.6 FUTURE DEVELOPMENTS

It is hard to see that there is much future scope either in the development of new anode alloys or the characterisation of existing materials. In terms of efficiency, for example, Al-Zn-In anodes exhibit values of around 95%, although codes such as DNV assume ~65%. This does not leave much scope for alloy improvement but more reliable characterisation may permit some relaxation of the codes.

These alloys are only likely to change if, like mercury, indium comes under an environmental cloud. If it is then prohibited, or becomes excessively expensive, the only recourse for exploiting aluminium as an anode would probably be to explore the Al-Zn or Al-Zn-Mg alloy options, or possibly Al-Ga with a higher level of Ga than 0.5%. This, in turn, would mean revisiting the CP design codes to accommodate their less negative operating potentials.

10.7 ELECTROCHEMICAL TESTING

10.7.1 Parameters measured

As we have already pointed out, there are two design parameters relevant to anodes: potential and capacity. We now also understand that these properties cannot be inferred simply from knowledge of the composition of the anode. They depend not only on the elemental make-up of the alloy, but also its metallurgical condition, and the anode's operational environment, temperature, current density and operating history. Accordingly, irrespective of the test configuration adopted, it is necessary to measure these two properties.

10.7.1.1 Potential

In principle, the measurement of potential is straightforward. All that is required is a reference half-cell and a high input impedance voltmeter, as we illustrated in Chapter 1. However, there are subtleties that need to be considered in the case of testing an anode when it is delivering current. The value we wish to determine is the potential difference across the anode electrolyte interface (which we simply call the "potential"). However, when the reference half-cell is placed any distance away from the surface of the anode, there will also be a contribution to the measured potential difference arising from the ohmic voltage drop of the current passing through the resistance of the seawater. This is known as the IR error. It works so as to make the operating potential of the anode (E_a as used in design calculations) appear less negative than it actually is. Usually, for anodes operating in the low resistivity environment of seawater, this error is not great. Nevertheless, the experimenter should always bear its effect in mind.

10.7.1.2 Capacity

There are two methods of determining the capacity of an anode: weight loss and hydrogen evolution.

The weight loss is the more obvious of the two, and the more straightforward, at least in principle. All that is required is that the specimen is weighed at the start of the test. Then, when the test is completed, it is cleaned and reweighed. The capacity is then determined as the quotient of the charged passed (Ah) divided by the weight lost (kg). Most of the test data in this chapter have been obtained in this way.

Unfortunately, that is not to say that such data are always as accurate as they might be. In almost none of the papers referenced did the authors record details of how they corrected their weight loss calculations for the effects of cleaning the sample. Certainly, none of them described a correction methodology using either control specimens, or else multiple replicate cleaning procedures of the type now described in ASTM G1. It may be that failure to remove scale effectively from anodes has resulted in underestimates of the weight loss and exaggerated estimates of the capacity.

This might explain some instances in the literature of reported anode capacities that are unusually high. For example, Lucas et al. [37] reported a capacity of 2958 Ah/kg, which equated to an efficiency >99.8% for the alloy tested. More dramatically, a study on Al-Ga anodes [69] reports a capacity of 3360 Ah/kg, which is about 12% higher than is theoretically possible. Similarly, in a "critical" review, Lemieux et al. [60] quoted capacities (measured by others) as 3435 Ah/kg, but omitted to point out that values this high contravene Faraday's law.

The alternative method of estimating the capacity of anodes is to measure the hydrogen evolved as it undergoes anodic dissolution. This method is essentially restricted to aluminium and magnesium alloy anodes and is only applicable to laboratory testing. It would be impracticable in field testing. It has the convenience that, in the case of an inefficient anode, it gives a result more quickly than a weight loss test. However, it is generally accepted as being less accurate. Although hydrogen measurement remains in NACE TM0190, the technique is rarely used nowadays. We need not concern ourselves with it here.

10.7.2 Testing modes and objectives

10.7.2.1 Screening tests

Screening tests form part of an anode alloy development programme. Their main objective is to eliminate materials that clearly do not warrant further investigation. For this reason, screening tests are devised by the developer to obtain rapid results and to give the candidate alloy a reasonable chance of making it through to the next phase of testing. An example is provided by the work of Reding and Newport [8] who evaluated some 2500 alloys. It would have been impractical to carry out a detailed evaluation of each alloy. For that reason, they conducted short-term (30-day) screening tests at an applied current density of $10.8\ A/m^2$. This was sufficient to select the most likely candidate alloys (and to get provisional patent applications filed). Longer-term testing on the more promising alloys came later.

Given that current Al-Zn-In alloy anodes offer acceptable operating potentials (around −1.1 V), and electrochemical efficiencies of ~95%, it is unlikely that the CP industry is going to trouble itself to seek an improved aluminium anode alloy. It would take something as radical as the banning of indium to force a resurgence of short-term screening tests.

10.7.2.2 Performance tests

From the standpoint of the CP designer, the relevant performance information on an anode alloy is provided in codes such as DNV-RP-B401. These data derive from historical laboratory testing or field trials. For the testing to be relevant, it would have needed to have been carried out over a long period of time (some argue that a year is not long enough), and the anode would have had to operate under conditions that reasonably reflect the real-life service. A case in point is the very extensive programme of testing sponsored by the US Navy and published by Murray and coworkers almost annually at the NACE Corrosion conference between 1993 [48] and 2004 [64].

However, even if an anode manufacturer tweaked the anode formulation, and really got on top of the quality of its production processes, such that performance testing demonstrated superior performance to that enshrined in the codes, it is unlikely that a CP designer would depart from the code guidance. We have already stressed the general reluctance in the CP industry to challenge the received wisdom of codes.

As with screening tests, it seems that performance testing has more relevance to history than to present day CP. The only exceptions are where it is anticipated that the anode will see service conditions outside the ordinary, and which are not fully anticipated in the codes. Examples include situations where the anode will be required to function where the environment is likely to be hot or polluted, or the water will be brackish or deep.

10.7.2.3 Deep water

CP practitioners have been interested in how anodes perform in deep waters for about half a century, although the definition of “deep” has changed progressively over that time. For example, in the early 1970s, with the burgeoning exploitation of the northern North Sea, there was interest in using anodes in water depths of up to 180 m, but to that point there had been little practical experience. Part of the reason for this lay in the practical difficulties of carrying out testing in such deep locations. Mackay [16,17] described testing in a sheltered Scottish Loch at depths of up to 137 m and found that Al-Zn-Hg and zinc anodes exhibited good efficiency down to these depths.

Various other tests were reported subsequently. This work was reviewed by Schrieber [39] in 1989, at a time when offshore structures were planned for installation in water depths of up to 780 m. He concluded that the spread of operating potentials and capacities between alloys of nominally the same composition exceeded the spread of results obtained when testing at different depths down to –1000 m.

In 2001, Fischer et al. [58] reached a slightly different conclusion on the basis of their review of the literature up to that date. They suggested that there were slight differences in the operating potential and efficiency of anodes at increased depth. However, their paper does not distinguish between anode performance changes resulting from changes in the current demand placed on the anodes by the cathodes and the functioning of the anode itself. With regard to the latter, there is no publication that demonstrates an effect of pressure alone on anode performance.

10.7.2.4 Elevated temperature

In contrast to hydrostatic pressure, elevated temperature adversely influences anode performance. Unlike some zinc anodes (discussed in Section 10.2.3), there is no evidence that aluminium alloy anodes undergo intergranular dissolution. Nevertheless, there is a convincing body of evidence that elevated temperature causes a reduction in capacity [26,29,43,67]. This accounts for the advice of most codes to down-rate the design capacity of Al-Zn-In anodes at elevated temperatures.

10.7.2.5 Biofouling

All seawater testing on anodes is affected by the fouling characteristics relevant to the locality. By definition, this makes the subject difficult to study. One of the few systematic attempts to study the effects of fouling was carried out on Al-Zn-Hg anodes in the late 1980s by Swain and Patrick-Maxwell [38]. Care must be exercised when interpreting their results. First, by definition, their observations can only be directly applicable to the fouling conditions at the location on the east coast of Florida where they carried out their testing. Second, it is interesting that they studied Al-Zn-Hg despite the fact it was by then widely supplanted by Al-Zn-In. The third problem lay in their experimental technique. They used galvanic coupling (i.e. "free-running") tests and ascribed all the differences in the cell current to changes in anode behaviour. They did not factor in the effect of changing cathodic current demand on their experiments.

Nevertheless, despite these experimental constraints, they produced a set of conclusions that are reasonable, and which accord with the experience of most marine CP practitioners.

An anode is more likely to acquire hard fouling (e.g. barnacles) if it is operated at a lower current density. Where an anode is covered with hard fouling, its resistance (R_a) increases in line with the extent of coverage of the fouling. Importantly, even if an anode is "100% covered" with hard fouling, it will nevertheless start to deliver current when called upon to do so. Furthermore, as anodic alloy dissolution increases, the barnacles detach.

The authors also expressed an opinion, which is also held by most CP practitioners, that soft fouling (e.g. seaweed) has little effect on anode performance. However, it should be noted that they did not actually study this.

10.7.2.6 Polluted environments

As a general rule, biofouling is severely inhibited in polluted seawater. So, if anodes are required to operate in polluted waters, we do not need to consider the effects of barnacles. However, we do need to consider whether the pollution itself impacts on the anode performance.

You will recall that in Chapter 1 we discussed accelerated low-water corrosion (ALWC) in the context of the nutrient-rich, or "slightly polluted" seawater encountered in harbours. CP is very effective in combatting ALWC. Indeed, the growing recognition of the ALWC problem by harbour engineers since the 1980s has resulted in good business for anode suppliers. I am not aware of any concerted investigation into anode performance in harbour environments. However, since the consensus is that sacrificial anodes have proven to be worthwhile investments, we can assume that if the slight levels of pollution influence anode performance, that influence is minimal.

10.7.2.7 Seabed sediments

Anodes tested in seabed mud perform less well than anodes tested in open, unpolluted seawater. However, there is no reason to view this fall-off in performance as being due to pollution. It can be explained by the chemical and electrochemical changes that develop due to the restricted ability of the anode corrosion products to be transported away from the anode surface.

There have been very few studies into the behaviour of anodes in polluted seabed sediments. An exception is the work of Felton et al. [61] who carried out work in connection with the Blue Stream pipeline project. This involved laying major pipelines in the bed of the Black Sea where, in some areas, very high levels of sulfide are encountered. Their work indicated that, in high sulfide (1000 ppm) seabed sediments, capacities for Al-Zn-In anodes were lower than 760 Ah/kg at 20°C. This compares with the recommended design value in DNV-RP-F103 of 1500 Ah/kg. These results confirmed much earlier work by Reding [12] who measured a capacity of 550 Ah/kg for an Al-Zn-Hg alloy in synthetic seawater purged with H_2S.

10.7.2.8 Estuarine waters

Aluminium alloy anodes require the presence of chloride in order to dissolve at usefully negative potentials. Chloride is plentiful in seawater, but not in fresh water. This explains why aluminium anodes work in the former, but not the latter. It also begs the question: how much chloride do we need in the environment for the anode to work?

Smith et al. [22], in summarising an extensive programme of laboratory testing, reported that diluting seawater to a resistivity of 4 Ωm, which corresponds to a seawater concentration of 5%, only has a modest impact on the behaviour of Al-Zn-In anodes. The operating potentials shifted from –1.08 V in full-strength seawater, to –1.03 V, and the capacity reduced slightly from 2550 Ah/kg to 2425 Ah/kg. This implies that aluminium anodes can be used in estuarine waters. However, it should also be recalled that, although the anode will perform, the increase in the environmental resistivity means that its resistance to the environment will also increase pro-rata. This needs to be factored into any CP design. Further information on the potential of Al-Zn-In anodes as a function of seawater resistivity was subsequently published by Schrieber and Murray [26]. Their results, which were obtained at an operating current density of 2.15 A/m^2 at 23°C, are reproduced in Figure 10.9. For convenience, this figure expresses seawater dilution in terms of its conductivity which is the reciprocal of resistivity. The potential in full strength seawater (conductivity ~5 S/m) is on the right hand side of the graph. The curve to the left shows the effect of progressively diluting the seawater on the anode's operating potential.

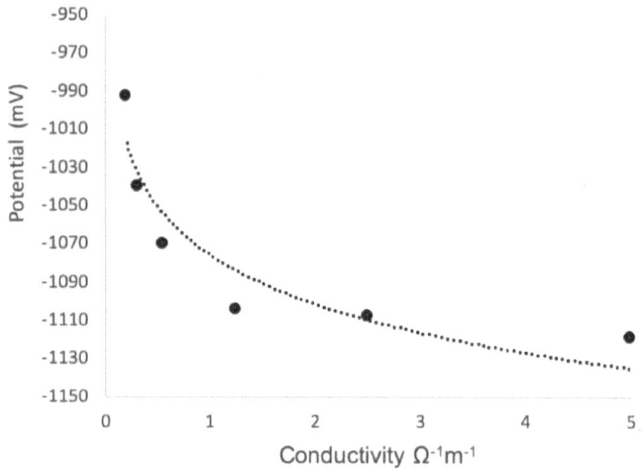

Figure 10.9 Al-Zn-In potential vs conductivity [26].

10.7.2.9 Pre-qualification and production testing

There is a belief that, for every major project requiring anodes, it is necessary to carry out pre-qualification electrochemical testing, followed by batch-by-batch testing to confirm the quality of the production run. Such tests are short-term and intended to produce a pass-fail result rather than data that can actually be incorporated into a design.

Crundwell, who had held senior positions in anode manufacturing, made the following comment in 2003 [63]. *In more than 25 years of practice in the use and manufacture of aluminium alloy sacrificial anodes the author of this paper has never heard of a report of the failure of an anode to operate as expected whose chemical composition was within specified tolerance. Thousands of electrochemical tests have been undertaken at great expense for no benefit whatsoever. It is to be hoped that the future will see realism and this waste of effort and resource consigned to history.*

I agree, but others do not.

10.7.3 Testing configurations

10.7.3.1 Constant current tests

It is simple to rewire a potentiostat, which we met in Chapter 5, to output a constant current as illustrated in Figure 10.10. In this configuration, the unit is termed a "galvanostat". Providing it has sufficient output power, it can be used to test multiple cells in series. This renders constant current testing of anodes relatively straightforward. A specimen of anode alloy, of a known surface area, is weighed. It is then placed in the test cell, and a known anodic current density is passed through it for a fixed time. The potential is recorded at intervals or continuously throughout the test. At the end of the test, the specimen is reweighed, and the current capacity can be determined as the quotient of the total charge (current x time) passed, divided by the measured weight loss.

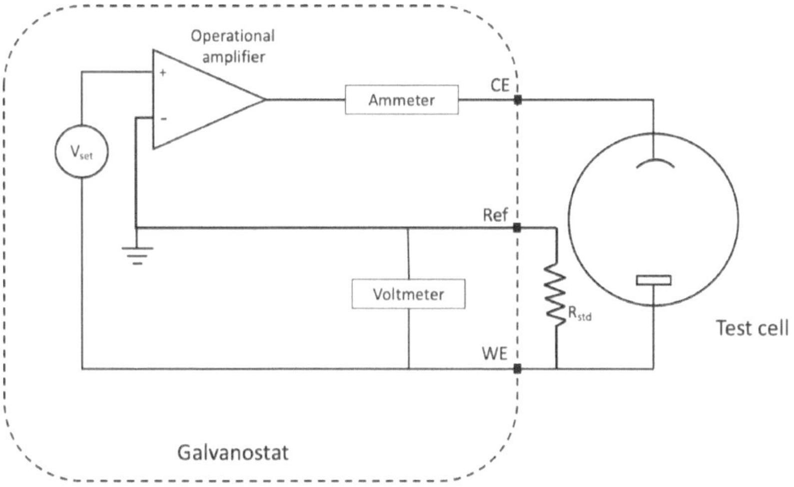

Figure 10.10 Constant current test – with a galvanostat.

Galvanostatic testing was a preferred screening test methodology during anode development. It was used, for example, in Reding and Newport's [8] seminal work in which they tested alloys at an anodic current density of 10.76 A/m^2 (1000 mA/ft^2) for 30 days.

Previous codes covered the method and, in some cases, prescribe acceptable results. For example, Appendix B of DNV-RP-B401 (2017) recommended a four-day test in which the anode current density is controlled at 15 A/m^2 for the first day, then 4 A/m^2, 40 A/m^2 and 15 A/m^2 for each of the following 3 days. The acceptance criteria for aluminium-based anodes are that operating potential must be –1.05 V or more negative on the fourth day, and that the capacity, as determined by weight loss, must be at least 2500 Ah/kg. The corresponding figures for zinc anodes are –1.00 V and 780 Ah/kg. The 2021 RP-B401, however, calls for test con ditio sn a nd results to be agreed between purchaser and manufacturer.

NACE TM0190 calls for a two-week test at an anodic current density of 6.2 A/m^2. Like DNV it does not commit itself to what should be an acceptable result for any type of alloy. This NACE method is referenced by EN 12496, which does not provide its own methodology. However, the EN makes the point that anodes do not operate galvanostatically and that capacity data obtained in high current densities are likely to be higher than obtained in service. It recommends that in order to provide an indication of likely anode performance the galvanostatic test can be performed at different current densities (0.003–10 A/m^2).

10.7.3.2 Constant potential tests

Constant potential testing has been used in the past for anode pre-qualification. This involves using a potentiostat to hold anode alloy specimen at a predetermined potential and recording its current output over a (relatively short) period. In practice, a series of tests would be carried out spanning a range of set potentials.

However, like the galvanostatic tests, potentiostatic tests do not mimic the way anodes actually operate. This is particularly so in the case of aluminium alloy anodes which operate in a pitting mode. When aluminium undergoes pitting, its natural potential fluctuates as pits initiate and then either re-passivate or propagate. The action of suppressing this natural behaviour of an anode, by artificially controlling the potential to a fixed value, does not make a lot of sense.

10.7.3.3 Free-running tests

Sacrificial anodes operate in neither the constant current nor constant potential mode. They form part of a galvanic couple in which they are usually connected to a much larger area of bare steel. The anode and cathode are then mutually polarised towards, but never reaching, a common potential value. Neither the potential nor the current output is constant. It varies in response to the development of active areas on the anode and the build-up of calcareous deposit on the cathode.

This galvanic coupling test is mimicked in so-called "free-running" tests. These more closely reflect the true electrochemical behaviour of a working anode. Figure 10.11 presents the results of one such laboratory test for a zinc anode in heated seawater [20].

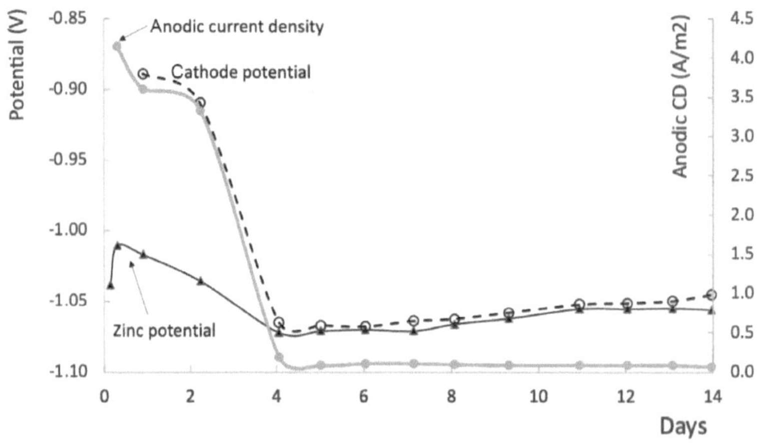

Figure 10.11 Example of a free-running test result.

Two approaches have been adopted. In the laboratory, anodes specimens have been coupled to steel plates with surface area ratios in the same ball-park as used in CP designs. Alternatively, where practicable, anodes have been coupled to large uncoated steel jetties [48].

One of the experimental problems with any free-running test is the measurement of charged passed. Historically, most experimental set-ups have involved connecting the anode to the structure via low-resistance shunt. The voltage drop across the shunt is measured at regular intervals (typically daily) and converted to a current by Ohm's law. Less archaically, the information has been recorded on digital data loggers. Digital or graphical integration of the current, over the test period, has then been used to estimate the total charge delivered. The capacity is then calculated from the estimated charge passed and the weight loss.

10.8 ANODE RESISTANCE

10.8.1 Relevance to design

We have already seen in Chapter 3 that, for a sacrificial anode CP design, the current output of an anode is assumed to be governed by Ohm's law:

$$I_a = (E_c - E_a) / R_a \tag{10.1}$$

where:

I_a = anode current output (A)

E_c = protected potential of the cathode (V)

E_a = operating potential of the anode (V)
R_a = anode resistance (Ω)

This was a slight oversimplification that we used to illustrate how a simple design is carried out. A more precise expression would be equation 10.2.

$$I_a = (E_c - E_a) / R_{total} \tag{10.2}$$

where:

R_{total} = circuit resistance (Ω)

R_{total} is the sum of all of the resistances in the CP circuit. These include the resistance of the anode to seawater, and of the structure to the seawater. It also includes the internal metallic resistances of the anodes and cathode themselves, together with the resistances of any cables or other connections between anode and structure.

R_{total} also includes the electrochemical resistance due to electrochemical charge transfer across the anode-seawater and seawater-cathode interfaces. All of these resistances are in series. So, if any is markedly greater than the others, it will dominate in controlling the CP current.

In practice, the anodes and their connections, and more particularly the structures, are substantial hunks of metal. This means that their internal resistances can usually be ignored.[4] It follows that most of the total resistance arises across the metal-seawater interfaces.

These anode and cathode resistances are dependent on surface area and geometry. Since the area of the cathode is invariably much greater than that of the anode, its resistance to the seawater is much lower. This means that, for practical purposes, R_{total} reduces the resistance of the anode to seawater. In other words, equation 10.2 becomes equation 10.1.

The anode resistance (R_a) is defined as its resistance to remote earth, or in this case remote seawater. Furthermore, the point has already been made that the EMF ($E_c - E_a$) available to drive the current against all of the series resistances in the CP circuit is typically only about 0.25 V. This means that R_a must be low (a fraction of an Ohm) in order that a useful current can be delivered.

10.8.2 How is R_a calculated?

The short answer to this question is simple. R_a is not calculated. It has to be estimated. We will now explore how these estimates have been arrived at.

Although R_a is the resistance of the anode to the remote environment, that is the environment at a distance of infinity, the greatest part of this resistance develops close to the anode's surface. It follows that, in an environment of any given resistivity, the value of R_a is determined largely by the shape and surface area of the anode.

Waldron and Peterson investigated the outputs of anode sizes in use on US Navy ships in the 1950s. They devised an experimental programme in which impressed current was applied between steel plates of the anodes' dimensions and a very large uncoated steel wall. They published [7] an Ohm's law relationship which, when converted to SI units, takes the form shown in equation 10.3.

[4] As we will see in Chapter 14, we cannot necessarily ignore the cathode resistance for pipelines.

$$\text{Log}\, I_a = 0.727\, \text{Log}\, A + \text{Log}\Delta E + 2.223 \quad (10.3)$$

where:

I_a=anode current output (A)
$\Delta = E_c - E_a$ (V)
A = anode surface area (m^2)

In the mid-1970s, Knuckey and Smith [19] revisited the Waldron and Peterson work, using zinc anodes instead of an applied current. They applied regression analysis to their results and derived a correlation similar in form to 10.3, but including a term for seawater resistivity.

Neither group of workers included any contribution from the shapes of the anode in their equations. Furthermore, although both groups reported that reducing the spacing between anodes had the effect of restricting the current output, neither factored that effect into their equation. Unsurprisingly, therefore, neither of these equations is used in modern-day CP design.

Nowadays, we use equations that take account of the anode shape as well as its surface area. These design equations do not have their origins in CP. They derive from earlier work on the earthing (or "grounding") of systems either for electrical safety or for providing earth return paths for currents powering electric trams or electric railways. However, irrespective of the motivation behind the original work, the objective is the same in CP as it is electrical power engineering. As Dwight [2] succinctly put it... *an electrical connection to the ground requires consideration of the engineering problem of obtaining the lowest number of ohms to ground for a given cost.*

For the most part, the calculation of resistance to earth has been carried on the basis of electrostatic principles. For example, the modified Dwight formula (10.5), which we first saw in Chapter 3, was originally derived from earlier work on the capacitance in air of a long cylinder. The formula Dwight arrived at, which ignores end effects, was:

$$R_a = \frac{\rho}{2\pi L_a}\left[\ln\left(\frac{2L_a}{r_a}\right) - 1\right] \quad (10.4)$$

where:

L_a=anode length (m)
r_a=anode radius (m)
ρ = seawater resistivity (Ωm)

In passing, it is worth mentioning that, although the modified version of the Dwight formula (see below) only became a common tool in offshore CP design in the 1980s, it had been used for decades previously in onshore CP. It was well established by the time it was featured in Peabody's classic book [10] in the mid-1960s.

We will not detail the origins of the various formulae here. The interested reader is directed to other publications that do better justice to the subject than I could. What is relevant is to present the various formulae in use within the CP industry and to consider how much reliance we may place upon them.

10.8.3 What the codes advise

10.8.3.1 Slender stand-off anodes

The early codes were non-prescriptive regarding the method of calculating R_a. In 1973, BS CP1021 referenced Dwight's paper [2] but gave no further information. However, by 1981,

DNV-TNA-703 was recommending the modified Dwight formula (10.5) for slender sacrificial anodes of the type used to protect offshore jackets; subject to a minimum stand-off distance of 0.3 m from the insert to the structure.

$$R_a = \frac{\rho}{2\pi L_a}\left[\ln\left(\frac{4L_a}{r_a}\right) - 1\right] \tag{10.5}$$

Although most authorities refer to this as the *modified Dwight* equation,[5] Cochran [33] has pointed out that it is more correctly termed the Dwight *Surface* equation. It defines R_a when one end of the anode is at the surface of the seawater. In the case of anodes, the *surface* is the submerged structure. The "modification" referred to is a doubling of the logarithmic term (compare 10.4 and 10.5). For a typical stand-off trapezoidal anode (1.6 m×0.16 m×0.16 m) in 0.25 Ωm seawater, the initial anode resistance (R_a) calculated according to the original Dwight formula would be 0.064 Ω. Calculated using the modified version of formula, as used in the codes, R_a would be 0.081 Ω.

All subsequent editions of RP-B401 have continued this advice, albeit the stand-off distance was increased from the 1986 edition onwards. Although the separation remained 0.3 m, it is now measured from the structure to the nearest face of the anode, not to the insert position as in the earlier edition. Further clarification was added in the 1993 edition where it was permitted to reduce the separation from 0.3 m to no less than 0.15 m, providing the calculated value of the resistance used for the design was uplifted by a factor of 30%.

Other codes, such as European Standards, have adopted the same resistance formula for stand-off anodes.

Before moving on, we should make a couple of important points concerning the formula shown in 10.5. It is a formula for calculating the resistance to remote earth of a single vertical cylindrical rod driven into a medium of known resistivity. Most offshore stand-off anodes are not cylindrical. They have a trapezoidal cross-section. This means that the CP designer has to calculate an equivalent anode radius (r_a). This is achieved by two simple methods. One is calculating the cross-sectional area of the anode (A) and then using relationship formula 10.6. The alternative method, given by DNV-RP-B401, uses the length of anode's cross-sectional periphery (equation 10.7).

$$r_a = \sqrt{\frac{A_a}{\pi}} \tag{10.6}$$

$$r_a = \frac{C}{2\pi} \tag{10.7}$$

where:

A_a = anode cross-sectional area (m^2)
C = anode cross-sectional periphery (m)

Equation 10.7 gives a slightly larger value of r_a, so leads to the prediction of a slightly lower value of the anode resistance. On this basis, using equation 10.6 in a CP design will understate the anode output and, thereby, add conservatism.

[5] As noted, it is a "formula" not an "equation". It only applies, to a reasonable accuracy, within a range of permissible ratios for anode length and radius.

10.8.3.2 Shorter stand-off anodes

DNV-RP-B401 gives formula 10.8 for shorter anodes (having $L_a < 4\ r_a$.). The symbols have the same meaning as in formula 10.5. In practice, however, it is very unusual to encounter a CP design based on stand-off anodes that are so short and stubby that $L_a < 4\ r_a$. Formula 10.8, which is referred to by Cochrane (30) as the *Sunde Surface Equation*, is therefore included here for completeness. It is rarely used in practice because such short stand-off anodes are themselves rare.

$$R_a = \frac{\rho}{2\pi L_a}\left[\ln\left(\frac{2L_a}{r_a}\left\{1+\sqrt{1+\left(\frac{r_a}{2L_a}\right)^2}\right\}\right)+\frac{r_a}{2L_a}-\sqrt{1+\left(\frac{r_a}{2L_a}\right)^2}\right] \tag{10.8}$$

10.8.3.3 Long flush-mounted anodes

There are many instances where there is simply not enough room to fit a stand-off anode, or else the stand-off profile would add unacceptable hydrodynamic drag. The classic example of this is a ship's hull for which flush-mounted anodes are required. A relatively simple resistance formula, known as the Lloyds formula (10.9), has been developed for these anodes.

$$R_a = \frac{\rho}{2S} \tag{10.9}$$

where:

S = arithmetic mean of the anode length and width.

As with the Dwight formula for stand-off anodes, the Lloyds formula is deemed to hold providing the anodes are long. In this case, an anode is considered as being long if its length is at least four times its width or its depth, whichever is the greater. Again, an anode that was only four times its width would not realistically be described as "long" in common parlance.

10.8.3.4 Short flush-mounted anodes and bracelets

Formula 10.10 is generally advised for calculating the resistance of short, flush-mounted anodes and, more particularly, pipeline bracelets. In some literature, this is referred to as the *McCoy* formula. This attribution appears to be due to a 1970 publication by the eponymous Australian corrosion engineer [11] who advocated its use. However, McCoy did not develop the formula. It had originally been published sixteen years earlier by Ryan [34]. It approximates to the formula that would be arrived at by using electrostatic principles to estimate the resistance of a disk in an infinite medium.

$$R_a = \frac{0.315\rho}{\sqrt{A_a}} \tag{10.10}$$

where:

A_a = the exposed surface area of the anode (m^2).

10.8.4 Validation of resistance formulae

In 1982, Cochrane [30] reviewed anode resistance formulae in use at the time. Whilst he was guilty of some exaggeration when he stated that there was… *a veritable deluge of equations*... in use, it is clear that the CP designer was not short of options. This point was illustrated by Strømmen [31] who listed six equations for stand-off anodes and a further eight for bracelets and flush-mounted anodes. It was also clear that ideas on the appropriate formulae differed on either side of the Atlantic. By 1978, DNV in Europe had reduced the list of acceptable formulae to three: Modified Dwight, Lloyds and McCoy [25].

Occasionally, workers have attempted to validate resistance formulae. It was easy enough to carry out desk studies to see how different formulae would compare with each other across ranges of anode sizes and geometries. However, the exercise would be sterile unless it could also be determined which, if any, of the formulae produced a result that was closest to reality. Resolving this point requires an independent assessment of anode resistance. Such an assessment can take one of three forms: in-service testing, laboratory scale modelling, or computer modelling.

10.8.4.1 In-service testing

There have been various attempts at in-service testing involving real CP systems on real structures. For example, Strømmen [31] cited earlier work by Wyatt in which monitored anodes (3.05 m×0.225 m×0.225 m) were installed on a structure. The objective was to convert the monitored current output to an anode resistance on the basis of the measured difference in potential (ΔE) between the anode and the adjacent structure. The results were mixed. For each anode, it was observed that the modified Dwight formula predicted a higher resistance, and therefore a lower current output, than was actually measured. This gave reassurance that the modified Dwight formula would force a conservative design. However, as observed by Eliassen (also cited by Strømmen [31]), the variations in the outputs of nominally the same anodes, on the same structure, indicated resistances that ranged from 0.013 to 0.064 Ω. This four-fold span of values, for what should be a single parameter, indicates how problematic it is trying to determine ΔE between some point on the anode, and some relevant point on the adjacent cathode.

As an alternative to the (very rare) installation of monitored anodes on real structures, anode current outputs have been estimated from measurements of the electric field gradient at the anode surface. Nisancioglu [32] has described work in which potential and electric field measurements were performed on a bracelet anode attached to a pipeline riser in the North Sea. The measurements of the field gradients were subject to computer analysis to produce an estimated anode output current of 0.615 A. Since the value of ΔE (0.036 V) had also been measured as part of the exercise, it was possible to apply Ohm's law to estimate the anode resistance (0.0585 Ω). Furthermore, since the surface area of the anode (1.56 m^2) and the seawater resistivity (0.3 Ωm) were also known, it was possible to calculate the anode resistance values according to the formulae in use. His results are reproduced in Table 10.6.

Table 10.6 Anode resistance results

Method		
Measured	*Estimated*	*Resistance (Ω)*
Field Data		0.058
	Hallén	0.053
	Modified Dwight (10.5)	0.037
	McCoy (10.10)	0.076
	Peterson	0.046
	Lloyds (10.9)	0.082

From [32].

He observed correctly that the Hallén formula gave the best agreement with the field data. However, from the perspective of a CP designer, the closeness of the agreement is less important than the observation that Hallén gave a non-conservative result (albeit only slightly so). Both Peterson and modified Dwight likewise return non-conservative results for this bracelet anode configuration. That leaves us with two formulae that are conservative that is their use will force a surplus of anode area into the design. These are the McCoy and Lloyds formula. Of these two, the McCoy is the less conservative. So, it is not surprising that it is this formula that is favoured by DNV, and other codes, for bracelet anodes.

10.8.5 Anode clustering

10.8.5.1 The problem

Typically, the codes offer a design value for the operating potential of the anode, and we then use this to calculate its current output when it is coupled to the cathode. It is implicit in this advice that we can regard the electrochemistry taking place on the surface of the anode as having essentially zero output resistance. In other words, we can draw as much current as we wish without materially affecting the anode's potential. Within the limits of most CP designs, this is a reasonable approximation because the dissolution kinetics of a well-formulated anode alloy in seawater are very rapid. For most offshore CP applications, it then becomes a reasonable design approach to estimate the anode output on the basis of its resistance to remote earth. We do this using one or more of the resistance formulae we have discussed above.

In itself, this presents a simple basis for an engineer to design a CP system. Once the output of the anode (i_a) has been estimated, all that needs to be calculated is the number of anodes (N) needed to satisfy the total current demand (I_a) using simple arithmetic:

$$N = \frac{I_a}{i_a} \tag{10.11}$$

There is, however, a problem implicit in equation 10.11. It assumes that i_a is independent of N. So, any value of I_a can be achieved simply by increasing N. This would seem logical to any engineer with a basic knowledge of electricity and an understanding of parallel resistors. Unfortunately, such an understanding is unhelpful when it comes to electrochemical systems. On a real structure, we cannot simply increase the total current output pro-rata to the increase in the number of anodes.

To understand why this is so, we need first to revisit equation 10.1:

$$I_a = \left(E_c - E_a\right) / R_a \tag{10.1}$$

This uses the anode potential E_a, for which a single value is given by codes. However, we may recall from Chapter 2 that *potential* in electrochemistry is the *difference* between the potential in the metal ($\varnothing_m$) and the potential in the adjacent solution ($\varnothing_s$).

$$E_a = \varnothing_m - \varnothing_s \tag{10.12}$$

Our use of the design value for E_a is only valid providing the potential in the solution ($\varnothing_s$) adjacent to the anode does not become noticeably more negative. However, this is just the situation which can arise when one anode is placed near another. In effect, the current output from an adjacent anode creates an electrical field which influences the value of $\varnothing_s$. This in turn effectively reduces the electrode potential difference between the anode and the seawater, and suppresses its output. The effect is, of course, mutual. Each of the two adjacent anodes influences each other in the same way. The situation is exacerbated if more anodes are placed in close proximity to each other.

This effective reduction of the potential difference between the anode and the seawater can also be viewed as an increase in the anode to seawater resistance. Both have the same effect of suppressing the anode's output. The CP industry adopts the resistance paradigm, not least because it relates more easily to anode resistance formulae. Nevertheless, the CP codes lack clarity on how a designer should allow for this mutual suppression of anode current output.

The fact that anode resistance increases if anodes are moved closer together has been well understood in the onshore CP industry. Dwight's 1936 paper [2] addresses the issue. It has also been recognised in the offshore CP industry for at least 40 years. For example, DNV highlighted the issue in a 1978 conference [25], stating that if anodes... *are grouped in arrays, interference between the anodes must be taken into account when calculating the anode resistance.* The paper went on to inform that *theoretical and practical investigations* were in progress for improving the use of the resistance formulae. It seems, however, that those investigations did not yield much that could usefully be included in any of the editions of the DNV recommended practices for CP. Up until, and including, the 2017 edition of RP-B401 the advice was that... *With the exception of very large anodes, shielding and interference effects become insignificant at a distance of about 0.5 m or more. If anodes are suspected to interfere, a conservative approach may be to consider two adjacent anodes as one long anode, or as one wide anode, depending on their location in relation to each other.* The 2021 edition withdrew any reference to minimum separation distances, replacing it with advice to carry out computer modelling to assess the reduction in output. We discuss computer modelling in Chapter 17.

10.8.5.2 The consequences

The pre-2021 DNV-RP-B401 advice was of limited value. Taken literally, it told designers that, with the exception of very large anodes, shielding and interference effects become insignificant at a separation of about 0.5 m. This implied, incorrectly, that installing anodes no less than 0.5 m apart would not compromise their outputs. This interpretation of the wording featured in at least one litigation in which I was involved.

To see the problem with this advice it is convenient to consider an artificial, but not unrealistic, situation of attempting to provide CP to a large diameter driven pile of the type that might be used to support the tower of an offshore wind turbine. Because anode attachments would shear off due to the piling vibrations, it is not possible to place anodes on the pile itself. One obvious solution is to install anodes attached to a collar which is subsequently installed at the top of the driven pile. This practice has been widely adopted in offshore wind farms, where the anodes are mounted on the transition piece which connects the pile to the tower.

If we base a design on a notional Al-Zn-In cylindrical stand-off anode 1 m long and 0.15 m diameter then, using the modified Dwight formula and taking the seawater resistivity as 0.3 Ωm, we estimate an anode resistance to remote earth of ~0.19 Ω. Using the difference between the values for protected steel potential (E_c −0.80 V) and anode operating voltage (E_a −1.05 V), we see that our driving voltage is 0.25 V. Ohm's law then tells us that our anode has an instantaneous output of ~1.3 A. Moreover, so long as we do not encroach on the 0.5 m limit on spacing suggested by DNV, we can simply add anodes around the circumference of the transition piece to increase the CP current output.

The process is illustrated by the solid markers in Figure 10.12. As the lateral anode spacing is reduced, more anodes can be fitted around the circumference and the cumulative output

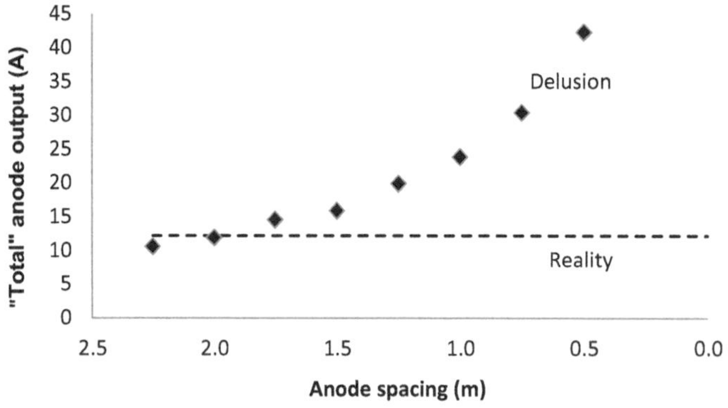

Figure 10.12 Notional outputs multiple stand-off anodes vs single bracelet.

increases. However, as noted in the figure, this approach is likely to lead to a "delusion". The problem is that by simply applying Ohm's law to the individual anodes, no account is taken of the mutual suppression of anode outputs. One way of providing a "reality check" is to adopt the view that the maximum output would be obtained if enough anodes were installed to produce a solid band around the circumference. This would constitute a bracelet, the resistance of which is most realistically estimated using the McCoy formula (10.9). The output of the equivalent solid bracelet is shown by the dotted line in the same figure, referred to on the figure as the "reality". Thus, we see that application of the Dwight equation to multiple anodes could lead the CP designer to overestimate the current output by a factor of up to four.

Indeed, exactly this problem has arisen on a number of offshore wind farms in European waters. We will return to this in Chapter 13.

10.8.5.3 The solution

There is no "solution" to the problem of mutual anode output suppression. It is simply a manifestation of the laws of physics. However, there is a need for a methodology to determine whether or not mutual suppression needs to be accounted for and, if it does, how this should be done.

It may well be the case that modifications to the anode resistance formulae used for offshore CP have not been developed because there has not been the need. The main market for sacrificial anodes has been space-frame jacket structures associated with petroleum production or submarine pipelines. A rule-of-thumb for a jacket might be to allocate one anode for every 30 m^2, or thereabouts, of steel surface. This ensures that anodes are reasonably well separated. Similarly, for pipelines, it would be unusual to have anodes spaced more closely than every twelve pipe joints. That places the anodes at least 146 m apart. These days, larger separations are the norm. This is sufficient to rule out any mutual output suppression.

More recently, however, the practical need to locate anodes in closer proximity to one another has developed. As noted, one example is the installation of anodes on wind farm monopile foundations. Other examples include the use of anodes mounted on sleds such as are frequently employed in CP retrofits. We consider retrofits for structures and pipelines in Chapters 13 and 14.

There has been comparatively little analytical work aimed at seeking modifications to the single anode resistance formulae to account for multiple anodes.

Knuckey and Smith [19] referred to the *current crowding* effect in 1977 when they investigated the effects of recent changes in the distribution of zinc anodes on the hulls of Royal Australian Navy ships. The change involved moving from widely distributed anodes to anodes placed in-line and closer together. Much of their paper focussed on adapting the Waldron and Peterson relationship (10.3). However, it is worth noting their tests in which the current outputs from two 0.15 m long flush-mounted zinc anodes were measured in 25 Ωm seawater at different linear separation distances. Their results, reworked in terms of anode current density and with anode spacing expressed in terms of multiples of anode length, are shown in Figure 10.13. Considering just two anodes, it appears that when the spacing needed exceeded the equivalent of ten anode lengths before mutual suppression of the current outputs could be discounted. Superficially, this observation suggests that the above-mentioned DNV advice for a minimum spacing of 0.5 m is non-conservative.

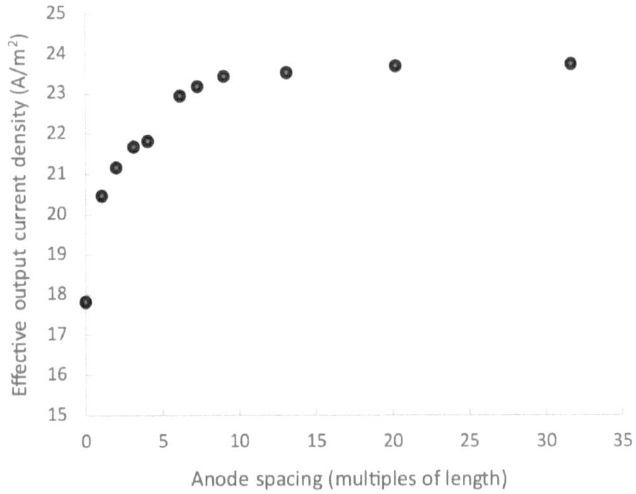

Figure 10.13 Effect of end-on spacing between two anodes [19].

A more detailed analysis of groups of offshore anodes was examined by Cochrane [30,33] in the early 1980s. He adopted the *Sunde Submerged Equation* (10.13) for a single anode which, it will be observed, is of a similar form to *Sunde Surface Equation* (10.8). He then incorporated Sunde's formula for the resistance of parallel equal length anodes in a circular arrangement (10.14).

$$R_a = \frac{\rho}{2\pi L_a}\left[\ln\left(\frac{L_a}{r_a}\left\{1+\sqrt{1+\left(\frac{r_a}{L_a}\right)^2}\right\}\right)+\frac{r_a}{2L_a}-\sqrt{1+\left(\frac{r_a}{L_a}\right)^2}\right] \tag{10.13}$$

$$R_{\text{array}} = \frac{1}{nR_a}+\sum_{m=1}^{n-1} R_a\left\{D\sin\left(\frac{m\pi}{n}\right)\right\} \tag{10.14}$$

where:

R_{array} = resistance of a circular array (Ω)
n = number of anodes in array
D = diameter of array (m)

The logic of Cochran's analysis was convincing, and the results produced in illustrative examples were intuitively reasonable. However, as with the other resistance formulae we have considered here, there has been no independent experimental investigation of the range of its applicability.

Subsequent to the mid-1980s, endeavours to determine the outputs of anodes and anode arrays, particularly when installed in congested spaces, have been pursued through computer modelling, rather than seeking to adapt classical resistance-to-earth formulae. We will return to this in Chapter 17.

10.9 ANODE DESIGN AND MANUFACTURE

10.9.1 Who does the design?

A prudent CP engineer produces a working estimate of the number, size and shape of the anodes required. However, they would be well advised to leave the detailed design of the anode to the anode foundry. After all, it is the foundry that possesses the knowledge of how best to manufacture an anode, and what is practicable and what is not. Unlike the electrochemical aspects of anode performance, the subtleties of the casting process are not widely disseminated in the corrosion literature. A notable exception is the paper by Warnock [51] which provides useful practical design information to aid the castability of both trapezoidal and bracelet anodes.

The same author [50] has also recorded a case where the CP design consultant... *had apparently made little attempt to discover the limiting factors concerning anode length as a function of anode diameter and thickness.* In my experience similar cases, which generally arise from an engineer slavishly following the design company's in-house spreadsheet, are not uncommon.

10.9.2 The anode specification

Let me tell you how a typical engineer drafts a sacrificial anode specification for a new offshore development. The first step is to dig out an anode specification from an earlier project. The second step is to cross out the name of that earlier project and client, and to write in the name of the new client and project. The third step is to complete the weekly time-sheet allocating 40 hours for the activity.

This probably explains the perpetuation of the requirement for electrochemical testing for the production quality control of anodes.

10.9.3 Anode inserts

10.9.3.1 Insert configuration

The terms *anode core* and *anode insert* are used interchangeably in the CP industry. With very few exceptions, all anodes are cast onto carbon steel inserts. For flush-mounted anodes, this will be a flat steel bar (Figure 10.14). The insert can either be welded or bolted to the structure. It is advisable to paint the underside of the anode to prevent dissolution from that face reaching the insert and causing detachment of the majority of the alloy.

For stand-off anodes up to about 80 kg, the insert can be a steel rod; and for larger stand-off anodes, it is likely to be tubular. Two configurations are in use as shown in Figure 10.15. Of the two, (a) is easier to cast because it does not involve fitting the core through penetrations in the mould. However, EN 12496 advises that, for stand-off anodes where divers or remotely operated vehicles (ROVs) are likely to operate, cores should protrude through the end face of the anodes. This is shown as configuration (b) and is often referred to as a "cow horn" anode. This eliminates the risk of divers' support umbilicals becoming snagged.

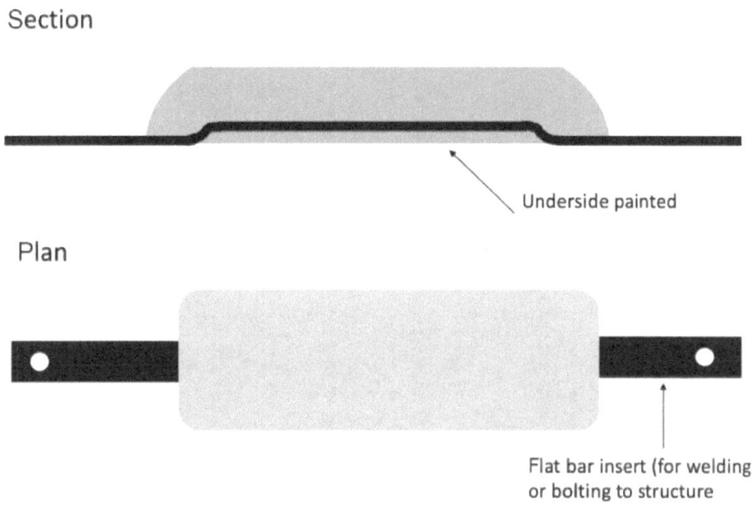

Figure 10.14 Schematic of flush-mounted anodes.

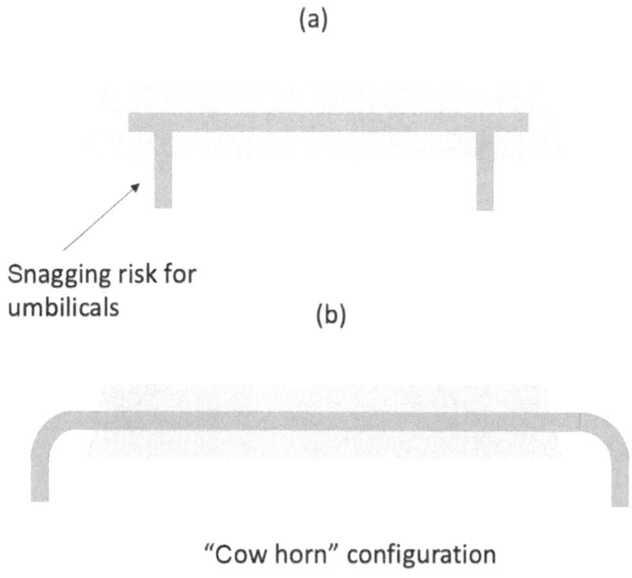

Figure 10.15 Stand-off anode insert configurations.

There are two modes of construction for bracelet anodes. One is for a group of anodes to be fixed by welding of their inserts to circumferential steel bands. This forms a segmented hinged bracket to fit around half of the circumference of the coated tubular (Figure 10.16). The profile of this arrangement renders it unsuitable for most pipe-lay methods. So, it is little used for pipelines.

Figure 10.16 Segmented bracelets (courtesy of Aberdeen Foundries).

More usually, pipeline anodes are cast in the form of half-shells (Figure 10.17). There are two options for closing the half-shells around the pipeline. From the installation point of view, the more convenient is usually to employ a hinged bracelet arrangement, which can then be secured by bolting. The less convenient alternative is to join the two halves by welding. Either way, the profile of the anode is designed so that it does not interfere with offshore pipe-laying.

Figure 10.17 Half-shell bracelets (courtesy of Aberdeen Foundries).

10.9.3.2 Insert surface preparation

Irrespective of the insert configuration used, it is vital that the anode alloy is well bonded to it. This is necessary not only to ensure electrical conductivity but also to prevent the physical detachment of hunks of alloy from the anode.

The standard method of preparing the steel insert for zinc or magnesium anodes is by hot-dip galvanising, although abrasive blast cleaned steel is also acceptable.

Galvanising is a common process carried out after the insert has been fabricated. It involves sequentially degreasing, pickling, rinsing and drying the inserts before immersing them for a prescribed period in a bath of molten zinc. The resulting galvanised layer is typically 50–100 µm thick. It comprises a series of zinc-iron alloys ranging from a thin layer, containing approximately 25% iron, at the interface with the steel, through progressively lower iron content alloys, to essentially pure zinc at the surface. Thus, galvanising is better described as surface alloying than a coating. When the molten zinc or magnesium is poured into the mould containing the galvanised insert, it fuses with the pure zinc in the outer surface of the galvanising layer creating a direct metallurgical bond between the anode and the steel insert.

I have never encountered, or heard tell of, any problems with the bonding of zinc anodes to galvanise steel inserts. However, this does not hold for aluminium.

DO NOT CAST ALUMINIUM ONTO GALVANISED STEEL.

There is a concern that the molten aluminium will melt the zinc with a resulting increase in the zinc, and iron, content of the anode alloy. More important: I have also been told independently by two individuals, with peerless experience in anode manufacture, that casting aluminium onto galvanising sets up an exothermic reaction that is so vigorous that molten alloy is ejected from the mould, and the steel partially melts!

Because aluminium, at any normal casting temperature, forms no metallurgical bond with steel, the adhesion of the alloy to the anode insert has to rely on intimate physical contact. This requires ensuring that the steel insert is free of rust. It also must have a roughened surface to maximise the steel to anode contact area. This steel surface condition is achieved by abrasive blasting.

10.9.4 The casting process

The sacrificial anode industry would not claim that there was anything sophisticated about the process of casting an anode. The steel insert is fixed in position in the mould. Then, as explained by Johnson [54], the task is to fill the mould as rapidly as possible but with a minimum amount of agitation or turbulence. The process is shown in Figure 10.18. Finally, more molten metal is added to compensate for the shrinkage that occurs as the alloy cools. Importantly, this additional topping up must be carried out before the bulk alloy has solidified.

When adequately cool, the anode is removed from the mould. For trapezoidal and flush-mounted anodes, this simply requires turning the moulds upside down. For half-shell bracelets, the cast anode is simply released when the two faces of the mould are separated.

For a while, there was an interest in semi-continuously casting anodes in water-cooled moulds [28]. This required a cylindrical mould orientated vertically with a mobile base which closes the mould at the beginning of the casting. A hole in the centre of the base allows for the insertion of the tubular steel insert. As the casting process starts, the mould is filled with liquid metal which solidifies when in contact with the base. The base is slowly lowered at the same speed as the insert. The mould, and moveable base, is cooled directly by water which runs around the complete circumference of the mould. As a result, the anode solidifies rapidly and reasonably symmetrically.

Figure 10.18 Anode casting (courtesy of Corrpro).

It was claimed that this method of casting produced a finer-grained alloy and more uniform distribution of alloying elements. Both of these claims were almost certainly true. However, there was no published evidence that indicated that the electrochemical properties of these semi-continuously cast anodes were measurably superior to those of anodes cast conventionally. A fine-grained anode microstructure does not necessarily translate into improved anode performance. Unsurprisingly, therefore, this more costly casting option was always at a disadvantage in the market. These anodes are no longer manufactured.

Nevertheless, there was at least one instance of these cylindrical anodes being selected ahead of the less expensive trapezoid shape. They were selected for the Oseberg "B" jacket [42] because the cylindrical profile would result in less hydrodynamic drag on the structure. Hence, it permitted a slight, but important, reduction in the steelwork weight. We will meet Oseberg "B" again in Chapter 13.

10.10 QUALITY CONTROL

An anode is not a precision-machined item. It is a casting and, as such, will inevitably exhibit some irregularities and defects. It is, therefore, important to define what is, and what is not, acceptable in a cast anode. The following guidance on this for stand-off anodes is given in NACE SP 0387. The advice in EN12496 mirrors this. In principle, these codes apply to either aluminium or zinc alloy anodes. However, it is clear that they were compiled with aluminium anodes in mind.

10.10.1 Sampling

A sampling plan will need to be agreed between the manufacturer and the purchaser. The advice given by SP 0387 on the intensity of sampling is summarised in Table 10.7. This applies to stand-off anodes because SP 0387 does not cover bracelets.

10.10.2 Dimensional and weight tolerance

The weight and dimensional tolerances advised by NACE SP 0387 are summarised in Table 10.8. The same figures are used for slender anodes in ISO-15889-2 and in EN 12496.

Table 10.7 Sampling of anode batches (per NACE SP 0387)

Inspection objective	*Sampling*	*Notes*
Non-destructive examination		
Individual anode weight	100%	anodes >140 kg
	10% min.	anodes< 140 kg
Batch weight of anodes	100%	
Dimensions (anode)	10% min.	
Dimensions (insert position)	10% min.	
Visual inspection of surfaces	100%	
Destructive examination		
Inspection for internal defects	Agreed between foundry and purchaser	

Table 10.8 Tolerance on stand-off anode dimensions

Dimension	*Tolerance*
Mass (individual anodes)	±3% or ±2 kg (whichever is the greater)
Mass (total anode batch)	0% to +2%
Length	The lesser of ±3% of nominal length or ±25 mm
Mean width	±5%
Mean depth	±10%
Deviation from straightness	<2% of the nominal length
Insert deviation from nominal position	<5% of length;< 5% of width;< 10% depth or subject to agreement between foundry and purchaser.

Essentially, the same guidance on casting quality for bracelet anodes is given by ISO 15889-2. It recommends checking at least 10% of the production run to confirm compliance with the dimensional requirements in Table 10.9. The same information is given in EN 12496.

Table 10.9 ISO/EN tolerance on bracelet anode dimensions

Dimension	*Pipe diameter*	*Tolerance*
Length	Any	The lesser of ±3% of nominal length or ±25 mm
Internal diameter	≦ 300° mm	0 to+4 mm
	>300° mm to ≦ 610° mm	0 to+6 mm
	>610° mm	0 to+1%
Thickness	Any	±3 mm
Bowing or twisting	Any	<2% of the nominal length

ISO 15889-2 also recommends confirming the dimensions of an agreed number of bracelet anodes from the production run by assembling them on an appropriately sized former. This is a useful check. I was involved in an instance in the 1980s where a batch of anodes failed to meet the specified dimensional tolerances and did not align properly on the formers. Some lusty blows with a sledgehammer rectified the misalignment and brought the dimensions back within tolerance.

10.10.3 Casting quality

10.10.3.1 Non-destructive examination

General appearance
The following is a list of dimensional limits on some typical features or blemishes which might be expected in commercially cast anodes. Meeting these limits should not be too onerous for any competent anode manufacturer.

- Shrinkage depressions should be limited to <10% of the nominal depth and must not expose the steel insert.
- Not more than 1% of the surface should contain non-metallic inclusions visible to the naked eye.
- Cold shuts or surface laps should not exceed a depth of 10 mm.
- Any protrusions that might pose a hazard to personnel handling the anodes should be removed.

External cracks
Neither zinc nor magnesium is prone to cracking. However, these are not normally used as stand-off anodes, so the codes need not address cracking.

Aluminium alloys, on the other hand, are susceptible to cracking. Indeed, it would be unusual for a sizeable cast aluminium alloy anode to be completely free from cracks. Some of the alloying ingredients that are known to benefit anode performance are also recognised as likely to cause cracks to develop. The propensity for cracking depends in a complex manner on the alloy formulation, the design of the insert and the casting process itself.

Cracks are of no consequence to the electrochemical functioning of the anode, unless they develop to the extent that alloy might become detached from the insert before it has dissolved to yield beneficial current. Unlike dimensional and weight tolerances, SP 0387 differs from EN 12496, and from ISO 15589-2, in its advice on the tolerable cracking in stand-off anodes. The EN and ISO advice may be summarised as follows:

Cracks are ***not*** accepted if:

- in the area where the anode material is not internally supported by the core, or
- penetrating to the steel core or penetrating through the anode, or
- in a section of the anode not supported by the core, or
- if they follow the longitudinal direction of long slender anodes (unless in the "topping-up" material).

Transverse cracks are acceptable providing:

- the length is <100 mm and the width <1 mm
- there are no more than 10 cracks per anode.

10.10.3.2 Destructive examination

For a large anode order, some destructive testing may be agreed upon. This might involve the sampling of one anode at random from the consignment. This would be sectioned typically at 25%, 33% and 50% of the nominal length. The cut faces should then be examined visually without magnification. SP 0387, EN 12496 and ISO 15589-2 offer essentially the same advice.

- No more than 2% of the total exposed surface area, and no more than 5% of any single surface, should exhibit gas holes or porosity. (ISO 15889-2 and EN 12496 add that no gas hole should exceed 100 mm^2.)
- No more than 1% of the total exposed surface area, and no more than 2% of any one surface, should exhibit non-metallic inclusions. (ISO 15889-2 and EN 12496 add that no non-metallic should exceed 100 mm^2.)
- No more than 10% of the total exposed perimeter of a tubular insert should have voids adjacent to the steel. For any individual exposed perimeter, the figure is 20%. If the insert is not tubular, it will be more difficult to control voids. Alternative acceptance criteria will need to be agreed.

10.11 ANODE INSTALLATION

The CP design, for both new and retrofit systems, should include a detailed anode installation specification. In particular, this will address the connection of the anode to the structure. Depending on circumstances, the connection options include:

- direct welding of the steel insert to the structure,
- direct welding of the steel insert to pre-welded pads or gusset plates,
- cable connection by pin brazing (direct or using brazed studs).

Thermite welding, which has been widely practiced for onshore CP work and is endorsed by various codes, is not always practicable in an offshore environment.

Under certain circumstances, mechanical (clamp type) connections may be used. Such allowable situations include: attaching lightweight anodes (<25 kg) in water ballast tanks (WBTs) by means of industry-standard "M" clamps, and the diver or ROV connection of cables from sub-sea retrofit sleds (see Chapters 13 and 14).

REFERENCES

1. H. Davy, *On the corrosion of copper sheathing by sea water, and on methods of preventing this effect; and on their application to ships of war and other ships*. Philosophical Transactions of the Royal Society A **114** 151–158, 242–246 (1824).
2. H.B. Dwight, *Calculation of Resistances to Ground* Electrical Engineering 1319 December 1936. (Reprinted in Materials Performance **22** (4) 23 (1983)).
3. H.A. Humble, *The Cathodic Protection of Steel Piling in Sea Water*. Proc. 5th Annual Corrosion Conf. NACE International (Houston) 1949.
4. D.B. Anderson and R.B. Teel, *The Effect of Minor Elements on the Current Output Characteristics of Zinc Galvanic Anodes in Seawater*. Proc. Navy-Industry Zinc Symp, *Cathodic Protection Zinc As A Galvanic Anode – Development, Application and Specification 21–22 April* 1955 Publ. Bureau of Ships Navy Department (Washington) 1955.
5. R.B. Teel and D.B. Anderson, *The effect of iron in galvanic zinc anodes in sea water*. Corrosion **12** 343–349 (1956).
6. J.T. Crennel and W.C.G. Wheeler, *Zinc anodes for use in sea water*. Journal of Applied Chemistry **6** 415 (1956).
7. L.J. Waldron and M.H. Peterson, *The current-voltage relationship of galvanic anode arrays in cathodic protection*. Corrosion **14** (6) 47 (1958).
8. J.T. Reding and J.J. Newport, The influence of alloying elements on aluminum anodes in sea water. *Materials Protection* **5** (12) 15 (1966).

9. G.L. Doremus and J.G. Davis, *Marine anodes: the old and the new – cathodic protection for offshore structures.* Materials Protection **6** (1) 30 (1967).
10. *Control of Pipeline Corrosion* A.W. Peabody publ. NACE (Houston) 1967.
11. J.E. McCoy, *Corrosion control by cathodic protection – theoretical and design concepts for marine applications.* Trans. Institute of Marine Engineers 82 (6) 1970.
12. J.T. Reding, *Sacrificial anodes for ocean bottom applications.* Materials Performance **10** (10) 17 (1971).
13. T.J. Lennox Jr. *Electrochemical properties of magnesium, zinc and aluminum galvanic anodes in sea water*, p. 176. Proc. 3rd Int. Cong. On Marine Corrosion and Fouling, 1972.
14. E.A. Lyublinskii et al., *Investigation and selection of cast aluminum protective alloys.* Journal of Applied Chemistry USSR **46** 823 (1973).
15. C.M. Grandstaff et al., *Performance Evaluation of Quality Aluminium Anodes in Seawater* Paper 110 CORROSION/74.
16. W.B. Mackay, *Deep Water Testing of Sacrificial Anodes*, Paper 114 CORROSION/74.
17. W.B. Mackay, *Deep Water Testing of Sacrificial Anodes (Part II)*, Paper 18 CORROSION/75.
18. S.N. Smith, C.F. Schrieber and R.L. Riley Jr. *Supplementary Studies of the Galvalum III Anode – Exposure Time and Low temperature*, Paper 35 CORROSION/77.
19. P.J. Knuckey and B.S. Smith, *Effects of array size and spacing on the calculation of galvanic anode current outputs.* Materials Performance **16** (5) 44 (1977).
20. C.G. Googan, *The Effect of Temperature on the Performance of Zinc Alloy Sacrificial Anodes in Sea Water.* MSc Dissertation, University of Manchester (1977).
21. F.Ø. Jensen, A. Rygg and O. Satre, *Testing of anodes for offshore buried pipelines at elevated temperature.* Materials Performance **17** (9) 9 (1978).
22. S.N. Smith, J.T. Reding and R.L. Riley Jr., *Development of a broad application saline water aluminum anode – "Galvalum" III.* Materials Performance **17** (3) 32 (1978).
23. C.G. Googan, *The Effect of Iron on the Performance of Aluminium Alloy Sacrificial Anodes.* PhD Thesis University of Manchester, 1979.
24. V. Ashworth, C.G. Googan and J.D. Scantlebury, *Intergranular dissolution of zinc alloy sacrificial anodes in sea water at elevated temperature.* British Corrosion Journal **14** (1) 46 (1979).
25. S. Eliassen and G. Valland, *Design rules for offshore cathodic protection systems.* Trans I Mar. Eng. **91** 43 (1979).
26. C.F. Schrieber and R.W. Murray. *Supplementary studies of the galvalum iii anode – hot saline mud and brine environments.* Materials Performance **30** (3) 19 (1981).
27. D.S. Ahmed et al. *An Examination of the Mechanism of Intergranular Attack in Zinc/Aluminium/Cadmium Sacrificial Anodes in Sea Water.* Proc. 8th Int. Congr. Metallic Corrosion, Vol. 2, p. 169 (1981).
28. M.C. Rebul, M.C. Delatte and J.M. Pieraerts, *Hydral 2 – A Semi-Continuously Cast Aluminium Anode* Paper 107 CORROSION/81.
29. C.J. Houghton and V. Ashworth, *The performance of commercial zinc and aluminium anodes in hot sea-bed mud.* Materials Performance **21** (7) 20 (1982).
30. J. Cochran, *A Correlation of Anode-to-electrolyte Resistance Equations used in Cathodic Protection*, Paper 169 CORROSION/82.
31. R. Strømmen, *Evaluation of Anode Resistance Formulas by Computer Analysis*, Paper 253 CORROSION/84.
32. K. Nisancioglu, *An analysis of resistance formulas for sacrificial anodes.* Materials Performance **23** (12) 36 (1984).
33. J. Cochran, *Anode-to-electrolyte resistance equations for offshore cathodic protection.* Materials Performance **24** (6) 39 (1985).
34. Editors' note in B.S. Wyatt, *Cathodic protection of fixed offshore structures*, in *Cathodic Protection – Theory and Practice*, Eds. V. Ashworth and C.J.L. Booker, Ellis Horwood (Chichester) 1986.
35. E.G. Haney, *Zinc sacrificial anode behaviour at elevated temperatures.* Materials Performance **25** (4) 31 (1986).
36. F.O. Jensen and T. Jore, *New Zinc Alloy for High Temperature Service.* Proc. UK Corrosion, 1986.

37. K.E. Lucas, M.H. Peterson and R.J. Guanti, *Long term Performance of Aluminum Anodes in Seawater*, Paper 277 CORROSION/89.
38. G.W. Swain and J. Patrick-Maxwell, *The Effect of Biofouling on the Performance of Al-Zn-Hg Sacrificial Anodes*, Paper 289 CORROSION/89.
39. C.F. Schrieber, *The Aluminum Anode in Deep Ocean Environments*, Paper 580 CORROSION/89.
40. D.S. Ahmed et al., *Mechanism of intergranular attack in Zn-Al-Cd sacrificial anodes at elevated temperature.* British Corrosion Journal **24** 149 (1989).
41. C.F. Schrieber and J.T. Reding, *Application Methods for Rapid Polarization of Offshore Structures*, Paper 381 CORROSION/90.
42. R.E. Lye, *A corrosion protection system for a North Sea jacket.* Materials Performance **29** (5) 13 (1990).
43. K.P. Fischer, *Anode Performance Data at High Pressure and High Temperature – A Laboratory Study in Natural Seawater*, Paper 235 CORROSION/91.
44. K.C. Lunden and T.M. Stastny, *Rapid polarization cathodic protection in the Rijn field.* Materials Performance, **30** (11) 24 (1991).
45. J.D. Burk, *Dualnode Field performance Evaluation Cathodic Protection of Offshore Structures*, Paper 309 CORROSION/91.
46. P.L. Bonora et al., *Improved Sacrificial Anode for the Protection of Offshore Structures.* Proc UK Corrosion, 1993.
47. K.J. Kennelley and M.W. Mateer, *Evaluation of the Performance of Bi-Metallic Anodes on a Deep Water Production Platform*, Paper 523 CORROSION/93.
48. J.N. Murray, R.A. Hays and K.E. Lucas, *Testing Indium Activated Aluminum Alloys Using NACE TM0190-90 and Long Term Exposure*, Paper 534 CORROSION/93.
49. C. de Waard, *Influence of anode cathode distance on sacrificial anode resistance.* Materials Performance **33** (2) 17 (1994).
50. P.A. Warnock, *Anode Design – User/Producer Collaboration for Profit. Some Case Histories*, Paper 493 CORROSION/94.
51. P.A. Warnock, *Offshore Sacrificial Anode Design – A Producers View of Limiting Factors for Success*, Paper 305 CORROSION/95.
52. S. Rossi et al., *Composite sacrificial anodes for offshore structures.* Materials Performance **35** (2) 29 (1996).
53. E. Aragon et al., *Influence of alloying elements on electrochemical behaviour of ternary Al-Zn-Ga alloys for sacrificial anodes.* British Corrosion Journal **32** (4) 263 (1997).
54. D.L. Johnson, *Anode Foundry Production Anomalies*, Paper 468 CORROSION/97.
55. J.-P. Pautasso, H. Le Guyader and V. Debout *Low Voltage Cathodic protection for High Strength Steel: Part 1- Definition of a New Aluminium Galvanic Anode Material* Paper 725 CORROSION/98.
56. S. Rossi, P.L. Bonora and M. Draghetti, *Cathodic protection revamping technology for offshore structures: the Agbarra platform.* Materials Performance **37** (3) 15 (1998).
57. J. Genescà and J. Juárez, *Development and testing of galvanic anodes for cathodic protection.* Contributions to Science **1** (3) 331 (2000).
58. K.P. Fischer, B. Espelid and B. Schel *A Review of CP Current Demand and Anode Performance for Deep Water*, Paper 01013 CORROSION/2001.
59. M.A. Talavera et al., *Development and Testing of Aluminum Sacrificial Anodes In/Hg Free*, Paper 01508 CORROSION/2001.
60. E. Lemieux, W.H. Hartt and K.E. Lucas, *A Critical Review of Aluminum Anode Activation, Dissolution Mechanisms, and Performance*, Paper 01509 CORROSION/2001.
61. P. Felton, J.W. Oldfield and M. Peet *Performance of Aluminium Anodes in Simulated Service Environments Containing Sour Sea Water and Sour Sediment*, Paper 02015 CORROSION/2002 NACE (Houston) 2002.
62. E. Lemieux et al., *Performance Evaluation of Low Voltage Anodes for Cathodic Protection*, Paper 02016 CORROSION/2002.
63. R. Crundwell, *The future of Sacrificial Anodes*, paper presented to a Cathodic Protection Symposium University of Manchester 2003.

64. J.N. Murray, *A Re-Investigation of the Testing Parameters for Evaluating the Initial Electrochemical Characteristics of Indium Activated Aluminum Alloy Sacrificial Anodes* Paper 04098 CORROSION/2004.
65. R. Orozco et al., *Electrochemical Characteristics of Al-Zn-Mg Alloys as Sacrificial Anodes in Sea Water*, Paper 05081 CORROSION/2005.
66. M.C. Quevedo and J. Genesca, *Flow-Induced Corrosion of Al-Zn-Mg Galvanic Anodes in Seawater*, Paper 08052 CORROSION/2008.
67. G. Gibson, *Behaviour of Al-Zn-In Anodes at Elevated Temperature*, Paper 10396 CORROSION/2010.
68. J. Genesca, M.C. Quevedo and V. Garcia, *Effect of flow on the corrosion Mechanism of Zn and Al Galvanic Anodes in Artificial Seawater*, Paper 11323 CORROSION/2011.
69. W. Monzel, A.P. Druschitz and M. Maxfield, *Development of New, Low-Voltage, Aluminium, Sacrificial Anode Chemistries*, Paper 4284 CORROSION/2014.
70. H. Abe et al., *Introduction to Anticorrosive Effect and Typical Performance of Al-Zn-In-Mg Alloys as Galvanic Anode*, Paper 55532 Eurocorr 2016.
71. D. Mim and L. Siqi, *Optimization cathodic protection design for offshore platform: Solution for the contradiction of big initial and small mean design current density by wing section sacrificial anode*, Paper 7303 CORROSION/2016.
72. Z. Wang et al. *Optimization of sacrificial anodes for one offshore jacket.* Materials Performance 55 (2) 20 (2016).
73. H. Aghajani, A.M. Atapour and R. Alibek, *Passivation of zinc anodes in marine conditions Materials.* Performance 55 (9) 34 (2016).
74. M. Kidd and A. Druschitz, *Understanding Al-Ga Sacrificial Anodes Via Simulation and Verification of Alloy Segregation*, Paper 11378 CORROSION/2018.

Chapter 11

Impressed current systems

"10,000 ways that won't work"

When Sir Humphry Davy [1] was wrestling with the problem of the corrosion of copper sheathing on wooden ships, he considered an approach which, had he pursued it, might have come to be known as impressed current cathodic protection (ICCP). He said... *I at first thought of using a Voltaic battery...* However, he quickly added... *but this could hardly be applicable in practice.*

Many years later, the prolific inventor Thomas Edison attempted to use a direct current to protect an iron hull. He was evidently unaware of, or undaunted by, Davy's conclusion. His experiment involved trailing copper cables from the ship and using them to pass a direct current into the water and back onto the hull. The system did not work because the copper quickly corroded through. As he once famously said... *I have not failed. I've just found 10,000 ways that won't work.* In the case of cathodic protection, it seems that he was not tempted to explore any of the remaining 9999 possibilities. Fortunately for the CP industry, others were tempted. Some were successful.

We met the essential features of ICCP systems in Figure 2.8. This shows that conventional (positive to negative) current leaves the positive terminal of the DC power supply and flows into the seawater via an impressed current anode. The current then returns from the environment onto the surface of the structure, thereby conferring protection. The circuit is completed by the current that has been collected on the surface of the cathode being returned to the negative terminal of the DC supply.

In this chapter, we examine the electrochemical reactions taking place in an ICCP system and their implications for the anode materials. We will then consider how these anodes are powered and become part of the system.

11.1 THE ELECTRODE REACTIONS

11.1.1 Cathodic reactions

The reactions taking place on the surface of the cathode are the same as when we apply CP using sacrificial anodes. The anodic dissolution of iron is suppressed in favour of the electron-consuming reduction of dissolved oxygen.

$$2H_2O + O_2 + 4e^- \rightarrow 4OH^- \quad (11.1)$$

We also noted that sacrificial anode systems can polarise the potential of the cathode sufficiently negatively to favour the hydrogen evolution reaction.

$$2H_2O + 2e^- \rightarrow H_2 + 2OH^- \quad (11.2)$$

DOI: 10.1201/97810032160702-11

In sacrificial anode systems using zinc or aluminium alloy anodes, the cathode potentials achieved are usually insufficiently negative to result in dramatic rates of hydrogen evolution, albeit the rate of the reaction can be sufficient to cause embrittlement of sensitive alloys (see Chapter 12).

Switching from sacrificial anodes to ICCP does not change the essential electrochemistry taking place on the surface of the cathode. However, because ICCP systems have the capability of achieving excessively negative potentials, the issues surrounding hydrogen evolution come more to the fore. These issues are principally cathodic disbondment of coatings (Chapter 9) and hydrogen embrittlement of susceptible alloys (Chapters 12 and 14).

11.1.2 Anodic reactions

11.1.2.1 Consumable anodes

Edison's results highlighted the main engineering challenge of transferring the electrical current from the positive side of the power supply into the seawater. Where the current flow changes from electronic (in the wire) to ionic (in the seawater), an electrode process has to take place. Where current flows from the positive terminal into the seawater, this electrode process has to be anodic. This is why Edison's copper cables failed. They simply corroded at an accelerated rate.

However, this does not necessarily prevent us from using a corrodible metal as an ICCP anode. In principle, we could use any electrically conducting material. For example, we could connect the sacrificial anodes discussed in the previous chapter to the positive terminal of the ICCP system. This would force them to yield more current. However, Faraday's law would still hold. These anodes would be consumed at an accelerated rate in producing the additional current.

The act of connecting a DC power supply into our CP circuit means that we are no longer constrained to select an anode that naturally exhibits a more negative potential than steel. There is no reason, for example, not to use iron or steel as the ICCP anode. The lack of any natural potential difference between it and the structure is irrelevant because the necessary EMF is provided by the DC power supply. In this arrangement, the iron anode would be anodically dissolved over time. However, if a sufficiently massive lump of scrap iron were used, it could have a useful life.

The use of iron anodes for offshore ICCP is almost unheard of. On the other hand, scrap iron or, more usually, iron alloyed with high levels of silicon, is extensively used in onshore CP.

11.1.2.2 "Non-consumable" anodes

Marine ICCP systems almost always employ anodes that are highly resistant to anodic dissolution in seawater. These are conventionally termed "non-consumable" anodes, although a more apt description would be "very slightly consumable". This means that the process of transferring the CP current into the seawater involves different electrode processes. The two candidates are the electrolysis of the water itself and the liberation of chlorine gas.

11.1.2.2.1 Electrolysis of seawater

We may remember an experiment from our school days. Figure 11.1 shows the Hoffman apparatus for the electrolysis of water. This is the classic experiment that demonstrates that water comprises two parts of hydrogen to one part of oxygen by volume.

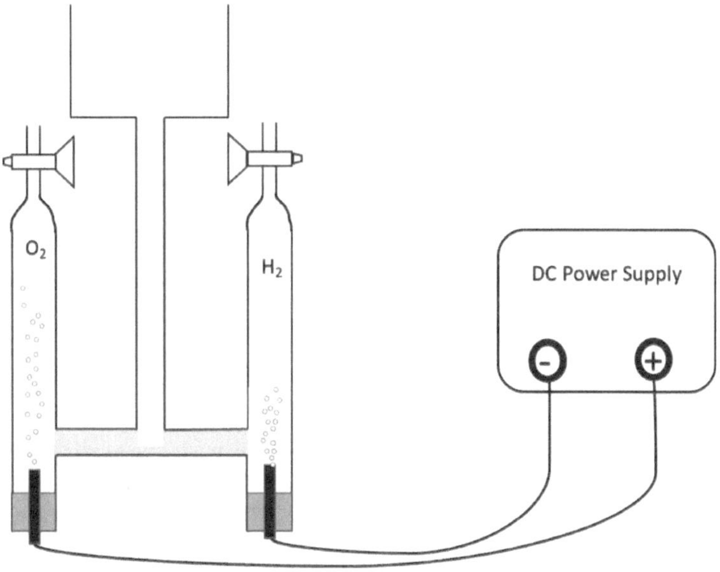

Figure 11.1 Hoffmann apparatus.

In this experiment, a direct electrical current is passed through the water[1] between two electrodes (usually platinum). At the electrode connected to the negative terminal (the cathode), hydrogen gas is evolved. This is equation 11.2. At the anode, the gas liberated is oxygen according to equation 11.3.

$$2H_2O \rightarrow 4H^+ + O_2 + 4e^- \tag{11.3}$$

This reaction is, by definition, anodic. It liberates electrons which flow towards the positive terminal of the DC supply.

You might also note that the negative terminal in the Hoffmann apparatus did not need to be made of platinum. It could be iron since it would be cathodically protected and would not corrode (at least while the apparatus was working). On the other hand, the positive electrode has to be an inert metal; otherwise, it would dissolve rather than serving as the substrate for the liberation of oxygen.

11.1.2.2.2 Chlorine liberation

Seawater contains a considerable quantity (~3.5%) of sodium chloride. This chloride gives rise to a second anodic reaction: the evolution of chlorine gas (equation 11.4).

$$2Cl^- \rightarrow Cl_2 + 2e^- \tag{11.4}$$

Like the liberation of oxygen (equation 11.3), this reaction also liberates electrons so it is anodic.

In practice, chlorine only has a transient existence in its elemental form. It reacts rapidly with the water to form hypochlorite. However, that need not concern us here. All that matters for ICCP is that anodic electrode processes are available in the seawater.

[1] In practice, the experiment uses water to which a little sulfuric acid water is added to aid conductivity.

11.2 ICCP ANODES

11.2.1 Requirements

It is clear from the above that the technical requirements for impressed current anodes are very demanding. They must be electrochemically very active so that they can deliver high currents into the seawater, but they must also be very resistant to electrochemical and chemical attack. As we can see from equations 11.3 and 11.4, the environment adjacent to an operating anode contains high levels of hydrogen ions (acidity) as well as the oxidising gases oxygen and chlorine (essentially bleach).

11.2.2 Onshore origins

Although the origins of CP lay in Humphry Davy's work on the protection of copper in seawater, marine applications of CP remained relatively limited for well over a century. ICCP developed, almost by accident, onshore. It coincided with the emergence of electrically powered tram systems in European and North American cities towards the end of the 19th century. Typically, these used DC current which was supplied to the vehicles via overhead conductors. This current then returned to the power supply facility via the rails. In compliance with the Law of Unintended Consequences, some current would leak from the rails into the ground and would find an alternative path back to the negative terminal of the power supplies. In so doing, the current might flow through local steel or cast iron gas or water mains as part of its return path. Where the current flowed from the soil onto the surface of the buried pipe it caused cathodic protection, even though the term did not exist at that time. However, where the current exited the pipeline to return to the power supply, it did so by an anodic electrode process. That anodic process was the dissolution of iron to form ferrous ions.

At the time, this was called "vagrant current". Lewis [16] has provided an interesting history of how this problem was addressed. It was found that installing drainage bonds to conduct the current back to the tram company's power supply could mitigate or eliminate the problem. "Vagrant current" is now more usually referred to as "stray current". Subsequently, instead of draining current from pipelines and returning it to the tram system power supply, it was drained via dedicated power supplies and dumped to earth. This reduced corrosion on the pipeline. ICCP was born.

There was little need for innovation in the development of anodes for land-based ICCP. The requirement was to provide a low resistance, robust and low-cost earth connection. For this reason, the anode arrangement became known as an anode ground bed. Typically, it would involve an array of scrap iron or, subsequently, high silicon iron electrodes. These were either buried directly in the earth or in a volume of conducting coke granules referred to as "carbonaceous backfill". ICCP ground beds being installed today are often little different from those installed almost a century ago.

11.2.3 Anode development

The massive ground beds employed for onshore CP are clearly unsuitable for a marine installation. More compact anode arrangements are required. In part, smaller anode dimensions are assisted by the fact that anodes working in seawater benefit from low environmental resistivities. However, against this we need to weigh the chemically aggressive nature of the environment generated adjacent to the anode in seawater.

To satisfy the demanding requirements of small anodes that resist the aggressive anodic environment, the offshore ICCP industry has made use of anodes developed for electrochemical industry applications such as the manufacture of caustic soda or hypochlorite.

11.2.3.1 Early ICCP anode alloys

11.2.3.1.1 Lead-based anodes

The earliest successful, so-called "non-consumable" anodes were based on lead alloys. The successful use of lead as an electrode in an aggressive electrolyte will not surprise anyone who has started a motorcar using a lead-acid battery.

The electrochemically active layer on a lead anode is lead dioxide (PbO_2), as it is on the positive plates of a charged lead-acid battery. The metallic lead, or lead alloy, essentially acts as a conductor to, and a physical support for, the PbO_2 layer. The development of lead anodes began in the late 19th century and was directed primarily at the industrial production of chorine and hypochlorite. Their first use as ICCP anodes for seawater was in the 1950s [7], and they were investigated extensively during the 1960s [8]. The alloying ingredients that were studied fell into two groups. The first was targeted at strengthening the inherently soft lead substrate, and the second group was aimed at increasing the electrochemical activity of the PbO_2 layer.

Two commercial lead-based ICCP anodes emerged [8,13]: Pb-6Sb-1Ag and lead-platinum "bi-electrodes". The latter consisted of lead with platinum micro-electrodes embedded into its surface. These two types were capable of delivering anodic current densities of 200 and 500 A/m^2, respectively. However, they had to run at a minimum of 50 A/m^2 to avoid passivation. A further observation was that these alloys were reported as exhibiting increased consumption rates at water depths greater than about 25 m. As far as I am aware, no further work was published as to what caused this increase.

Burgbacher [3] described Shell's use of lead-based ICCP anodes on structures in the Gulf of Mexico prior to 1968. Previously, lead-silver anodes had been used. These were simply deployed overboard, suspended by cables or ropes, after the structure had been installed. However, this approach had given way to Shell's first "permanent" anode installation in 1966. In this case, the ICCP anodes and their conduits were fixed in place during the fabrication of the structure. The lead-silver anodes were replaced by lead-platinum.

By the end of the 1960s, however, lead-based anodes were disappearing from the market. For example, one major North American ICCP anode producer discontinued its production around 1971, turning its attention to anodes manufactured by winding a platinum wire onto a niobium rod.

11.2.3.1.2 Platinum-based anodes

Platinum has long been recognised as a desirable electrode for carrying out electrochemical reactions. It is very resistant to corrosion or electrochemical dissolution, yet its surface is electrochemically very active. This provides the opportunity for the ICCP system designer to extract very high current densities and to operate at high output voltages, if necessary.

Platinum's obvious drawback is that it is a precious metal. So, to avoid excessive cost, and to minimise the risk of pilfering, it has to be used sparingly. This prompted research targeted at maximising the surface area of platinum exposed to the seawater whilst minimising its weight and cost. This resulted in a 1958 patent (cited in [8]) for electroplating thin (typically <10 μm) layers of platinum onto titanium. Similar patents also included niobium or tantalum substrates. Subsequently, other approaches included extruding, rather than electroplating, the platinum.

The thin layer of platinum is the electrochemically active surface that supports the anodic reactions (equations 11.3 and 11.4). The titanium, niobium or tantalum, which are often referred to as "valve metals", provide the mechanical support for, as well as the electrical conduction to, the platinum. The valve metals are all characterised by passive behaviour in seawater. This is important because the thin platinum film will inevitably contain defects or be subject to mechanical damage that would expose the substrate. If a non-passive conductor, such as copper, were used for the substrate, then this would fail rapidly wherever such exposure occurred. The valve metals resist this anodic dissolution up to the potential at which their passive oxide film breaks down in seawater. This means that, if we change the substrate from titanium to niobium or tantalum, we produce anodes that can operate at higher voltages.

Unsurprisingly, the benefit of increasing operating voltage is matched by increasing cost. However, since tantalum offers no benefit over the less expensive niobium, platinised tantalum did not catch on. Furthermore, for an environment with good conductivity such as seawater, it would be unusual to design an ICCP system with such a high output voltage that any substrate other than titanium was required. For this reason, most offshore ICCP platinum-based anodes have used platinised titanium.

Although platinum is very highly resistant to corrosion, platinised titanium anodes do not last indefinitely. Although they are termed "non-consumable", this term is an approximation. There is a progressive, albeit very slow, loss of platinum over periods of operation. This loss is more properly referred to as "wear" or "attrition" rather than "corrosion". An approximate figure for this wear rate, together with other operating parameters for platinised anodes, is given in Table 11.1.

Table 11.1 Properties of some platinised anodes in seawater

Electrode	*Platinum*		
Substrate	*Titanium*	*Niobium*	*Tantalum*
Maximum current density (A/m²)	1000	2000	
Maximum operating voltage (V)	9	>100	
Consumption rate (kg/A year)	$8–18 \times 10^{-6}$		

From [8,13].

There are technical reasons why is not possible to extend the life of a platinised anode indefinitely by simply increasing the thickness of the platinum layer. Increasing the thickness results in the cohesive forces within the platinum overcoming the weak adhesive forces that bond it to the substrate. Increasing the thickness beyond about 12.5 µm increases the likelihood that the platinum would exfoliate.

Platinised titanium was first used in a marine ICCP system in 1959 on a jetty in the Thames Estuary. That material, and to a lesser extent platinised niobium, dominated the market for marine ICCP anodes from the 1970s to the 1990s. That is not to say, however, that these materials were without their problems; two of which were fouling and AC ripple.

Normally, the conditions around an operating marine ICCP anode, particularly the low pH and the generation of hypochlorite, are inhibitive to marine growth. However, if the anodes spend time switched-off, fouling can develop. This could be damaging to platinised anodes.

Within a few years of the first application of platinised titanium ICCP anodes, Juchniewicz [2] reported a series of experiments where he examined the effect of AC ripple currents on working ICCP anode materials. He found no problem with the commercial lead-based alloy systems then in use, but there were issues with platinum and, in particular, with platinised titanium. For example, in one experiment, platinised titanium was polarised at a DC anodic current density of 700 A/m² in 3% NaCl for 2000 hours. No corrosion was observed. However, when a 50 Hz AC current of 70 A/m² (10% ripple) was superimposed on the DC polarisation, pitting corrosion was evident after as little as 18 hours. The implication of this was that it was judged necessary to install chokes to smooth out AC ripple in the power supplies energising platinised titanium anodes.

It was subsequently shown that the pitting reduces with increasing frequency. Warne and Hayfield [5] reported that... *there was no evidence for the dissolution rate being significantly accelerated by ripple frequencies of 100 Hz or higher.* This meant that, in practice, single-phase and three-phase full-wave rectifiers produced output waveforms suitable for feeding these anodes. Hence, the propensity for AC corrosion of platinised titanium did not seriously restrict its success in the ICCP market over the last four decades of the 20th century.

11.2.3.2 Mixed metal oxide ("MMO") anodes

In their mid-1980s review of ICCP anodes, Shreir and Hayfield [8] reproduced a table first published in 1976 by Brand. That table gives the anode properties of... *thermally deposited noble metal oxide electro catalyst on titanium.* However, the narrative of their review makes no mention of these materials. This is unsurprising since these materials had found very little application in either onshore or offshore CP prior to that time.

In a follow-up review presented in 1989 (published in 1993), Moreland and Howell [14] include a section dealing with these materials under the heading *Metal oxide anodes (catalysed)*. In it, they describe anodes as *mixed metal oxide*-coated titanium. These anode materials are now universally referred to by the abbreviation "MMO".

MMO anodes were patented in the mid-1960s. During the 1970s, they completely replaced the extant graphite anodes in the chlor-alkali industry [11]. Although there are references to their use in offshore CP as early as 1971, they were first reported on substantial projects in the early 1980s [6]. Since that time, they have effectively displaced lead alloy and platinised titanium anodes from the marine ICCP market.

MMO anodes are manufactured by applying paints containing salts of platinum group metals, usually ruthenium, iridium or combinations of the two, to the pre-etched titanium substrate. The painted component is then thermally treated at a selected temperature between 350°C and 650°C in a controlled atmosphere. The organic components of the paint breakdown and volatilise. At the same time, the salts decompose to oxides which become incorporated in the TiO_2 film on the substrate. The resulting mixed metal oxide is electrically conducting and electrochemically active, and so can function effectively as an ICCP anode.

As a generalisation, the ruthenium-containing anodes were developed for chor-alkali service, so are suitable for seawater. The materials containing the more expensive iridium were developed for electrolytic oxygen generation.

Moreland and Howell [14] tell us that these can deliver 500 A/m^2 in seawater and would be expected to operate for in excess of 20 years. Other authorities claim a maximum operating current density of 600 A/m^2 [13], with a consumption rate of 0.5–1.0 mg/Ay [10], which means that the life is dependent on the thickness of the MMO layer which for largely technical reasons cannot exceed about 12 µm. In practice, this means that designing anodes with lives of up to 40 years should not be a problem. I am aware of one (onshore) installation based on MMO anodes where the anode design life was 60 years.

Unlike platinised anodes, MMO anodes are untroubled by AC ripple, irrespective of the frequency. Furthermore, since both types of anode are usually supported on titanium substrates, the maximum operating voltage would be restricted to the breakdown voltage of the titanium dioxide passive film. For platinised titanium, this is in the region of 10 V [4], although codes are more cautious. For example, ISO 13174 advises a limit of 8 V but adds that higher voltages may be used with... *MMO coated anodes or in less saline environments.*

This code relaxation in favour of MMO takes account of the fact that the fused oxides modify the passive film on titanium, with the result that the breakdown voltage is higher. Unfortunately, the codes give no clue as to how much higher. However, the actual breakdown voltage is probably not relevant. Funahashi and Wu [18] showed results of polarisation tests in

3% NaCl where, even drawing a current of ~5000 A/m², did not cause an MMO-coated titanium specimen to polarise to more than about +5V. In other words, the notional breakdown voltage, if it exists on MMO-coated titanium, will never be reached in any real ICCP system.

11.2.4 Anode configuration and resistance

In principle, the same anode resistance formulae that we discussed in Chapter 10 in the context of sacrificial anodes will apply to ICCP anodes. In practice, however, the range of geometries employed for ICCP anodes is fairly limited. The customary configuration is a strip of anode mounted in a polymeric anode housing. This assembly is surrounded by a dielectric shield (see Section 11.2.5).

This arrangement is always used for the ICCP of the external hulls of ships and other floating installations. Variations of this arrangement are also usually employed when ICCP systems are installed on fixed structures such as jetties and petroleum production platforms. Where the anode is long and slender, as shown in Figure 11.2, it is customary to use the Lloyds formula (11.5) for estimating the anode resistance R_a in seawater of resistivity (ρ).

$$R_a = \frac{\rho}{\text{Length} + \text{width}} \tag{11.5}$$

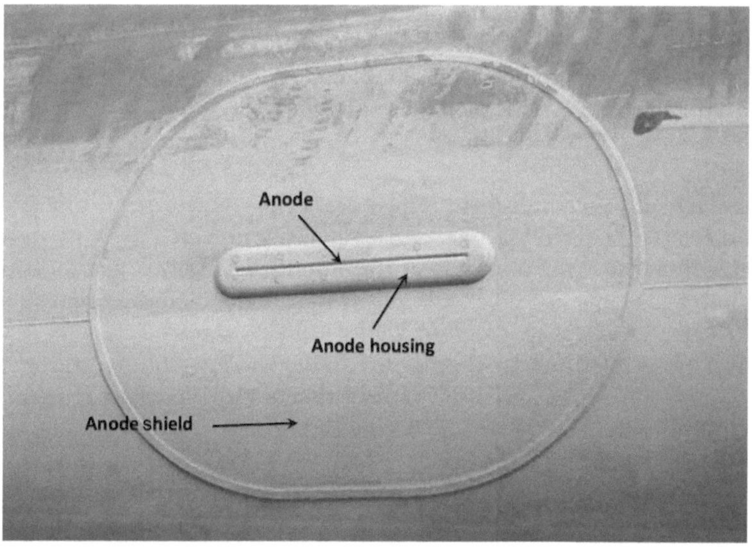

Figure 11.2 Flush-mounted ICCP anode on a cargo ship.

Less frequently, rod-style anodes that protrude into the seawater have been used. The resistance for these can be estimated from the modified Dwight formula, albeit that this is likely to provide a slightly conservative result. Whilst the use of a rod-style cantilevered anode might have a benefit in providing a lower resistance to earth than the same area of flush-mounted anode, it brings with it the practical problems of providing physical protection to a vulnerable anode in a hostile environment. It has, however, been used successfully on some offshore projects (see Chapter 13).

11.2.5 Anode shields

11.2.5.1 *Why do we need anode shields?*

An ICCP system is a high-energy current source. This has the profound advantage that we need far fewer ICCP anodes than sacrificial anodes to protect any given structure. The downside is that the electric current it introduces into the seawater will always seek the lowest resistance path back to the negative terminal of the power supply. With an ICCP anode mounted on the protected structure, this means that the steel nearer to the anode will receive a disproportionately higher current density than more remote parts of the structure.

This excessive polarisation may in turn lead to excessive cathodic polarisation adjacent to the anode. As noted in Chapter 10, this can accelerate the disbondment of coatings. It may also result in hydrogen embrittlement of high-strength steels or other susceptible alloys, and it may have undesirable implications for the fatigue life (see Chapter 12). Taking these threats together, therefore, it is customary to limit the degree of polarisation adjacent to the anode by installing an anode shield such as is shown in Figure 11.2.

The anode shield, which is also referred to as a dielectric shield, comprises a substantial thickness of non-conducting material: usually glass-reinforced polymer (GRP). It is more mechanically robust than the hull paint system. The shield effectively displaces the nearest area of exposed steel cathode to a position that is somewhat more remote from the anode. This attenuates the potential thereby reducing the propensity towards cathodic disbondment of the coating. It also has the effect of improving the overall balance of CP current over the surface of the structure.

11.2.5.2 *Anode shield size*

The radius of an anode shield is usually calculated on the basis that the working anode acts as a point current source generating a hemispherical pattern of potential gradients. A so-called tension hill calculation is then carried out using equation 11.6

$$r = \rho \frac{I}{2\pi\eta} \tag{11.6}$$

where

η = overpotential at the edge of the anode shield (V)
ρ = seawater resistivity (Ωm)
r = shield radius (m)
I = anode output current (A)

If we assume an anode operating with an output $I = 50$ A, operating in seawater of resistivity ($\rho = 0.25$ m), and if we wish to limit the polarisation of the steel at the edge of the anode shield to ($\eta = 1$ V), then we see that the required anode shield radius (r) is ~2 m.

11.3 BASIC DESIGN

11.3.1 Cathodic current demand

The cathode has no knowledge of whether the current it is taking comes from an ICCP system or sacrificial anodes, so the calculation of the cathodic current demand for a structure follows the same process that we described in Chapter 3. However, since the current output capacity of the ICCP system remains constant throughout its life, there is no need to carry out separate initial, mean and final cathodic current demand calculations. Marine ICCP systems are sized to supply the maximum predicted cathodic current demand. For an uncoated structure, this will be the initial current needed to achieve polarisation. For a coated structure, this will be the final demand when the extent of coating breakdown is at its maximum.

We will discuss calculating the cathodic current demand for ships, where ICCP systems are frequently used, in Chapter 15. We will also examine some other situations such as retrofits in Chapter 13.

11.3.2 System output calculations

11.3.2.1 Current

The required current output of the system should match the predicted maximum current demand. However, since a key benefit of an ICCP system is that it uses far fewer, much higher output, anodes than a sacrificial system, it follows that the current from an ICCP system is likely to be less uniformly distributed over the cathode. To compensate for this, it is customary to design a system where the output current capacity exceeds the maximum demand. A margin of around 20% would be typical.

11.3.2.2 Voltage

The DC output voltage of an ICCP system is not restricted to the potential difference between a working anode alloy and a protected cathode (typically ~0.25 V). It is determined by the lower the permitted rating of the DC power supply or the breakdown voltage of the anode substrate. In this context, the power supply unit must have sufficient output voltage to overcome:

a. the voltage drop across the anode seawater interface, plus
b. the voltage drop in the anode feed cables, plus
c. the back EMF.

11.3.2.2.1 Anode voltage drop

The voltage drop across the anode is calculated as the product of the anode resistance (R_a) multiplied by the anode current (I_a). The process is analogous to the calculations we carried out in Chapter 3 for sacrificial anode systems. For example, if it is intended to use a flat disc-shaped anode with a surface area of 0.2 m² to deliver current into 0.3 Ωm seawater, we can calculate R_a using the McCoy formula:

$$R_a = \frac{0.315\rho}{\sqrt{A}} \qquad (11.7)$$

where:
A = total exposed area of anode (= 0.2 m²)
ρ = seawater resistivity (0.3 Ωm)

gives a value of ~0.21 Ω. If the design calls for a maximum anodic current of 40 A from each anode, the voltage drop anticipated at the anode will be ~8.4 V. This is close to the breakdown limit for a platinised titanium but does not pose a problem for MMO.

11.3.2.2.2 *Anode cable voltage drop*

We consider anode cables in Section 11.6. A typical calculation of the voltage drop in the cable might be as follows. If the cable size has been selected as 16 mm^2 cross-section, and the cable route from the power supply positive terminal to its underwater anode is 250 m, as might arise on a jetty CP system, then the conductor resistance would be 250×1150 μΩ = 0.2875 Ω. Rounding this up to 0.3 Ω tells us that the voltage drop in the cable would be 0.3 Ω×40 A = 12 V. Thus, in addition to provide enough voltage to drive the current through the anode to remote earth, the T/R would need to have at least 12 V additional to the 8.4 V calculated above.

11.3.2.2.3 *Back EMF*

If we placed an MMO anode in the seawater near to a steel cathode, but without any connection between them, there would initially be a potential difference between the two. The anode, which would not be corroding, would be acting as a dissolved oxygen electrode. Depending on the temperature and dissolved oxygen concentration in the water, this could be at a potential as positive as +1 V, although +0.6 to +0.8 V would be more typical. The steel would be corroding at somewhere in the region of −0.65 V. Thus, the potential difference would be up to (+1 V−(−0.65 V) = 1.65 V.

When we connect the two via the ICCP circuitry this 1.65 V, which we refer to as the back EMF, needs to be overcome before any protection is conferred on the steel. As we then apply CP current, the potential of the steel moves in a negative direction to (say) −1.0 V, and the potential of the anode becomes more positive as we start to liberate oxygen (and chlorine) on its surface. Thus, the back EMF generated by a working ICCP system based on MMO anodes is likely to be around 2–2.5 V.

11.3.2.2.4 *T/R voltage output selection*

On the basis of this calculation, the minimum required output voltage of the power supply would be 8.4 V (to accommodate the anode resistance) plus 12 V (for the cable resistance) plus 2 V (to overcome the back EMF). This gives a total of 22.4 V. This total assumes that the cathode structure itself is so massive, and of such a large area, that its resistance is low enough to be ignored.

Typically, T/R units are supplied with 12, 24 or 48 V DC outputs. Whereas a 24 V unit fits the bill, it is likely that a CP designer might opt for 48 V if local regulations permitted. This would be in keeping with the conservatism inherent in offshore CP design practice.

11.3.3 Design calculation process

An ICCP design exercise will involve some degree of iteration in which:

a. The number of anodes is provisionally selected. For many structures, this is usually a rule-of-thumb exercise. For example, a design which sets out to provide 600 A might be based on twelve 50 A anodes.
b. A trial anode design is examined and its resistance is calculated using the formula appropriate to its shape (see Chapter 10). Further calculations would then be carried out to ensure the following hold.

i. The surface area of the anode ensures that the maximum operating current density for the alloy (typically set to below 500 A/m^2) is not exceeded. In other words, for a 50 A anode, we would be looking for at least 0.1 m^2 of anode surface area.
ii. A further calculation might then be carried out to check that the breakdown voltage of the anode material would not be exceeded. This would be relevant for platinised titanium anode but unnecessary for an MMO anode.

   If either check (i) or (ii) failed, then the design would need to be reworked based on more or larger anodes.

c. The anode resistance then needs to be added to the series resistance of the anode cables (see Section 11.6). This combined resistance is then multiplied by the maximum anticipated anode current in order to determine the minimum required DC output voltage of the power supply.
d. To this voltage, we then need to add an allowance for the back.EMF (typically 2 V).
e. Finally, as is the case with the designed current capacity, it would be normal to ensure that there was a margin in the voltage availability to ensure that the system runs comfortably below its maximum rating.

It might require several iterations of this process to come up with a workable configuration of anodes. It would then be necessary to determine exactly where the anodes would go. We return to this when we discuss structures (Chapter 13), ships and floating installations (Chapter 15) and modelling (Chapter 17).

Although the above design process has a logic to it, we should be aware that at least one major offshore operator dispenses with the above steps in favour of a modularised design approach. Its standard practice for offshore platforms, of which it has many, is simply to calculate the current demand, based on its in-house design current densities. It then installs the requisite number of its standard design 50 A anode arrays, mounted on seabed sleds, around the periphery of the structure. Elsewhere, for example in ship-building, CP calculations are often dispensed with, and systems are installed according to established in-house practice.

11.3.4 Anode locations

Many years ago, when I was learning about CP, I imagined the options for illuminating a prestigious municipal building. Two approaches might be considered. Either the building could be festooned with a large number of low-power light bulbs, or else a small number of high-powered floodlights could be positioned some distance away and aimed at it. I regarded the first approach as a useful analogy for sacrificial anode CP, with a large number of low-power anodes distributed more-or-less uniformly over the surface. The second approach could be thought of as representing ICCP, with a much smaller number of more powerful current sources.

In many ways, this is a handy analogy, particularly when trying to get to grips with how CP works. However, it does embody a serious shortcoming. It is all too easy to over-interpret the analogy with regard to the floodlights and ICCP systems. In particular, the fact that the floodlight arrangement will cast shadows in re-entrant corners of the building might lead to the interpretation that CP depends on line-of-sight. Indeed, I still occasionally meet CP practitioners who believe that this is true.

This is not the case. The flow of electrolytic current through a conducting medium is quite different from the propagation of light. It does not travel in straight lines. That is not to say, of course, that a high-powered ICCP anode will polarise (or using the analogy: illuminate) all parts of the structure uniformly. We will return to the subtleties of the distribution of CP current through the seawater surrounding a structure when we discuss modelling in

Chapter 17. For now, all we need to be aware of is that when we choose to polarise a structure using impressed current anodes, either on the structure itself or on remote sleds, we should get a result that works. How well it will work depends on numerous other factors.

11.4 POWER SUPPLIES

11.4.1 What we need to know

I am sitting writing this on a well-known, somewhat pricey brand of laptop computer running a no less well-known word processor application. It may not surprise you to learn that I know next to nothing about how these items of hardware and software actually work. Nevertheless, my ignorance of such matters does not prevent me from generating a result. All that I really need to know is that, if I type the letters in the correct order, the phrase "cathodic protection" appears on the screen.

By analogy, the same applies in regard to CP practitioners' attitude to the DC electrical power source for an ICCP system. They need not be overly concerned with how the output DC voltage and current are produced. At the basest level, they merely need to be able to tell the equipment supplier how many volts and amps will be required.

With this in mind, the following coverage of ICCP power supplies is circumscribed by both the need-to-know of the CP practitioner, and by the fact that my own expertise in this area is on a par with my knowledge of computer science. Like most CP workers, I can twiddle the knobs and read the dials, but it would be unsafe for me to open the box and fiddle around inside!

11.4.2 The basics

With very few exceptions, offshore ICCP systems are installed in facilities where there is a source of AC electrical power. Examples include fixed and floating production and hydrocarbon processing facilities, ships and jetties. These installations generate power for their day-to-day operational and domestic needs. Providing there is sufficient spare power capacity, ICCP is an option. Furthermore, this demand for onboard electrical power for normal operations usually means that the installation carries personnel who are qualified to maintain the equipment.

To contribute to ICCP, the onboard power supply voltage, which is usually either single or three-phase AC, has to be *transformed* to a lower value and *rectified* to DC. For this reason, ICCP power supplies are generally referred to as transformer-rectifiers (or "T/R's"). Most engineers will have covered the principles of transforming and rectifying AC voltage during physics lessons in their school days, and the schematic circuit sketched out in Figure 11.3 might jog the memory.

The transformer part of the T/R (on the left-hand side of Figure 11.3) works on the basis of Faraday's Law of Electromagnetic Induction (Yes: him again). The incoming (primary) supply is fed through a coiled conductor wound around a soft iron core. This AC current induces an oscillating magnetic field in the core. This oscillating field induces an AC current in the secondary coil. The ratio of the primary to secondary voltage is approximately equal to the ratio of the number of coils in the primary and secondary windings. Typically, ICCP power supplies output at 12, 24 or 48 V. Since the primary voltage is higher (typically 240 or 110 V AC), the units are generally known as "step-down" transformers.

The rectifier part of the T/R (on the right-hand side of Figure 11.3) comprises a diode bridge. The diodes serve to block the reverse part of the signal resulting in a conversion from

AC to DC. However, as indicated in Figure 11.4, although the output rectified voltage is now DC in that its polarity does not cycle, the waveform is not smooth. Depending on its design, the T/R may also include smoothing circuits.

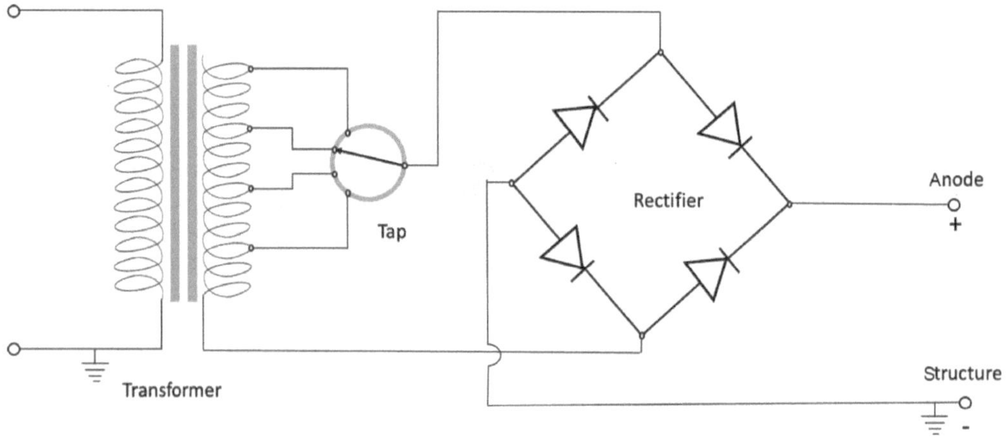

Figure 11.3 Transformer/rectifier – basic circuit.

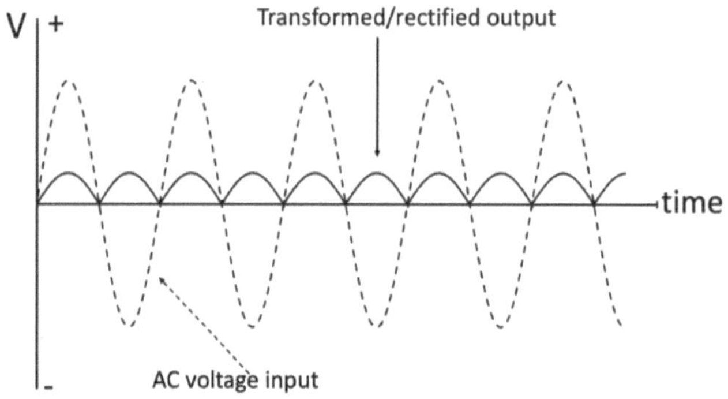

Figure 11.4 Transformer/rectifier – input and unsmoothed output.

Ultimately, what is on the inside of a T/R, or any other DC power supply unit, need not concern the CP engineer any more than the inner workings of a computer. All that matters is what it does. Providing that is understood, it can legitimately be regarded as an electrical box as is illustrated in Figure 11.5.[2]

11.4.3 Manual control

As noted, ICCP systems were developed for the protection of onshore pipelines some decades before they were introduced into offshore or marine situations. The output voltages of early onshore T/R's were adjusted by altering the number of transformer coils on the secondary side using a tapping arrangement. This tapping is also illustrated schematically in Figure 11.3. Effectively, these taps enabled the CP operator to control, albeit coarsely, the AC voltage fed to the rectifier.

[2] This unit is switched off because the vessel is in dry-dock.

Figure 11.5 Shipboard T/R unit.

This arrangement is generally satisfactory for many onshore pipelines, for which the CP current demand only varies relatively slowly over the operational life. However, it has two drawbacks. First, it requires monitoring, typically every 6 months; and, if required, occasional changes of the tap settings. Second, the primary coil stays fully energised, consuming power and generating heat.

On the plus side, the arrangement is simple and robust. Many such T/R units remain in service today. The lack of day-to-day control over the output might be deemed a disadvantage. However, in practice, this proves to be less of a problem than one might imagine. Although there is no facility for adjusting the output voltage in line with variations in the day-to-day current demand of the installation, there is often some measure of "self-adjustment". On a jetty, for example, the cathodic current demand might be higher at high tide when more of the steelwork is immersed. However, for the same reason, the CP circuit will have a lower resistivity, so the "fixed" output voltage will result in the delivery of more cathodic current.

11.4.4 Automatic control

Fixed output voltage T/Rs are being superseded in marine and offshore applications by systems that lend themselves to automatic control. In principle, these units can be set to control the potential of the structure at a pre-set value (say −0.8 V) and will automatically hold this potential by adjusting the power output in response to the changing cathodic current demand. In this respect, they may be regarded as large-scale versions of the laboratory potentiostat that we met in Chapter 7.

Automatic control of a T/R requires a control input and an interface that uses that input signal to adjust the voltage output of the unit. The control input will be the potential difference between the structure and a reference half-cell (see Section 11.5). Two types of control circuitry are found in ICCP T/R's: thyristors or switch-mode power supplies.

Thyristors, which are also known as silicon-controlled rectifiers,[3] are used instead of conventional diodes in the rectification stack. The thyristor differs from the diode in that it has an additional control input or "gate". The signal pulse applied to this gate triggers the thyristor into its conducting mode. This chopping action of the thyristor in response to the timing of the gate pulse controls the output of the unit. As above, the unit will usually include circuitry to smooth out this chopped waveform.

Thyristor-controlled T/R's obviously have advantages over manually controlled units in controlling the level of polarisation of the structure. However, they are not without their drawbacks. In particular, although they adjust the output voltage of the unit, they do not affect the AC input. The result is that the transformer remains permanently energised which is both inefficient and, in some instances, can lead to issues with the heat generated.

A more energy-efficient system is to replace the transformer with a switched-mode[4] power supply [15,19]. These types of power supplies have gained prominence in a wide range of domestic and industrial electrical appliances since the 1970s. As with thyristor control, the output of the switch-mode device is controlled by the interface with the signal from the reference half-cell.

11.5 CONTROL INPUTS

11.5.1 Ag|AgCl|seawater

Thus far we have used the Ag|AgCl|seawater half-cell as our reference potential. This is in keeping with marine CP practice. The half-cell is manufactured by inserting a piece of silver wire into molten silver chloride salt (>455°C) causing the wire to be coated with a layer of the salt. This becomes an Ag|AgCl electrode, the potential of which depends on the temperature and the chloride ion concentration of the seawater. This is described by the Nernst equation which we met in Chapter 4.

In practice, the salinity and temperature of open seawater vary comparatively little around the world. This means that the reference potential exhibited by these Ag|AgCl|seawater half-cells is sufficiently predictable and stable to be used with confidence for CP measurements.

These references can be made even more precise if, instead of exposing the electrode directly to seawater, it is encapsulated in a solution, or an aqueous gel, containing a fixed concentration of chloride ions. Contact with the external environment is then achieved via a porous plug, usually made of sintered glass.

The use of such encapsulated half-cells, often incorrectly still referred to as Ag|AgCl|seawater, means that they can be used at locations where the environmental chloride concentration is lower (or higher) than 19000 mg/L, which is the typical figure for open seawater.

Although a newly manufactured Ag|AgCl electrode provides a precise reference in seawater or its encapsulating chloride solution, it is sensitive to other anions present in the seawater. In addition, as students of photography will be aware, light causes the silver chloride salt to decompose to metallic silver and chlorine. Although modern encapsulated half-cells restrict the inward diffusion of contaminants and are intentionally non-transparent to light, it is impractical to create perfect barriers to these influences. For this reason, Ag|AgCl|seawater half-cells will become less reliable over time.

[3] Silicon controlled rectifier is a General Electric Corp. trade name.

[4] Switch-mode is a Motorola trade name.

This is not a problem when using portable reference half-cells for carrying out periodic CP surveys. The portable references can, and should, be calibrated against certified laboratory references at the beginning and end of each survey session. Replacement of references that have fallen outside their specification, for example, if they differ from a prescribed standard value by more than (say) 10 mV, is straightforward.

However, this option of calibration and periodic replacement is not available when it comes to selecting permanent references intended to work with automatically controlled ICCP units.

11.5.2 Zinc reference electrodes

The reference electrode of choice for many automatic ICCP systems will, therefore, be a block of zinc. Unlike Ag|AgCl, zinc is not a reversible electrode. It corrodes.

Its use as a reference depends on the fact that, once it achieves a "steady state" corrosion condition in seawater, the reference potential exhibited by zinc, which is actually its corrosion potential, is relatively stable. Over time, it settles down to a value within a few tens of millivolts of –1000 mV (as measured against Ag|AgCl|seawater). Although potentials measured with respect to zinc will lack the precision of those measured against a new Ag|AgCl|seawater reference, this lack of precision has no practical consequence for corrosion control. The advantage of zinc is its long-term durability. This provides an advantageous trade-off against the early life precision of Ag|AgCl|seawater.

An additional characteristic of zinc electrodes is that their potential in seawater can be made slightly more stable by permitting a small anodic current to flow on the surface. This current is arranged to be so low as not to noticeably corrode the zinc block over its designed working life.

When reviewing ICCP system performance data, we must always be aware of which reference is being quoted. I still remember the initial shock I had when, after some years of reviewing CP data showing potentials in the range –800 to –1000 mV, I was confronted with my first shipboard ICCP system which was set to control at +200 mV. My horror soon subsided when I realised that the (unquoted) reference was zinc.

The moral here is that we should always record the reference

11.5.3 Dual references

In some CP designs, the combination of an accurate reference half-cell (Ag|AgCl|seawater), for the early stages of immersed service when the ICCP system is under more frequent adjustment, and the more robust zinc reference electrode for later in life when things are more settled, makes sense. We discuss dual references in more detail when we consider CP system management in Chapter 18.

11.6 CABLES

11.6.1 Conductors

With the exception of anode sleds, which we will return to in Chapters 13 and 14, sacrificial anodes are connected to the structure by a short length of thick section anode core. The ohmic voltage drop in the anode connections need not be considered. On the other hand, ICCP systems may require the anode cables to be deployed at some distance from the T/R. The ICCP design will have to allow for the additional voltage needed to drive the current through the resistance of the anode cable. By contrast, less attention is directed at the negative cable since this is usually just a short run to a convenient earthing point on the structure or a ship's hull.

The cable selection will need to follow the installation owner's in-house electrical standards which, in turn, should comply with the appropriate electrical codes. Table 11.2 shows the dimensions and resistances of some typical conductors that might be considered for connecting the positive output terminal of a T/R to its anode. More details on cable sizes (metric and AWG) can be found in numerous electrical references.

Table 11.2 Some copper conductors

Size (metric)	*No of strands*	*Strand diameter (mm)*	*Cross sectional area of conductor (mm²)*	*Conductor weight (kg/m)*	*Conductor resistance (μ/m)*	*Typical DC current rating (A)*
6.0	7	1.04	5.95	0.054	3080	31
10	7	1.35	10.02	0.091	1830	42
16	7	1.70	15.89	0.145	1150	56
25	7	2.14	25.18	0.229	727	73
35	19	1.53	34.93	0.317	524	90
50	19	1.78	47.28	0.429	387	145
70	19	2.14	68.34	0.620	268	185
95	19	2.52	94.76	0.860	193	230

For example, an ICCP design for an offshore installation might elect to use anodes with a (maximum) estimated output of 40 A. This would be (just) within the current rating of a size 10 (nominally 10 mm² cross-section) conductor. However, as a matter of conventional practice, most CP designers would opt for a higher-rated conductor: 16 mm² nominal cross-section. This reflects the natural tendency towards conservatism that pervades the offshore CP industry.

Test cables are not intended to carry CP current; indeed, it is important that they are prevented from doing so. Hence, the selection of the cable gauge is determined solely by the mechanical robustness required.

11.6.2 Insulation

11.6.2.1 What is required?

The cable insulation has to satisfy one basic requirement: to provide complete electrical isolation between the conductor and either the metallic structure or, in some instances, the seawater. In itself, this is not a particularly onerous requirement. As we have seen, the DC side of an ICCP system generally operates at no more than a few tens of volts. A covering of any non-conducting polymeric compound would suffice.

However, the requirement becomes markedly more demanding when we consider that the insulation needs to be mechanically robust enough to avoid any damage during the cable installation process. Even the slightest nick in the insulation is likely to lead to system failure in very short order. Furthermore, once in place, the insulation must retain its integrity for the duration of the ICCP system's life, which may be measured in decades. In particular, it will need to resist, or be protected from, day-to-day mechanical damage; and it must be able to resist degradation by the various environmental circumstances to which it will be exposed during its life.

Finally, it will need to comply with the various electrical codes applicable to the installation.

11.6.2.2 *What do the codes say?*

Not much.

The main international CP codes (e.g. ISO and EN) advise selecting cable insulation that is mechanically robust and resistant to the anticipated environment. However, they are non-specific with regard to the insulation material itself. For example, Section 12.5 in ISO 15589-1 states that the... *insulating materials shall be resistant to chlorine, hydrocarbons and other deleterious chemicals*; but it offers no further information. Similarly, Section 6.3.5.6 in ISO 20313 advises that the... *insulation materials shall be resistant to their environmental conditions and satisfy classification requirements.* Again, no further details are provided.

11.6.2.3 *Candidate materials*

Early electrical cables were insulated with natural rubber. From the 1950s onwards, natural rubber was progressively replaced by synthetic thermoplastic or elastomeric polymers. You may recall that we encountered these basic polymer science terms when we discussed protective coatings in Chapter 9.

In the 1980s, an Institute of Corrosion/NACE joint task group worked on a document to provide advice on cable insulation and sheathing for marine ICCP systems. However, although the document was approved by the Institute in 1990 [12], it does not appear to have been formally published. It considered six types of insulation/sheathing material. The following is based on the content of that document.

11.6.2.3.1 Ethylene propylene rubber (EPR)

This material is based on a polyethylene/polypropylene copolymer. It was developed to provide a material intermediate in properties between polyethylene and the more rigid polypropylene. Further developments led to the incorporation of diene monomer which contributes some elastomeric (rubber-like) quality to the otherwise thermoplastic copolymer.

It serves well as cable insulation. It exhibits reasonable resistance to heat and various chemicals, and it exhibits very good low-temperature flexibility. Its main drawbacks are sensitivity to hydrocarbon solvents and burning fairly readily. Hence, if it is to be considered as a cable insulation, it needs to be sheathed with a hydrocarbon- and fire-resistant material.

11.6.2.3.2 Chlorsulfonated polyethylene (CSP)

This elastomeric polymer is more readily known by its registered Dupont tradename "Hypalon". It has good flame resisting properties and has been described as "self-extinguishing" after exposure to flame. It has reasonably good electrical properties and good mechanical properties, except at low temperature where it is inferior to EPR. It is generally used as a sheathing layer rather than the primary insulator.

11.6.2.3.3 Polyvinyl chloride (PVC)

Many of us are familiar with PVC in the form of window and door frames or building facias. In its solid form, where it is often referred to as unplasticised PVC (u-PVC), it is too brittle to serve as a cable insulator. However, flexibility is achieved by blending it with any or a range of lower viscosity resins known collectively as plasticisers. Plasticised PVC finds very widespread use, including as a fabric and cable insulation. Its insulating properties, whilst not as good as some other candidate materials, are more than adequate for the low-voltage cables employed in ICCP. It has generally good resistance to seawater and chlorine and is reasonably flame retardant.

A significant drawback of PVC is a phenomenon known as plasticiser migration. The plasticising resins are not chemically bound to the PVC. So, over time, these molecules will tend to diffuse outwards to the surface leaving an embrittled matrix behind. This embrittlement is accelerated by elevated temperatures or the influence of some organic solvents.

11.6.2.3.4 Polyethylene (PE)

PE is a thermoplastic polymer which has very good electrical insulation properties and excellent resistance to water and some chemicals and solvents. It is easy to process. When used as insulation, it is often pigmented with carbon black to improve its weathering resistance. It has a limited maximum operating temperature (<70°C) and only possesses moderate resistance to oils. It is also flammable, a fact which excludes it from service in numerous situations such as ships or some hydrocarbon production facilities.

11.6.2.3.5 Cross-linked polyethylene (XLPE)

PE, which is a non-cross-linked thermoplastic polymer, can be "artificially" cross-linked, making XLPE, by treating it with strongly oxidising chemicals or by irradiation. XLPE retains the good electrical insulation properties of PE but with improved thermal stability. It nevertheless retains the disadvantage of being flammable.

11.6.2.3.6 Polyvinylidene fluoride (PVDF)

The task group's summary also included one fluorinated thermoplastic polymer: polyvinylidene fluoride (PVDF).[5] However, the document itself omitted any narrative description of this material.

PVDF is a thermoplastic polymer with the formula $[\text{-CF}_2\text{-CH}_2\text{-}]_n$. It may be considered as intermediate between the non-fluorinated polyethylene $[\text{-CH}_2\text{-CH}_2\text{-}]_n$ and the fully fluorinated material PTFE[6] $[\text{-CF}_2\text{-CF}_2\text{-}]_n$. Thus, its properties lie between those of PE, which is easily processed but lacking in flame resistance, and PTFE which is highly resistant to flame and most forms of chemical attack but is brittle.

11.6.2.3.7 Comparative properties

The task group produced a summary table ranking the properties of the candidate materials. Table 11.3 reproduces a selection of that summary. It should be noted that the ranking results reflect the experience and opinions of the task group members. They are not the outcome of any form of independent testing or trials. It is clear, however, that the fluorinated polymer was much admired. However, it remains relatively expensive.

11.6.2.4 Current practice

Nobody is in the business of developing a new insulation material for the sole benefit of the relatively small ICCP market. The designer has to select one of the proprietary types marketed by major cable manufacturers. Usually, the designer would be wise to select a cable from the range normally stocked by CP equipment suppliers.

Thermoplastic insulation is normally preferred for cables which, once placed, are not expected to move or be disturbed. With the heavier duty cables, it is common to have a sheathing layer outside the main conductor to provide additional mechanical protection

[5] Also known as Kynar®.

[6] Poly(tetrafluorethylene) or Teflon™ was trade-marked by Dupont in 1938.

and, in some cases, to confer some flame-retardant benefit. Thus, PVC sheathing is sometimes provided with PVC, PE or XLPE insulation.

Table 11.3 Insulation performance rankings

	EPR	*CSP*	*PVC*	*PE*	*XLPE*	*PVDF*
Electrical insulation	8	3	4	9	8	9
Heat resistance	7	7	4	3	7	9
Cold temperature	8	5	2	9	8	8
Flame resistance	1	8	7	0	1	5
Oil resistance	3	7	5	5	5	6
Solvent resistance	4	6	4	9	9	8
Water resistance	7	6	6	9	9	9
Abrasion resistance	3	7	6	3	6	9

Scale: 0 (very poor) to 9 (excellent).

From [12].

Where elastomeric insulation is selected, the insulation of choice is usually ERP. Where sheathing is required, particularly for flame retardation, CSP presents the obvious elastomeric option.

11.6.2.5 Mechanical protection of cables

The protection of ICCP cables from mechanical damage in the offshore environment is of paramount importance. This protection can take two forms: steel wire armouring (SWA) and steel conduiting. It is unlikely that a cable that is going to be provided with mechanical protection will be required to be flexible, so a typical cable construction scheme might be XLPE (insulation)/PVC (inner sheathing)/SWA/PVC (outer sheathing).

Experience suggests that, if a cable is required to transit through the splash zone and waterline, the only durable approach is to run it through a robust rigid conduit.

11.6.2.6 Cables connections

ICCP anodes for subsea use are invariably provided with factory-made connections to an appropriately insulated and protected cable. The cable length should be specified to ensure that there is no need to install splices that will be exposed to seawater. The cable can be run, with the appropriate mechanical protection to the positive terminal of the T/R which should ideally be placed in a weatherproofed location.

In the case of ships' external hulls (see Chapter 15), a hull penetration has to be made using a cofferdam. The anode to cable connection is then made in the dry environment within the cofferdam.

In marine CP, the cable-to-structure connection rarely involves anything more than running a cable from the negative terminal of the power supply to the nearest convenient earthing point on the metal structure.

11.7 STRAY CURRENT INTERFERENCE

In Chapter 2, we made the point that one of the advantages of ICCP was also a potential disadvantage. Its ability to provide a high-power option for delivering CP current into the seawater is a clear advantage over the use of low-power, bulky sacrificial anodes. The corresponding disadvantage is that the intense electrical fields generated in the vicinity of ICCP anodes, and the high currents produced, can cause stray current interference damage to steelwork that is not part of the same CP system. In the past, this has caused accelerated corrosion damage on offshore pipelines adjacent to, but isolated from, fixed structures protected by ICCP.

A possible stray current interference situation is illustrated schematically in Figure 11.6. In this scenario, a fixed offshore structure is protected by an ICCP system delivering current from remote subsea sleds. The position of one of the sleds is close to a pipeline which, although not connected to the structure, passes close to it. In this scenario, the pipeline coating has some areas of damage such that bare steel is in contact with the seawater. This scenario also assumes that the pipeline's own sacrificial bracelet anodes are widely spaced, and none is close to either the ICCP anode or its host structure.

The stray current problem then arises because the pipeline acts as a conductor intersecting the electric field generated by the anode sled. Where the pipeline runs close to the sled, it picks up current. At this location, the direction of the current is from the seawater onto the pipe. So, the current is protective. Ultimately, however, this current must return to the negative terminal of the power supply. To do so, it must exit the pipeline. It does this at a location nearer to the structure. As we have learned, where current exits, it does so by an anodic process. This causes intense corrosion, termed stray current electrolysis, at this location on the pipe.

We will return to stray current issues in Chapters 13 and 14.

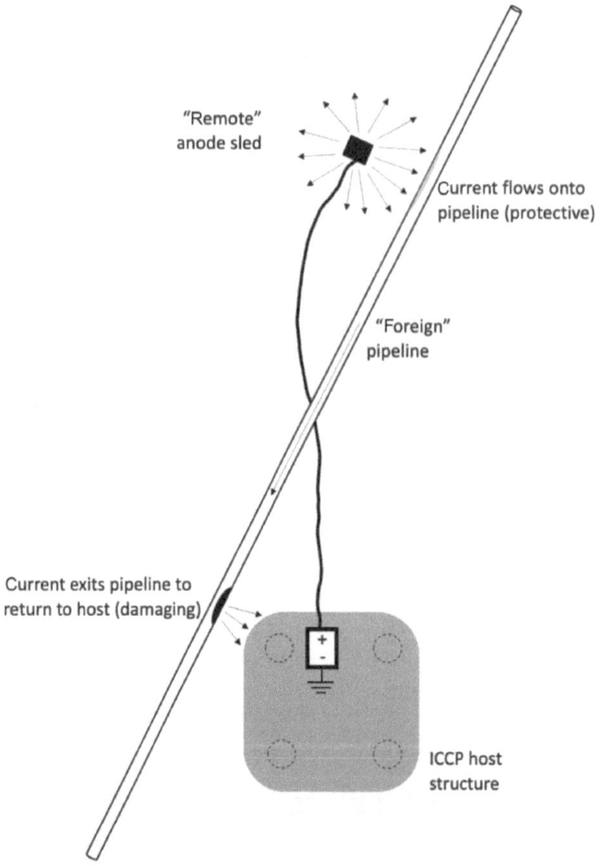

Figure 11.6 Stray current corrosion situation.

11.8 ICCP SYSTEM SAFETY

11.8.1 Transformer-rectifiers

Most CP professionals are not qualified electrical engineers. A CP design engineer should be able to specify what is required from the T/R. This includes the mains supply (voltage and phase), the output DC voltage and current range, manual or automatic control, control discrimination, the number of meters and indicator lamps, and so on. It is neither necessary nor desirable for the CP engineer to design the unit itself. That should be left to the experts employed by the equipment manufacturers. Importantly, the manufacturer will be responsible for ensuring that the equipment complies with all of the relevant electrical safety regulations.

On the other hand, few if any ships or offshore installations include a trained CP technician among the crew. It follows that the day-to-day operation of offshore and marine ICCP systems falls under the aegis of the electrical department. It has to be assumed that these individuals are adequately trained in electrical safety, although as a matter of experience, they are rarely very knowledgeable about CP itself.

11.8.2 Diver safety

Most are aware that the combination of electricity and salt water is potentially hazardous. It is perhaps unsurprising that, in 1974, the UK Department of Energy issued a Diving Safety Memorandum which called for ICCP systems to be isolated when divers were working in the vicinity. Although the regulations were revoked in 1981, it has remained customary for ICCP systems to be switched off whenever divers work on an offshore installation [9].

It has to be conceded, however, that the evidential basis behind this concern is thin. We are broadly aware of the effects of an electric current on the human body. For example, in the 1980s, Moulton [9] reviewed the physiological effects of DC electrical currents (see Table 11.4). He also provides information on the higher currents that lead to respiratory failure and ventricular fibrillation. However, these are just medical terms for death, and since the uncontrolled muscle contraction would render a diver at risk of getting into an unrecoverable situation, the real limits have to be set at a figure well below the figure that causes death directly.

Table 11.4 Physiological effects of electric current

	Current (mA)	
Effect	*AC*	*DC*
Perception	1–3	5
Pain	6–9	40–50
Muscle contraction	9–16	70–80

From [9].

Authoritative guidance is now provided by IMCA[7] D 045 [17]. This advises a maximum system voltage output of 24 V DC and 6 V AC (rms). Although this document provides more detail than Moulton's earlier paper about the tolerable safe AC and DC current limits, the figures have remained broadly the same.

However, it is difficult to impose a current limit because it is very problematic to calculate how much current would go through a diver if he (and it invariably is a "he") should pass close to, or touch, a working anode. Moulton provides some examples of estimates based on

[7] IMCA: International Marine Contractors Association.

divers of certain sizes and shapes intersecting the electric field created by typical platform-mounted ICCP anodes. He concluded that... *the probability of a diver receiving a serious shock from a local impressed current anode is low.* Nevertheless, he also acknowledges that we cannot award a safety carte blanche since that involves the philosophical conundrum of proving a negative. As he says, we... *cannot conclude that impressed current anodes do not present a hazard to divers.* On this basis, it seems that the current practice of turning ICCP systems off when divers are in the water is destined to persist.

Indeed, the attitude of the offshore industry has been, and sometimes remains, equivocal on this point. For example, one major operator's standard on diving operations calls for the diving contractor's operating manual to... *specify safety cautions to be deployed based on a risk assessment...* when working close to impressed current cathodic protection systems. However, this apparent permission to work (with care) close to a working anode is then overridden later in the same standard where it requires that... *all items of plant or equipment requiring physical isolation has been closed down and locked (including any impressed current cathodic protection systems) when working in the vicinity of the structure.*

The only solid conclusion that we can draw from this is that safety is paramount. The very existence of doubt, therefore, forces the decision to switch ICCP systems off when divers are in the water nearby. Since divers are only present for a limited period, switching the ICCP off will not impact the integrity of the structure.

Needless to say, however, it is imperative that the ICCP is switched back on when the divers have finished!

REFERENCES

1. H. Davy, *On the corrosion of copper sheathing by sea water, and on methods of preventing this effect; and on their application to ships of war and other ships.* Philosophical Transactions of the Royal Society **114**, 151–158, 242–246 (1824).
2. R. Juchniewicz, *The influence of alternating current on the anodic behaviour of metals.* Proceedings of 1st International Congress on *Metallic Corrosion*, London, 1961.
3. J.A. Burgbacher, *Cathodic protection of offshore structures.* Materials Protection 7(4), 26 (1968).
4. W.R. Jacob, *Substrate materials for platinised anodes.* Corrosion Prevention and Control 22 (8) (1975).
5. M.A. Warne and P.C.S. Hayfield, *Platinized titanium anodes for use in cathodic protection.* Materials Performance **15** (3), 39 (1976).
6. C.F. Schrieber and G.L. Mussinelli, *The Lida® impressed current system: performance in natural waters and sea muds.* Proceedings of UK Corrosion, 1984.
7. N.D.S. Hill, *Galvanostatic evaluation of lead magnetite composite anodes in simulated seawater.* Materials Performance **23** (10), 35 (1984).
8. L.L. Shreir and P.C.S. Hayfield, *Impressed current anodes*, Chapter 6 in *Cathodic Protection: Theory and Practice,* Eds. V. Ashworth and C.J.L. Booker. Ellis Horwood (Chichester) 1986.
9. R.J. Moulton, *Hazards to divers from cathodic protection systems,* Chapter 21 in *Cathodic Protection: Theory and Practice,* Eds. V. Ashworth and C.J.L. Booker. Ellis Horwood (Chichester) 1986.
10. V.F. Hock et al., *Structure, chemistry, and properties of mixed metal oxides*, Paper no. 230 CORROSION/88.
11. J.T. Reding, *Performance of mixed metal oxide activated titanium anodes in deep groundbeds,* Paper no, 9 CORROSION/88.
12. UK Corrosion Control,Engineering joint venture Task Group E4-4 document, *impressed current system cables for marine use*, 1990.
13. J.W.L.F. Brand and P. Lydon, *Impressed-current anodes*, Chapter 10.3 in *Corrosion*, vol. 2, 3rd edition, Eds. L.L. Shreir, R.A. Jarman and G.T. Burstein. Butterworth-Heinemann (Oxford) 1993.

14. P.J. Moreland and K.M. Howell, *Impressed current anodes, old and new,* Chapter 11 in *Cathodic Protection: Theory and Practice*, Eds. V. Ashworth and C. Googan. Ellis Horwood (London) 1993.
15. G. McMillan, *Developments in cathodic protection rectifiers: parts 1 and 2.* Materials Performance **32** (11, 12), 31 (1993).
16. M. Lewis, *How 'Vagrant current' became impressed current cathodic protection parts 1 & 2.* Materials Performance **47** (11), 36 & **47**(12), 34 (2008).
17. IMCA D 045, *Code of practice for the safe use of electricity under water*, 2010.
18. M. Funahashi and H. Wu, *What you need to know about MMO coated metal anodes*, Paper no. 2107 CORROSION/2013.
19. G. Mulkahy, *Successfully adapting high frequency switch mode power supply technology to impressed current cathodic protection.* Paper no. 5437 CORROSION/2015.

Chapter 12

The effect of CP on mechanical properties

12.1 INTRODUCTION

12.1.1 Outline

In Chapter 1, we came up with a ball-park figure of <0.1 mm/year for the long-term corrosion rate for steel fully submerged in seawater. Even lower rates are generally observed in the buried zone. This might suggest that underwater corrosion need not be a relevant integrity threat to a massive steel structure with a design life of (say) 30 years. Simply adding a few millimetres of corrosion allowance to the steel thickness should suffice. CP would then be irrelevant. Indeed, there are more than likely to be some marine structures fitted with entirely functional, but ultimately pointless, CP systems. Even if these structures were allowed to corrode naturally, they would see out their intended lives without falling over as a result of corrosion.

There is, of course, the economic counterargument that adding a few mm to the thickness of a large structure adds to the cost, not just in respect of the added material but also in increased fabrication, transport and installation costs. There is also a relevant environmental issue: the production of each additional tonne of steel would generate about three tonnes of CO_2 emissions.

In addition, for many structures the relevant integrity benefit of its CP system will not be its ability simply to control metal thinning caused by corrosion. It is more likely to be the influence it exerts on fatigue or, more correctly, corrosion fatigue. Accordingly, in this chapter, we start by exploring fatigue and corrosion fatigue of typical structural steels. We will then examine how corrosion fatigue is mitigated by CP.

From this, our attention moves to higher strength steels which, like their lower strength counterparts, are also prone to corrosion fatigue. However, with the higher strength, we have to consider an additional factor: the role of atomic hydrogen. As we shall see, hydrogen atoms, and the way they interact with the metallurgical structure of the steel, introduce complexity and risk to the issue of marine CP. We will also discover that the role of hydrogen is not limited to the high-strength steels. It can also exert troublesome influences on some of the CRAs which may find themselves under CP.

12.1.2 Some basics

When we think of metals, we think in terms of the properties which we experience in our everyday interaction with metallic objects: strength, stiffness, elasticity, toughness, ductility and so on. In this chapter, we explore how CP might influence these mechanical properties. In order to get to this, it is worthwhile revisiting some of the basic concepts taught in the school physics lab.

DOI: 10.1201/97810032160702-12

We can start by considering the 17th century work of Robert Hooke who recorded the extension of wires to which he attached weights. The weight or applied load is termed the stress and is defined as the load applied per unit cross-sectional area of the specimen. It is usually given the Greek symbol σ. Its SI system unit is the pascal (Pa) or, more usually, the megapascal (MPa). The extension, also referred to as the (tensile) strain, and often given the symbol ε, is a dimensionless quantity. It is the ratio of the extension of the stressed specimen divided by its original length. For convenience, this extension is often expressed as a percentage.

Figure 12.1 illustrates the relationship between the applied stress and the resulting strain for two metals. For "A", we see that an increase in applied stress produces an increase in strain. Up to a point, as we increase the stress, the strain increases linearly in proportion. This behaviour is known as Hooke's law.

The ratio of the stress to the strain (σ/ε) is known as Young's modulus (E). This linear part of the stress-strain diagram depicts *elastic* behaviour. If we remove the applied stress, the strain disappears and the specimen returns to its original length. Viewed simplistically on an atomic scale, the effect of the stress is to stretch the interatomic bonds in the metal. Removing the stress permits these bonds to return to their original length.

As the applied stress continues to increase, small deviations from this elastic behaviour become apparent. The elastic limit of the material is reached, which means that the material now does not return precisely to its original dimensions. There is some permanent deformation. This is referred to as *plastic* deformation. Keeping with our simplistic view, not only have the interatomic bonds become stretched, but planes of atoms have actually slipped over one another producing permanent distortion. Metallurgists refer to this change as the movement of dislocations within the metal.

In defining the characteristics of metals, the stress at which the change from elastic to plastic behaviour becomes perceptible is referred to as the *yield stress* or *yield strength*. In practice, the exact point at which the material starts to yield is difficult to determine. For this reason, it is customary to prescribe an easily measurable level of plastic strain (usually 0.2%), which is slightly higher than the yield stress. This is referred to as the 0.2% proof stress. The strength of alloys is usually characterised by their specified minimum yield stress (SYMS). For most practical purposes, this is the same as the proof stress.

Pure iron has a yield stress of between 80 and 100 MPa. Alloying iron with carbon to form steel increases the strength. Depending on the alloy composition and thermal treatment, steel yield stresses can be anywhere from ~200 MPa to over 1500 MPa. However, most (but not all) of the steels used in the construction of ships, offshore structures and pipelines have yield stresses in the range of 350–500 MPa.

Continuing to increase the applied stress over and above the yield stress produces further permanent plastic deformation. This is not necessarily a bad thing. For example, our ambition might be to press the metal into a desired shape. However, if the stress continues to be increased, the specimen will break. At this point, we have reached the ultimate tensile stress (UTS). For the lower strength steels, the UTS might exceed the yield stress by as much as 80%. For higher strength steels, the UTS is likely to be little more than 10% above the yield stress.

Figure 12.1 also shows a stress-strain diagram for material B. Young's modulus is greater showing that B is stiffer than A, and both the yield strength and UTS are higher, showing it to be stronger. These observations might tempt inexperienced structure designers to conclude that B is superior to A since it is stronger.

Unfortunately, life is not so simple.

We may recall from our school physics lessons that the work done, or energy expended, is the product of the force times its displacement. This means that the area under the stress-strain curves in Figure 12.1 represents the energy absorbed in breaking the specimens. Since the area under curve A is greater than that under curve B, we can conclude that material A will absorb more energy before it breaks. Material A is termed *ductile*, whereas the stronger material B is *brittle*. In practice, the development of steel formulations has often involved a trade-off between achieving the mutually exclusive attributes of strength (or hardness) and ductility.

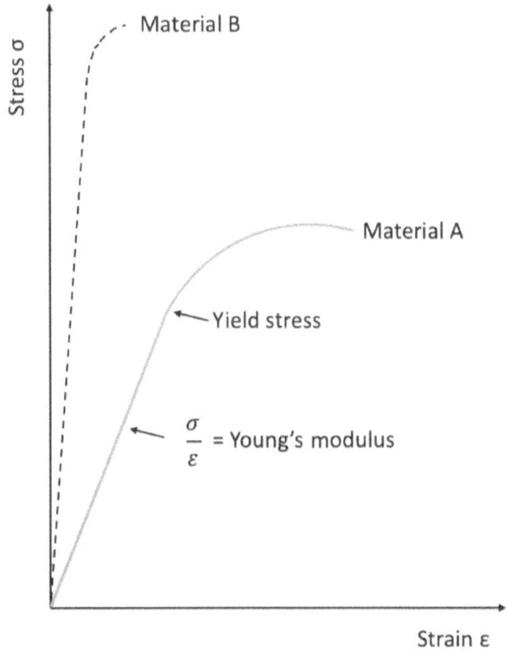

Figure 12.1 Schematic stress-strain curves.

This distinction between A and B is apparent if we examine the specimens at the end of the stress-strain test. As indicated in Figure 12.2, the more ductile specimen will have extended more in length, and it will have exhibited more pronounced "necking" at the fracture location than the brittle material. As we shall see, this necking, or reduction in cross-sectional area (RA), expressed as a percentage, is a useful parameter when studying the mechanical properties of structural or pipeline steels.

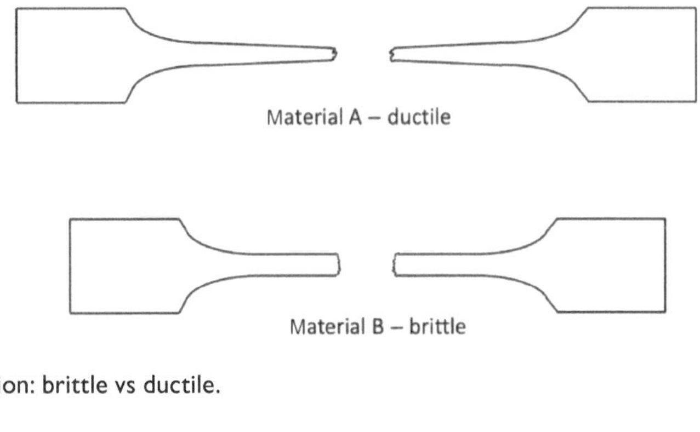

Figure 12.2 Elongation: brittle vs ductile.

$$\text{RA} = \frac{A_o - A_f}{A_o} \qquad (12.1)$$

where

A_o = is the original cross-sectional area of the specimen
A_f = is the cross-sectional area of the specimen at the point of fracture

12.2 MATERIALS OF INTEREST

12.2.1 Structures

The material of choice for the construction of an offshore structure is usually carbon steel (also referred to as carbon-manganese steel). This is a generic term covering a multitude of materials of various chemical compositions, microstructures, thermal histories and mechanical properties. To add to this complexity, we need also to be aware that any one steel might have a bewildering number of designations deriving from multiple national and international standards.

12.2.1.1 Medium-strength steels (SMSY < 550 MPa)

Historically, the personnel and facilities involved in the design and construction of offshore structures had links to shipbuilding. The grades of steel commonly employed for ship construction have, therefore, predominated. For example, almost all of the structural steelwork in the North Sea has conformed to BS 4360 grade 50D.[1] It has reasonable strength (minimum yield stress around 345 MPa) and good ductility (it extends by around 20% before fracturing). It is also easily welded.

It follows that the majority of the mechanical testing that is relevant to offshore construction has been carried out on carbon steels of this grade, or on materials that are very similar. However, even for a material with an established composition range, the improvements in steel-making processes over time have resulted in significant changes within that range. Taking 50D as an example, the specified composition sets a limit of 0.04% by weight for phosphorus and sulphur impurities. Versions of this steel produced in the early 1970s contained 0.024% sulphur and 0.038% phosphorous [22]. A decade or so later, a much cleaner version of the same steel grade, levels of sulphur and phosphorus were 0.003% and 0.005% respectively [15], was in common use. There has been little or no systematic study to examine the effect of such compositional changes on mechanical properties. Nevertheless, it is reasonable to assume that cleaner steels possess better characteristics than their less pure progenitors.

12.2.1.2 Higher strength steels (>550 MPa)

In the last quarter of the 20th century, there was an increasing use of higher strength steels for some structural elements. These materials exhibit very similar corrosion behaviour to their lower strength cousins and are equally amenable to corrosion control. However, as we shall see, their mechanical properties are susceptible to being influenced by CP.

[1] This British Standard is now superseded by EN 10025-3, for which the relevant grade designation is (approximately) S355N. This material has equivalent designations according to other standards throughout the world, but it has not been assigned a UNS designation.

12.2.2 Pipelines

As with structures, the first-choice material for a pipeline designer will be steel. For many pipeline projects, a grade of similar strength to ship plate steel, such as ISO 3183 Grade L360 [54], fits the bill. This material has a specified minimum yield stress of 360 MPa and is similar to grade 50D. However, pipeline engineers remain more firmly wedded to US customary than to SI units. Grade L360 is still commonly referred to as API 5L grade X52 (minimum yield stress of 52 ksi[2]). ISO 3183 and API 5L are essentially the same document. As with structural steels, pipeline designers will opt for higher strength steels if need be. Nevertheless, it is rare to encounter an offshore carbon steel pipeline fabricated from a grade stronger than ISO 3138 grade L485. That is ~485 MPa yield strength (equivalent to API 5L grade X70).

Latterly, the need to transport more corrosive fluids subsea occasionally brings with it the need to construct flowlines in CRAs such as the duplex or martensitic stainless steels which we discussed in Chapter 8. As we will discover, the combination of such novel CRAs with CP has not been without its problems.

12.2.3 Equipment

Where practicable subsea equipment is also fabricated in medium-strength carbon steel, although the bolting is usually a high-strength low-alloy (HSLA) steel. However, mechanical requirements, or the need to resist the corrosivity of a contained fluid, frequently force the material selection towards CRAs. As we shall see, the selection of such alloys also needs to take account of their compatibility with the CP systems to which they will be connected.

12.3 FATIGUE

12.3.1 What is it?

Figure 12.1 might tempt us to conclude that all we need to do to design an offshore structural component would be to calculate the maximum load it must bear, select a grade of steel with the appropriate yield strength and then calculate the thickness of metal needed to take that load. We might also add a few millimetres of thickness as a corrosion allowance. Unfortunately, such a simplistic engineering approach might not result in a structure with the desired service life.

As mentioned, even if corrosion were not controlled, comparatively few structures installed in seawater would fail because the steel had thinned to the extent that they could no longer sustain the deadweight load. In most instances, failure would come earlier as a result of fatigue or, more correctly, corrosion fatigue.

Fatigue has been recognised as a failure mode for metals since the early part of the nineteenth century. As long ago as 1843, William Rankine reported to the UK Institute of Civil Engineers[3] on *the causes of the unexpected breakage of the journals of railway axles* [1]. He did not use the word "fatigue", but he reported that the observed fractures... *commenced with minute fissures... which had gradually penetrated from the surface towards the centre.* He argued that these breakages were... *owing to a tendency of the abrupt change in thickness... to increase the effect of shocks at that point.*

[2] The unit "ksi" is thousand pounds per square inch.

[3] You may be wondering why Rankine presented his findings at a civil engineering meeting. The reason is simple. The Institute of Mechanical Engineers was not founded until four years later.

Eleven years later Rankine was in attendance when Frederick Braithwaite delivered a paper to the same institute on the fatigue and fracture of metals [2]. The word "fatigue" had been introduced in the interim by French workers who had also been investigating fractures in railway locomotives. They had mused that the metal had become "tired" ("fatigué" in French) of taking the load. The tone of Braithwaite's presentation echoed this anthropomorphism. He concluded that a metal... *in a state of rest, although sustaining a heavy pressure... will continue to bear that pressure without fracture, so long as its rest is not disturbed, or the same strain is not too often repeated.*

This early paper contains the seeds of our current understanding of metal fatigue. At its simplest, fatigue is failure caused by the cumulative action of cyclic stresses, the individual magnitudes of which are below the yield stress. This means that, when designing structures, it is important to know not only the yield strength of the metal but also its fatigue behaviour. As a matter of experience, many more mechanical failures of metallic structures have been caused by fatigue than by simple over-loading.

It will not surprise you to learn that fatigue is a complex phenomenon that has generated a very extensive body of technical literature. There is a limit to how deeply into this oeuvre it is reasonable for a corrosion or CP engineer to delve. So, in this book, we will limit our coverage to what a competent CP practitioner needs to appreciate in order to understand the objectives of other engineering disciplines involved with any project. If you are interested in drilling down deeper into the subject then, in the first instance, I suggest the following references [11,24,35,38,46]. What follows below is an expurgated treatment of the subject.

Fatigue affects all metals, but we will concentrate our attention on structural steels. The phenomenon is generally studied either by S-N experiments or by applying fracture mechanic principles to the growth of fatigue cracks.

12.3.2 S-N testing

12.3.2.1 Plain specimens

Since the middle of the nineteenth century, the fatigue behaviour of metals has been characterised by conducting experiments in which a specimen is subjected to cyclic stress of amplitude (S). The number of cycles to failure (N) is then recorded. The results are plotted on a graph of S (or log S) against log N. A schematic diagram, using published test data for a commercial steel tested in air, is given in Figure 12.3. In this figure we see that, at relatively high applied cyclic stresses (around 70% of the yield stress), the test specimens fracture at a little over 10^4 cycles. This means that, if the testing was carried out at (say) 0.5 Hz, the specimens would have failed in less than a day.

Moving to the right along the graph we see that, as the magnitude of the applied stress (S) is reduced, the number of cycles (N) needed to produce failure increases. Below a certain minimum stress, in this example around 45% of the yield stress, the shape of the S-N curve tends towards a horizontal line. This horizontal line is termed the *fatigue limit*. For most steels, irrespective of their actual strength level, it happens to be somewhere close to 50% of yield strength.

Examples of such S-N diagrams for steel and other metals are to be found in the literature (e.g. [11]). Examinations of individual results in such publications usually reveal the inherent variability of fatigue testing. This means that fatigue curve exemplified in Figure 12.3 is not an analytical line. Rather, it is a boundary between populations of specimens that failed, and populations that did not fail, during the test.

It is worth mentioning here that not all metals exhibit a fatigue limit. Aluminium is a case in point. This will eventually fracture by fatigue even at very low-stress amplitudes

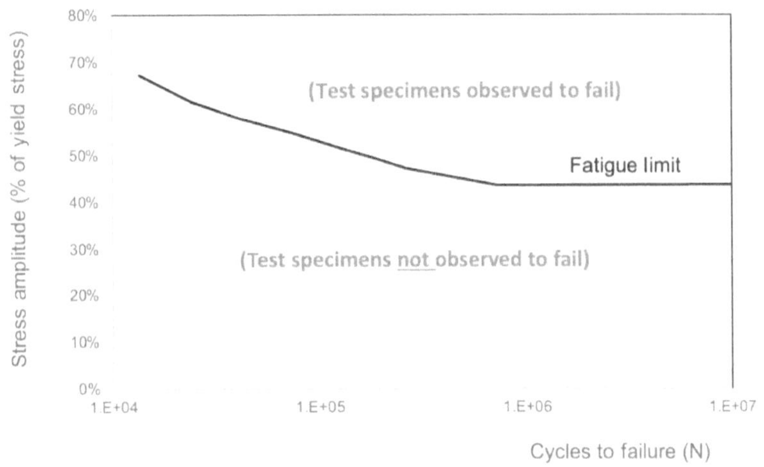

Figure 12.3 Schematic S-N curve for steel in air.

providing one has the patience to submit it to a large enough number of cycles. For this reason, the fatigue behaviour of aluminium is usually quantified in terms of its *fatigue strength* rather than its fatigue limit. The fatigue strength, which is also termed the *endurance limit*, is defined as the maximum value of a completely reversed stress that a material can withstand for a finite number of cycles. Typically, N will be set as 10^6 or 10^7.

We should also point out that our description of N in Figure 12.3 makes no attempt to characterise the processes that are taking place as the specimen proceeds towards failure. In particular, S-N curves do not distinguish between the cycles that are involved in initiating a crack and those which are subsequently involved in crack growth. Indeed, the distinction between crack initiation and crack propagation remains the subject of debate between fatigue specialists. Some reason that the initiation is a distinct process. Others argue that it is simply a period of crack growth, during which the cracks are too small to be identified.

12.3.2.2 Notched specimens

Up until the 1960s, researchers on fatigue almost always used smooth, sometimes polished, specimens. In principle, this made the calculation of the stress in the S-N experiment very straightforward. S is simply the applied load, or load range, divided by the cross-sectional area of the specimen. However, in the real world, steel structures are not smooth. They contain irregularities ranging from intended changes in profile, through notches to surface cracks. The effect of these geometric irregularities, which are also termed *stress-raisers*, is to concentrate the stress locally. Thus, if we carry out S-N fatigue testing at any chosen value of S, we would expect a specimen containing a stress raiser to fail in fewer cycles. This is essentially what Rankine had reported as early as 1843.

One way of characterising the effect of a stress raiser is its *stress concentration factor* (SCF). The SCF is defined as the ratio of the maximum stress in the vicinity of the stress raiser divided by the average, or nominal, stress (i.e. the total applied load divided by the total cross-sectional area). Pioneering work in this area dates back over a century to Charles Inglis at Cambridge. He was perhaps the first to articulate that stress should be viewed not as a static force, but as an influence that "flowed" through the metal, rather than just being the load applied at its extremities.

It is helpful to consider an analogy with lines of magnetic flux when a specimen is magnetised. Figure 12.4 illustrates this concept of stress flow for a material under simple tensile loading. Ideally, the "flux" or intensity of the stress would be uniform through the specimen. However, the presence of a notch, or other geometrical irregularity, means that the stress field is distorted such that there is a concentration of the stress.

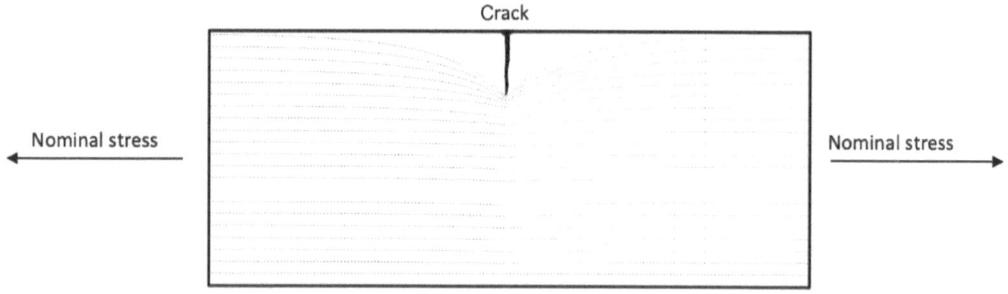

Figure 12.4 Stress concentration.

There has been a considerable body of work on stress concentration published since that time in the journals of mechanical engineering. The topic itself falls outside the domain of the CP engineer. However, we do need to have some awareness of it. In particular, we should appreciate that, although the size of the stress raiser exerts some influence on the SCF, it is the shape that is critical. The sharper the notch or more acutely angled the profile change, the greater the SCF it will produce.

This is important when it comes to the consideration of marine structures which, in practice, are never smooth. They comprise welded steel plates. Inevitably, when the weld metal cools and shrinks, tensile stresses develop. For example, this means that the area referred to as the "toe" of the weld, which is where the visible face of the completed weld (termed the weld "cap") meets the plate metal, will contain microcracks. In summarising research work carried out prior to 1979, Hartt et al. [10] pointed out that... *the geometrical discontinuity at the weld toe and the resultant stress concentration are fundamentally important with regard to fatigue life.*

In the case of welded structures, therefore, crack initiation may be thought of as already having taken place. This explains the observation that, other parameters being equal, fewer cycles are required to crack welded specimens than smooth ones. It also explains why the vast majority of fatigue cracks observed in practice are associated with welds.

For this reason, most of the fatigue testing and analysis that underpins the various industry codes has been carried out on welded specimens. In addition, the reality that a welded structure contains pre-existing cracks means that, in the view of many experts in fracture and fatigue, it is more relevant to analyse fatigue and corrosion fatigue using fracture mechanics.

12.3.3 Fracture mechanics

12.3.3.1 Basics

Although they are widely used, S-N curves have a number of limitations. The first is that there are only two possible outcomes: "failure" or "no failure". This means that it is usually necessary to test for a very large number of cycles in order to obtain a result. Another limitation of S-N data is that they do not provide any information on whether, or at what rate, an existing known crack will propagate under a given fatigue loading. However, it is reasonable to assume that all real structures, particularly those that are welded, contain crack-like defects. Accordingly, the fatigue assessment of these merits a fracture mechanics approach in accordance with an internationally recognised code such as BS 7910 [34].

Fracture mechanics deals with predicting whether or not a crack, or any other flaw, will give rise to fracture when stress is applied. As with other scientific disciplines that we touch on in this book, there is a limit to how far a CP professional might reasonably delve into this subject. We need to be able to appreciate how fracture mechanics considerations are applied to fatigue, but we must also remain aware that it is a discipline populated with its own experts.

Since fracture mechanics applies to all materials, we might look in the first instance at glass. This is a very strong material, but that strength comes at a price. As we have all experienced at one time or another, glass is very brittle. Glaziers frequently make use of this brittleness when cutting a pane of glass to size. A straight scratch is marked onto the surface using a diamond scribe. Then, with the application of a little pressure, and more skill than I have ever mustered, the pane snaps cleanly along the intended line.

What interests us here is the fact that a small induced notch or crack, in combination with an applied stress, which is very much lower than the tensile strength of the glass itself, leads to rapid fracture. This fracture of glass was first studied by Griffiths around the time of World War I. He was aware of Inglis' work on stress concentration but found that applying SCF formulae produced unhelpful results when it came to a sharp crack. In the extreme, if the crack tip were perfectly sharp, the SCF was calculated to be infinite.

Accordingly, he reasoned that a crack could only grow if the elastic strain energy released was greater than the energy consumed in creating a new crack surface area. This led to an explanation as to why minute cracks in glass can result in it fracturing at applied stresses that are only a small fraction of the tensile strength.

Fortunately, metals are less brittle than glass. This means that, as the applied stress is increased beyond the elastic limit, some plastic deformation takes place before the final rupture. We illustrated this in the case of ductile material A in Figure 12.1. In that case, only a part of the strain energy released as the crack grows is involved in creating new surfaces. The majority is absorbed in causing deformation in the plastic zone at the head of the crack tip.

It is not actually possible to measure this stress intensity. Nor, strictly speaking, is it possible to calculate it with complete accuracy. Nevertheless, fracture mechanic specialists have derived a workable construct: the *stress intensity factor* (K). This is an important fracture mechanics parameter which seeks to describes the state of stress at the crack tip. K can be estimated based on a consideration of the mode of cracking, the nominal applied stress, and the crack size and geometry. There are three cracking modes including. These are described in detail in the literature (e.g. BS 7608 [42]), but I prefer the simplistic descriptions of pulling, sliding or twisting. Here we will content ourselves with considering only pulling or, more prosaically, tensile stress. This is referred to as mode I crack growth. It should properly be described as tensile crack opening with the applied stress perpendicular to the direction of crack growth. Most tensile testing and most fatigue experiments involve mode *I* cracking.

The value of K associated with this mode is often designated K_I. Information on calculating K_I for different crack sizes and geometries can be found in BS 7910 and various online sources.[4] One of the simpler examples is for an edge crack in a semi-infinite plate.[5]

$$K_I = 1.12\sigma_\infty\sqrt{\pi a} \tag{12.2}$$

[4] www.fracturemechanics.org

[5] The area of the plate is very much larger than the area of the crack.

where

K_I is the stress intensity factor (MPa$\sqrt{\text{m}}$)
σ_∞ is the nominal stress (MPa)
a is the crack length (m)

12.3.3.2 The Paris law

The key link between fracture mechanics and fatigue, or corrosion fatigue, is given by the so-called Paris' law which states:

$$\frac{da}{dN} = A(\Delta K)^m \quad (12.3)$$

where

$\frac{da}{dN}$ is the rate of increase in crack length per fatigue cycle
ΔK is the stress intensity factor range
A and m are constants

As an aside, you may have noticed that I have referred to this relationship as the so-called Paris law. In my view, the term "law" implies a stronger scientific foundation than is actually provided by Paris. "Paris Analogy" would be more fitting. Nevertheless, in keeping with the fatigue fraternity, I shall refer to it as a "law".

A simplified representation of the Paris law is shown in Figure 12.5 for a metal undergoing fatigue in air. Below a certain value of the stress intensity factor range (ΔK), the fatigue crack will not grow. This threshold value may reasonably be likened to the fatigue limit. If this threshold is exceeded, then the crack will grow by fatigue according to relationship (12.3). The use of a log-log plot results in a linear relationship. Of course, as the crack grows by fatigue, ΔK increases (for example, according to 12.2). Ultimately, ΔK reaches the point where the fracture toughness of the material is exceeded and the test specimen breaks.

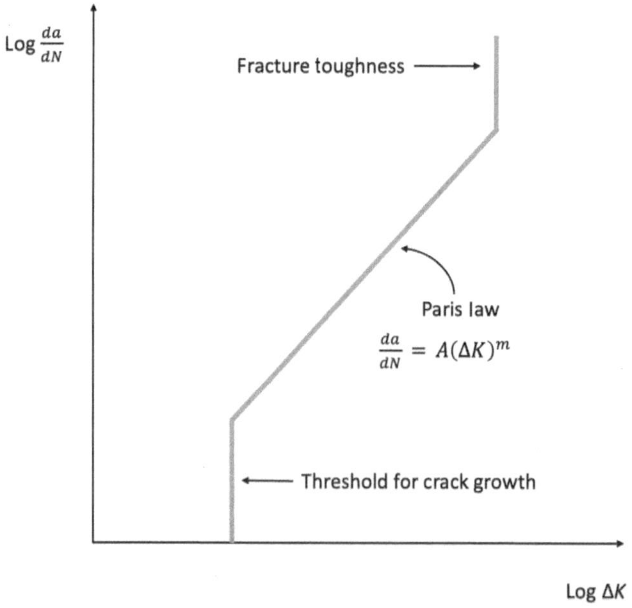

Figure 12.5 Paris' law schematic.

Actual fatigue crack growth data never adhere as neatly to the Paris law as Figure 12.5 might imply. The data will always contain a considerable degree of scatter. More to the point, evidence suggests that there is a two-stage crack growth process for steel, with the exponent m undergoing a reduction at a ΔK value somewhere around 6 MPa$\sqrt{m}$.

Before moving on, we need to have a quick word about the units of K. You will note that, as with the rest of this book, I have converted the units in source publications to their SI equivalents. Hence, we see MPa$\sqrt{m}$ for stress intensity. If you choose to refer to original publications, you need to be aware that you will find other units such as the imperial ksi $\sqrt{\text{inch}}$ or the metric unit N/mm$^{3/2}$. As it happens, ksi$\sqrt{\text{inch}}$ is numerically only about 10% above MPa$\sqrt{m}$. More confusion occurs around the use of N/mm$^{3/2}$ instead of MPa$\sqrt{m}$, which differs numerically by a factor of $\sqrt{1000}$. Even though the pascal (Pa) was officially adopted as the SI unit for stress as long ago as 1971, the use of the newton (N) has proved difficult to expunge from stress intensity calculations.

In air, and irrespective of which units we choose, the values of the Paris law constants A and m depend solely upon the characteristics of the metal. It is useful to put some numbers to these constants and to see the form of the relationship they produce. We can take the mean values for these constants from BS 7910. Since the standard is compiled on the basis of experimental data from various sources, it helpfully adds values of these constants that are two standard deviations on the conservative side of the mean.

We can apply these crack growth values to the schematic plot in Figure 12.5. The result is presented in Figure 12.6, in which the continuous line represents the mean curve and the dashed line represents the curve drawn to capture data points falling within two standard deviations of the mean value.

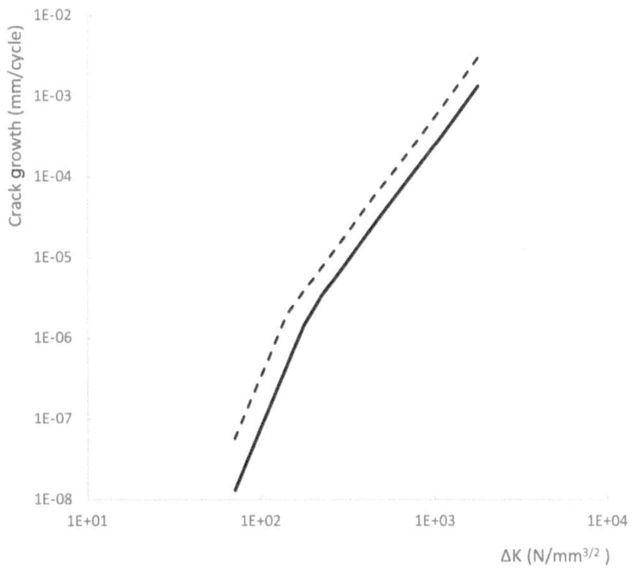

Figure 12.6 Schematic Paris' law fatigue crack growth for a welded joint in air.

Figure 12.6 embodies a threshold value of ΔK of ~2 MPa$\sqrt{m}$. In most tests, it is found that a crack will not grow if its ΔK value is lower than this. This figure may be thought of as the fracture mechanics analogy of the fatigue limit. Where the fatigue loading causes a ΔK value that is greater than this threshold, the Paris law tells us that the crack will grow. Furthermore, since the crack grows with each fatigue cycle it endures, the value of ΔK

increases slightly (in accordance with equation 12.2, for example). So, the next cycle of the fatigue load produces a slightly greater extension of the crack and a further slight increase in ΔK for the next cycle.

We should understand here that, where I am using the expression "slightly" in respect of the increments in ΔK and crack growth per cycle, I really do mean very "slightly" indeed. We can get some feel for how "slightly" by conjuring up an artificial, but not unrealistic, example along the following lines.

Let us assume that we have a welded plate containing a crack. If there is no obvious crack, it is customary to assume that one exists with a depth that is just below the limit of detection of the inspection equipment. A typical figure might be 0.05 mm. If we then arbitrarily assume that the nominal cyclic stress range is a rather punishing 200 MPa, then we estimate the initial ΔK as ~2.8 $\mathrm{MPa}\sqrt{\mathrm{m}}$ (from equation 12.2). Applying the Paris law relation, with the mean crack growth parameters from BS7910 (converted from $\mathrm{MN/mm^{3/2}}$ to $\mathrm{MPa}\sqrt{\mathrm{m}}$), we find an incremental crack growth of $\sim 4.1\times10^{-8}$ mm on the first cycle. This seems a reasonable, and an agreeably small, figure. However, it presents a problem when we consider that the interatomic spacing between iron atoms in steel is about 3×10^{-7} mm. In other words, fracture mechanics anticipates a crack growth increment that is too small to reconcile with atomic theory!

This academic point has little relevance when compared with two more telling practical issues. The first is that an incremental crack growth of $\sim 4.1\times10^{-8}$ mm is far too small to measure using any device available to a fatigue laboratory. The second is that it is not obvious from Figure 12.6, how many cycles, at this nominal stress, the welded joint might be expected to endure before failure.

In practical testing, there is never any attempt to measure the crack growth in an individual cycle. The normal practice is to measure the crack length extension after a block of (usually many thousands) of cycles during which it is assumed that ΔK remains approximately constant.

The prediction of fatigue life from fracture mechanics data is illustrated by continuing with the example of the 0.05 mm crack in a welded plate subject to a 200 MPa nominal fatigue stress. Numerical analysis[6] produces the predicted profile of the growth in crack length versus number of cycles shown in Figure 12.7. As can be seen from this simplified, and intentionally over-loaded, example, we would be very concerned if the joint was approaching 10^5 fatigue cycles.

12.3.4 Reliability of testing

We shall make use of both S-N curves and Paris law plots in our discussions of corrosion fatigue and the effect of CP. Before we do this, however, it is important to say a few words about the reliability of such plots. Figure 12.3 shows a reasonably smooth curve drawn through a set of test data. Although I have transcribed the data points in Figure 12.3 accurately from a source document, I have to admit that I find the closeness-of-fit to an idealised S-N curve rather unusual. In my experience S-N data, even when obtained by diligent workers, are generally much more scattered. The possibility of Figure 12.3 data having been "edited" in the original publication cannot be ruled out. The Paris law plot in Figure 12.5, on the other hand, shows no data points whatsoever. The lines in the figure are idealised because, in practice, crack growth results usually exhibit a wide degree of scatter.

Similarly, a cursory glance at Figure 12.6 might suggest that the closeness of the two curves indicates a high degree of agreement among the source data. However, we need to

[6] By numerical analysis, I mean playing around with a (large) spreadsheet.

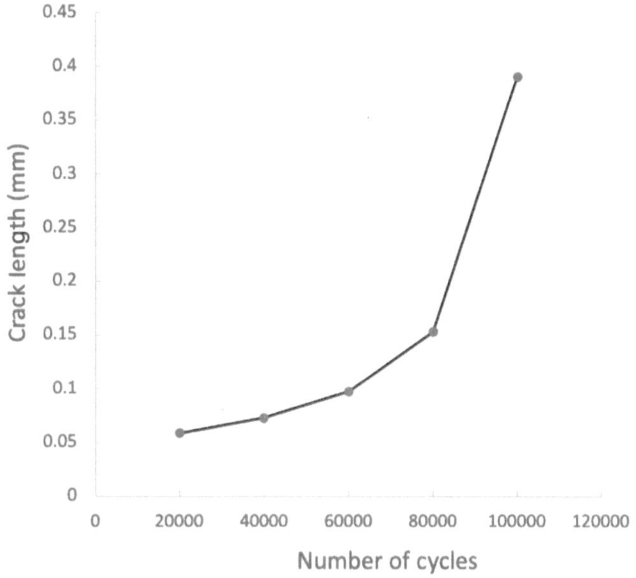

Figure 12.7 Crack growth: welded steel plate in air (stress range 200 MPa).

remember that this is a log-log plot. It spans numerous orders of magnitude, particularly on the crack growth rate axis. Closer examination shows that the curves are separated by a factor of about 3. In any fatigue assessment, therefore, it would be prudent to use the more conservative line.

The reason that I raise these testing difficulties here is because we will need to bear them in mind when we use S-N or crack growth rate data to evaluate the interactions of corrosion and CP with fatigue. In addition, we need also to be aware of how the fatigue design codes have accommodated the intrinsic experimental variations by erring on the side of caution.

12.4 CORROSION FATIGUE

12.4.1 Discovery

Early fatigue experiments, which were all of the S-N type, were generally carried out in dry air. The results obtained have, therefore, been determined solely by the metallurgical properties of the steel and should not have been influenced by corrosion.

An important feature of fatigue experiments carried out in air is that the number of cycles to failure (N) is, as far as we can tell, independent of the frequency at which the cyclic load is applied. This brings about the important benefit that tests can be accelerated by simply increasing the frequency of the load cycling. There is, however, a practical limit to the acceleration that can be gained. This is because it is difficult to increase the frequency beyond a certain point without heating the specimen, and thereby altering its fatigue characteristics.

Early attempts to counteract this heating involved the simple expedient of spraying water onto the test piece to cool it. Although this achieved the desired cooling, it was soon realised that the effect of the cooling water was actually to lower the number of cycles to failure. An additional facet of fatigue had been discovered: *corrosion fatigue*.

12.4.2 Characterisation

In practice, steel structures are almost always exposed to an environment that is more corrosive than dry laboratory air. We are interested in structures immersed in seawater which is corrosive. This combination of exposure to a corrosive environment and fatigue results in corrosion fatigue. Viewed in the context of S-N experiments, its main effect is a reduction in the number of cycles to failure at any given applied stress amplitude. A schematic example of this is given in Figure 12.8. Again, we need to be aware that the fatigue line indicated exhibit masks the considerable degree of scatter that would be expected in actual experimental results.

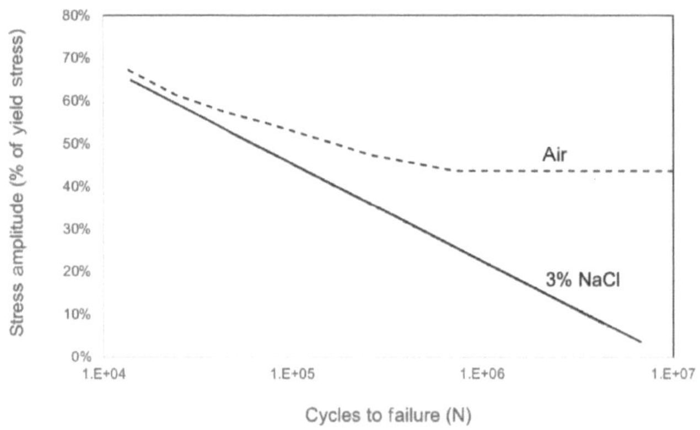

Figure 12.8 Schematic S-N curve for a steel in 3% NaCl.

An obvious additional effect of corrosion is to remove the fatigue limit (the horizontal part of the curve in the dry air test). For this reason, S-N data for steels tested in corrosive environments are often reported in terms of fatigue strength or endurance limit.

By way of an alternative perspective, the effect of corrosion on fatigue is shown schematically in a fracture mechanic crack growth type diagram in Figure 12.9. As we will see later (Figure 12.11), the agreement of the real results with the schematic can, sometimes, be less than wholly convincing. Nevertheless, both S-N and Paris law type tests are unequivocal in demonstrating that corrosion has a deleterious effect on the fatigue resistance of steel.

12.4.3 Theories

There are various theories to account for the interaction of corrosion and fatigue. The simplest is along the following lines. The crack opens during the tensile half of the stress cycle. This causes a small increment in the length of the fatigue crack. In addition, where a corrosive medium is present, this is drawn into the crack as it opens. The contact of this fresh corrodent with the freshly exposed metal surface at the crack tip produces a transient episode of rapid corrosion. Thus, the crack grows by a combination of mechanical growth (fatigue) and corrosion during each cycle. This is followed by the compression phase of the stress cycle during which the "spent" corrodent is expelled. The cycle then repeats.

An important point emerges from this simple model. Whereas fatigue is independent of the frequency at which the cyclic load is applied, corrosion fatigue is not. Where corrosion is contributing, lower frequencies result in a reduced number of cycles to failures, albeit that the lower frequency means that the actual time to failure might be long. There are two possible explanations for this dependence on frequency.

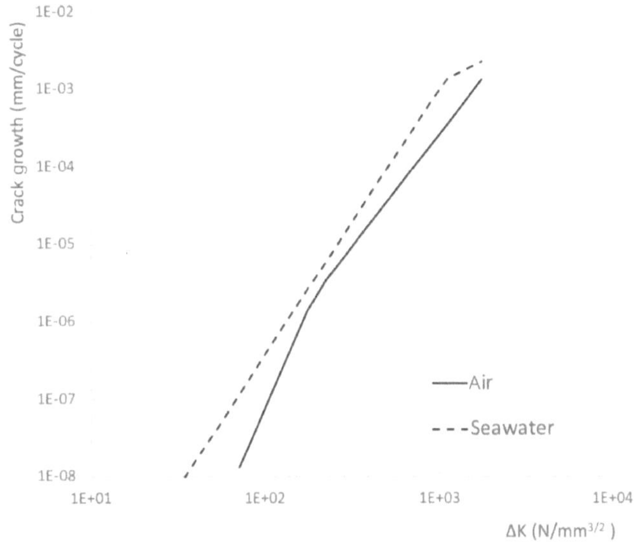

Figure 12.9 Fracture mechanics – schematic depiction of corrosion fatigue.

The first explanation is intuitive. At low frequency, the time of corrosion exposure per cycle is increased. Thus, as the frequency reduces, the contribution of corrosion to crack growth, relative to the purely metallic fatigue component, increases. The result is a reduction in the number of cycles that need to be completed before the specimen fails.

An alternative explanation [17] hinges on the observation that, in the case of a pre-cracked pipeline steel (~450 MPa yield strength) in 3.5% NaCl, detailed fractography indicated that accelerated crack growth rates, compared to air, were the result of hydrogen embrittlement (HE). We will return to the subject of HE in Section 12.7. In passing, we should note that the posited mechanisms of accelerated corrosion at the crack tip, and HE are not in conflict with each other.

12.4.4 Stress ratio (R-value)

At this point, we need to introduce another stress-related parameter that is relevant to fatigue. This is the stress ratio, which workers in the area often refer to as the R-value. It is a parameter which is simple enough to define. It is the ratio between the minimum stress and the maximum stress (equation 12.4).

$$R = \frac{K_{\text{min}}}{K_{\text{max}}} \qquad (12.4)$$

You will recall that ΔK is the difference between K_{max} and K_{min}. However, it now becomes clear that we can achieve the same value for ΔK over a range of R values. This is the same as saying that we can obtain the same stress amplitude (S) in an S-N experiment over a range of values for the minimum and maximum stress in the fatigue cycle. If K_{min} is zero[7], then so is R. However, in practice, it is rare to have any structure that has a mean stress of zero. Even if applied loads are ignored, a welded structure, for example, will contain residual

[7] A negative value of K_{min} indicates compression and fatigue crack closure.

stresses. For this reason, the data from Table 4 of BS7910, which have been used to compile Figure 12.6, are based on a *R*-value of >0.5.

The *R*-value is a measure of the extent to which the mean stress (or stress intensity) is removed from zero. For some materials, particularly aluminium, increasing the value of *R* markedly increases the crack growth rate for any given ΔK value. The effect of *R* on steel, on the other hand, is relatively limited [35]. So, there is no need for us to consider it further.

12.5 THE EFFECT OF CP

12.5.1 Information from S-N testing

Given that it has been long known that CP controls corrosion, it was obvious to ask the question: could it also control corrosion fatigue? This is a topic that has been studied since the 1960s. The early work was almost exclusively carried out using S-N testing.

The experimental difficulties faced when carrying out, or interpreting, long-term fatigue testing on specimens under CP should not be understated. Furthermore, in some instances, workers have made experimental simplifications that, with the benefit of hindsight, were not always justified. For example, a number of early investigators used 3% NaCl solution instead of seawater. This might sound reasonable since the two solutions have almost identical salinities and conductivities. However, it fails to take account of the benefits of ionic species, such as calcium, magnesium and bicarbonate, which are naturally present in seawater. As we saw in Chapter 7, these ions benefit the application of CP. It also emerges that they exert a beneficial effect on the interaction of CP with corrosion fatigue.

Aside from the variations in electrolyte composition, in particular its dissolved oxygen content and temperature, there is a wide range of parameters to consider in any testing programme. The variables in steel grade and metallurgical condition, fatigue loading and frequency, stress ratio (*R*-value), and levels of applied polarisation, taken in combination, mean that there is essentially an infinite number of experiments that could be conducted. Hence, we must be circumspect in drawing conclusions from the body of work reported in the literature.

Early experimentation clearly demonstrated that CP does indeed control corrosion fatigue. For example, in 1968 Duquette and Uhlig observed that not only did corrosion accelerate fatigue but also that the more corrosive the environment the greater the acceleration. This prompted them to investigate the reasonable, but as yet untested, proposition that reducing the corrosivity also reduced corrosion fatigue. In their initial experiments, they found that deaerating the seawater restores the fatigue limit for mild steel to its in-air value [3]. Subsequently, in 1971 Uhlig [5] also reported that... *cathodic protection to* $-0.49\ V_{(S.H.E)}$ *accomplishes the same result.* That potential is equivalent to −0.74 V versus Ag|AgCl|seawater.

This figure agrees well with the other contemporaneous work. For example, Hudgins et al. also reported [4] that in-air fatigue performance could be attained in seawater if the specimens were polarised to −0.75 V. Indeed, examination of their tabulated results indicates that the fatigue performance under CP in seawater was actually superior to the in-air performance, albeit that the authors did not comment on this at the time.

In addition, based on four tests at a potential of −0.7 V, the same authors were also able to report that... *even small potential shifts in the cathodic direction greatly increases the specimen life.* This is important because it tells us that, like its effect on corrosion itself, a little CP is much better than none at all. There is no "threshold" potential that has to be attained before there is a benefit in reducing corrosion fatigue. It is also worth observing that these potentials are less negative than the "threshold" potentials for full CP advised in modern-day codes (see Chapter 6).

Hudgins et al. repeated some of their testing on notched specimens and reported that... *a potential of – 0.75v vs Ag/AgCl will prevent corrosion fatigue on notched specimens also...*

12.5.2 The interaction with CP

As we have mentioned, according to one school of thought, the overall fatigue or corrosion-fatigue process is made up of two time-dependent processes: crack initiation and crack propagation. In practice, there have been numerous instances of corrosion-fatigue cracks having initiated at corrosion pits, the latter having acted as stress-raisers. On this basis, therefore, it might be tempting to suggest that the role of CP in combatting corrosion fatigue is to extend the initiation period back to what it would be in the absence of corrosion.

Unfortunately, any such simple model of CP returning corrosion-fatigue behaviour to that of pure fatigue by preventing early crack initiation does not work. In the first place, there are instances where corrosion fatigue cracks have originated on the surface in the absence of any corrosion pits. Secondly, it would be hard to reconcile this model with the observation that the benefit of CP in controlling corrosion fatigue is progressive as the potential is made more negative. Thirdly, and most tellingly, the fact that CP has been demonstrated by Hudgins et al. to be effective on notched specimens means that a mechanism involving the prevention of crack initiation is unlikely to provide a sufficient explanation.

Hooper and Hartt [9], using an experimental set-up described earlier [6], augmented the work of Hudgins et al. on notched specimens. They characterised the endurance limit of notched UNS G10180 (~345 MPa yield strength) steel specimens exposed to 10^7 cycles of fatigue loading whilst polarised potentiostatically in seawater. They also measured the endurance limit, over the same number of cycles, for the same steel in air. Their results showed that not only does CP in seawater return the endurance limit to the in-air value, it actually enhances it considerably. Importantly, this beneficial effect is not observed in experiments in which the seawater is replaced by 3.5% NaCl.

Furthermore, this enhancement in fatigue performance in seawater under CP is observed to be potential dependent. In these experiments, the optimum fatigue performance was obtained within a potential range between −1.05 and −1.25 V [9].

If the role of CP were simply to return the corrosion fatigue behaviour to the in-air behaviour, then we might reasonably conclude that it simply removed the "corrosion" component from corrosion fatigue. However, since CP provides an enhancement on the in-air performance, the mechanism is evidently more nuanced. Hooper and Hartt posited that the benefit of CP is intimately linked to the role of calcareous deposits. As we discussed in Chapter 7, these form on bare steel when it is under CP in seawater. They are advantageous in that they reduce the cathodic current demand and, as they develop, they help to spread the polarisation over the surface. The nature and rate of formation of these deposits vary in a complex manner depending on the seawater chemistry (particularly oxygen content), temperature, polarisation current density and previous polarisation history.

Two reasons for the benefit of calcareous deposits in reducing fatigue have been advanced.

a. They form down the sides of a crack, so reducing the cathodic current demand on the existing crack surface. This reduces the attenuation of potential down the crack. This aids the "throw" of protection to the developing crack tip.
b. The physical presence of the deposit restricts the crack opening and closing amplitude, effectively propping the crack open. This means that it effectively reduces the stress range at the crack tip.

Both these benefits are credible, but neither has been fully characterised.

At the time, Hooper and Hartt [9] linked the observed fall-off in enhancement at more potentials to the onset of the evolution of hydrogen and its known antagonistic effect on calcareous deposits. However, in subsequent publications, they, and other workers, considered the possibility that the role of hydrogen was more profound than simply disrupting calcareous deposits. We return to the issue of hydrogen in Section 12.7.

12.5.3 The fracture mechanics perspective

Figure 12.10 compares the Paris law relationships for steel undergoing corrosion fatigue (as seen in Figure 12.9) and the change brought about by applying CP at two levels of polarisation[8] −0.85 and −1.1 V. These are the mean curves from BS7910. They are based on best fits to a range of test data. We see that, at lower values of ΔK, the code advises that the effect of CP is to return the corrosion-fatigue characteristic in seawater back to the in-air fatigue value. At higher values of ΔK, however, the position appears not to be so encouraging. Under these conditions, the effect of CP appears to be antagonistic, giving crack growth rates similar to those expected in seawater in the absence of CP. Furthermore, the code advises that polarising to the more negative potential (−1.1 V) produces an antagonistic effect.

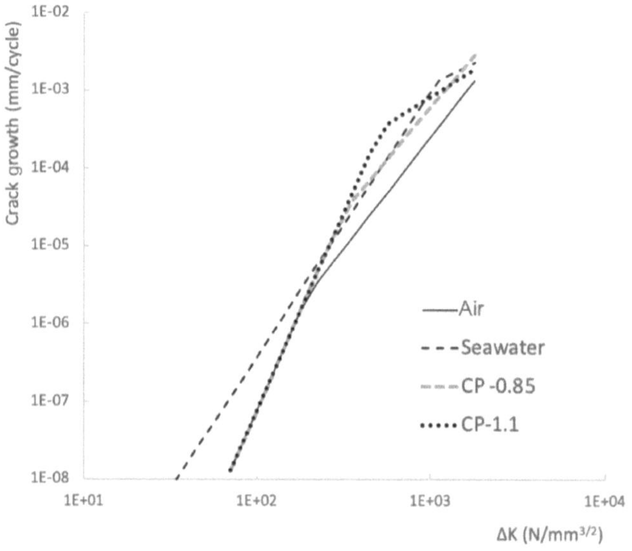

Figure 12.10 Fracture mechanics schematic: CP and corrosion fatigue.

In viewing these graphs, we need to be aware that fracture mechanics is not telling us to omit CP! In any real fatigue situation, the majority of the life of the specimen is spent in the region of slow crack growth. It follows that the effect of CP returning the corrosion-fatigue behaviour to the in-air value for the majority of the life is very beneficial.

12.5.4 S-N testing versus crack growth rate data

Since both fracture mechanics and S-N testing are just different ways of looking at the same phenomenon, we might reasonably expect complete agreement between the two

[8] I assume that the potentials are referenced to Ag|AgCl|seawater; but BS 7910 [6] omits this information.

methodologies. However, this does not always seem to have been the case. Based on information available in the early 1980s, Procter [17] concluded... *S-N data indicate that normal cathodic protection restores the corrosion fatigue properties to those observed in air. However, crack growth rate data indicate that cathodic protection, and particularly, overprotection, can result in further marked acceleration of the corrosion fatigue crack growth rates.*

My own view is that Procter was correct at the time to highlight the apparent mismatch between data obtained from fracture mechanics and S-N tests. However, we need to be careful not to read too much into this. We have already seen that fracture mechanics testing provides a wide spread of data for (nominally) identical experiments. For example, Figure 12.11, used samples of the same material, exposed to the same loading frequency at the same stress ratio, in the same seawater solution at approximately the same temperature. Even then, there is considerable scatter at each of the applied potentials studied. Furthermore, the use of a log-log plot inevitably serves to downplay this scatter. It follows that it is very difficult to attempt to compare reported S-N and fracture mechanics data which have been compiled in different laboratories on different steel grades, or different compositions within a grade, exposed at different frequencies and stress ratios, in different solutions (seawater and 3.5% NaCl) and at a range of temperatures.

12.6 THE CODES

12.6.1 Code development

We have already dipped into the codes during our discussion of fatigue and corrosion fatigue. It is now timely to outline how these codes have been developed and how they are utilised.

The object of any industry fatigue code is to produce a set of guidelines that the engineer can rely on to design a structure that will have a sufficient fatigue life. To that end, committees of the wise and the well-intentioned have gathered under the auspices of industry regulators to review the available evidence (both published and unpublished). The results of these deliberations have then been published, usually in the form of standardised S-N curves.

In the early days of code development, the information available to such committees would have been the results of comparatively few short-term laboratory tests on small-scale specimens of relatively simple geometry. There will have been considerable uncertainties involved in scaling up such laboratory tests to reflect a massive structure installed offshore. As with all code drafting committees, the understandable and appropriate response to such uncertainties will have been to err on the side of caution, at least until reliable field data became available to permit relaxation.

What was needed, therefore, was reliable long-term data from real structures, or at least from real structural components.

12.6.1.1 Laboratory testing

It is not the role of a CP textbook to review the very substantial body of corrosion fatigue test data. This has been done comprehensively by others. For example, Jaske et al. [11] reviewed test results prior to 1980, not all of which were in the public domain. This review encompasses both fracture mechanics and S-N data.

For our purposes, it is sufficient to take a look at some real test data to get an appreciation of the inherent variability. By way of an example, I have selected data produced by Scott and Silvester [8] issued to the UK Department of Energy in 1977 and included figures 54–56 in

Jaske's review. This is useful because it uses a common structural material (BS 4360 grade 50D) and examines the crack growth rates at a range of stress ratios and at three levels of polarisation: −0.65, −0.85 and −1.1 V. The −0.65 V data represent the typical steady-state free corrosion for steel in seawater. The potentials of −0.85 and −1.1 V were selected to represent respectively the normal levels of CP and the maximum level of CP achievable with a sacrificial anode system. The tests were conducted in seawater at 5°C–10°C, and at a frequency of 0.1 Hz; both of which are good approximations to real exposure in an environment such as the North Sea.

As is common to fracture mechanic testing, the Scott and Silvester study produced a great deal of data. For convenience, we can look just at the results obtained at a stress ratio (*R*) of 0.5. The relevant results, extracted from the three figures in [11], are consolidated in Figure 12.11. This figure helps us to appreciate just how much scatter is to be expected in fatigue experiments, even when carried out with diligent control exercised over the key variables. We should always remember that code-generated line graphs, such as those shown in Figure 12.10, are constructed by drawing upper bounds around data sets such as are shown in Figure 12.11. This alone should tell us that fatigue prediction is far from a precise art!

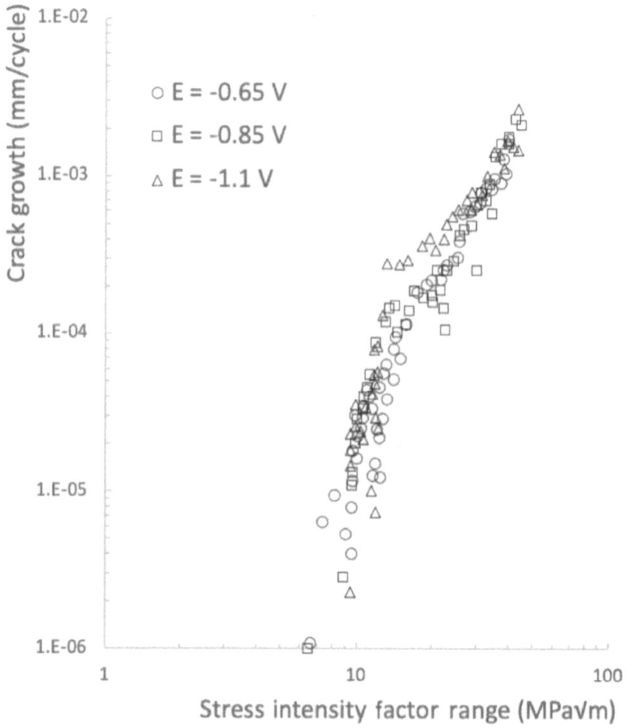

Figure 12.11 Examples of fracture mechanics data. (Transposed from Ref. [11].)

12.6.1.2 The Cognac fatigue "experiment"

In 1977, the same year as the data presented in Figure 12.11 were issued, the base section of Shell's Cognac platform was installed in a water depth of 306 m in the Mississippi Canyon of the Gulf of Mexico. At the time, it was the world's largest offshore structure.

Its fatigue design followed the contemporary code recommendations. The CP system had a target design life of 25 years [27]. Unusually for an offshore structure, it was fitted with monitoring equipment to measure oceanographic parameters, movement of the structure, and strains in select subsea structural components. This was to provide feedback to the project team on the validity (or otherwise) of their design assumptions. It was also intended to provide fatigue data to which end a set of pre-cracked and welded fatigue specimens was attached to one of the instrumented braces at a depth of 52 m [14].

These specimens were earthed to the structure, so were incorporated into its CP system. The intention was to run the experiment for 20 years which, as far as we know, would have been a record for a fatigue testing programme. However, despite the planned test duration, it was still deemed necessary to accelerate the rate of fatigue damage in order to produce useful information within that time frame. With that in mind, the fatigue specimens were attached in a manner designed to amplify the stresses in the brace on which they were located.

The first set of specimens was recovered after 1 year, and the second after 3 years. In the event, there was no fatigue crack growth and no fatigue crack initiation. This was unexpected. Particularly in the view of the experiment strategy which provided for stress amplification. Two reasons were posited for this. First, the in-service stresses actually experienced during the first 3 years of service were lower than allowed for in the structure's design. Second, the American Welding Society fatigue curves used for the design were conservative.

The lack of observable fatigue in 1981 would have doubtless been re-assuring to workers residing on the platform. On the other hand, it was presumably a disappointment for the researchers who sponsored the testing. It confirms the old adage that you do not really know a material until you have succeeded in breaking it.

The absence of any subsequent publications suggests that the planned 20-year Cognac fatigue experiment was shelved after the third year. However, the structure re-emerges in the CP literature in 1997 when Goolsby and McGuire [27] (rather belatedly) described a CP retrofit that had been installed in the early 1980s to remedy shortcomings in the performance of the original system. We will deal with retrofits in the next chapter. In 2014, some years after its originally intended end of life, the Cognac processing facilities were refurbished and the operating life was extended. This also indicates how conservative the original fatigue design had been.

12.6.1.3 Further testing

There has been no shortage of fatigue testing since Jaske et al. published their 1981 review. One of the outcomes of this further testing is that the conclusion that CP increases the fatigue resistance to above that of the "in-air" value needs to be modified, at least in so far as the higher strength steels used in offshore construction are concerned. Whilst the conclusion remains valid for long fatigue lives (high cycle fatigue $>10^7$ cycles), the benefit of CP is not so clear-cut at short lives (low cycle fatigue $<10^7$ cycles). This point is illustrated in Figure 12.12, which shows the S-N design curve recommended by DNV [45] for a tubular joint.

A reasonable interpretation of this graph is that the low cycle fatigue case involves a considerable degree of plastic deformation ahead of the growing crack tip, for which CP might be antagonistic (we return to this when we discuss hydrogen embrittlement). This point is probably moot, however, since most marine structures are designed for longer fatigue lives ($>10^8$ cycles), so are designed not to take the high fatigue stresses seen to the left-hand side of Figure 12.12.

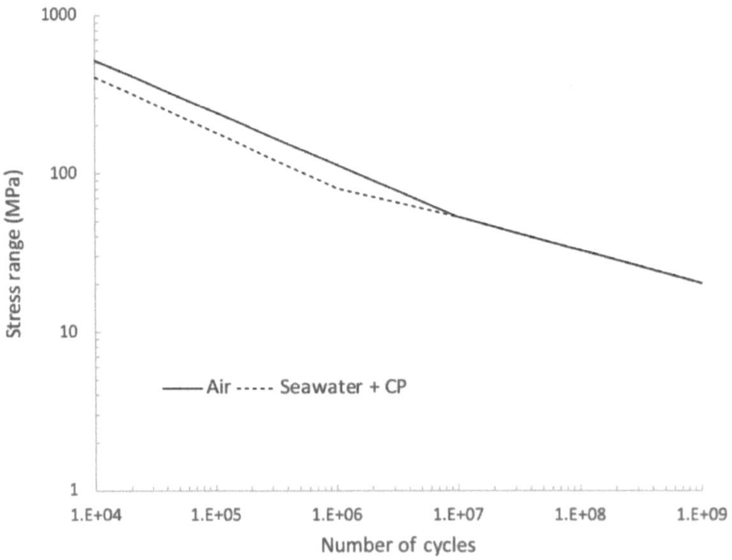

Figure 12.12 S-N curves for tubular joints. (Courtesy of DNV Ref. [45].)

12.6.2 Using the codes

12.6.2.1 Overview

It would be unusual for a CP engineer to be involved in the fatigue assessment of a structure. That activity falls squarely within the domain of the structural or mechanical engineer. Nevertheless, as we have already suggested, it is helpful for the CP engineer at least to be aware of the process. To that end, we will give a brief sketch of how a fatigue assessment for an offshore structure is carried out. We will assume that the structure is to be fabricated in a carbon steel with a SMYS below 550 MPa. We will also assume that the assessment is to be carried out in accordance with a code such as DNV-RP-C203 [45]. This steers the designer towards basing the fatigue assessment on S-N data, rather than fracture mechanics.

At its most basic, carrying out a fatigue assessment according to this recommended practice requires the designer to carry out the following activities: select the critical structural elements and estimate the fatigue loadings, select the appropriate S-N curve and then assess the fatigue life.

12.6.2.2 Elements and fatigue loadings

The design of the structure is analysed to determine which elements are fatigue critical. This is firmly in the domain of the structural engineer, so there is not much that we can say about it here. Nevertheless, we can make some obvious generalisations. For example, in the case of jacket-type structures, which we considered in Chapter 3, the fatigue assessment is likely to be at the major welded nodes linking the bracing members to the main legs. On the other hand, for offshore wind turbine monopile foundations, which we will discuss in Chapter 13 (external protection) and Chapter 16 (internal protection), the maximum fatigue loading is likely to be at the first circumferential weld above the seabed.

For each such element, the fatigue loadings to which it will be subject over the course of its lifetime are estimated. Unlike a typical laboratory fatigue test, in which the specimen is usually subject to a predetermined stress range (S), real structures are exposed to more complex patterns of *S*, depending on the prevailing circumstances. Real structures also have stress-raisers, so it is also necessary to apply an appropriate value for the SCF.

If the dominant fatigue influence is (say) the wave loading, the value of *S* will vary across a spectrum of values depending on whether the conditions are calm or stormy. Such variable loads are customarily analysed using "rainflow" counting. This is a process whereby the stresses anticipated over a period of time are analysed to produce a "count" of the number of cycles expected within each stress range.

The output of the rainflow analysis, which may be visualised as a histogram, can then be combined using Miner's rule.[9] This rule is based on the proposition, broadly supported by testing [13], that fatigue damage is simply cumulative. It enables the combination of the various fatigue loadings in proportion to their contribution to the overall rain flow histogram. This, in turn, permits the fatigue to be analysed on the basis of a single S-N curve.

As an aside, it emerges from this type of analysis that structures in the North Sea have to be designed to achieve a prescribed fatigue life. On the other hand, fatigue calculations have less relevance for structures near-shore in the Gulf of Mexico. These are designed to withstand the extreme, but comparatively rare, loadings arising from hurricanes [12].

The design detail being assessed is then analysed to determine the relevant SCF to apply to the nominal stress. Appendices B and C of DNV-RP-C203 give information as to how this should be done. Like the selection of the critical elements, and the calculation of the nominal stress range, this activity is firmly in the domain of the structural engineer.

12.6.2.3 Select the S-N curve

The engineer then decides upon which family of S-N curves the fatigue analysis is to be based. Like all such codes, C203 offers a binary choice for structures immersed in seawater: "freely corroding" or "cathodically protected". Understandably, this DNV document focusses almost exclusively on structures that are under CP.

However, unlike BS7910, which is based on fracture mechanics, DNV-RP-C203 does not tell us what potential has to be achieved for a structure to be deemed cathodically protected. For example, it does not cite its companion document DNV-RP-B401. However, even if it did, that would not help. Although RP-B401 tells us the potential to be used in CP design calculations, it is not a performance standard. It does not tell us what potential actually needs to be achieved. BS 7608, on the other hand, specifies cathodic protection as being in the range of −0.85 to −1.1 V.

Accordingly, most workers adopt −0.8 V as being the relevant least negative potential. This is a reasonable interpretation of CP codes more generally. However, it needs to be recognised that, like all other aspects of fatigue design, this choice of potential adds conservatism. As we have already seen, early work has shown potentials of −0.75 V to be acceptable.

The appropriate individual S-N curve is then selected. Using RP-C203, this requires the relevant classification of the structural detail to be selected from its Appendix B. Fifteen options are available: ranging from a non-welded tubular section (B1) which exhibits the most fatigue-resistant behaviour, to... *Weld metal in partial penetration or fillet welded joints around a penetration through the wall of a member on a plane essentially parallel to the plane of stress* (W3). This exhibits the least fatigue-resistant behaviour of the configurations considered. Again, this selection lies firmly within the work scope of the fatigue specialist.

[9] Sometimes referred to as the Palmgren-Miner rule. Miner developed the mathematical formalization of the concept, first proposed by Palmgren, that fatigue damage is cumulative.

12.6.2.4 Assessment

If the minimum number of cycles to failure (*N*) indicated by the application of the code exceeds the number of cycles within the design life of the structure, taking into account the various recommended design (safety) factors, then the fatigue life is judged acceptable.

On the other hand, if the predicted fatigue life arrived at by this methodology is inadequate, there are a number of options available. Obviously, the first is to modify the design of the element of the structure being assessed either to lower the nominal stress or, usually more rewardingly, to lower the SCF.

Additionally, or alternatively, the fatigue life can be extended by grinding the welds. It might seem counter-intuitive to improve fatigue life by grinding away metal and thereby increasing, even if only slightly, the nominal stress. However, the pivotal benefit of grinding is that it removes the stress-raising cracks in the weld toe. For this reason, RP-C203 permits an increase in the stress range (S) of 30% for a ground weld compared to one in the unground state.

Other ways to achieve fatigue life extension during fabrication would be to carry out hammer peening or to apply a stripe coat of paint to the weld.

12.6.3 The role of the CP engineer

It would be remiss to write a marine CP textbook without including some discussion on corrosion fatigue and the effect of CP. Hence the above discussion. However, I have also made the point that CP engineers, unless they are particularly versatile, do not normally get involved in the assessment of fatigue, either in the design of new structures or the evaluation of the remaining life of existing ones. This, therefore, prompts the question: how deeply should a CP engineer get involved in matters relating to fatigue? There is no simple answer to this. All that I can do is offer some personal thoughts.

In my experience, where problems arise, they can be traced to the attitude to engineers (of either the fatigue of the corrosion variety) to numbers printed in standards. Many engineers have been brought up in the hard-nosed school of compliance. For these workers, life is black-and-white. A structure either complies or does not.

Unfortunately, this approach can result in battles that never need to be fought. For example, I have twice been involved in multi-million dollar litigations centred on fatigue and CP. The cases and their outcomes remain confidential, but I can disclose the following. Each instance resulted from an owner's rigid interpretation of standards. In both cases, fatigue was the life-limiting factor, and the fatigue design required CP to be effective. The problems arose after installation when inspection indicated that not all parts of the structures had, in the opinion of the owners, reached an appropriate level of polarisation within an acceptable period of time.

In the first case, the owner elected to retrofit supplementary anodes, at very considerable expense, because of a professed view that not all of the measured potentials were in the range −850 to −1100 mV, so the "in-air" fatigue design assumptions could not be relied on. In fact, the least negative potential observed was recorded at −802 mV at the relevant time. Whilst this figure was, strictly speaking, outside the −850 to −1100 mV target range, it was negative enough to ensure that corrosion was not a problem. More important, the historical evidence shows that −802 mV is more than adequate for ensuring the restitution of "in-air" fatigue characteristics.

Ironically, subsequent fatigue analysis based on measured loading demonstrated that, even assuming free-corrosion behaviour, the required fatigue life was assured.

The second case was similar, but common sense prevailed before the owner indulged in a pointless CP retrofit.

12.7 HYDROGEN EMBRITTLEMENT

12.7.1 The problem

We have already mentioned HE in connection with its possible involvement in corrosion fatigue. We now need to say more about this interesting phenomenon which, incidentally, is the most widely researched topic in all of metallurgy (see for example, Sir Harry Badeshia's review [44]). HE is something of a catch-all term that covers the reduction in ductility (i.e. embrittlement) of a metal that can be brought about by the presence of atomic hydrogen within its structure. Although the critical process is the embrittlement, there is only a practical problem if this actually leads to a crack and, more particularly, a failure of the component. Such HE cracking failures have received names such as hydrogen-induced stress cracking (HISC) or hydrogen stress cracking (HSC). In this chapter, we shall simply refer to HE.

It may be convenient to think of HE in terms of a Venn diagram (Figure 12.13). The three inter-related factors which need to be operative in order to bring about an HE failure are:

- a susceptible material,
- a tensile stress and
- a source of atomic hydrogen.

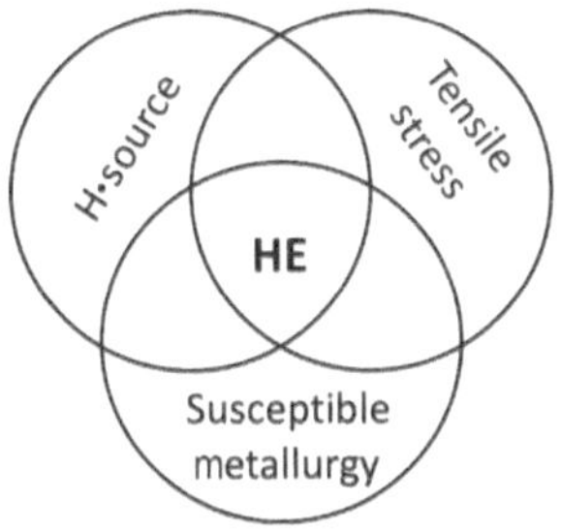

Figure 12.13 Risk factors for HE.

Any material might be subject to tensile stress. Indeed, it is hard to conceive of a circumstance where this is not the case. Accordingly, we can regard this element of Figure 12.13 as a given. However, there will usually be a threshold level of tensile stress (or, rather, stress intensity), for any particular material and circumstance of exposure, below which HE does not lead to cracking. We return to this point when we look at individual materials below. But first, we need to look at the sources of the hydrogen and consider how this hydrogen leads to the loss of ductility.

We will begin by looking at these questions in the context of carbon steels of the types widely used in offshore construction. That is, principally, the carbon-manganese steels with a specified minimum yield stress (SYMS) typically below 500 MPa. We will then consider the higher strength steels, as might be employed for some critical components, and finally, we will examine the interaction of hydrogen with certain CRAs.

12.7.2 What is the source of atomic hydrogen?

We have already encountered two sources of hydrogen. In Chapter 1, we noted that the cathodic corrosion process involves the reduction of hydrogen ions (equation 12.5).

$$2H^+ + 2e^- \rightarrow H_2 \tag{12.5}$$

You also may recall that, at the pH of natural seawater (~8), this reaction only has a minor role to play in the corrosion of steel. Nevertheless, it cannot be ignored. The natural corrosion processes that occur in seawater can provide sufficient hydrogen for certain very susceptible materials to suffer HE and to crack, even in the absence of an externally applied stress.

In Chapter 11 we also considered the cathodic electrode process that occurs when ICCP is applied. The result can be the electrolysis of water (equation 12.6). Depending on the level of potential achieved, the rate of this reaction can easily be so high that bubbles of hydrogen are observed fizzing off the surface of the steel.

$$2H_2O + 2e^- \rightarrow H_2 + 2OH^- \tag{12.6}$$

Interesting as these two equations are, they do not indicate that any hydrogen is actually entering the steel. In each equation, the cathodic reaction product is given as molecular hydrogen gas (H_2) which is discharged to the environment. We now need to look at these processes in more detail.

For example, if we apply our critical faculties, we see that, as written, equation 12.5 involves a four-particle collision. Two hydrogen ions (H^+) come together with two electrons (e^-) to produce one molecule of hydrogen gas (H_2). As a matter of experience, instantaneous four-particle collisions are improbable enough in any circumstance. They become even more so when we realise that the hydrogen ion is actually a proton. This means that, as written, equation 12.5 implies simultaneous collisions of four subatomic particles. It would be a mind-bogglingly improbable nuclear event!

This might suggest that the experimental observation defies the laws of physics. Fortunately, this is not the case. In fact, this problem is easy to resolve. We simply recognise that equation 12.5, in common with all chemical and electrochemical equations, is just a statement of what we start with and what we end up with. It lists the reactants (protons and electrons) and the product (hydrogen gas). Crucially, it makes no attempt to explain the mechanism by which the reaction actually proceeds. An insight into this mechanism emerges if we break equation 12.5 down into the following three subprocesses.

$$H^+ \rightarrow H^+_{ads} \tag{12.5a}$$

$$H^+_{ads} + e^- \rightarrow H^\bullet_{ads} \tag{12.5b}$$

$$2H^\bullet_{ads} \rightarrow H_2(gas) \tag{12.5c}$$

The first step (12.5a) depicts a hydrogen ion in the solution arriving at the surface of the steel and adsorbing onto it. Since the hydrogen ion is very mobile, and the steel surface presents a relatively enormous target area, there is nothing improbable about this process. The second step (12.5b) indicates the positively charged hydrogen ion receiving an electron from the steel to form a hydrogen atom ($H^\bullet$), which remains adsorbed on the surface of the steel. Again, there is no impediment to this process since the metal can be viewed as an almost infinite source of electrons which will be attracted to the positive hydrogen ion.

Although an atom is the smallest species that can be regarded as characteristic of a chemical element, it is not necessarily the smallest unit of matter that can have an independent existence. The smallest species that can exist independently is the molecule. A hydrogen molecule comprises two hydrogen atoms combined chemically. This combination of two unstable atoms to form a stable hydrogen molecule is indicated in equation 12.5c. Hydrogen atoms adsorbed on the steel surface are mobile. When two collide, with sufficient energy, a molecule of hydrogen gas is formed. This process requires a two-particle collision. However, since there are numerous hydrogen atoms (which are much bigger than protons), and they are limited to two-dimensional movement on the surface, there is nothing improbable about such collisions.

Of the three reaction steps making up the overall hydrogen evolution reaction, (12.5c) is the slowest. In chemistry, it is termed the rate-determining step. This may be thought of as a bottleneck in the overall process. It causes a build-up of the concentration of adsorbed hydrogen atoms on the surface. A side effect of this build-up of surface concentration is that some hydrogen atoms can diffuse into the steel instead of combining to form gas at the surface. The reason that this is possible is the relatively small dimensions of the hydrogen atom, which has a diameter of $\sim 0.11 \times 10^{-9}$ m, compared to the iron atom (diameter $\sim 0.31 \times 10^{-9}$ m).

This difference in atomic sizes means that hydrogen atoms are small enough to fit, or rather "squeeze", into the interstices between the iron atoms, which, at all temperatures above absolute zero, will be vibrating. This permits the inward diffusion of hydrogen atoms driven by the concentration of adsorbed surface hydrogen atoms. Molecular hydrogen (i.e. hydrogen gas) does not diffuse into the metal.

There are two factors that can affect the rate of atomic hydrogen entry into any alloy: the potential and the environment. In the case of some materials, the control of potential can be critical, and it is possible to design CP systems to achieve this control. However, since any surface immersed in seawater is prone to fouling, it is not usually possible to control the environment. This has prompted work to investigate the interaction between microbiological activity and hydrogen uptake. Overall, this work has been interesting but far from definitive.

Robinson et al. [19] studied the effect of microbiological activity on the hydrogen uptake of cathodically protected BS 4630 50D steel. They concluded that... *there may be a synergistic effect between the SRB and the applied cathodic potential that results in enhanced hydrogen absorption by the steel.* However, in a subsequent review [30], Robinson and Kilgallon revised this view slightly, stating that... *it appears that hydrogen uptake is promoted by sulphides rather than by a direct effect of bacteria.* That review contained a wealth of tabulated information on hydrogen absorption which Batt and Robinson subsequently summarised pictorially [32].

In related investigations, Edyvean et al. [29] reported the effect of biofilms on both hydrogen uptake and fatigue crack growth rates. They observed that CP at –1000 mV[10] resulted in an observable hydrogen flux, and this flux was markedly increased as a result of the presence of a biofilm. However, bearing in mind the view expressed by Robinson and Kilgallon, we should be cautious in ascribing this increase to the sulfide in the biofilm, or to the microorganisms it contains.

On the other hand, Edyvean et al. also reported that, in corrosion-fatigue experiments in seawater-sulfide mixtures, crack growth rates were lower in the presence of a biofilm. The latter point is confusing because Benson and Edyvean later reported [31] the same results, but this time interpreted it as showing that... *anaerobic biological activity, generating hydrogen sulphide, considerably enhances crack growth rates.* Edyvean et al. [29] concluded, not unreasonably, that... *the effect of biological activity on structures susceptible to HE and corrosion fatigue are not going to be easy to predict.* However, since they compared hydrogen flux measurements on a 355 MPa yield stress material under CP, and crack growth

[10]These workers [50] omitted to record what reference half-cell they used. However, members of the same group later reported [52] that they used the saturated calomel reference. This is reasonably close to Ag|AgCl|seawater.

rate experiments on a 690 MPa steel in the absence of applied CP, it is perhaps understandable that the two sets of results were difficult to reconcile.

In short, therefore, there is clear evidence that microbiological activity enhances hydrogen uptake in cathodically polarised steel, although the details of the mechanism by which this occurs are unclear.

Another factor which has been demonstrated to increase hydrogen absorption by steel is water depth. Equation 12.4c indicates that the combination of hydrogen atoms on the surface forms hydrogen gas. Any increase in pressure, therefore, increases the energy that needs to be expended to make the reaction happen. A corollary is that increasing pressure will reduce the rate of hydrogen gas discharge, thereby causing a build-up of atomic hydrogen adsorbed on the steel surface. This in turn will accelerate the entry of atomic hydrogen into the steel matrix. This has been experimentally confirmed by Festy and Tigges [23] in polarisation tests carried out in a pressurised tank.

12.7.3 What does the atomic hydrogen do?

There are various forms of damage that can be caused by hydrogen when it is in the metal. However, the only one that we need to be concerned with is HE. At its simplest, hydrogen atoms harden (i.e. strengthen) steel, but they can only achieve this by lowering the ductility. The mechanism by which the hydrogen atoms interact with the steel to lower the ductility is complex, widely researched but incompletely understood. Since this is a CP textbook, not a treatise on HE, I develop my description of the effect of hydrogen in two stages. First, we will adopt a simplistic view, but one which serves to communicate something of what is involved. Then we take a view that is a bit more complex. It serves to introduce some of the subtleties of the phenomenon, and to give an insight as to why HE is still the subject of research.

12.7.3.1 A simplistic view

Consider a snooker table (or a pool table if you must) completely covered in snooker (or pool) balls. These balls represent a plane of iron atoms in a crystal of steel. We will ignore the other minor component atoms, such as carbon and manganese. Now picture another full layer of balls sitting on top of the first. Each ball will sit in the space created by three adjacent balls in the first row. We now have a model of two planes of iron atoms.

Next, imagine taking a straight-edged piece of wood and aligning it along one edge of the upper plane of balls. Then, very gently, we start to apply pressure. At first, nothing happens; the plane of balls resists the pressure (or stress) that we are applying. However, as we gradually increase the pressure, there comes a point when the entire upper plane of balls moves. The balls simultaneously bobble along to the adjacent positions.

In this simplistic model, the balls represent iron atoms, and the slippage of one plane of atoms over another represents the way in which the metal reacts to applied loads above the yield stress. The metal-to-metal bonds do not break. Rather, planes of metal atoms slip over one another. In other words, the metal starts to yield and plastic deformation takes place (see Figure 12.1). This is a manifestation of the metal's ductility.

Now, consider the same scenario of two planes of balls representing iron atoms; but with the addition of a few smaller balls, about one-third the diameter, representing hydrogen atoms, wedged into spaces between the two planes. It can be envisaged that these hydrogen atoms will introduce some distortion to the planarity of the layers of atoms.

Maintaining our snooker/pool ball analogy, if we repeat the exercise of applying pressure to one edge of the upper plane of balls, we will find that the effect of the smaller balls is to increase the resistance of the upper plane of balls to being displaced. We have to apply more pressure (or stress) in order to get the upper row to move. In terms of our analogy, the hydrogen atoms in the steel matrix have increased the strength of the steel or hardened it.

This can be viewed in metallurgical terms as interstitial hardening. The hydrogen produces an effect that is broadly similar to the addition of carbon to (soft) iron to make (hard) steel. Superficially, therefore, we might assume that the presence of atomic hydrogen in the steel would be beneficial since it increases its strength. However, we need to remember that specifying the mechanical properties of a structural alloy such as steel requires compromise. There is always some degree of trade-off between its strength and its ductility. If the mechanical engineer has selected a grade of steel that offers a suitable compromise between strength and ductility, it is unlikely to be beneficial if the environmental exposure causes an alteration in these properties.

12.7.3.2 A less simplistic view

The above simplistic view conveys something of an idea of how hydrogen atoms might lead to the embrittlement of steel. However, there are some problems with the analogy.

The first problem is that hydrogen concentrations as low as about 1 part per million (ppm) by weight can give observable HE problems. Viewed another way, and recognising that an iron atom is ~56 times heavier than a hydrogen atom, this suggests that as few as one hydrogen atom for every ~15,000 to 20,000 iron atoms can cause HE.

The second problem is that the observation of an embrittlement effect depends on the way in which we try to measure that effect. If we apply a high strain rate, we do not see a difference in ductility between a hydrogen-containing steel specimen, and a comparable specimen that is essentially free of hydrogen. For example, the Charpy impact test [47], which has been in use since 1900, is a high strain-rate test that determines the amount of energy absorbed by a material during fracture. As already noted, this can be thought of as equivalent to the area under the stress-strain curve. The more energy absorbed, the tougher the material. This in turn indicates higher ductility. However, Charpy testing does not reveal whether or not a material is subject to HE.

HE only becomes apparent in slow strain rate tensile testing. In a typical test, a specimen is extended by the application of a progressively, but slowly, increasing applied load. At first, the extension is reversible. That is, the specimen returns to its original length after the applied load is removed. As we have already noted, this is elastic behaviour. However, as the load is increased further it exceeds the yield stress, beyond which permanent plastic extension is produced. Further increasing the load results in the specimen continuing to deform. It elongates, and its cross-sectional area is reduced. Eventually, the applied load exceeds the ultimate tensile strength of the remaining cross-section and the specimen fractures.

As we have mentioned, a common means of assessing the relative ductility in tensile tests of this type is to measure the cross-sectional area of the specimen at the point of fracture and to compare it with the original cross-sectional area of the specimen. The greater the reduction in area (RA) (see equation 12.1), usually expressed as a percentage, then the greater is the ductility. The effect of HE in such a test is to reduce the percentage RA in comparison with specimens of the same material that are not embrittled.

The fact that the effect of hydrogen atoms depends on the rate at which the steel is strained tells us that the role of hydrogen is subtler than the snooker ball analogy suggests. The relationship with strain rate confirms that not only is the presence of atomic hydrogen important, but so too is the movement of that atomic hydrogen within the structure of the metal.

Over a period of time, much of the atomic hydrogen atoms will diffuse to sites such as grain boundaries, interfaces between the iron alloy matrix and precipitates and other discontinuities in the crystal matrix of the metal. These sites can be thought of as low-energy sinks. Once the hydrogen arrives there it is effectively trapped, and it plays little or no observable role in producing HE.

On the other hand, the hydrogen that remains in solution in the iron alloy matrix is able to diffuse. Crucially, its role in HE is connected to the ability of this hydrogen to diffuse to the high stress intensity plastic deformation zone ahead of a crack tip. This is where the embrittlement effects become apparent.

We now turn to the metallic materials commonly, and not so commonly, used in subsea engineering; some of which are more susceptible than others to HE.

12.8 LOW- AND MEDIUM-STRENGTH CARBON STEELS

Thus far in this chapter, we have considered only low- to medium-strength carbon steel of the type used for most offshore structures and pipelines. These are steels with a SMYS below 550 MPa. As a matter of experience, these materials do not suffer HE failures as a result of the application of CP, even when excess levels of cathodic polarisation are applied.

That is not to say that these materials do not undergo some degree of HE. For example, if we conduct a tensile test (as in Figure 12.1) on a sample of a typical offshore structural steel, e.g. HD50, we can observe a reduction in area (see equation 12.1); in the region of 40%. If we now re-run the test with the specimen immersed in seawater and polarised to around −1100 mV, we obtain a lower value of RA (say 30%). This shows that the CP causes some reduction in ductility, albeit only observable at high levels of applied stress. Nevertheless, the excellent performance of this material in service under CP tells us that, at the design stresses permitted by the codes, any embrittlement that may occur will be inconsequential.

12.9 HIGH-STRENGTH LOW-ALLOY STEELS

12.9.1 General

Although "ship plate" steel is the most widely used material of construction for subsea structures, its modest strength renders it inadequate for some applications. In these circumstances, the next "step up" in terms of alloy strength (and cost) is the high-strength low-alloy (HSLA) steels. These materials are alloyed with low levels of chromium and molybdenum: typically, up to ~1% and 0.5% respectively. These levels of alloying additions improve the strength of the alloys compared with carbon steel but add nothing to the corrosion resistance. CP is still required.

There are two areas in particular where these types of alloys are used subsea: fasteners[11] and structures.

12.9.2 Fasteners

HSLA fasteners are very widely used for connecting substantial items of carbon steel subsea hardware, such as pipeline flanges and wellhead equipment. They are also sometimes

[11] A fastener is the catch-all term for studs, bolts and nuts.

employed on CRA equipment. A project's materials selection exercise will invariably select the grade of HSLA in accordance with experienced-based codes and standards. In hydrocarbon production, for example, the Operator's in-house materials selection guide will be adopted which, in turn, is likely to cite guidelines such as ISO 21457 [39] or EEMUA[12] 194 [41].

The ISO is only of moderate use. It tells us that, if CP cannot be guaranteed, it is necessary to use a seawater-resistant CRA. It also states that fasteners... *submerged in seawater shall be resistant to hydrogen embrittlement. The material strength of carbon, low alloy and SS should therefore not exceed a SMYS of 725 MPa.* The EEUMA document is more prescriptive. It informs us that HSLA bolting material used for subsea connections conforms, almost universally, to either ASTM A193 [55] or ASTM A320 [56] where low-temperature impact testing is deemed necessary. The usual materials used are A193 grade B7/B7M or A320 grade L7/L7M (<50 mm diameter) or grade L43 (<100 mm). It also reminds us that CP can cause HE, suggesting that... *it is preferable to specify bolting and nut materials with the 'M' designation (e.g. B7M, L7M, 2HM), limiting the hardness of the material to 22 HRC.*[13] However, it also advises that, if stronger bolts are needed, A193 grade B7 material resists HE up to a hardness of 34 HRC. Some of the relevant information from these two ASTM standards is presented in Table 12.1.

Ayama et al. [48] pointed out that the actual failure rate of subsea fasteners connected into CP systems is low, albeit that the consequences of such failures can be dramatic. However, they cite a case where 11 out of 20 grade L43 riser bolts failed soon after installation. Investigation revealed that, although the bolts were supplied with a manufacturer's material test report, they had actual yield strength values of up to 1165 MPa. As can be seen from Table 12.1, this is considerably in excess of the upper limit of 755 MPa recommended in the ASTM standard. It has to be concluded that the certification was fraudulent.

Table 12.1 HSLA bolting materials compatible with CP

Standard	*ASTM A193/A193M*		*ASTM A320/A320M*		
Grade	*B7*	*B7M*	*L7*	*L7M*	*L43*
Chemical composition (wt%)					
C	0.38–0.48				0.38–0.43
Cr	0.80–1.10				0.70–0.90
Mo	0.15–0.25				0.20–0.30
Ni					1.65–2.00
SMYS (MPa)					
<64 mm dia.	720	550	725	550	725
65–100 mm	655				755
100–180 mm	515	515			

[12]The Engineering Equipment and Materials Users Association.

[13]HRC is hardness measured on the Rockwell "C" scale. 22 HRC approximates to a yield strength of 790 MPa, 34 HRC corresponds to about 1050 MPa.

12.10 CORROSION-RESISTANT ALLOYS

12.10.1 Stainless steels

12.10.1.1 Classes

We encountered the main crystalline classes of stainless steel: austenitic, ferritic and martensitic, together with the important twin-phase austenitic-ferritic (duplex), in Chapter 8. We learned that irrespective of whether or not these materials required CP in their own right, they were likely to be connected into the CP system installed to protect the carbon steel structure. As we discussed in Chapter 8, designing a CP system to protect the carbon steel involves also providing sufficient current to polarise any attached (usually uncoated) CRA. However, we also need to be mindful of the possibility that cathodically generated hydrogen might influence the properties of certain types of stainless steel.

12.10.1.2 Austenitic stainless steels

As far as we know, fully austenitic materials such as the 300 series stainless steels do not suffer HE in seawater. This is irrespective of the level of cathodic polarisation. In this respect, the –1100 mV limit imposed by ISO 15589 (see Chapter 8) seems unduly restrictive. The solubility of hydrogen in austenitic stainless steels is quite high compared with the other types. However, its rate of diffusion is about 10^4 times lower than for a ferritic stainless steel [28]. It seems, therefore, that the resistance of this material to HE stems from the fact that atomic hydrogen is unable to diffuse sufficiently rapidly to the susceptible plastic zone ahead of the crack tip.

12.10.1.3 Ferritic stainless steels

We might whimsically offer two rules for the CP of ferritic stainless steels:

- Rule 1: do not apply CP to ferritic stainless steels.
- Rule 2 : do not even think about applying CP to them.

HE and, consequently, HISC are pretty much guaranteed if we breach Rule 1. Breaching Rule 2 is a step in the wrong direction.

However, since ferritic stainless steels are not employed subsea, you may be wondering why I have troubled you with these two rules. The reason has to do with the fact that we can, and do, apply CP to duplex stainless steels (see following). That is fine. But we also need to be aware that, if we get the duplex metallurgy wrong, we could accidentally find ourselves applying CP to a zone of ferritic material within the duplex.

12.10.1.4 Duplex stainless steels

We discussed duplex stainless steels in Chapter 8. These alloys are manufactured using very careful thermal control in order to produce two-phase materials comprising grains of both ferrite and austenite. The material is widely used for the construction of subsea flowlines carrying highly corrosive crude hydrocarbons containing both chlorides and sulfides. A somewhat oversimplified (but not entirely incorrect) view is that the ferrite phase resists the cracking threat due to chloride whilst the austenite phase resists the cracking threat due to the sulfides. The material also has high strength, and because it has a lower nickel content than the austenitic stainless steels, it is attractively priced. Such attributes explain their popularity in subsea pipeline engineering. There are innumerable duplex stainless steel subsea systems installed worldwide. For the most part, their durability and reliability have been entirely satisfactory. Failures have been very rare.

These materials require polarising to –500 mV for the prevention of crevice corrosion. In practice, however, these lines are likely to be connected to carbon steel structures which are intentionally polarised to more negative potentials. This raises a question as to if, or under what circumstances, the materials might be at risk of HE.

In 1996 a newly installed subsea hydrocarbon production manifold in the UK sector of the North Sea failed when hydro-tested. Upon recovering the manifold to the surface, it was found that locations on the forged 25% duplex stainless steel connector hubs had cracked [33]. Some similar instances followed [36], also in the North Sea. In each case, the cracking was described as HISC, which is a manifestation of HE. These incidents prompted two strands of investigation: into the metallurgy of the hubs, and into the sacrificial anode CP.

A considerable effort was directed at modifying the CP design so as to limit the extent of cathodic polarisation achieved on the duplex components. This involved investigation into applying CP at limited potentials; a topic we covered in Chapter 10.

In the event, however, it was the examination of the metallurgy of the failed hubs that provided the most rewarding route to remedying the problem. Ideally, a duplex stainless steel is thermally processed to ensure a fine-grained microstructure comprising approximately 25%–65% ferrite, preferably as near to 50% as can be achieved. If the product is not cooled sufficiently quickly, the ferrite grains will grow at the expense of the austenite. In these cases, areas of the duplex hubs contained thinner-walled appendages which, as a result of the heat flux from the more massive hub forgings, suffered conversion of austenite to ferrite. At thinner-walled locations, this transformation was so complete that an essentially continuous ferrite path existed through the full wall thickness.

Thus, although the hub was nominally duplex, small, but critical, zones of it were actually ferritic stainless steel. The intentional application of CP to the duplex was, therefore, an unintentional breach of rule 1: *do not apply CP to ferritic stainless steels!*

Lessons were learned, and the importance of achieving an austenite-ferrite phase balance in duplex stainless steels is now well established. However, we should also note that control of metallurgy is not the complete answer. EEMUA 194 contains a neat summary of the findings of this work [41]. Some of its findings are as follows.

- Given the appropriate circumstances, any duplex stainless steels might be susceptible to cracking when subject to the usual levels of CP in seawater (–800 mV or more negative).
- There is little difference between the 22% Cr duplex and the 25% Cr superduplex materials in this respect.
- The threshold stresses for crack initiation become lower as the potential becomes more negative. If the potential reaches –1100 mV, which is near the negative limit achievable with a CP system based on Al-alloy anodes, crack initiation occurs at stress levels producing ~0.5% strain.
- Small (ferrite) grain size is beneficial. As the grain size reduces, the level of strain needed for crack growth increases.
- Coarse grains, particularly with the grains aligned in the through-thickness direction, are particularly susceptible to cracking. (This is what occurred in the above-mentioned connection hub failures.)

In addition to the failure of these hubs, other HISC failures of subsea duplex equipment, including flowlines, were observed in the 1990s. In some of these cases, the problems were ascribed to welding which, if not properly controlled, can likewise lead to ferrite grain growth. In some instances, however, failures occurred which could not be attributed simply to this. These latter failures involved flowlines and equipment that had to be subject to an excessive degree of strain.

Thus, the key to the successful use of duplex stainless steel lies in a combination of the appropriate thermal processing to give a reasonable balance of austenite and ferrite grains, and judicious mechanical design to avoid over-straining. In 2008, these requirements were formalised in DNV-RP-F112 [37]. The available evidence is that, providing this guideline is followed, conventional CP design methodologies may be adopted. Although there have been subsequent reports of hydrogen stress cracking failures of duplex stainless steel under marine CP conditions, investigations have shown the root cause to be non-compliance with the DNV requirements [50].

Of course, it may not be easy to ensure that a duplex stainless steel always does receive the appropriate thermal processing, particularly when it comes to welding. In this respect, some recent Brazilian work [52] is interesting because it discusses a friction stir welding technique for a 25% Cr superduplex material which results in the weld zone exhibiting greater fracture toughness under CP than does the parent plate.

Conversely, if the microstructural and mechanical loading factors cannot be guaranteed to be eliminated, the only remaining option would lie in control of the CP potential.

12.10.1.5 Martensitic stainless steels

As noted in Chapter 8, weldable 13% Cr martensitic stainless steels have been developed mainly to fill a niche in the flow line market between carbon steel and duplex stainless steel. Although less intrinsically resistant to corrosion by crude hydrocarbons than the 22% Cr duplex materials, they are sufficiently durable to handle fluids of modest corrosivity. They are also stronger and, crucially, cheaper than duplex.

These materials have low PRE_N values (typically below 15). As such, they need to be provided with CP for subsea service. However, like duplex, they can be subject to cathodic hydrogen charging. During 2002, two similar leaks were found on 13% Cr flow lines that had been newly installed in the Norwegian sector of the North Sea. In each case, the leaks originated at the 25% Cr fillet weld connecting a type 316 stainless steel doubler plate to the flowlines. The doubler plates had been installed to accommodate the welded cable connection from the bracelet anodes. The cracking progressed from the doubler plate fillet weld through the pipe wall. The failures were detected during pressure testing which, for various reasons, took place about 2 years after the flowlines were installed.

A considerable amount of failure investigation work was carried out which led to the conclusion that these failures were due to HE. It was further concluded that, in these cases, the source of the hydrogen (measured at around 5 ppm) was the flowlines' CP systems. As result, the EEMUA 194 recommendation for the CP of 13% CR pipelines is to try to avoid bracelet anodes and doubler plates. We return to this in Chapter 14.

12.10.2 Nickel alloys

12.10.2.1 Solid solution alloys

If you keep adding nickel to an austenitic stainless steel (UNS SXXXXX series), you eventually have to call it a nickel alloy (UNS NXXXXX), and you have to pay more for it. The two most widely used nickel alloys employed subsea are the solid solution materials: Alloy 825 and Alloy 625. We discussed both of these in Chapter 8. These alloys are fully austenitic and, as such, do not suffer HE problems. We have no concerns if equipment, such as flowlines or small-bore manifold piping, fabricated in either of these materials is connected to the host structure's CP system.

Alloy 625 is recommended in Norsok M-001 [43] as a subsea fastener material if CP cannot be provided.

12.10.2.2 Precipitation-hardened alloys

12.10.2.2.1 Alloy K-500

A precipitation-hardened alloy is formulated to be heat-treated to produce very fine (submicroscopic) intermetallic precipitates. These precipitates strengthen (or harden) the alloy. An example of such an alloy is the Ni-Cu-Al alloy K-500 (UNS N05500). This material is formulated to combine the high seawater corrosion resistance of the Ni-Cu alloy system with the high strength that derives from the presence of intermetallic Ni-Al precipitates. Such properties suggest that this material would be a good choice for subsea fasteners. Indeed, it is widely used for just that purpose.

There was, however, a series of high-profile offshore failures of K-500 riser bolts in the early 1980s [16] that were attributed to HE. All of the bolts failed soon after entering service. Laboratory investigations of the recovered components revealed that they had suffered brittle fracture and contained high levels of hydrogen (up to 18 ppm) near the fractures. Although the bolts were within the appropriate specification with regard to chemical composition, there were regions of high hardness adjacent to the thread. In addition, the bolts had all been highly stressed and had been polarised to potentials in the region of –1000 mV. Thus, all of the risk factors for HE indicated in Figure 12.13 were in play.

In this instance, the operator resolved the problem by introducing a two-stage heat treatment. First, after the threads were machined, the bolts were given a full anneal (at ~1000°C for 10 minutes followed by a water quench) to reduce the local hardness. This was followed by the precipitation hardening heat treatment (600°C for 16 hours). At the time of reporting, replacement bolts manufactured according to this revised process had been in service for a couple of years without problems. In addition, K-500 fasteners are still used subsea in combination with CP.

12.10.2.2.2 Other alloys

Recently the effect of CP, at potentials of –1100 and –1400 mV in 3.5% NaCl, on five other precipitation-hardened nickel alloys, in the yield strength range of 867–1161 MPa, has been studied [49]. The alloys studied all showed fracture toughness that reduced as potential was made more negative. However, only one of these materials (N07718) is recognised as a subsea fastener by Norsok M-001, albeit that it is not often used as such. This material has suffered a well-publicised hydrogen stress cracking failure when highly stressed and exposed to a sour hydrocarbon well fluid. As far as we are aware, there has been no reports of it suffering HISC as a result of the combined action of seawater exposure and CP. Samples of N07718 have also been tested by another group [53] who, likewise, found reduced fracture toughness as a result of increasing cathodic polarisation.

It should be understood that, although these experiments mimicked excessive CP polarisation conditions, the object of the testing was to examine differences between the fracture toughness of alloys rather than to qualify the materials for subsea service. Thus, although the experiments showed that increasing cathodic polarisation caused reduction in fracture toughness in slow strain rate testing, which was attributable to HE, it lacked direct relevance to practical marine CP applications. Nevertheless, taken together, these papers tell us that we would need to be cautious if designing a CP system for subsea equipment which included precipitation-hardened nickel fasteners. In particular, it would be important to guarantee that these items were neither highly stressed nor exposed to excessively negative potentials.

12.10.3 Copper alloys

As far as we know, copper alloys are immune to HE-induced failures. This also includes precipitation-hardened copper alloys. For example, cupro-nickel alloy C72420 which, being a fastener alloy, is one of the very few copper-based alloys used subsea both under tension and in combination with CP [40]. Similarly, it has been recently determined [51] that the combination of marine exposure and CP does not affect the fracture toughness of a precipitation-hardened copper-beryllium alloy (C17200) used for split lock rings and trunnions.

12.10.4 Titanium

To date, there has been comparatively little use of titanium, or its alloys, in subsea construction. Nevertheless, its strength-to-weight ratio renders it a potentially attractive material for some components of offshore hydrocarbon production systems, e.g. risers. It also has the advantage of being highly resistant to corrosion in seawater, even up to elevated temperatures. For example, Schutz [40] points out that, by 2012, there were 1.98×10^8 m of commercial purity titanium heat exchanger tubes in seawater-cooled steam condensers at coastal power plants worldwide. In 40 years of service, none of these condensers had experienced a seawater corrosion failure. Similarly, there has never been a reported instance of titanium suffering from MIC.

The material itself, therefore, clearly has no need of CP. However, where titanium is contemplated for subsea use, it is likely that it will be deliberately or adventitiously connected into cathodically protected steel structures or equipment. This may present a problem because, as ISO 21457 [39] states, *...Titanium-based alloys should not be used for applications involving exposure to cathodic protection.*

The reason for this is that CP introduces a risk because titanium can be embrittled by cathodically generated hydrogen. In this instance, however, the mechanism is distinct from the mechanisms that we have discussed above which involve atomic hydrogen atoms hindering plastic deformation at the tips of growing cracks. In the case of titanium, the issue involves the results of chemical reaction between the titanium and atomic hydrogen that results in the formation of titanium hydride.

In considering how to mitigate the problem of hydride formation, Gartland et al. [26] reviewed work published prior to 1997. Most of that work focused on grade 2 material, which is widely used in desalination plant seawater heat exchanger tubing. This unalloyed, commercially pure grade possesses a hexagonal close-packed crystallographic structure. It is referred to as an alpha-type alloy.

In general, the degree of hydride formation in this grade of material is a function of the temperature, and the degree and duration of cathodic polarisation. However, not all of the early work on hydride formation yielded unequivocal information. For, example Lee, Chung and Tsai [18] found 33 ppm of absorbed hydrogen, and no hydrides after polarisation at room temperature for 8 months at –1.0 V. Conversely, over 18 months of testing, Lee, Lin and Oung observed hydrides to a depth of 66–70 µm in a specimen polarised at –1.0 V. On the basis of this, they advised that the potential must be restricted to more positive than –0.75 V [21]. This advice was similar to that offered by the Kobe Steel Co. [7] which designated –0.75 V as the threshold for hydrogen absorption. Conversely, Schutz and Grauman [20] suggested a more relaxed negative limit of –1.0 V, based on their testing at 45°C in near-neutral brine for 6 months.

Venkataraman and Goolsby [25] conducted a number of laboratory and field studies into titanium embrittlement in the context of offshore construction, including in deep waters. They concluded that the alpha-material, which has a low hydrogen solubility and, therefore, lower threshold for hydride formation, is more susceptible than beta-alloys (body-centred cubic) which appeared the more attractive option for extended subsea service applications.

REFERENCES

1. W.J.M. Rankine, *On the causes of the unexpected breakage of the journals of railway axles; and on means of preventing such accidents by observing the Laws of Continuity in their construction.* Paper no. 596, Institute of Civil Engineers (London) 7 March 1843.
2. F. Braithwaite, *On the Fatigue and consequent fracture of metals.* Paper no. 915, Institute of Civil Engineers (London) 16 May 1854.
3. D.J. Duquette and H.H. Uhlig, *Effect of dissolved oxygen and NaCl on corrosion fatigue of 0.18% carbon steel.* Trans ASM **61**, 449 (1968).
4. C.M. Hudgins et al., *The effect of cathodic protection on the corrosion fatigue behaviour of carbon steel in synthetic seawater.* Journal of Petroleum Technology **23** (3) 283 (1971).
5. H.H. Uhlig, *Corrosion and Corrosion Control*, 2nd edition. John Wiley & Sons (New York) 1971, p. 150 (ISBN 0-471-89563-6).
6. W.H. Hartt, J.E. Fluett and T.E. Henke, *Cathodic protection criteria for notched mild steel undergoing corrosion fatigue in sea water,* Paper no. OTC 2380, Offshore Technology Conference, 1975.
7. Kobe Steel Ltd., Titanium Dept., *On the cathodic protection of titanium tubed turbine condenser - experimental results and some considerations*, Technical note No. 004, 1975.
8. P.M Scott and D.R.V. Silvester, *The influence of mean tensile stress on corrosion fatigue crack growth in structural steel immersed in seawater*, UK Department of Energy Interim Technical Report UKSORP 3/02, 1977.
9. W.C. Hooper and W.H. Hartt, *The influence of cathodic polarization upon fatigue of notched structural steel in sea water.* Corrosion **34** (9), 320 (1978).
10. W.H. Hartt, P.E. Martin and W.C. Hooper, *Endurance limit enhancement of structural steel in sea water by cathodic protection,* Paper no. OTC 3511, Offshore Technology Conference, 1979.
11. C.E. Jaske, J.H. Payer and V.S. Balint, *Corrosion Fatigue of Metals in Marine Environments*, 1981 (ISBN 0-935470-07-7).
12. W.H. Hartt, *Cathodic protection and fatigue of offshore structures.* Materials Performance **20** (11), 50 (1981).
13. G.S. Booth, *Corrosion fatigue of welded steel joints in sea water, Proceedings of UK Corrosion*, 1982.
14. P.W. Marshall et al., *The cognac fatigue experiment*, Paper no. OTC 4522, Offshore Technology Conference, 1983.
15. J.W.C. Thompson, *Phenomenological investigation of the influence of cathodic protection on corrosion fatigue crack propagation behaviour, in a BS 4360 50D type structural steel and associated weldment microstructures, in a marine environment*, PhD Thesis, Cranfield Institute of Technology, 1984.
16. K.D. Efird, *Failure of Monel N-Cu-Al K-500 bolts in seawater.* Materials Performance **27** (4), 37 (1985).
17. R.P.M. Proctor, *Detrimental effects of cathodic protection: embrittlement and cracking phenomena,* in *Cathodic Protection Theory and Practice,* Eds. V. Ashworth and C.J.L Booker. Ellis Horwood (Chichester) 1986, pp. 293–313.
18. J.-I. Lee, P. Chung and C.-H. Tsai, *A study of hydriding of titanium in seawater under cathodic polarization*, Paper no. 259 CORROSION/86.

19. M.J. Robinson, C.H.J. Parker and K.J. Seal, *The influence of sulphate reducing bacteria on hydrogen absorption by cathodically protected steel,* Proceedings of UK Corrosion, 1987.
20. R.W. Schutz and J.S. Grauman, *Determination of cathodic potential limits for prevention of titanium tube hydride embrittlement in saltwater*, Paper no. 110 CORROSION/89.
21. J.-I. Lee, J.-C. Lin and J.C. Oung, *Hydrogen problems in cathodically polarized titanium tubes in seawater*, Paper no. 115 CORROSION/89.
22. A.S. Dolphin and D.R. Tice, *Protection potential and stress ratio on fatigue thresholds of structured steels in sea water*, HSE Offshore Technology Report OTH 92 369, 1992.
23. D. Festy and D. Tigges, *Hydrogen embrittlement of high strength low alloy steel cathodically protected in seawater: effect of hydrostatic pressure*, Paper 421 CORROSION/92.
24. P.M. Scott, *Corrosion fatigue*, in *Corrosion*, 3rd edition, Eds. L.L. Shreir, R.A. Jarman and G.T. Burstein. Butterworth-Heinemann (Oxford) 1993, pp. 8:143–8:183.
25. G. Venkataraman and A.D. Goolsby, *Embrittlement in titanium alloys from cathodic polarization in offshore environments and its mitigation*, Paper no. 554 CORROSION/96.
26. P.O. Gartland, F. Bjørnås and R.W. Schutz, *Prevention of hydrogen damage of offshore titanium alloy components by cathodic protection systems*, Paper no. 477 CORROSION,/97.
27. A.T. Smith and W.D. Dover, Corrosion fatigue of API 5L grade X85 welded tubular joints with applied cathodic protection of -1000 mV (Vs Ag/AgCl), HSE Offshore Technology Report OTO 96 034, 1997.
27. A.D. Goolsby and D.P. McGuire, *Cathodic protection upgrade of the 1,050' water depth cognac platform,* Paper 472 CORROSION/97.
28. R.N. Gunn (ed.), *Duplex Stainless Steels.* Abingdon Press (Cambridge) 1997.
29. R.G.J. Edyvean et al., *Biological influences on hydrogen effects in steel in seawater.* Materials Performance **37** (4), 40 (1998).
30. M.J. Robinson and P.J. Kilgallon, *A review of the effects of sulphate reducing bacteria in the marine environment on the corrosion fatigue and hydrogen embrittlement of high strength steels*, UK Health & Safety Executive Report OTH 555, 1998.
31. J. Benson and R.G.J. Edyvean, *Biologically influenced hydrogen entry into steel in seawater*, Corrosion Management, p. 14, May/June 1999.
32. C.L. Batt and M.J. Robinson, *Optimising cathodic protection requirements for high strength steels under marine biofilms*, Corrosion Management, p. 13, September/October 1999.
33. T.S. Taylor, T. Pendlington and R. Bird, *Foinaven super duplex materials cracking investigation,* Paper no. OTC 10965, Offshore Technology Conference, 1999.
34. BS 7910 (formerly PD 6493), *Guide to methods for assessing the acceptability of flaws in metallic structures,* 2005 (re-issued in 2019).
35. Z. Ahmed, *Corrosion fatigue*, in *Principles of Corrosion Engineering and Corrosion Control.* Elsevier Science & Technology Books (Amsterdam) 2006, pp. 221–241.
36. S. Huizinga et al., Failure of subsea super duplex manifold hub by HISC and implication for design, Paper no. 06145, CORROSION/2006.
37. DNV-RP-F112, *Design of duplex stainless steel subsea equipment subject to cathodic protection*, 2008. (Re-issued *Duplex stainless steel design against hydrogen induced stress cracking* 2019).
38. R. Akid, Corrosion fatigue, in *Shreir's Corrosion,* Eds. J.A. Richardson et al. Elsevier (Amsterdam) 2010, pp. 929–951.
39. ISO 21457, *Petroleum, petrochemical and natural gas industries: materials selection and corrosion control for oil and gas production systems*, 2010.
40. R.W. Schutz, *Titanium alloys*, Chapter 7, in: *The Corrosion Performance of Metals for the Marine Environment: A Basic Guide,* Eds. C. Powell and R. Francis. EFC Publication No. 63, CRC Press (London) 2012, pp. 13–16.
41. EEMUA Publication 194, *Guidelines for materials selection and corrosion control for subsea oil and gas production equipment*, 2012.
42. BS 7608, *Guide to fatigue design and assessment of steel products*, 2014.
43. Norsok M-001 Materials Selection, 2014.

44. H.K.D.H. Bhadeshia, *Prevention of hydrogen embrittlement in steels.* ISJI International **56,** 24–36 (2016).
45. DNVGL-RP-C203, *Fatigue design for offshore steel structures*, 2016.
46. P. Pedeferri, Section 13.8, in *Corrosion Science and Engineering*, Eds. L. Lazzari and M. Pedeferri. Springer Nature Switzerland AG (Cham) 2018, pp. 243–273.
47. ASTM E-28, *Standard test methods for notched bar impact testing of metallic materials*, 2018.
48. H.E. Ayama, B. Fahimi and R. Thodla, *Hydrogen embrittlement of ASTM A320 L7 & L43 grade carbon alloy steel bolts in subsea environments,* Paper no. 12785 CORROSION/2019.
49. J. Feiger et al., *Fracture toughness of precipitation hardened nickel alloys under cathodic polarisation environments,* Paper no. 12849, CORROSION/2019.
50. M.S. Hazarabedian et al., *Hydrogen induced stress cracking of super duplex stainless steel UNS S32760: A root cause failure investigation*, Paper no. 12963 CORROSION/2019.
51. A. Bajvani et al., *Fracture toughness of wrought copper beryllium alloy in seawater with cathodic protection conditions for subsea applications,* Paper no. 13010 CORROSION/2019.
52. C.E.F. Kwietniewski et al., *Hydrogen embrittlement under cathodic protection of friction stir welded UNS S32760 super duplex stainless steel,* Paper no. 12994 CORROSION/2019.
53. X. Li, R. Thodia and G.B. Viswanathan, *Hydrogen embrittlement study of three heats of UNS N07718 in subsea applications,* Paper no. 13057 CORROSION/2019.
54. ISO 3138, *Petroleum and natural gas industries: steel pipe for pipeline transportation systems,* 2019.
55. ASTM A193/A193M, *Standard specification for alloy-steel and stainless steel bolting for high temperature or high pressure service and other special purpose applications*, 2019.
56. ASTM A320/A320M, *Standard specification for alloy-steel and stainless steel bolting materials for low-temperature service*, 2021.

Chapter 13

Fixed steel structures

13.1 STRUCTURES FOR HYDROCARBON PRODUCTION

13.1.1 Early sacrificial anode systems

This chapter is concerned with structures that are fixed in place, either in the sea or at its edge. We have already covered the CP of such structures up to a basic level in Chapter 3, where we followed the CP design codes to produce sacrificial anode designs. Now it is time to ask a few more questions such as: how did these codes come about and, more importantly, what do we do in situations where they are silent? We will try and answer these questions. In so doing, we will see that the development of the codes followed, rather than led, the evolution of offshore CP itself. Furthermore, and up until recently, that evolution was almost entirely driven by developments in the offshore hydrocarbon industry.

In the USA, oil and gas production took its first tentative steps offshore in the late 1940s. The first structure was installed in the shallow nearshore waters of the Gulf of Mexico in 1948 [44]. Those early structures were a far cry from the behemoths that were being launched 30 or 40 years later. In many cases, the installations amounted to a single wellhead in a couple of metres of water in the Mississippi Delta. The term "jacket" came into the offshore lexicon because the support structure for the well platform literally formed a jacket around the well conductor. Indeed, some of the earliest jackets were little more than simple wooden frames.

The hydrocarbon industry was already aware of the benefits of CP in controlling corrosion. It had been practiced onshore for pipelines since the 1920s, where both impressed current and sacrificial magnesium anode systems were in common use. Elsewhere, CP was also occasionally used to protect harbour jetties and piling. That application also usually featured magnesium anodes [1].

The early small offshore steel installations were generally protected using a handful of magnesium anodes, of the type in common use for onshore pipelines. These were simply lobbed off the deck and suspended from the structure by their connecting cables. There was little incentive to develop "permanent" CP systems. The magnesium anodes were simply regarded as consumable items. Their expected life was no more than a couple of years, at the end of which the cables were hauled up and the depleted anodes discarded and replaced.

13.1.2 Early ICCP systems

Initially, little thought was given to the robustness of ICCP systems. Brandt [28] tells us that the first offshore ICCP systems simply used conventional land-based graphite or silicon-iron anodes connected to pad-eyes and suspended over the side. The normal practice was to switch the power off and haul the anodes up onto the deck, whenever a storm blew through. Somebody might then get around to lobbing the anodes back over the side and switching

DOI: 10.1201/9781003216070-13

the power back on. For example, the "Emmy" platform, which was installed in 1962 in ~15 m of water [22] off the Californian coast, was protected by 15 lead-silver anodes dangling from polyethylene cables. Each cable was deployed from a manual winch to enable the anodes to be hauled out during storms. This system was not entirely reliable. Sometimes the cables chafed against structure causing a direct short and blowing the circuit breakers on the power supplies. Occasionally, divers found severed anodes at the base of the platform. These were simply recovered, reconnected and redeployed. Thus, just as there was no incentive to design permanent sacrificial anode systems, there appears to have been some tolerance of frail ICCP systems. Persuading divers to work in shallow waters off the California coast was, we may assume, not too expensive an exercise.

13.1.3 Deeper waters

Things started to change after 1962 [37] when the US federal government issued licences for production in deeper waters (up to 61 m). This required larger structures, and it soon became obvious that deploying, and regularly replacing, the required number of magnesium anodes would be very onerous. The obvious route was to adapt the onshore ICCP technology to the offshore setting. For the most part, early ICCP systems on structures continued with the practice of swinging anodes on cables or installing heavier anodes on the seabed. The reliability of such systems was no better than for "Emmy", but the move to deeper waters meant that the costs associated with that unreliability increased.

Attempts were made to engineer more robust ICCP systems with the anodes fixed to the structure and to provide some protection for the anode cables which needed to be connected back to the power supplies on the deck. Shell's first "permanent" ICCP system was installed in the Gulf of Mexico in 1966 [5]. It used "high output" lead-platinum anodes and had the anode cables brought to the surface through protective conduits. These anodes achieved full polarisation of the uncoated structure to −0.8 V within a week. By 1970, the inventory of Shell platforms protected by ICCP had increased to nine [6], in water depths up to 103 m.

Despite these engineering efforts, these early impressed current systems still generally did not achieve the desired level of reliability. A 1979 paper by Wyatt [9] referred to an earlier retrofit by Shell of a dozen structures in the Gulf of Mexico. It seems likely that some of these platforms may have been those with the early "permanent" CP systems. There were reports of... *various mechanical failures.*

Through the late 1960s and into the 1970s, therefore, sacrificial anodes became the "go to" option for the CP of offshore oil and gas structures. This was reinforced by the development of the aluminium alloys described in Chapter 10. By 1963, Shell was using Al-Zn-Sn anodes [5]. From the mid-1960s onwards, the more reliable and efficient Al-Zn-Hg materials were coming into play. The high current capacity of these aluminium alloys, compared with zinc or magnesium, and their useful operating potentials, particularly compared with the over-active magnesium alloys, meant that Al-Zn-Hg quickly came to dominate the sacrificial anode market for offshore structures.

The combination of larger structures in deeper water and the availability of reliable high current capacity aluminium anodes meant that there was both an incentive and an opportunity to develop long-life sacrificial anode designs. Instead of hooking anodes to the structure after it was installed, larger anodes were welded on in the fabrication yard.

The attachment of these "permanent" anodes required an element of design. In turn, this required definition of the target protection potentials and guidance on the current density needed to polarise steel to that potential. The protection potential was established early on. As we discussed in Chapter 6, a figure of −0.8 V Ag|AgCl|seawater entered the codes at the end of the 1960s, and this figure has not been challenged since.

The required cathodic current density was a less straightforward parameter to tie down. This design figure varies according to location and circumstances. Since much of the development of offshore CP in the 1960s related to waters <100 m deep in the Gulf of Mexico, that was the area for which design cathodic current densities were first documented, at least in the English language technical literature. Chapter 7 contains a synopsis of the evolution of recommended current densities for offshore CP.

13.1.4 To coat or not?

It has always been recognised that CP is ineffective in the splash zone and offers only limited benefit in the inter-tidal zone. As a consequence, all offshore steel structures are coated down to at least 1 m below the level of lowest astronomical tide (LAT). Below LAT, however, operators historically split into two camps. For example, in the 1960s, BP's Arabian Gulf and southern North Sea jackets were painted with coal tar epoxy and cathodically protected with zinc anodes. Indeed, in the 1980s, BP simply purchased its preferred coal tar epoxy paint manufacturer. On the other hand, the majority of other operators opted for uncoated jackets and, later into the 1960s, for aluminium alloy anodes.

The protective coatings industry used its best endeavours to persuade operators to coat immersed steelwork. However, the industry's resistance to the concept arose as a result of a number of considerations. The main counterargument to coatings was the cost of their application and, in particular, the cost of the necessary surface preparation of the steel. More importantly, the impact that the extra staging, grit blasting, paint application and inspection would have on the construction schedule was usually regarded as an insurmountable obstacle. Time was of the essence when it came to getting the hydrocarbon flowing.

A second consideration weighing against coatings was the uncertainty on how they would perform in the long term. As we saw in Chapter 9, early versions of the CP codes offered very pessimistic predictions of the rate of coating breakdown. Broadly speaking, the designers of jackets, with required lives of up to 40 years, were obliged to assume that the coating would have completely disappeared by the end of life. This conservatism was, perhaps, understandable since the epoxy paints which were being offered as candidate coatings for these structures had, by that time, only been in existence for about 20 years.

Thus, even though coating, or at least partial coating, was known to reduce the weight of anodes required, there was uncertainty about how much weight reduction could reasonably be justified. Hence, once the use of uncoated structures had become the standard way of doing things, intellectual inertia meant that it was likely to stay that way. This was despite the fact that various authors, some even without any affiliation to the paint industry, pointed out the technical and economic benefits of coating. For example, Garner [16] argued that... *to ensure that the construction schedule is not impacted the most economic method is to coat the jacket in the fabrication yard and acknowledge that the coating systems may be neither 100% or of high quality*. This pragmatic attitude encapsulates an important reality of which we should never lose sight. Corrosion engineering is not, and never should be, a counsel of perfection. The object of our profession is to keep things working for as long as is required. We should leave the cosmetics to others.

13.1.5 Weight saving

13.1.5.1 Same problem – different solutions

I am not attempting to write a history of offshore CP. So, with apologies to the authors, I am ignoring very many worthy journal and conference papers that describe the CP installations on offshore jackets. Instead, I will highlight two North Sea installations that illustrate diametrically different approaches to essentially the same problem: Conoco's Murchison platform [11,13], and Norsk Hydro's Oseberg "B" [19].

As we see in Table 13.1, the two platforms were similar in terms of their size and geographic proximity. Furthermore, an important design constraint that applied to both structures was that they were to be launched from a specially constructed barge. This imposed a limit on the maximum launch weight. It follows that weight reduction was an important driver in the overall jacket designs, including their corrosion protection systems.

Table 13.1 Murchison and Oseberg "B"

Structure	*Murchison*	*Oseberg "B"*
Location	Northern North Sea	Northern North Sea
Water depth	156 m	109 m
Installed	1979	1987
Installation method	Barge Launch	Barge Launch
Immersed and buried steel area	70,300 m^2	71,857 m^2
Wells (planned)	38	27
Coated	No	Yes
CP system	ICCP plus sacrificial anodes	Sacrificial anodes

Nevertheless, despite what appears to have been very similar corrosion control requirements, the two design teams came up with quite different approaches for reducing the weight of the CP systems. The Murchison designers opted for an uncoated structure and a hybrid CP system combining a (much reduced) sacrificial anode burden with an ICCP system. The Oseberg "B" team, on the other hand, opted for reducing the weight of sacrificial anodes by fully coating the structure. It is informative to examine the thinking behind the approaches adopted by each team.

13.1.5.1.1 Murchison

The Murchison design team considered painting the structure as a route to saving weight. However, it rejected the approach of coating the entire jacket. It considered that it would be... *difficult to get a good coating job because of the size and complexity of the jacket, and the weather conditions in Scotland where the jacket was to be built* [13]. The team went on to say that... *coating failure is one of the major recurring maintenance problems in the North Sea.*

With hindsight, there was scope for further discussion here. The comments on the difficulties in coating the jacket were reasonable, and few would contest their assessment of the Scottish weather. The reference to *a good coating job*, however, warrants scrutiny. The relevant question would not have been: how do we ensure a good coating job? It should have been: how good would the coating need to be in order permit the required reduction in the weight of anodes and associated steelwork?

The comment on the coating maintenance problems in the North Sea was correct. Indeed, paint deterioration remains an endemic problem throughout the offshore industry worldwide. However, the comment was irrelevant when it comes to underwater coatings. These are never maintained. It is simply assumed that they will deteriorate progressively over time, and the CP system is designed to accommodate this.

Be that as it may, the decision not to paint the structure was very much in keeping with established practice of the time. It was the decision to employ ICCP that bucked the trend. The design team was well aware of the received wisdom that impressed current systems were inherently unreliable. However, the decision to avoid coating forced a need to mitigate those historical sources of unreliability. The objective was to produce... *a robust design of the underwater parts of the system to ensure maximum integrity for the life of the platform together with facilities for retrofitting additional anodes if required.*

The result may be regarded as a masterclass in how to go about designing an offshore ICCP system. The installation comprised 216 rod type (20 mm diameter 1 m long) platinised niobium anodes, each rated at 50 A. 100% redundancy was achieved by installing the anodes in one branch of a "Y" fitting [11,13]. Furthermore, the threads of the screwed anode connection in the fitting were platinum coated. This was, in part, to ensure ease of replacement by divers. Further system robustness was ensured by providing ample potential monitoring points and by running all cables in welded steel conduit.

In addition to the ICCP installation, a total of 596 Al-Zn-In sacrificial anodes (318 kg each) were also installed such that all critical joints would receive protection whenever the ICCP system was off. These sacrificial anodes were regarded as essential, not least because there would be a gap of around a year between the installation of the jacket and the energising of the ICCP system. I do not know if the corrosion or fatigue implications of a year without CP were analysed. They received no mention in the public-domain corrosion literature.

13.1.5.1.2 Oseberg "B"

Oseberg "B" followed a few years after Murchison, and it had a longer design life: 40 years compared with 30 years for Murchison. However, the problems faced by the designers were very similar. In particular, the upfront decision to launch the structure from a barge meant that, like Murchison, the weight of the jacket had to be carefully controlled. However, whereas the Murchison team opted for ICCP because it regarded coatings as unreliable, the Oseberg "B" team opted for coating the jacket [19]; in part, because it regarded ICCP as unreliable. It argued that... *no impressed current system with a proven design life of 40 years in a North Sea environment is available.* This statement was incontrovertible; but a cynic might reasonably have also pointed out that, in 1988, there was neither a sacrificial anode system nor a coating system with a 40-year track record.

Nevertheless, the project team opted for fully coating the jacket with 300 µm of coal tar epoxy, and a white epoxy top coat (200 µm) and Al-Zn-In anodes. They calculated that, because of the benefit of the coating, they were able to reduce the installed anode mass from 1526 tonnes, needed for an uncoated structure, to 763 tonnes. A further reduction in weight was achieved by using cylindrical profile anodes, which reduced the fatigue due to wave loadings, and further permitted some trimming of steelwork weight. This 50% reduction in anode mass, and the reduction in supporting steelwork weight was critical in meeting the structure's upper weight limit for the barge launch.

Unusually for a structure protected by sacrificial anodes, Oseberg "B" was fitted with permanent reference cells and monitored anodes. Even more unusually, a data collection unit was pre-installed such that it was in place and operational as soon as the structure was launched. When the data were subsequently recovered, it was found that the coated jacket had polarised fully within a day.

Having made the case for coating Oseberg "B", it is unsurprising that the neighbouring Oseberg "C", launched 3 years later, was also coated to save anode weight. This was even though its design life was a more modest 25 years [25]. However, unlike "B", the "C" structure was not fully coated. A predesign study had examined the options of coating all the straight structural members and leaving the nodes bare, and the converse of only coating the nodes. We are told that a CP computer model showed that coating the straight members would result in faster polarisation and better anode distribution. No details of the model are given, but it seems that its results endorsed the intuitively obvious. Coating the straight members covered about 80% of the steel surface, whereas the nodes only covered 20%. It is little surprise that the CP models indicated that the option with a much larger area coated was easier to polarise. Accordingly, the design opted for only coating the straight members.

Oseberg “C” also introduced another innovation. Bracelet sacrificial anodes were installed on coated members down to a depth of −45 m. These reduced the hydrodynamic loadings on the structure and, according to the modelling, provided adequate protection.

13.1.5.2 Comment

Murchison and Oseberg “B” illustrate the point that there is no single “correct” answer when it comes to designing a CP system. Both systems were successful. They achieved the weight reduction needed for launch, and, crucially, they protected the structures for their intended lives. Murchison was decommissioned by its last owners (CNR) in 2015. At the time of writing (2021), Oseberg “B” is still in service. It is operated by Equinor ASA (a company formed by a merger between Norsk Hydro and Statoil).

They also exemplify the old adage that nobody challenges a CP design that works. Now, long after the event, we might reasonably point out that each CP concept was arrived at with some lack of rigour. The Murchison team discounted coatings, and the Oseberg team eschewed ICCP, in each case without any report of a critical evaluation. In the world of hydrocarbon production, however, the result matters more than the journey to that result.

Although it would now be a sterile exercise, re-examination of the CP design processes would probably reveal costs that might have been safely trimmed on either project. However, any such evaluation would need to be viewed in the context of the general levels of risk associated with such pioneering projects. As DNV subsequently argued [33], there is a sound case for some measure of design conservatism. The cost of a yard-installed CP system for an offshore structure is typically somewhere between 0.5% and 1% of the overall project fabrication cost. It follows that there would never be much incentive to seek a marginal saving on a marginal cost.

Whereas conservatism is inexpensive for yard applied CP systems, it can be an expensive indulgence for CP systems retrofitted offshore. We will return to this point in Section 13.5.

13.1.6 Into the sunset?

At the risk of oversimplification, the history of the first half-century of the offshore hydrocarbon industry has been one of seeking to exploit reserves in ever more inhospitable locations. This has required installation of steel or concrete structures in deeper and more hostile waters. To a detached observer, these decades saw oil companies competing against each other in ever more macho offshore projects. The spin-off of this endeavour was a progressive improvement in the quality of CP hardware and of CP design standards and practices.

It now looks like future offshore oil and gas production will tail off. This is due to the combined pressures of harder to find reserves and, much more relevantly, the gathering global impetus towards greener energy. We can reasonably ask ourselves: is there any more that we need to say about the CP of fixed marine structures?

The short answer is: yes.

Even if we adopt the premature view that offshore hydrocarbon production is a "sunset" endeavour, we are still left with numerous instances where we need to protect fixed steel structures in the sea. Some of these instances, such as jetties and harbours, are well established. We also see rapidly emerging technologies such as offshore wind energy (see below). Tidal power projects may gain more traction whilst other offshore endeavours, such as seabed mining, are just becoming visible over the horizon. There is little doubt that steel, or other potentially corrodible metals, will form the backbone of the subsea equipment that will be needed for these endeavours.

We also need to bear in mind that there is an accumulation of marine hardware and infrastructure, much of which is now required to remain in service for many years beyond its original end-of-life date. It will occasionally be necessary to retrofit CP to these existing structures.

13.2 OFFSHORE WIND FARMS

13.2.1 Development

Unfortunately, our exploitation of fossil fuels, including offshore hydrocarbons, has had a significant hand in the global warming crisis that we now face. This crisis has, in turn, prompted the quest for alternative sources of energy; a notable example of which is the exploitation of offshore wind.

The first offshore wind farm was commissioned in Danish waters in 1991 [56]. By 2019, the installed and commissioned offshore wind-generating capacity, most of which is in NW Europe, was about 25 GW. This is enough to replace about 50 conventional coal or gas-burning power stations. No doubt, this figure will have increased further by the time you are reading this. As an island, the UK is handily placed to exploit wind power. As of 2019, about one-third of the global offshore wind energy capacity was in its waters. The growth of the UK's offshore wind capacity is illustrated in Figure 13.1. This shows the number of wind turbines installed offshore in the first two decades of the 21st century. The global growth mirrors this pattern.

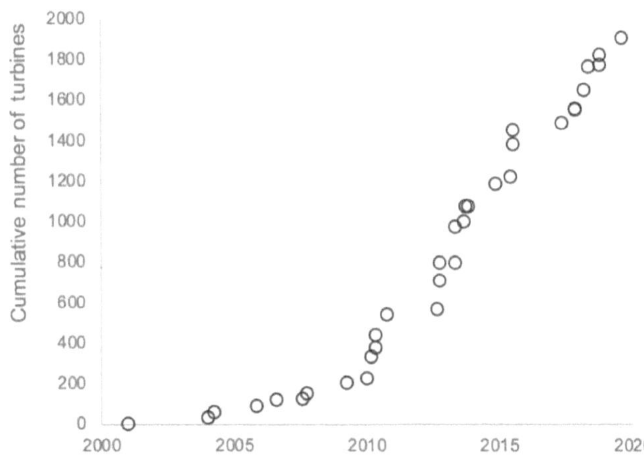

Figure 13.1 Development of offshore wind power in the UK.

13.2.2 Foundation options

In the UK, over 1900 offshore wind turbines had been installed by the end of 2019. With the exception of a handful of experimental floating installations, all of these installations are fixed in place and mounted on a foundation structure. The vast majority of these foundations (~1800) are steel monopiles; most of the remainder being steel jacket-type substructures. Accordingly, we will focus on monopile foundations. As of 2020, these are employed up to a maximum water depth of 32 m, although the average depth is a little over 20 m.

Depending on the local circumstances, a typical monopile foundation will be 5–7 m in diameter, with a steel grade and wall thickness designed to take both the piling impacts and the through-life fatigue loads. Once piled into place, typically >20 m into the seabed, a transition piece is installed onto the monopile. This transition piece is levelled to a high degree of precision before being either bolted or, more usually, grouted into place. The turbine tower is then bolted onto the transition piece.

As a matter of routine, these monopile foundations are protected externally from corrosion by protective coating above the waterline and CP below it. Conventionally, the submerged steelwork is not coated. Up until comparatively recently, these CP systems have been designed using the methodologies and guidelines developed within the oil and gas industry, typically using the current density guidelines in DNV-RP-B401. Initially, this posed little problem. However, as wind farms developments migrated further offshore, and foundations grew to support larger towers in deeper water, it became apparent that the design guidelines developed for space-frame type hydrocarbon structures were not always appropriate.

13.2.3 Monopiles

All monopile foundations are exposed to seawater both externally and internally. In some instances, the internal seawater can exchange, either deliberately or accidentally, with external seawater. For this reason, in some cases, some operators deem it appropriate to install CP internally in monopiles. Because this application of CP introduces some unique challenges, we cover it in Chapter 16. In this chapter, we will only look at the CP of the external surfaces.

13.2.3.1 Is external CP needed?

Ideally, the question of whether or not a monopile needs to be protected from corrosion should be addressed by carrying out a corrosion engineering assessment. The thickness of the steel plate used in their construction, usually >70 mm, is dictated in the first instance by the need to take the piling loads. Based on what we learned about marine corrosion rates in Chapter 1, there is no realistic likelihood of the plate thinning to the extent that it can no longer support the deadweight load imposed by the tower and turbine. Based on a typical monopile design life of ~25 years, an average steel thickness loss of no more than a few millimetres would be anticipated. This would not cause a tower to topple over. Indeed, it has been remarked [51] that at least one monopile design resulted in a 50 mm corrosion allowance once the pile was set in place, and there was no longer a need to withstand the piling loads.

In practice, however, the life-determining factor for monopile foundations will be corrosion fatigue. This reality has been summed up by Lichtenstein[1] who described a wind turbine as... *a machine built to destroy its foundations*. Thus, in a rational world, the need

[1] J. Lichtenstein, (DNVGL) presentation at EuroCorr, 2020.

or otherwise for external CP should be based on the fatigue and corrosion-fatigue analysis for the structure. In practice, this would reduce to a fatigue analysis of the most critical area. If the result of the analysis was that even under free corrosion the fatigue life was adequate, then there would be no rationale for CP.

The world is not always rational, and the decision to apply external protection is usually taken as read. As far as I am aware, there has never been an assessment to determine if it is actually necessary to apply CP. This is exemplified by DNV-RP-0416, Section 4.4.2 of which states that it... *is mandatory that external surfaces of the submerged zone shall have cathodic protection.*

It is interesting that, in this respect, this *recommended practice* document is in the paradoxical position of setting a *mandatory* requirement. There is scope for some confusion because Section 4.16.2.1 of DNV-ST-0126, which is the relevant standard that invokes DNV-RP-0416, refers to... *structures for which a cathodic protection (CP) design is applied.* This wording suggests that the standard, which is superior in the document hierarchy to the recommended practice, envisages circumstances where CP might not be applied.

Be that as it may, and since DNV is likely to be the entity certifying the foundation design, no wind farm corrosion engineer is going to gainsay the imperative in RP-0416. As a matter of routine, therefore, offshore wind turbine foundations are protected externally from corrosion by a combination of protective coating above the waterline and CP below it.

13.2.3.2 CP design – codes

Up until around 2015, most monopile foundations in European waters, and many elsewhere in the world, have been fitted with CP systems designed using the methodologies and guidelines developed within the oil and gas industry, in particular DNV-RP-B401. There is some logic to this since, as with the majority of oil and gas structures, nearly all monopile foundations thus far installed have been protected by sacrificial anodes. However, as we pointed out in Chapter 3, and as reiterated by DNV [53], this recommended practice is intended for jacket-type structures associated with offshore hydrocarbon production, not for monopiles.

In 2016 DNV issued RP-0416 which addresses corrosion in general but which has limited coverage of CP. It bases its approach to sacrificial anodes CP on RP-B401, but with some modifications to account for the particular circumstances of windfarms. ISO 24656 is under development which focusses more on the CP requirements. This includes the innovative approach to calculating the design current densities from metocean[2] data. Based on the supporting publication [62], however, it is not clear precisely how these data can be transcribed into usable design parameters. Both the DNV and the ISO documents also cover ICCP systems for wind turbine foundations. Indeed, there have been some ICCP systems installed. Nevertheless, the vast majority of wind farm CP has thus far been by sacrificial anodes.

13.2.3.3 CP design – challenges

With respect to CP design, there are two fundamental differences between a jacket and a wind turbine monopile foundation. The first is the fact that most offshore wind farm structures are located closer to shore than petroleum production structures. The second is that anodes are distributed differently on a monopile foundation.

[2] "Metocean" is derived from meteorological and oceanographic.

13.2.3.3.1 Nearshore locations

In recent years, much has been made of the fact that waters nearer to shore are subject to greater tidal variations in water depth and, more particularly, higher tidal flows than encountered further offshore. This factor has been linked to the observation that a number of wind farms have been observed to polarise disappointingly slowly.

This coincidence of nearshore location and slow polarisation has been interpreted as indicating that the cathodic current densities given in DNV-RP-B401 for offshore petroleum production structures are inadequate for nearshore monopile foundations.

In response to this perception, a guidance note appended to Section 5.2.2 of DNV-RP-0416 calls for a 50% uplift on design current densities compared with the values recommended in RP-B401. This advice to increase the current provision will inevitably bring about faster polarisation of the monopiles since it obliges the CP designer to increase the current output of the system. In a world where it is the result that matters, this is fine. However, it is probably a case of getting the desired outcome for the wrong reason.

This is illustrated by a recent case that went as far as litigation. The owners of a new offshore wind farm averred that the CP system provided by the design contractor was inadequate. It took a long time, in some instances over a year, for some of the monopiles to polarise adequately over their entire submerged length. One reason advanced for this alleged inadequate CP performance was the insufficiency of the DNV-RP-B401 current density recommendations for the location.

However, this allegation had a couple of flaws. The first was that the wind farm in question was in a similar water depth, and approximately similar distance offshore, as a gas field a modest distance further along the coast. The gas field structures had been protected adequately for many decades using sacrificial anode CP systems designed using less conservative cathodic current densities in force at the time.

More relevantly, instrumental monitoring deployed on the seabed at the wind farm location demonstrated that the protection current densities required were lower than the intentionally conservative figures offered by DNV-RP-B401.

As mentioned, the forthcoming (2021) ISO document has also been developed on the presumption that the design current densities in codes such as RP-B401 are inadequate for nearshore monopiles. It will advise higher design current densities. As noted, it also recommends generating site-specific current densities from metocean data. However, as far as I am aware, important details of just how this should be done are not included in the standard.

13.2.3.3.2 Monopiles and jackets

The more telling reason, in my view, for the observed instances of slow polarisation arises from the fundamental differences in the way in which monopile foundations and jacket structures are designed and installed. The most important of these differences is that, in the case of a jacket, the sacrificial anodes are installed directly onto the structure in the construction yard. This means that the anodes can be distributed more or less uniformly over the surface of the structure as shown in Figure 13.2.

On the other hand, it is not possible to fix anodes to the monopile itself because the shock forces experienced during the piling process would cause the welds between the anode core and the monopile to shear. This means that, in the majority of designs, the anodes have to be fixed onto the small area of the skirt of the transition piece that will be below the waterline.

This, in turn, has four important consequences, each of which is antagonistic in respect of anode performance.

Figure 13.2 Anodes on a frame structure. (Courtesy of Aberdeen Foundries.)

a. Cable connections are required between the transition piece and the monopile.
b. Anodes are positioned close to each other.
c. Anodes are positioned some distance away from some areas of the monopile.
d. There will be a delay between the installation of the monopile and the attachment of its anodes.

We consider these four points in order below.

a. Cable connections

Since the transition piece is usually grouted, rather than welded, to the monopile, it is necessary to install electrical bonding cables to connect the monopile into the CP circuit. As we saw in Chapters 3 and 10, sacrificial anodes are low voltage power sources, so any additional resistance in the CP circuit will supress the levels of protection achievable. Unless these straps are adequately specified, competently installed and inspected periodically through their life, there is a risk that they will degrade in the humid atmosphere inside the monopile and will impose an unforeseen resistance between the anodes on the transition piece and the monopile which they are intended to protect.

b. Anode clustering

In keeping with the general ethos of offshore engineering, there is an incentive to minimise the as-installed weight of the foundation. Since the production of each tonne of steel involves the liberation of about three tonnes of CO_2 into the atmosphere, it would seem counterintuitive to be liberal with the steel on a "green" energy project. One option for trimming weight is by limiting the area of the transition piece skirt that extends below the waterline. A consequence of this is that anodes must be clustered in close proximity to one another.

This proximity of anodes to each other suppresses their current outputs. The closer they are, the greater this degree of mutual suppression. As we discussed in Chapter 10, this is an inevitable consequence of the electrochemical behaviour of the anodes [50]. Unfortunately, however, very few engineers involved in producing CP designs in accordance with recommended practice documents are versed in electrochemistry.

Similarly, the various editions of DNV-RP-B401 have provided little guidance on this topic. This is unsurprising since, as seen in Figure 13.2, anodes on the structures of the type for which RP-B401 is intended are usually well-spaced.

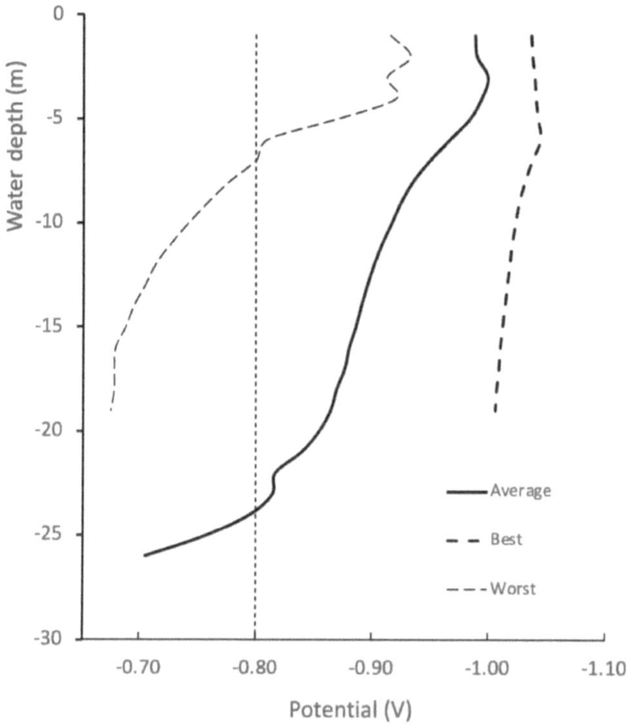

Figure 13.3 Potential versus depth for monopiles.

c. **Anode distribution**

We are aware from Chapter 7 that polarisation is not an instantaneous process, particularly for an uncoated structure. Even on a jacket, with well-distributed anodes, it can take months for the entire surface of the structure to polarise to the target protection potential (−0.8 V). In the case of a monopile, it can take considerably longer for full polarisation to be achieved down to seabed level. This is illustrated in Figure 13.3 which shows the variation of potential versus depth for monopile foundations on a windfarm offshore NW Europe. The potentials were measured by dropping a weighted reference half-cell on an anchored guide wire positioned close to the monopile. The solid black line shows the average curve for the entire population of 88 foundations. There is a clear trend from very negative potentials near the water's surface, with potentials becoming less negative with increasing depth.

This potential attenuation down the pile is simple enough to explain when we recognise that, since the anodes are located on the transition piece, the upper levels of the monopile will initially receive most current and polarise preferentially. Figure 13.3 also shows the individual potential versus depth profiles for the "best" and "worst" polarised monopiles. The best is fully polarised to more negative than −1000 mV down its entire length, whilst the worst is only fully protected (more negative than −800 mV) down to a depth of 7 m.

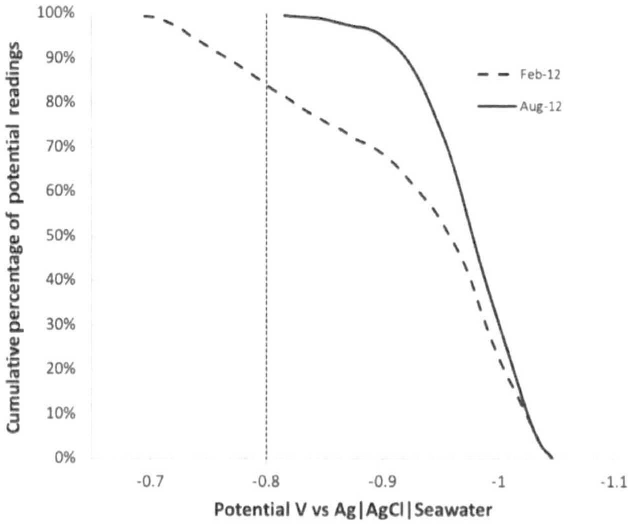

Figure 13.4 Potential versus time for monopiles.

In viewing these potential profiles, it needs to be understood that Figure 13.3 does not depict the final state of polarisation of the foundations. It is a "snapshot" of the situation taken during a single month (February 2012). At that time, the "best" monopile had been installed for over a year and had had time to polarise fully. The "worst" monopile had only been recently installed, and polarisation was just getting underway.

The polarisation was continuing. A repeat drop-cell survey 6 months later revealed that all foundations were fully protected along their entire immersed lengths. This is illustrated in Figure 13.4 which presents cumulative distribution plots for all 2000+ potential readings. The increasing levels of overall polarisation are indicated by the displacement of the distribution curve to the right. All potentials were more negative by August. Furthermore, sample surveys carried out in May of the following year showed a continuing displacement of the curves to the right, indicating the achievement of further polarisation.

d. Delay in anode installation

When a piece of unprotected steel is first immersed in seawater, its potential is observed to be relatively positive (more positive than –300 mV). However, its potential moves progressively towards more negative values as air-formed oxide films on its surface are reduced. Depending on circumstances, and usually within a day or two, a steady-state potential somewhere around −650 mV, becomes established.

If the steel is immersed with anodes attached, the high initial potential difference between the steel (more positive than –300 mV) and the anode (more negative than –1000 mV) means that the actual EMF driving the CP current is very much higher than the "cookbook" value of ~250 mV envisaged in the codes. It follows that the initial cathodic current density on the steel is high. All the available evidence points to high initial current densities being very beneficial in calcareous deposit formation.

In the case of a wind turbine monopile, however, the anodes are only attached after the steel has reached a steady-state corrosion condition (E_{corr}~−650 mV). This means that the initial current densities that the anodes will produce on the steel are lower, so initial calcareous deposit formation will be less pronounced. It is likely that this factor will also delay polarisation of a monopile compared to a structure launched with anodes pre-attached.

13.3 HARBOUR STRUCTURES

13.3.1 Historical background

From the nineteenth century onwards, driven steel sheet piling started to replace stonework, or even wood, as the material of choice for the construction of harbour wharfs. The need to handle deeper draft vessels brought with it the need to extend jetties into deeper water. This caused a change of construction from sheet piling to tubular piles to support decks and walkways.

In as much as any corrosion engineering was applied in the early years, it was assumed that the steel sheet piling was thick enough to tolerate corrosion damage for several decades. For example, BS 6349-1 [40], which is restricted in its scope to temperate waters, quotes a "mean" typical corrosion rate of only 0.04 mm/side/year for immersed steel piling, and 0.17 mm/side/year as a "maximum". This longevity was increased if a bituminous coating was applied to the steel.

The advent of tubular piling on larger jetties coincided with an increased awareness of the corrosion control benefits of installing CP. Humble's 1949 paper [1] had demonstrated the success of CP using magnesium anodes. He found that current densities of 32–65 mA/m^2 could protect at the North Carolina coast. He had previously emphasised the importance of calcareous deposits and had demonstrated the benefits of polarising at high current densities for the first 5 days in order to improve the protective quality of the deposits.

It is informative to compare the differences between harbour piling and offshore structures in so far as they impact CP design. Offshore structures are constructed onshore and transported to site, the environment is usually full salinity seawater, and the sea states can be severe in storm conditions. Harbours, on the other hand, are constructed in-situ. They are frequently in estuarine locations, where the water can vary between brackish and full salinity. Although harbours are likely to have a higher tidal range than offshore locations, they are usually either protected by breakwaters or located so as to minimise wave action. Harbours also tend to be more polluted than offshore locations, although, happily, less so in modern times.

Other differences between harbours and offshore installations that are relevant to CP arise from the manner in which they are operated. The seabed around an offshore petroleum production platform is a no-go area for shipping. A 500 m exclusion zone is typical. This is to avoid subsea hardware being damaged, for example by dragged anchors, and to minimise the risk of a ship impacting the structure. On the other hand, harbours encourage shipping activity. Whilst this should all be carried out in an orderly manner, occasional mishaps, such as vessel groundings, do occur.

Finally, most national legislation requires offshore petroleum production installations to be cathodically protected. For most harbours, CP has historically been optional.

13.3.2 Current densities

The differences in the CP approach to nearshore harbour structures and offshore installations are indicated by the fact that CEN developed separate standards for the two applications: ISO 13174 (formerly EN 13174) for harbours and EN 12495 for offshore installations. However, although the two documents address separate CP applications, there is a very great deal of commonality between the two. Nevertheless, they differ in their recommendations of design cathodic current densities. Both documents offer values for initial, mean and final (or repolarisation) current densities. However, whereas EN 12495 is similar to RP-B401 in selecting values according to geographic location, ISO 13174 advises values that depend on water flow rates, levels of aeration and, significantly, on whether there is a perceived threat of accelerated low water corrosion (see Chapter 1). The ISO 13174 values are set out in Table 13.2.

Table 13.2 Current densities (mA/m²) for harbour structures (ISO 13174)

	Initial		Mean		Repolarisation	
Level of aeration	**Poor**	**Good**	**Poor**	**Good**	**Poor**	**Good**
Seawater immersed (no established ALWC or other MIC)						
Tidal flow <0.5 m/s	80–100	120–150	50–65	65–80	60–80	80–100
Tidal flow >0.5 m/s	120–150	170–300	60–80	80–100	80–100	100–130
Seawater immersed (ALWC or MIC established)						
Any flow	170–200		60–100		80–130	
Burial in mud						
No MIC	25		20			
MIC	30–50		25–30			

At first sight, the current densities in ISO 13174, which are "informative" rather than "normative", seem straightforward enough. However, to make use of these figures in a CP design requires decisions to be made.

- There is no definition as to what constitutes "poorly" or "well" aerated waters. No dissolved oxygen concentration is specified. However, since most harbours are tidal, the CP designer is likely to lean towards conservatism and to assume that tidal exchange will maintain aeration.
- Similarly, there is no guidance on whether the cut-off rate of tidal flow (0.5 m/s), which is just under 1 knot in nautical parlance, is an average value over the tide cycle, or the maximum rate that might be encountered. Broadly speaking, for harbours protected by breakwaters, flows in excess of 0.5 m/s are unlikely, but they are likely where piers and jetties are in estuarine waters.
- The standard also requires the CP designer to make a decision about whether ALWC, or other forms of MIC, are established; and higher current densities need to be provided. This means that arguing for the lower current densities incurs the classical philosophical conundrum of trying to prove a negative. It will be less trouble to simply assume that ALWC may be present somewhere in the harbour and to select the design current density accordingly.
- Finally, the standard gives a range of current densities for each situation but offers no advice as to how to pick a design value within this range. Again, the natural conservatism of CP designers, which is never challenged by CP equipment suppliers, will be to opt for the top end of the range.

Thus, interesting as the guidance figures in ISO 13174 may be, CP designers are likely simply to select 200, 100 and 130 mA/m² respectively for the initial, mean and repolarisation current densities for immersed steelwork. It is probably no coincidence that these are the same figures that DNV-RP-B401 advises for steelwork in shallow (<30 m) temperate (7°C–12°C) seawater.

The same logic will steer designers to values of 50 mA/m² (initial) and 30 mA/m² (mean and repolarisation) for the mud zones.

Similarly, the advice given by ISO 13174 for coating breakdown still leaves the CP designer with some head-scratching to do. This point is illustrated in Table 13.3 which gives the "best-case" and "worst-case" design average and final coating breakdown values for a "high durability" coating. In this context, a "high durability" coating is defined in ISO 12944-1 [40] as having a life to first maintenance in excess of 15 years, although that ISO does not indicate how much breakdown would have to occur before maintenance was required. As can be seen from the table, the designer has a very wide range of coating breakdown figures within which to work. However, since conservatism has always pervaded CP design exercises, it is likely that, for most projects, a pessimistic view of coating performance is likely to be adopted.

Table 13.3 Forty-year coating breakdown – high durability coating (ISO 13174)

Exposure condition	*Best case (%)*	*Worst case (%)*
Immersed	21	42
Buried (pile-driven)	45	90
Buried (back-filled)	25	65

13.3.3 Sacrificial anodes

For many of the small dockside applications involving interlocking sheet piling, sacrificial anodes provide a straightforward answer to providing protection. Anode sleds may be placed on the dock floor, although the designer needs to be aware both of the potential loss of output as a result of silting and the risk of mechanical damage to cabling.

A more robust solution, albeit more time consuming to install, is to fix long sacrificial anodes (see Figure 13.5) in the rebates of the piling profile where they will be physically protected from damage by vessels.

Figure 13.5 Sacrificial anodes for interlocking sheet piling. (Courtesy of Aberdeen Foundries.)

13.3.4 ICCP systems

13.3.4.1 Seabed anodes

Irrespective of the coating breakdown factors and current densities selected for the design, the large exposed steelwork area in harbours often favour an ICCP option over sacrificial anodes. This is further supported by the fact that mains electricity will be available, and most sizeable harbours employ electricians who can take care of the maintenance of ICCP equipment. As we will see when we discuss pipelines in the next chapter, jetty piling in the Arabian Gulf was first protected by ICCP before World War II, albeit that was a side effect of applying CP to a non-isolated pipeline. A brief 1964 paper by Cherry [2] refers to the early impressed current anodes that were in use at the time: including graphite, silicon iron and scrap steel. Platinised-titanium anodes had just been introduced (at Thames Haven in the UK), and there was a growing use of T/R_s with fully automatic control. The use of graphite electrodes persisted into the 1970s, with 2-tonne seabed anodes being used to protect a Brazilian harbour [8] constructed on an artificial island.

The literature (e.g. [24]) contains examples of anodes deployed on the seabed being damaged, or completely destroyed, as a result of vessel groundings or propeller wash, particularly from tug boats. Nevertheless, the arrangement is still used in some instances, not least because it represents the simplest way to provide the large amounts of CP current often required for jetty structures. For example, Huck and Javia [55] have described a system installed in 2016 to protect 247 coated steel piles of a harbour structure in Indonesia. The arrangement comprised six weighted sleds, deployed by cranes from the jetty. Each sled contains two tubular MMO anodes (1.5 m long by 25.4 mm diameter), given each sled a total output of 175 A.

13.3.4.2 Pile-mounted anodes

There are a number of options for mounting ICCP anodes on tubular jetty support piles. The challenge, as with all marine ICCP systems, is to engineer arrangements that offer sufficient robustness. On that basis, it is generally a good idea to ensure that the anodes are orientated inwards, so that they are not at risk from damage by workboats or other vessels. As discussed in Chapter 11, It is also necessary to install an adequate anode shield. To this end, anodes are often pre-assembled in GRP supports that can be readily fastened around the tubular piles by divers.

In the case of sheet piling, the profile of the steel can be used to provide protection for piles mounted on the seaward face. Again, the option would be for the anode elements to be integrated into their GRP anode shields, which are preformed to fit into the rebate of the piling.

13.3.4.3 Conventional "onshore" anodes

Mounting anodes in the rebates of interlocking sheet piling will protect the seaward face of the piling. Usually, this is all that is required. Providing a corrosion engineering assessment has been carried out, and that this has determined that the corrosion of the rear of the piling, from the land-side, is tolerable, then this is all well-and-good. On the other hand, it is not so reasonable to ignore rear-face pile corrosion simply because it would be too difficult to install CP on the land side.

If it is required also to protect the landward side of sheet piling, there are two options. Either a separate anode installation (ICCP or sacrificial) needs to be installed in the soil or backfill on the landward side, or a "deep-well" anode ground bed arrangement is required to protect both the seaward and landward faces of the piling.

A land-based CP system needs to be integrated into the design and construction of a new harbour. If this is done, then future replacement of anodes should be achievable. For example, Davis [39] was able to report a successful like-for-like anode replacement in eight harbour-side deep-well groundbeds that had originally been installed 23 years earlier.

However, once a harbour has been constructed, the piling capped in concrete, and the decking, roadways and other harbour facilities constructed, it is very difficult indeed to gain access to areas on the landward side. This makes it very problematical indeed to install any anodes in the ground on the landward side of the piling.

Where existing groundbeds on the shore-side are not available, it may well be that the most economical way forward will be to install deep-well anode groundbeds on the seaward side. An example of this approach in the early 1990s was described by Tinker [24]. The harbour wall was constructed in 1954 and consisted of interlocking sheet piling. No consideration was given to CP during construction; structural integrity relied on the corrosion allowance. However, with the consumption of the corrosion allowance over the years, and the realisation that the future design life was indeterminate, CP became necessary. Calculations showed that the current requirement would be 1000 A.

The approach adopted was to drill ten deep-well groundbeds immediately in front of the sheet piling using a portable drilling rig. Each groundbed was 61 m deep and contained MMO anode strings sized to give an output of 110 A per string. Thus, an essentially onshore CP technique was applied to a marine location.

13.3.4.4 Harbour structures versus platforms

As noted above, it is generally more convenient to install and manage an ICCP system on a shore-side jetty than on an offshore platform. However, there is one aspect in which the jetty is at a disadvantage. Whereas the offshore installation is a single continuous cathode and is relatively compact, a jetty structure may be discontinuous and can be very spread out. Some jetties are kilometres long. This can present challenges when it comes to delivering the necessary CP current.

Vukcevic et al. [48] describe an interesting approach to dealing with this type of situation on the Swanson Dock in Melbourne. This had been constructed in stages between 1974 and 1988 without, it would seem, any CP. The decision was taken to apply CP to the mix of tubular and sheet piling. With an estimated total current demand of 3500 A, the only sensible option was ICCP. However, there was a design requirement that no hardware could be located on the top of the deck. This meant that all cabling had to be led under the deck, and, importantly, the *T/R* units could only be installed at the extremities of the wharfs. This meant that some anodes would have to be up to 600 m from the *T/R*. With conventional *T/R* outputs of ~20 V DC, anode feed cables would have had to have been up to 120 mm^2. This in turn would have required 36 tonnes of copper conductor. Aside from the cost of the cable, the challenges of physically running this under the deck would have been particularly onerous.

They circumvented this problem with a *current multiplier system*. The anode side current from the *T/R* was delivered at 300 V DC, with the consequent lower current permitting much lighter gauge anode feed cabling. On the under-deck immediately prior to the anode, the voltage was stepped down to 10 V DC enabling a maximum feed of 45 A to each MMO rod anode.

13.3.4.5 How not to do it

A 2015 example of anode failures in a harbour ICCP system makes it clear that the ICCP lessons of the 1970s have long since been forgotten by some. The case involved a jetty ICCP system [52]. The anode cables were set in cantilevered tubes that transited the splash zone. Most failed within a year. The authors' review of the design point to the lack of protection of the tubes. They attributed the rapid corrosion to the combined effects of the severe splash zone environment, combined with the effect of chlorine generation at the anode. Both causes are reasonable.

However, the authors did not consider the vulnerability to fatigue of the style of anode support. More interestingly, they made no mention of the clear fact that the anodes were installed near the mid-tide level. So, they would have spent about half of their (short) service lives not delivering any CP current at all, but with their support tubes corroding!

13.4 ALLOWANCES FOR CURRENT DRAINAGE

13.4.1 Simple rules

Sometimes, we cannot be sure if the structure we wish to protect will be connected, intentionally or otherwise, to other structures. There are two simple rules.

- If connection to another structure would be undesirable, then assume that such a connection will exist and design the CP system accordingly.
- If we intend to supply CP from our host structure to an ancillary component, that might or might not be connected to it, then we must assume that no adventitious connection will exist. We must either ensure electrical bonding to the cathodically protected host or, usually more sensibly, ensure a separate and sufficient CP provision for the ancillary component.

We now turn to the subject of metalwork that will drain current from our CP system, even though it may not be our intention to protect it.

13.4.2 Well casings

The only reason the oil industry has installed platforms in the sea is to enable the extraction of hydrocarbon fluids. This requires wells to be drilled and pipe to be installed into the drilled holes. The outer pipe in any well is termed the casing. Its main function is to prevent the well bore from collapsing. When the well has been drilled, part of the "well completion" activity is to fill the annulus between the casing and the well bore with cement. If this has been done effectively, the external corrosion rate of the casing will be too low to be of any practical concern.

Thus, offshore structure CP systems are not designed to protect the well casings. As explained by Gartland and Bjønaas [35], CP... *is frequently applied to well casings of onshore wells to prevent detrimental corrosion and leaks...* but... *subsea wells are not defined as critical objects that require cathodic protection.* So, it is rarely intended to apply CP to wells. Nevertheless, a well casing will be earthed to its wellhead and tree, and thence to its host structure. As we have already discovered in this book, it is one thing not to need CP current to go somewhere but quite another to stop it going there. In practice, therefore, some of the current intended for the structure will inevitably find its way onto the subsurface well casing. It follows that any CP design for a drilling platform needs to include an allowance for current train to the wells.

Traditionally, this allowance has been arbitrarily assigned. A current allowance of 5A per well is recommended by DNV-RP-B401. This figure is echoed by Norsok M-503 for wells drilled from a platform. However, it recommends 8 A per well for a subsea well for the first 30 years, reducing to 5 A per well thereafter. No explanation is given for this difference in recommendation between the two documents, or for the abrupt change in requirement after 30 years. Indeed, it is hard to see where any of these figures arise. They appear to have evolved historically, more as a result of conservative committee consensus than on the basis of any site investigations.

Where measurements have been attempted, lower figures have been reported. For example, Thomason et al. [27] estimated a drain of 0.6 A per well in the Gulf of Mexico. They arrived at this figure by recording the change in output of a single monitored anode on a newly installed small platform. They assumed that the output of the monitored anode was typical of the other 67 on the structure. On this basis, they were able to estimate the overall increase in current demand as a total of 13 wells were drilled, and casings were installed. They recommended that a current drainage allowance of 1 A per well was sufficient for this area, indicating that the DNV and Norsok codes are, as ever, conservative in their recommendations.

Up until the end of the 1970s, unlike other areas, external casing leaks due to corrosion were something of a problem in the Arabian Gulf. Hamberg et al. [17] suggested that this was due to... *multiple casing strings and more elaborate cementing programmes often found*... elsewhere. The susceptibility of early offshore well casings in the Arabian Gulf to corrosion leaks, probably as a result of cementing problems, has meant that wells are not simply regarded as a current drain but as parts of the structure that need to be protected. On platforms equipped with ICCP, some installations use pulsed current to throw protection further down the wells. On small wellhead structures that did not have electrical power, large block-type magnesium anodes [46] were tried with the object of providing additional current to the well.

13.4.3 Other buried steelwork

Whether or not they have wells, most items of hydrocarbon production hardware, from major platforms to remote manifold structures, positioned on the seabed will have steel designed either to prevent sinking into the sediments or lateral movement due to tides and currents. This steelwork, including piles and mud mats, is usually of sufficient thickness that corrosion will be inconsequential within the project lifetime. Nevertheless, these components, which are invariably uncoated, will drain current from the host CP system. It follows that the CP designer has to incorporate sufficient additional current capacity to accommodate the current drains. The codes provide some assistance in this matter. For example, the DNV-RP-B401 (2021) recommends... *a design current density (initial/final and average) of 0.020 A/m²... irrespective of geographical location and depth.* This is the same as for Norsok M-503. The advice has changed little since 1981 when DNV TNA 703 recommended 0.025, 0.020 and 0.015 A/m^2 for the initial, mean and final current demand respectively. These figures are also replicated in EN 12495. NACE SP 0176 states that typical... *mud zone protective current densities are 11 to 33 mA/m²* but provides no indication of how to settle on a design value within this range.

Interestingly, but unhelpfully, none of these standards suggest that there will be any reduction in current demand with depth of burial. In view of the overall lack of research on what currents are actually taken by steelwork in mud, and at what depths of burial, there is little else that the authors of these codes can offer in the way of advice. Once again, as is their wont, the codes are conservative.

13.4.4 Concrete reinforcement

Reinforced concrete structures have been in service offshore since World War II when the so-called Maunsell forts were placed in the estuary of the River Thames. After more than seven decades, some of the atmospherically exposed parts of these structures are collapsing. However, the underwater sections remain intact. This points to the fact that steel reinforcement that is provided with an adequate cover good quality concrete is remarkably resistant to corrosion when immersed in seawater.

It is little surprise, therefore, that offshore design engineers turned to reinforced concrete as a material of construction for some of the larger offshore petroleum production structures. In the North Sea, Mobil's Beryl "A", installed in 1975, was the first "condeep": an acronym coined from **con**crete **deep**water structure. Some notable other condeeps, such as Statfjord "B" and "C", and Ninian Central followed in due course. Even though these structures did not require CP in their own right, some CP nevertheless remained necessary. As Saetre and Jensen [12] pointed out in 1982, the aim of CP was not primarily to protect the rebar. It was required to protect external steel appurtenances which may have been in contact with the rebar. They observed that it was necessary for the CP design for these appurtenances to include an allowance for the current drain to the rebar. However, although research was in progress in Denmark and Norway, they were not able to offer a view at that time on what this allowance should be.

Ultimately, the research bore fruit, to the extent that standards now offer current drainage values for concrete reinforcement. The 1986 edition of DNV-RP-B401 contained the advice that... *an average current density of 0.5 to 1 mA/m² for the outer reinforcement layer is normally sufficient*. By the 1993 edition, this advice had changed to include different current densities for different depths and seawater temperature zones. The figures ranged from 0.8 mA/m² for cold deep water to 3 mA/m² for near-surface tropical waters. Editions from 2005 onwards have advised very slightly lower current densities as indicated in Table 13.4.

Table 13.4 Current densities (mA/m²) for concrete reinforcement (DNV-RP-B401)

	Temperature			
Depth (m)	*>20°C*	*12°C–20°C*	*7°C–12°C*	*<7°C*
0–30	2.5	1.5	1.0	0.8
30–100	2.0	1.0	0.8	0.6
>100	1.0	0.8	0.6	0.6

It should be recognised that these figures relate to the area of the steel reinforcement, not simply to the surface area of the concrete. This may pose a problem for the CP designer, particularly when dealing with older structures where detailed reinforcement drawings are unavailable, or difficult to interpret. In such cases, it is always worth getting the advice of an engineer familiar with reinforced concrete construction.

13.5 CP RETROFITS

13.5.1 What is a "retrofit"?

The prefix "retro-" in English implies "going back". On that basis, we might reasonably define any act of going back to an offshore structure, after it has been launched, to install anodes as a CP "retrofit". By that definition, just about every offshore CP installation up to the early 1960s was a retrofit since the anodes (sacrificial or impressed current) were lobbed over the side after the structure had been launched. By the same token, the routine periodic replacement of those suspended anodes could also be termed retrofits. In a similar vein, the installation of CP on offshore wind farm foundations (Section 13.2) and harbour structures (Section 13.3) are also retrofits. However, in this book, we will restrict the term "retrofit" to the installation of a CP system to a structure after the original system is failing (or has failed) or is approaching (or has passed) the end of its design life.

The first "retrofits" were carried out when the decision was taken to install "permanent" CP systems to existing installations. For instance, in 1968 Shell issued a single contract to fit "permanent" sacrificial anodes to 516 Gulf of Mexico structures. Now, before any reader in the business of supplying CP retrofits starts to salivate at the thought of a contract involving 516 structures, we need to be clear that these were small structures. The majority were single-well jackets, some of which were even made of wood. The objective was limited to providing (some) CP to the well casings. All structures were all in shallow water. The maximum depth was 21 m, and many were at depths of <5 m. Each structure was protected by retrofitting between seven and twenty-four 120 kg Al-Zn-Hg anodes.

I regard this as the earliest documented offshore CP retrofit, although the term "retrofit" is absent from the paper [7] describing the work. The word only gained traction in the 1970s when it was applied to CP installations designed to replace or supplement exiting "permanent" CP systems that were underperforming or had failed. Few if any had reached the end of their intended lives by that time.

Importantly, the retrofit technique used, which involved one or two anodes welded to a clamp that was mechanically attached by divers to a tubular member (Figure 13.6) set a pattern that persisted for some time. The approach was common into the late 1980s, was still the predominant retrofit technique into the mid-1990s [28] and was still used routinely by some operators through into the second decade of the 21st century. For example, Chevron used the approach to supplement a problematic ICCP system on the Ninian North platform [21]. Similarly, a Petrobras paper [20] describes a typical application but mentions in passing that... *the literature anticipates a working life of 4 to 5 years for this type of retrofit system.* Unfortunately, despite invoking "the literature", the authors did not provide any specific references. In my own experience, they understated the longevity of this type of arrangement.

Figure 13.6 Retrofit arrangement for a tubular. (Courtesy of Aberdeen Foundries.)

13.5.2 Information on retrofits

The corrosion literature is not short of references to CP retrofits. A list of some papers is given in Table 13.5. As can be seen, there are documented instances of impressed current systems having been retrofitted with further ICCP or sacrificial anodes, and sacrificial systems supplemented with more sacrificial anodes or ICCP.

In viewing this list of references, we need to be cautious. First, we have to remember that there is a commercial incentive to presenting a conference paper or getting an item published in a recognised journal. Accordingly, we are more likely to see enthusiastic write-ups of impressed current retrofit systems from ICCP equipment suppliers, and offerings in praise of sacrificial anodes from companies with foundries. The same marketing pressures also dictate that any newsworthy retrofit should be written up as soon as possible. Most of the papers cited in Table 13.5 report a recently installed retrofit CP system, invariably regarded by the authors as having been a success. None of these papers tells us whether, or to what extent, the retrofits outperformed the systems they replaced.

Table 13.5 Some retrofit references

Original system	*Retrofit system*	*References*
Impressed current	Impressed current	[9,22,31]
	Sacrificial anodes	[21,30,31]
Sacrificial anodes	Impressed current	[10,18,23,26,34,41,43,45,49,58,60]
	Sacrificial anodes	[4,5,7,9,15,20,29,32,36,38,42,47,54,57,59]

Indeed, I am aware that one of the cited retrofits literally fell off its host structure within a few years of installation. That event was not lauded in the corrosion literature. Commercial entities are understandably reluctant to publicise failures.

13.5.3 Do we need to retrofit?

By far the most economical way of providing CP to an offshore structure is to install the system, either sacrificial anodes or impressed current, in the construction yard. The relative cost of installing CP onto a structure when it is offshore greatly exceeds the cost of doing it on dry land. Nevertheless, there are some reasons why it might be necessary to retrofit CP to a structure after it has been installed offshore. If we ignore the situations where the CP is intentionally retrofitted after a structure is set in place, there are two principal reasons for retrofitting, design mishaps and life extension.

13.5.3.1 Design mishaps

Many instances of design mishaps involved ICCP systems which, particularly in the early days, often lacked the mechanical resilience for extended service in harsh offshore environments. The industry view was summed up by ffrench-Mullen in 1980 [10], saying that the ICCP method was little used because... *in spite of numerous attempts, virtually all impressed current systems have proved to be mechanically unsatisfactory. They have broken down, almost invariably from failure of the cable conduit or cable to anode connections, and a general inability to withstand storm damage.*

Less commonly, there were fundamental design errors in sacrificial anode systems. A rare published example of the latter, from the operator's perspective, is given in a 1997 paper by Goolsby and McGuire [37]. This describes a retrofit carried out on the Gulf of Mexico Cognac structure in the early 1980s. In that case, the designers of the original sacrificial anode system had overlooked the current drain to the well conductors. Similar instances, albeit not so widely reported, also occurred in the early development of the southern North Sea.

Another example of a retrofit-waiting-to-happen was the Hondo platform [42] installed in 1976 in 260 m off the California coast. As early as 1984, it was observed to be depolarising towards the −0.8 V criterion. Such a short life might have seemed surprising until it was realised that, to save installation weight, only a 10-year system had been fitted in the first instance. Furthermore, it transpired that there was more bracing and there were more piles on the structure than had been disclosed to the CP designers. Finally, for reasons that were not divulged, up to 10% of the allocated anodes had not actually been installed. On mature reflection, the CP system's achievement of an 8-year life has to be regarded as a triumph.

13.5.3.2 Life extension

Nowadays, the predominant reason for retrofitting is to extend the life of a CP system beyond its original design target. This has usually been because successful developments in enhanced hydrocarbon recovery, and in exploiting marginal fields, have required the life of the structure itself to be extended.

13.5.3.3 Unnecessary retrofits

All of the above are valid justifications for the expenditure associated with retrofitting CP. However, there have also been some instances where retrofit CP systems have been installed, at considerable expense, which by any realistic analysis were unnecessary. For the most part, these unwarranted retrofits have remained undocumented in the public-domain literature. An exception is the case of retrofits on two platforms carried out by Arco in 1991 [29]. The structures were 19 and 22 years old respectively, and each was required to remain in service for a further 10 years. Divers had estimated that only around 20% of the anodes remained. Thus, with the 80% of the anodes having been consumed in 65–70% of the operating life, there seemed to be a reasonable case that a retrofit would be needed.

The authors were aware that designing the retrofit on the basis of code current densities for new structures would result in an over-design of the retrofit system. Accordingly, they attempted a novel approach of determining the actual current density being taken by the structures. The approach they adopted was as follows.

1. They used divers to make detailed measurements of the dimensions of a representative sample of the existing anodes.
2. The divers also made potential measurements of the anodes (E_a) and of the adjacent steel structure (E_c).

From the anode dimensions, and knowledge of the seawater resistivity, they were able to estimate the anode resistance (R_a) using the modified Dwight equation (which we met in Chapters 3 and 10). This enabled them to calculate the average current output of each anode (I_a) using Ohm's law equation we first met in Chapter 3.

$$I_a = (E_c - E_a)/R_a \quad (13.1)$$

From the known number of anodes (N_a), and the known surface area of the structure, they were able to estimate the cathodic current density (i_c).

$$i_c = I_a N / A \quad (13.2)$$

The results were surprisingly low: 3.4 and 8.8 mA/m^2 respectively for the two structures. They stated that... *the only explanation for these high*[3] *structure potentials and low current demand was the presence of very thick calcareous deposits, which must have developed late in the life of the structure. Evidently the long exposure time and chemistry of the water combined to produce a very protective deposit layer.*

[3] In common with many CP practitioners, the authors used "high" in the context of "highly negative".

This low measured current demand meant that the remaining 20% of the original anodes would be more than sufficient to see out the remaining 10 years of operation. On that basis, one might reasonably have expected them to shelve the plans for the retrofit. However, that is not what happened. Instead, the company went ahead with the retrofit, which it designed on the basis of a current density of 32 mA/m^2 (seawater) and 5 mA/m^2 (mud).

Thus, a technically competent operator elected to go ahead with an offshore CP retrofit even though it was in possession of relevant data from its own structures that demonstrated that the retrofit was unnecessary. This was above, and beyond, accepting the conservatism inherent in CP design codes. This was simply a waste of money.

That it occurred three decades ago might allow us to dismiss it as a historical anomaly. However, it is not an isolated case. Within the last 5 years, I have been involved in cases where other technically competent operators have squandered multiple-millions of dollars on retrofits, even when their own CP inspection data demonstrated that no retrofit was required.

13.5.4 Retrofit requirements

Unlike the cases mentioned above, most retrofits are actually needed. The designers must then come up with a retrofit system that is robust, so avoiding the need for a future retrofit of the retrofit, and which provides the appropriate current and current distribution to the structure. More to the point, the dominant cost of any offshore retrofit will not be the cost of the CP hardware, be it sacrificial anodes or impressed current. It will be the cost of the offshore installation work; in particular, the cost of any diving or ROV time that will be needed.

Thus, any offshore retrofit design is arrived at as a result of a two-pronged approach. First, the current demand is estimated. Then the least expensive method of installing that current capacity offshore is sought.

13.5.5 Current demand

As we saw when we carried out some routine CP designs in Chapter 3, the preliminary task of CP system designers is to estimate the cathodic current demand of the structure. For new installations, they are guided by codes such as DNV-RP-B401 in the allocation of initial, mean and final current demands. These codes are intentionally conservative. As has already been pointed out, this conservatism for new structures is easily justified. The cost of CP installed on dry land is usually modest, particularly when judged against the capital cost of the structure. A smidgeon of over-design is, therefore, justified. It can be thought of as insurance against system underperformance and, in the extreme, a future need for an expensive offshore retrofit.

Once we start to contemplate retrofitting CP offshore, the cost per amp of installed current capacity increases dramatically. However, at the time of writing (2021), there are no recognised codes for designing retrofits. This leaves the designers with problems in selecting the design current densities. In some instances, design houses have circumvented the lack of relevant codes by simply opting for the values in existing codes for new structures.

For example, a 1987 Amoco paper [15] describes the CP retrofit design for four platforms originally installed in 1973 and protected by Al-Zn-Sn anodes. Interestingly, the designers... *did not attempt to take credit for the existing calcareous scale, but used the initial values. This has resulted in a more conservative design.* This approach is arguably naïve and unnecessarily expensive. This is illustrated by the much more recent case of a retrofit programme for a population of platforms in the Arabian Gulf that we discussed in Section 7.7.2.

13.5.6 Retrofit strategies

13.5.6.1 ICCP vs sacrificial anodes

Having decided how much current provision needs to be retrofitted, the next decision is whether it should be provided with sacrificial anodes or an ICCP system. In some cases, the decision requires no analysis since the scale of the retrofit forces the ICCP option. For example, the recent life extension of the North Cormorant platform [58,60] involved the retrofitting of over 7000 A current capacity. This was challenging enough for the ICCP system adopted but would have been wholly impracticable with sacrificial anodes.

In addition to determining how much retrofit current is needed, and how it is to be provided, there is the additional issue of current distribution. This becomes increasingly problematic as the structures to be retrofitted become larger and of more complex geometry. There is also a measure of mutual exclusivity between the desire to minimise installation costs, by installing a small number of high current sources, and the objective of uniform current distribution, which is favoured by multiple smaller current sources. As we have explained, CP does not obey "line-of-sight" rules. That is not to say, however, that a few high-powered current sources will give an even current distribution and uniform levels of polarisation. The reality is more subtle. This is fertile territory for computer modelling, to which we return in Chapter 17.

A key driver in the design of any offshore CP retrofit is the requirement to minimise the installation cost. This, in turn, means minimising the amount of time for which divers and support vessels are needed. Other things being equal, the higher the current requirement, the more likely that ICCP will win out over sacrificial anodes.

In the real world, however, "other things" are very rarely "equal". This means that there is no simple answer to the selection of ICCP, sacrificial anodes or a mixture of the two when it comes to retrofits. Any retrofit project should begin with an independent conceptual study to determine:

- if a retrofit is actually needed (and, if so when by),
- the current demand and distribution, and, crucially,
- whether the system should be impressed, sacrificial or hybrid.

A full ICCP retrofit might well be the technically desirable choice. However, it could only work if there is sufficient spare electrical power available, and enough space to install the DC power supply and control units. Similarly, any decision to opt for impressed current on a normally unmanned installation would necessitate a commitment to telemetric remote control and monitoring. Thus, even though ICCP might often seem an obvious choice, the "fit-and-forget" attributes of sacrificial anodes might hold sway.

13.5.7 Retrofit implementation

13.5.7.1 Sacrificial anodes

13.5.7.1.1 Sleds and pipes

Rather than crane single or double anodes onto a structural member and rely on divers to complete the mechanical installation, it is usually much more efficient to prefabricate sleds containing a rack of multiple anodes and deploy those sleds around the periphery of the target structure. These are then cable connected back to it. Figure 13.7 illustrates a small example of such a sled.

Figure 13.7 Retrofit sacrificial anode sled.

This approach, which remains widely used, has not been without its problems. For example, designers of sleds have failed to account for the mutual suppression of outputs that arises from placing anodes close together. We discussed this in Chapter 10. I have encountered several instances of CP designs that assumed that placing *N* anodes on a sled would give *N* times the output of a single anode. Occasionally, and even quite recently, the disclosure of overly congested sleds has found its way into the literature. In 1997, Brandt published an example of a sled with particularly closely spaced anodes [32]. However, the inclusion of some magnesium anodes in the array would have gone some way to boosting the instantaneous current output. More recently Agel et al. [57] described a particularly congested double-banked array of closely spaced anodes which, unsurprisingly, yielded particularly disappointing outputs.

In addition to the anode clustering issue, I have also encountered cases where the designers had evidently not considered the effect of the resistance of the cable connecting the sled to the structure. This omission was explained by the fact that DNV-RP-B401 explained how to calculate the anode to seawater resistance (R_a) for a single anode, but made no mention of any other circuit resistances.

A variation of the sled, which has been used in a small number of retrofit applications, is the "pipe anode" (Figure 13.8). This simply involves installing bracelet anodes on a length of coated line-pipe. The anode resistance is usually computed from the McCoy formula which we encountered in Chapter 10, with the anode length taken as the aggregate length of the bracelets. As far as I am aware, such anode pipes have been considered for offshore structures [42] but have not actually been installed.

The approach has some advantages in simplicity of manufacture, but its design renders it likely to become buried in seabed sediments. The example shown in Figure 13.8, which was installed on a pipeline, uses magnesium anodes.

13.5.7.1.2 Pods

In addition to the suppression of anode outputs that results from placing anodes in close proximity on a sled, there will be an added problem should the sled eventually become buried in seabed sediment. In the case of Al-alloy anodes, there will be a slight positive shift in the anodes operating potential (E_a), thereby reducing the voltage ($E_c - E_a$) available to drive the CP current. More important, there is likely to be at least a four-fold increase in the environment resistivity, and a corresponding further drop in the (already low) sled current output.

Figure 13.8 Retrofit "pipe anode".

Redesigning the sled into a pod-type arrangement (Figure 13.9) goes some way towards increasing the spacing of the anodes and, to a large degree, reduces the possibility of the anodes being silted over.

These sacrificial anode pod arrangements are now widely used for offshore retrofits. However, despite their advantages over congested flat sled arrangements, even anode pods are likely to prove inadequate to satisfy retrofit current demands of the larger structures. The number of pods that would need to be installed on the relatively small area of vacant real estate around the base of a major platform is often prohibitive. For example, the above-mentioned North Cormorant retrofit would have required the installation of over 600 gross tonnes of aluminium alloy anodes [58]. That would have needed an awful lot of pods!

Figure 13.9 Retropod®. (Drawing courtesy of Deepwater Corrosion Services Inc.)

13.5.7.1.3 Strings

Any arrangement in which anodes are simply deployed over the side of the structure is reminiscent of the pioneering days of the 1940s and 1950s. Nowadays, its use with sacrificial anodes is mainly limited to the internal protection of wind farm monopile foundations. We will return to this in Chapter 16. It has also been used occasionally as a technique for (temporarily) attaching magnesium anodes to accelerate the polarisation of the external surfaces of monopiles. In these instances, and in contrast to the early days when anodes were suspended by their copper connecting cables, the anode weight is taken by polypropylene rope to minimise the risk of the cable failure.

13.5.7.2 Impressed current

By contrast, the string deployment of anodes has been more rewarding when applied to impressed current anodes. There were some early applications in which platinised-titanium rods were deployed on strings However, the technique gained considerable traction with the development of tubular titanium anodes coated with electrochemically active mixed metal oxide (MMO) coatings (see Chapter 11). These materials permitted multiple anodes to be sealed onto a single insulated copper conductor. The first application was termed LIDA®, an acronym for linear distributed anode.

This type of anode arrangement is suited to deployment as a string anchored to a deadweight on the seabed. An early reported example of this retrofit approach was in 1987 where it was used to replace the original (1976) sacrificial anode system on four platforms in 85 m of water in the Loango field offshore the Congo [18]. The system was referred to as "the tensioned string" concept. It comprised ten strings, each consisting of eleven 1 m long × 25.4 mm diameter MMO tubular anodes. Nine strings were around the periphery and one down through the centre of the jacket. Each string was secured to a tensioned polymer-coated steel hawser, which was secured to a concrete block on the sea floor. Six power cables were used to feed the anodes on each string (five supplying a pair of anodes, and one supplying a single). Each anode was rated at 55 A. The paper unsurprisingly concludes that the retrofit exercise was successful.

Examples of other tensioned anode string type retrofit applications are to be found in the literature [22,23,34]. All were reported as being successful. This illustrates a relevant point about technical papers relating to all retrofits, and to CP systems more generally. The lead authors are usually representatives of the system suppliers. They inevitably, and entirely understandably, have a commercial motivation for extolling the virtues of their employers' systems. There are plenty of conference papers that we can bowdlerised as follows: the original CP system failed or expired, we installed our proprietary retrofit system, and it works really, really well.

What is missing from oeuvre are the reports of how the retrofits are performing 10 or 20 years down the line. The nearest we find to this is a note on the Loango retrofit. We are told [26] that after 5 years of service, 98% of the anodes were still working successfully and providing complete protection to the structures. It is a moot point whether or not 98% is a good "score" as far as retrofit anodes are concerned. Clearly, the loss of a couple of anodes per platform was not a problem as far as the levels of protection were concerned. However, if these anode failures were just the first indicators of a more profound system vulnerability, then it might suggest that, at 5 years, the retrofit was well into its service life. We simply do not know because there is a lack of follow-up publications. Based on offshore experience more generally, there must always be a concern about running power cables, with little or no mechanical protection, through the splash zone.

Inevitably, the issue of running cables through the splash zone presents a challenge to all retrofit ICCP systems. Unless a means of physical protection is provided, as was the case, for example, when steel conduits were welded to the Murchison platform in the construction yard, the life expectancy of an ICCP retrofit is likely to be limited.

Fortunately, there have been some instances where a platform has unused J-tubes integrated with the structure. If present, these J-tubes offer an opportunity for pulling-in cables that have been factory-connected to remote ICCP sleds. This means that no connection needs to be made underwater, and the cables are protected through the splash zone. If an available J-tube does not exist, then consideration should be given to installing one if a long-life ICCP retrofit is required.

Assuming the cabling issues can be securely resolved, remote sleds offer a (relatively) inexpensive option for installing ICCP retrofits. Numerous sled designs have been tried. In recent years, all of these have been based on MMO anodes, mostly of the tubular variety. Britton [41,43] has pointed out the benefit of attaching the anode to underwater buoys so that they are orientated vertically above the sled. This greatly reduces the prospect of their becoming silted over. An example of this arrangement, known as a Retrobuoy®, is shown in Figure 13.10. This shows the equipment complete with frame and cable. The MMO anodes can be seen along the lengths of the floats which will hold them free of sediment when the unit is deployed.

13.5.7.3 Connections

The US Navy's Atlantic Underwater Test and Evaluation Center (AUTEC) is located in a deepwater area near the Bahamas known as the Tongue of the Ocean. In 1962 Waldron and Peterson, two pioneers of marine CP, attempted to protect the newly laid AUTEC deepwater mooring cables using magnesium anodes [3]. However, when the moor was inspected in 1964, divers found that all of the magnesium had disappeared. Replacement anodes were installed in 1965, but the mooring cables failed in 1966. Groover examined some recovered failed sections the following year and remarked that they had suffered from *serious corrosion* [4]. As a result of this investigation, he was obliged to conclude that the installation of the replacement anodes... *was accomplished by divers who apparently were not aware of the technical requirement of providing electrical contact between the anode and the moor.*

Despite his restrained choice of words, we can feel Groover's exasperation. This early and embarrassing CP failure highlights the point that the expense involved in any retrofit is wasted if the connections fail.

Figure 13.10 Retrobuoy®. (Courtesy of Deepwater Corrosion Services Inc.)

In the early days of offshore CP retrofits, anodes were installed individually or in pairs, and their cores fixed to the structure by underwater arc welding. This technique was pioneered by Khrenov in the Soviet Union in the 1930s. Surprisingly to non-welders, the technique works reasonably well. It remains quite widely used. However, it is not without its practical problems, and ensuring weld quality has never been easy.

Quite early in the history of offshore retrofits, the technique was superseded by mechanical connection. Typically, a short length of structural member, usually a horizontal bracing, was water-jetted to remove fouling. Sacrificial anodes mounted singly or in pairs, on a hinged steel carrier, were then lowered from the deck and clamped in place by divers. The anode-to-structure connection sometimes relied on the clamping of the frame. More usually, however, the frame included a simple volcano bolt contact. This was tightened by the diver to produce a more secure electrical connection.

As the scale of retrofits increased, it was no longer efficient to install numerous individual sacrificial anodes on the structure itself. For impressed current retrofits, there was no need to make an electrical connection to the structure below the waterline. However, with sacrificial anode pods or sleds, a direct connection is required between the anode array and the structure. There are two options:

a. the anode cable can be brought above the waterline and the connection made to the structure at a convenient topside location, or
b. it can be led to the nearest convenient point on the structure, and the connection made underwater.

Option a) has the advantage that the connection can be made under near-ideal conditions, and it remains inspectable and maintainable. However, it requires a cable to be routed through the inter-tidal and splash zone. The need to ensure mechanical protection in the zone is well established in the offshore CP industry. More to the point, this approach incurs the disadvantage that the cable adds considerably to the ohmic resistance between the anodes and the structure. In some sacrificial anode CP designs, it has proved necessary to switch from Al-alloy anodes to magnesium to achieve the voltage necessary to drive the CP current against this resistance.

Option b) mitigates the problem of the electrical resistance of the longer cable lengths, and it obviates the issue of protecting vulnerable cables transiting the splash zone. However, it retains the historic requirement to make an underwater connection to the structure. To this end, mechanical devices are available to provide positive clamping to a cylindrical structural member. An example is shown in Figure 13.11.

As an alternative, a threaded stud can be friction welded to the structure and cable lugs attached by bolting. Underwater friction welding is not particularly new. It was described for the purpose of fixing anodes in 1984 [14]. The technique involves a stud being rotated at high speed whilst being held at pressure against the steel structure. The friction created generates heat at the metal surfaces which causes plastic flow. The surface impurities are expelled by the rotational forces, and the two surfaces are forged together as a result of the compressive force. The process is usually completed in under 10 s. The temperatures involved are lower than those involved in fusion welding. The two metals being friction welded do not actually melt.

There have recently been some successful applications of friction welding to monopile foundations of offshore wind turbines, for which large-diameter encircling clamps would be impracticable. Although the details are not yet in the public domain, I am aware that the results have been entirely satisfactory.

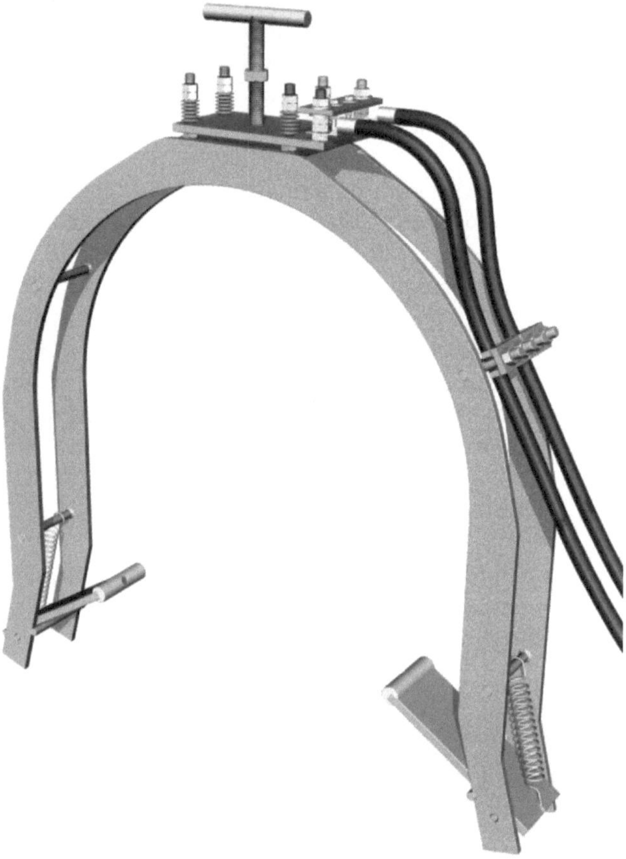

Figure 13.11 Retroclamp®. (Courtesy of Deepwater Corrosion Services Inc.)

13.6 THE FUTURE

It has been no easy task to try to sum up the essential history of the CP of offshore structures and to catalogue some of the lessons that have been learned on the way. It is an even taller order to attempt to predict the way this element of the CP industry will develop in the future. Nevertheless, I think that we can take an educated guess. It is reasonable to assume, for example, that the future will see far fewer massive offshore steel structures dedicated to the recovery of hydrocarbons. Nevertheless, although the world is weening itself off petroleum, its final demise is still some way over the horizon. Even if its contribution to satisfying the global energy demand diminishes, its role as a petrochemical feedstock for manufacturing a vast range of products from plastics to pharmaceuticals is destined to persist. This means that hydrocarbons will still need to be produced offshore. It follows that there will be a need to protect new equipment and to keep existing structures serviceable beyond the ends of their design lives. Indeed, some existing installations may need to be repurposed for new roles in carbon capture.

We can also assume that world's marine trade will continue to grow, so existing harbours will need to be kept operational, and new harbour facilities will continue to spring up. It is hard to see a change from the dominant construction practice of using steel piling.

Similarly, it is hard to see the trend in offshore wind farm development, illustrated in Figure 13.1, taking a downturn anytime soon. Again, it is also difficult to picture steel being replaced as the material of choice for the turbine foundations. However, we may see a move away from monopiles. These require a considerable mass of steel, not least in the wall thickness needed to accommodate the piling loads. As already noted, steel production is a major contributor to atmospheric CO_2. It is ironic that an industry dedicated to reducing CO_2 production through its operations remains rather profligate when it comes to its construction work. There is some blue skies thinking going on regarding how to construct lighter weight foundations. For example, Lomholt et al. [61] describe a new jacket-type foundation concept. This is based on coated standard line-pipe joints. The mass-produced joints are assembled to form structure subcomponents which, in turn, can be bolted together at the dockside to produce a complete lattice frame foundation. If this potentially cost-saving idea gains traction, it will require some new approaches to CP, but this should prove less problematic than the monopile foundations presently in vogue.

Elsewhere, promising new technologies such as tidal and wave power may contribute usefully to future energy generation. Some of these installations may have to be integrated into the new coastal defence structures that will be needed to defend against the rising sea levels that we must expect as a result of our previous exploitation of hydrocarbons. Other technologies, such as subsea mining, are just about visible over the horizon.

What we can reasonably predict is that a very large slice of the infrastructure needed for these future projects will be in carbon or low alloy steel. Corrosion control will be required and cathodic protection will feature as a part of that. If history teaches us anything, the past debates about impressed current versus sacrificial anodes, and between coated and uncoated steelwork, will rerun for new projects. It is also a safe bet that the historical lessons learned will, once again, be forgotten.

REFERENCES

1. H.A. Humble, *The cathodic protection of steel piling in sea water.* Corrosion **5**, 294 (1949).
2. P. Cherry, *Cathodic protection of jetties*, Proceedings of Symposium *Recent Advances in Cathodic Protection*, Marston Excelsior Ltd. (Wolverhampton) 1964.
3. L.J. Waldron and M.H. Petersen, *Unique cathodic protection system for a deep sea moor.* Materials Protection **4** (8), 63 (1965).
4. R.E. Groover, *Analysis of the failure of the AUTEC TOTO II deep sea moor and the performance of its cathodic protection system*, NRL Memorandum Report 1950 (1968).
5. J.A. Burgbacher, *Cathodic protection of offshore structures.* Materials Protection 7 (4), 26 (1968).
6. W.L. Berry, *Cathodic protection of deep water platforms with impressed current*, Paper no. OTC 1273 Offshore Technology Conference, 1970.
7. E.P. Doremus and R.B. Pass, *Cathodic protection of 516 offshore structures: engineering design and anode performance.* Materials Protection & Performance **10** (5), 23 (1971).
8. A.C. Dutra, *Corrosion prevention of an offshore Brazilian Artificial Island.* Materials Performance **16** (6), 9 (1977).
9. B.S. Wyatt, *Design of retrofit cathodic protection systems for deep water offshore platforms*, Proceedings of International Symposium *Corrosion and Protection Offshore* CEFRACOR, Paris, 1979.
10. T. ffrench-Mullen, *An impressed current system for the protection of offshore platforms.* Materials Performance **19** (11), 15 (1980).
11. W.M. McKie, E. Levings and R. Strømmen, *Muchison cathodic protection system*, Proceeding of UK Corrosion, 1982.
12. O. Saetre and F. Jensen, *Developments in cathodic protection of offshore concrete structures.* Materials Performance **21** (5), 30 (1982).
13. R.M. Vennett, R.W. Seager and M.A. Warne, *Design of cathodic protection system for Conoco's North Sea Murchison Platform.* Materials Performance **22**(2), 22 (1983).

14. E.D. Nicholas, *Underwater friction welding for electrical coupling of sacrificial anodes*, OTC 4741 Offshore Technology Conference, 1984.
15. J.S. Smart III, *Optimized design of anode retrofits on offshore platforms*, Paper no. 61 CORROSION/87.
16. A.F. Garner, *The synergism of coatings and cathodic protection of fixed offshore structures*, Proceedings of UK Corrosion, 1988.
17. A. Hamberg, M.D. Orton and S.N. Smith, *Cathodic protection of offshore well casing.* Materials Performance **27** (3), 26 (1988).
18. R.L. Cabe et al., *Impressed current cathodic protection systems of offshore structures*, Paper no. 285 CORROSIO/89.
19. R.E. Lye, *A corrosion protection system for a North Sea Jacket.* Materials Performance **29** (5), 13 (1990).
20. R. de Oliveira Vianna and C.R. da Fonseca, *Anode replacement on offshore platforms by Bracelet attachment.* Materials Performance **29** (5), 19 (1990).
21. W.J. Cochrane et al., *A computerized CP retrofit design of the Ninian Northern platform*, Paper no. 375 CORROSION/90.
22. C.J. Mudd, J.E. Flake and W.B. Wilder, *Offshore platform cathodic protection design and installation*, Paper no. 382 CORROSION/90.
23. J. Edwards et al., *Retrofitting offshore structures with tubular metal oxide activated titanium anodes*, Paper no. 240 CORROSION/91.
24. W.A. Tinker, *Cathodic protection for steel bulkheads under relieving platforms.* Materials Performance **30** (11), 30 (1991).
25. R.E. Lye and F.Ø. Jensen, *Cathodic protection with reduced wave load features for an offshore steel jacket.* Materials Performance **31** (4), 20 (1992).
26. M. Tettamanti, U. Caterini and E.G. Field, *Cost effective impressed current cathodic protection systems for retrofitting of offshore platforms*, Proceedings of UK Corrosion, 1993.
27. W.H. Thomason, K.P. Fischer and I.J. Rippon, *In service measurement of the performance of a CP/coating system on a Gulf of Mexico jacket*, Paper no. 516 CORROSION/93.
28. J.A. Brandt, *Optimized retrofitting of offshore cathodic protection systems*, Paper no. 518 CORROSION/93.
29. M.W. Mateer and K.J. Kennelley, *Designing anode retrofits for offshore platforms.* Materials Performance **33** (1), 32 (1994).
30. J. Jelinek et al., *Current density surveys for optimizing offshore anode retrofit design.* Materials Performance **35** (4), 19 (1996).
31. S.P. Turnispeed, *Offshore platform cathodic protection retrofits.* Materials Performance **35**(10), 11 (1996).
32. J.A. Brandt, *Sacrificial anode retrofits: innovative solutions for the challenge of deepwater and other complex marine environments*, Paper no. 479 CORROSION/97.
33. T. Sydberger, J.D. Edwards and J.B. Tiller, *Conservatism in cathodic protection design.* Materials Performance **36** (2), 27 (1997).
34. M.L. Smith and C.P. Weldon, *Impressed current tensioned anode strings for offshore structures.* Materials Performance **37** (4), 19 (1998).
35. P.O. Gartland and F. Bjønaas, *Computer modelling studies of current drain to subsea wells in the North Sea.* Paper no. 555 CORROSIO/98.
36. S. Rossi, P.L. Bonora and M. Draghetti, *Cathodic protection revamping technology for offshore structures: the Agbarra platform.* Materials Performance **37**(3), 15 (1998).
37. A.D. Goolsby and D.P. McGuire, *Cathodic protection upgrade of the Cognac platform.* Materials Performance **37** (4), 13 (1998).
38. J.H. Kiefer, W.H. Thomason and N.G. Alansari, *Retrofitting sacrificial anodes in the Arabian Gulf.* Materials Performance **38** (8), 24 (1999).
39. J.G. Davis, *Long-term evaluation of CP for interlocking sheet pile bulkheads.* Materials Performance **39** (3), 38 (2000).
40. BS 6349-1, *Maritime structures: code of practice for general criteria*, 2000.

41. J. Britton, *Cost saving offshore cathodic protection retrofit methods*, Proceedings of UK Corrosion, 2001.
42. S.N. Smith, *A new approach to the design of retrofit sacrificial anodes,* Paper no. 01504 CORROSION/2001.
43. J. Britton, *Impressed current retrofits on offshore platforms the good, the bad and the ugly,* Paper no. 01505 CORROSION/2001.
44. S.N. Smith and H.R. Hanson, *The history of NACE RP-0176*, Paper no. 06100 CORROSION/2006.
45. M. Tettamanti et al., *20years of impressed current cathodic protection retrofit of offshore platforms using MMO tensioned string anodes*, Paper no. 07049 CORROSION/2007.
46. M.A. Al-Arfaj, S.A. Al-Zubail and R.T. Aguinaldo, *Performance of large size magnesium anodes used for protection of Wellcasings at non-electrified offshore fields*, Paper no. 08010, Proceedings of *12th Middle East Corrosion Conference*, Bahrain, 2008.
47. D.K. Flanery et al., *Installation of ROV-friendly cathodic protection systems on two deepwater subsea developments*, Paper no. 09073 CORROSION/2009.
48. R. Vukcevic, J. Furstenberg and I. Godson, *Resolving CP design restrictions at an Australian Wharf: a novel marine impressed current CP system*, Paper no. 10031 CORROSION/2010.
49. D.K. Flannery, *Cathodic protection retrofit of a spar platform*, Paper no. 11053 CORROSION/2011.
50. C. Googan, *Subsea CP: electrochemistry meets engineering*, Proceedings of ICorr London Branch Meeting: *Offshore Cathodic Protection- Structures and Pipelines*, Institute of Corrosion (Northampton) June 2013.
51. S. Ayyar, J. Jansson and R. Sørensen, *Cathodic protection design for offshore wind turbine foundations.* Materials Performance **53** (9), 26 (2014).
52. A. Aghajani and A. Saatchi, *Failure of impressed current anode supports on jetty piles.* Materials Performance **54** (3), 36 (2015).
53. L. Lichtenstein, *DNV GL standard harmonization: recommended practice of corrosion protection for wind turbines*, Paper no. 69255 EuroCorr, 2016.
54. J. Vittonato and M.A. Pellet, *Platform cathodic protection retrofit with anodes racks and subsea current measurement,* Paper no. 7615 CORROSION/2016.
55. T. Huck and S. Javia, *Impressed current anode systems for jetty piling protection.* Materials Performance **55** (1), 22 (2016).
56. B.B. Jensen, *Specifying corrosion protection for the offshore wind turbine industry*, Paper no. 9091 CORROSION/2017.
57. E. Agel et al, *Feedback from revamping of cathodic protection for offshore jackets*, Paper no. 95062 EuroCorr, 2017.
58. E. Rodriguez, A. Delwiche and T. Queen, *Strategy and results of an impressed current cathodic protection retrofit in the North Sea*, Paper no. 8957 CORROSION/2017.
59. H.B. Liu et al., *Optimizing an offshore platform cathodic protection system retrofit.* Materials Performance **56** (4), 56 (2017).
60. J. Britton and A. Delwiche, *Continued operation of a large offshore platform requires innovative new cathodic protection systems to be retrofitted.* Corrosion Management (137), 14 (2017).
61. T.N. Lomholt et al., *Unification of corrosion protection for offshore wind farms: collaboration in partnerships*, Paper no. 11170 CORROSION/2018
62. B.S. Wyatt, J. Preston and W.R. Jacob, *Cathodic protection of offshore renewable energy infrastructure – parts 1 and 2.* Corrosion Management issue 156 (p. 19) and issue 157 (p. 16) (2020).

Chapter 14

Submarine pipelines

"Just metal tubes!"

A while back, I was chatting to a pipeline engineer. He recounted a tête-à-tête with his wife the previous evening. I gathered at the time that this may not have been one of their most harmonious conversations. At some point, she had uttered the following unflattering appreciation of his career: "Anyway, what's so special about pipelines? They're just metal tubes!" If you are a pipeline engineer feeling offended by these words, I share your pain.

I have been around enough pipeline projects to know that there is a lot more to them than simply being metal tubes. There is nothing trivial about designing and constructing an offshore pipeline, which is essentially a pressure vessel perhaps hundreds of kilometres long. It is even more demanding to lay it safely on, or under, the sea floor, and then to ensure that it will operate safely for decades into the future.

Historically, applying CP has usually proved to be one of the least challenging parts of any overall offshore pipeline project. The performance of CP systems on subsea pipelines, particularly on carbon steel lines, has been exemplary. I am not aware of any external corrosion damage on fully submerged subsea pipelines in the North Sea, despite some pipelines having been in service for over half a century. More relevantly, Roche has summarised the worldwide experience of the French oil company Total [39]. The position in 2005 was that... *the only external corrosion cases encountered occurred in the splash zone or aerial zones of risers. No corrosion problems under disbonded coatings* have *ever been recorded so far in the submerged zones.*

This experience accords with a 1981 DNV paper [6] which concluded that... *cathodic protection seems to prevent corrosion in crevices filled with stagnant seawater.* Likewise, I am not aware of any subsequent reports that would alter the DNV conclusion or Total's offshore experience. Even so, as we will see in this chapter, pipeline CP has not been entirely without its problems.

14.1 EARLY SUBMARINE PIPELINES

It is difficult to sketch a history of subsea pipelines without mentioning the "PLUTO" project of World War II. The acronym was originally contrived from "pipe-line underwater transport of oil"; but this soon morphed into the more memorable, but inaccurate, "pipeline under the ocean". PLUTO involved various individual lines. Some were lead telegraph cable sheaths, but without the copper cables. Others were 76 mm diameter welded steel pipe. All were laid across the English Channel in the late summer of 1944, to supply fuel to allied forces liberating Normandy.

DOI: 10.1201/9781003216070-14

Without doubt, PLUTO was a major achievement of military engineering. It made use of the ship-mounted rotating drum method of laying telegraph cables across oceans. It was the first time that a reeled steel pipeline had been laid. This is a technique that has been widely used since [36]. Ultimately, however, PLUTO had little relevance. It only supplied 8% of the fuel used by allied forces, and none at all in the critical early stages of the campaign. It was interesting, but only as a footnote to history. The pipelines were not provided with CP. About 90% of the total inventory had been recovered for its scrap value by the end of 1946.

In fact, PLUTO was not even the first hydrocarbon-carrying steel pipeline laid subsea. Duncan and Haines [7] describe six loading pipelines that were laid between Sitra Island and the loading wharf in Bahrain in 1936, about 5 years after oil was first discovered on the island. The loading lines ran for ~1 km underwater. They were initially unprotected but were observed to be suffering severe corrosion after 2½ years. This led to the first pipeline CP installation in the Middle East, and one of the first outside the USA. The original installation was an impressed current arrangement using consumable steel anodes and motor-driven generators. The system was modified at various times over the years. In 1952 magnesium sacrificial anodes were used. In 1956, these were replaced by another ICCP system, this time using graphite anodes. Platinised-titanium anodes were introduced in 1961, and power supply units were replaced from time-to-time as required. At the time they prepared their paper (1981), the original six sea lines, together with various other submerged pipelines that had been added over the years, had remained protected.

14.2 PIPELINE TYPES

14.2.1 Flowlines

14.2.1.1 General

Formerly flowlines were not particularly long, usually no more than some hundreds of metres, and rarely more than a few kilometres. With the onset of satellite field developments, the lengths have increased, in some instances to tens of kilometres. They are of relatively small diameter (<0.3 m) so they possess sufficient negative buoyancy not to require a concrete weight coating. Furthermore, because they are mostly within the area of restricted navigation around an offshore petroleum installation, they do not usually need to be buried or trenched to protect them from trawling or anchor damage.

14.2.1.2 Production flowlines

Flowlines transport fluids either to or from subsea wells, either directly or else via gathering or distribution manifolds. A "production" flowline transports crude hydrocarbon from a well. The transported fluids may contain water and acid gases (CO_2 and H_2S) so will be potentially corrosive. Operators' preference will be to construct in carbon steel, with an appropriate corrosion allowance incorporated into the pipe-wall thickness. However, some reservoir fluids are so corrosive that the flowline will be constructed in a CRA: 13% Cr supermartensitic, 22% Cr duplex or 25% Cr superduplex stainless steel. As we discussed in Chapter 8, these materials are either not resistant to corrosion in ambient temperature seawater (13% Cr and 22% Cr) or at best only marginally resistant (25% Cr). It follows that even CRA flowlines need to be cathodically protected.

As offshore hydrocarbon exploration has progressed to deeper reservoirs, the temperatures of the crudes have generally also increased. If flow assurance calculations show a need to maintain the crude fluid at high temperature, the flowline will need to be insulated or, in some instances, heated. These factors have needed to be considered by CP designers and by the authors of the pipeline CP design codes.

14.2.1.3 Injection and gas lift flowlines

Fluids sent to subsea wells in flowlines are involved in either enhanced oil recovery or effluent disposal. The fluids transported will be either lift gas or water for reservoir injection. In addition, there is now an increasing interest in disposing of carbon dioxide by injecting it into hydrocarbon reservoirs. It would be very unusual to construct an injection or gas lift flowline in anything other than carbon steel.

14.2.2 Trunk and service lines

14.2.2.1 Export lines

Export lines are the pipelines that transport hydrocarbons either between offshore platforms or from offshore platforms to onshore terminals. Since the hydrocarbons are processed on the platform to reduce their corrosivity, trunk lines are always constructed in carbon steel. They are generally longer than flowlines, sometimes extending for hundreds of kilometres. They are also of larger diameter (>0.75 m). As a result, they will inevitably require a concrete weight coating to provide negative buoyancy.

Because of the risk of damage by trawling, or a dragged anchor, trunk lines are often trenched or buried. In some instances, they also need to be rock-dumped for stability.

14.2.2.2 Sea lines

Also falling under the same description as trunk lines are much shorter lengths of large diameter pipeline that transport stabilised crude oil or refined products from onshore facilities to loading jetties. These lines are variously referred to as sea lines, loading lines, transfer or shipping lines. The transported products are non-corrosive, so the pipelines are carbon steel. Their large diameters (>0.75 m) mean that they have to be concrete-coated. Since these lines originate in an onshore facility, they are usually protected by incorporation into the site ICCP systems.

14.2.2.3 Service pipelines

The production of wet natural gas from offshore fields usually requires a supply of chemicals such as methanol or glycol which suppress the formation of solid hydrates in the pipelines. Where large chemical consumption is required, the fluids are delivered from the onshore facility to the offshore platform in carbon steel pipelines. Typically, these are relatively small diameter (<0.125 m), non-concrete-coated, pipelines laid along the same route as the incoming export lines. Sometimes, to economise on pipelay vessel time, the service line is piggy-backed to the trunk line. In this case, the two pipelines share the same CP system, with the anodes being installed on the larger diameter trunk line. This arrangement requires electrical bonding straps to be installed between the two pipelines at intervals along their length.

14.2.3 Risers

The term "riser" is applied to a section of mainly vertical pipeline that connects the pipeline on the sea floor to the processing facilities on the platform. Confusingly, the term riser applies to both pipes that take fluids up to the processing facility ("import" risers), and to the pipes that take fluids down to the seabed section of the pipelines ("export" risers).

The CP applied to a riser will, of course, only provide corrosion protection to its immersed section. However, the riser also transits the severely corrosive inter-tidal and splash zones. Because of the potential vulnerability to corrosion in these zones, risers have additional pipe-wall thicknesses and a particularly high integrity protective coating system, such as 25 mm of neoprene rubber. Some risers are externally clad with a corrosion-resistant alloy through the splash zone.

Steel risers are either rigid or catenary. Rigid risers are fixed in place and supported by clamps on a fixed structure. Steel catenary risers (SCRs) are often to be found in deeper water. They connect a floating installation, such as a floating production storage and offloading (FPSO) vessel to subsea manifolds or pipeline terminations. The name catenary, which derives from the Latin word for "chain", refers to the looping shape these risers adopt between the structure and the seabed. In some instances, the riser is tethered to a floatation buoy near its mid-depth, so forming a double catenary.

14.3 CODE-BASED CP DESIGN

14.3.1 Methodology

In Chapter 3, we discovered that it was a relatively straightforward "cookbook" exercise to produce a workable CP design for an uncomplicated offshore structure. The same applies for pipelines. The essential steps are set out below. Although pipelines and structures obviously differ in detail, the overall CP design processes are quite similar. The main difference is that the topic of anode spacing needs to be addressed on pipelines whereas it is rarely a consideration for simple structures.

Step 1: System type and code selection

As with all CP systems, there is a basic choice to be made between a sacrificial anode and an ICCP system. However, with the exception of comparatively short pipelines linked to a facility with existing ICCP systems, the only sensible option will be sacrificial anodes. The codes reflect this reality.

It is likely that any offshore pipeline CP design will be based on either ISO-15589-2 or DNV-RP-F103 which mirrors much of the ISO document.

Step 2: Interfaces

Pipelines do not exist in isolation. They transport fluid between two points. These may be items of hardware such as subsea valves, tie-ins, pipeline end manifolds (PLEMs) and pipeline end terminations (PLETs), or they link offshore installations with offshore or onshore facilities. Usually, but not always, these terminal installations will also be provided with their own CP systems, but these will have been designed to other standards. This leaves the designer of the pipeline CP system with the obligation to consider any possible interactions with the terminal structures or their CP systems.

As a consequence, the pipeline CP engineer might need to consider arranging to isolate the pipeline electrically from these installations, or else to cater for the possibility that they might draw current from the pipeline. We return to these issues when we discuss the issue of isolation (see Section 14.5).

Step 3: Design factor

The pipeline owner might require the CP designer to incorporate a *design factor.* Alternatively, the CP designer might recommend that a design factor is included and might offer an opinion on what its value might be. This design factor is designated as "*k*" in a guidance note in DNV-RP-F103, which states... *that any additional conservatism for CP design is introduced by use of a design factor, k, rather than modification of one or more of the CP design parameters...* As explained by DNV [**49**], the use of this design factor is one of the key differences between the DNV code and ISO 15589-2.

This effectively means uplifting guideline CP current densities as a catch-all means of compensating for unknown future shortcomings in, for example, the performance of the coating or the functioning of the sacrificial anodes. Interestingly, DNV does not venture any opinion as to what the value of *k* should be.

Step 4: Anode spacing

A core part of the design exercise is to calculate the spacing between the anodes. We are at liberty to adopt the advice in DNV-RP-F103 that there is no need to carry out such calculations providing the anodes are no more than 300 m apart. This is as near as makes no difference to one anode for every 25 standard pipe joints, where the pipe joint length is nominally 12.2 m (40 feet).

However, this should not be taken to imply that, if we opt for a spacing in excess of 300 m, achieving protection will become a problem. With modern high impedance pipeline coating systems, it is possible for anodes on the terminal structures to protect many kilometres of pipeline. This point is acknowledged in both DNV-RP-F103 and ISO 15589-2. By way of a practical example, Surkein et al. [**28**] comfortably protected a 2-mile long hot pipeline only using anodes at each end. We will return to the topic of anode spacing below.

Step 5: Area calculation

The calculation of the surface area of the pipeline is very straightforward. All we have to do is remember the formula for the surface area of a cylinder.

$$\text{Area} = \pi \times \text{diameter} \times \text{length}$$

The position becomes a little more complex if we also have to allow for the surface area of valves or flanges and, in particular, subsea flange bolts. For now, we shall ignore these items. We can simply take note of the fact that, as a "belt-and-braces" measure, many CP designers simply "double up" on the anode provision towards the ends of the pipeline to cater for any such increase in the current demand.

Step 6: Protection potential

Subsea pipelines are often buried, intentionally or otherwise. There is a view held by many CP practitioners, and promulgated by some codes, that this service merits a more negative protection potential (−0.9 V) than a structure exposed to open seawater (−0.8 V). The added 100 mV of cathodic polarisation is (incorrectly) believed to be necessary to counteract the threat of MIC in some seabed sediments. We explained in Chapter 6 why this perceived need for a more negative protection potential is incorrect. DNV-RP-F103 (unlike ISO 15589-2) correctly sets the design protection potential for pipelines at −0.8 V, irrespective of whether or not they will be exposed to seabed mud or to the threat of MIC.

Step 7: Bare steel current densities

DNV-RP-F103 gives the current densities shown in Table 14.1 for pipelines, irrespective of whether they are constructed in carbon steel or CRA. It should be noted that these values are without the addition of any design factor (*k*).

Table 14.1 DNV-RP-F103 current densities

	Internal fluid temperature (°C)				
	<25	*25–50*	*50–80*	*80–120*	*>120*
Pipe coated to DNV-RP-F106 and F-102					
Non-buried	0.050	0.060	0.075	0.100	0.130
Buried	0.020	0.030	0.040	0.060	0.080

Step 8: Coating breakdown

Subsea pipelines are always coated. Allocating coating breakdown values for a pipeline CP design follows exactly the same process as for a coated structure. However, the quality of factory-applied pipeline coating is now expected to be so high that we use different coating breakdown guidelines than advised for structures. Examples of breakdown figures for factory-applied coatings are given in Table 14.2 where 'a' is the initial breakdown and 'b' is the annual rate of breakdown. These figures reflect the generally superior performance of factory-applied pipeline coating systems. A somewhat lower expectation of durability is recommended for the field joint coating systems (see Table 14.3).

Table 14.2 Some pipeline coating breakdown factors (from DNV-RP-F103)

Coating (refer to DNV-RP-F106 for details)			
No.	*Description*	*a*	*b*
1	FBE with concrete weight coating	0.03	0.0003
2	3-layer (FBE/adhesive/PE) with or without concrete weight coating	0.001	0.00003
3	3-layer (FBE/copolymer adhesive/PP) with or without concrete weight coating		
5	Polychloroprene (neoprene)	0.010	0.001

Table 14.3 Some field joint coating breakdown factors (from DNV-RP-F103)

Field joint coating type (refer to DNV-RP-F102 for details)	*Infill type*	*a*	*b*
Heat shrink sleeve: backing+adhesive in polyethylene or polypropylene, liquid epoxy primer	None or moulded polyurethane	0.03	0.003
FBE	None	0.10	0.010
	Moulded polyurethane	0.03	0.003
FBE with polyethylene heat shrink sleeve	None or moulded polyurethane	0.01	0.0003
FBE with PP heat shrink sleeve	None	0.01	0.0003
Polychloroprene (neoprene)	None	0.03	0.001

Step 9: Current demand

Initial current

There is no need to calculate the initial current demand for a pipeline with a newly applied modern coating system.

Mean and final current demands

The mean and final current demands are calculated in an analogous manner to the coated structure we considered above. Each figure is the product of the surface area, the current density (Table 14.1) and breakdown factors for the factory-applied line pipe and field joint coating (Tables 14.2 and 14.3). The coating breakdown factors are applied pro-rata to the relative lengths of line pipe and field joint coatings. This estimated current is then multiplied by the contingency factor *k*.

Step 10: Anode alloy

As with structures, the selected anode material will either be a zinc or an aluminium alloy. However, anodes on structures only experience ambient conditions. Pipeline anodes may operate at elevated temperatures. Thus, DNV-RP-F103 contains additional information on anode performance at elevated temperature (Table 14.4).

Table 14.4 Anode alloy properties (DNV-RP-F103)

Alloy type	*Environment*	*Temperature (°C)*	*Capacity (C) (Ah/kg)*	*Potential (E_a) (V)*
Al-Zn-In	Seawater	≦30	2000	−1.05
		60	1500	−1.05
		80	720	−1.00
	Sediments	≦30	1500	1.00
		60	680	1.00
		80	320	1.00
Zn	Seawater	≦30	780	−1.03
		30–50	780	−1.03
	Sediments	≦30	750	−0.98
		30–50	580	−0.98

Step 11: Anode design

The options for pipeline anode design are limited. Most pipelines are protected using either half-shell or segmented bracelet anodes of the type we saw in Chapter 10. A bracelet anode for installation on a concrete-coated pipeline will normally be square-ended to abut to the concrete layer. Bracelet anodes for non-weight-coated pipelines are usually tapered to enable the anode to pass over the stinger on the lay-barge (see Figure 10.17).

The internal diameter of the bracelet must equal the combined outer diameter of the pipe and its anticorrosion coating. The outer diameter is controlled by pipelay considerations. If the pipeline is concrete-coated, then the anode thickness will need to match that of the concrete. This means that the only significant design variable available to the CP designer will be the anode's length. However, this is not without its constraints. The practicalities of casting bracelet half-shells often limit the length to about 1 m.

In some cases, there are options to use sled-mounted anodes cable connected to the pipeline, or else to satisfy the pipeline current demand using the CP provision already installed on one or both terminal structures. We consider these approaches in Section 14.4.2.4.

Step 12: Instantaneous current output

The resistance of a short flush-mounted anode, which includes pipeline bracelet type anodes is given by the McCoy formula (14.1).

$$R_a = \frac{0.315\rho}{\sqrt{A}} \tag{14.1}$$

where
A=total exposed area of the anode (m^2)

It is conservative to assume that a pipeline's anodes will be fully buried in seabed sediment. The environmental resistivity (ρ) of sediments is not always easy to quantify. DNV recommends a design figure of 1.5 Ωm if other data are not to hand.

R_a is calculated on the end-of-life anode dimensions to derive the final current output. For a bracelet anode, it is customary to assume that the anode length remains the same throughout the life. The surface area, therefore, only reduces very slightly through life due to the slight reduction in the outer diameter of the bracelet.

14.3.2 Example calculation

14.3.2.1 Pipeline condition

For the purpose of this design, we will assume that the pipeline is intentionally buried.

14.3.2.2 Design factor

There are no rules governing what the design factor should be. For this exercise, we will adopt a contingency of 50%, making k=1.5.

14.3.2.3 Current demand

Our input data for a notional pipeline are given in Table 14.5. This table also includes the wall thickness (WT) of the pipeline steel. This is not something that concerned us when we considered CP designs for simple structures in Chapter 3. We return to this point when we discuss anode spacing and potential attenuation in Section 14.4.

Table 14.5 Pipeline design example – design data

Parameter	*Value*	*Units*	*Remarks*
Design life	40	years	
Internal fluid temperature	40	°C	
Pipeline outside diameter (*D*)	0.762	m	
Pipeline corrosion allowance (CA)	3	mm	Allocated for internal corrosion
Pipeline wall thickness (WT)	15	mm	
Pipeline inside diameter (*d*)	0.738	m	D−2×(WT−CA)
Sea bed resistivity (ρ)	1.5	Ωm	Assumed in the absence of measurements
Pipeline condition	Buried		
Pipeline length per anode	620	m	50 pipe joints (trial value)
Pipe joint length	12.2	m	Typical value (40′ pipe joint)
Field joint cut-back	0.6	m	0.3 m at the end of each joint
Line pipe coating (type)	DNV-RP-F106 No. 2 (3LPE)		
Line pipe coating (thickness)	3.5	mm	
Field joint coating	DNV-RP-F102 No. 2b(2) (FBE+PE heat shrink sleeve+polyurethane infill)		
Concrete weight coating	40	mm	Decided by pipeline design team

Table 14.6 CP design parameters

Parameter	*Value*	*Units*	*Remarks*
Design factor (k)	1.5		50% additional design margin
Protection potential (E_c)	−0.9	V	Owner's conservative requirement (over-rides DNV value of −0.8 V)
Initial current density (CD) for bare steel	n.a.		Not needed for coated pipeline
Area protected per anode	1460	m^2	Area of 50 pipe joints
Mean CD for bare steel	0.045	A/m^2	0.030 A (Table 14.1) uplifted by $k = 1.50$
Final CD for bare steel			

Table 14.7 Design coating breakdown factors (40 year life)

Parameter	*Value*	*Remarks*
Factory coating		
Initial breakdown (a)	0.001	Table 14.2
Annual breakdown (b)	0.00003	
Mean coating breakdown	0.0016	Refer to Chapter 3 for details of calculating coating breakdown factors
Final coating breakdown	0.0022	
Field joint coating		
Initial breakdown (a)	0.01	Table 14.3
Annual breakdown (b)	0.0003	
Mean coating breakdown	0.016	Refer to Chapter 3 for details of calculating coating breakdown factors
Final coating breakdown	0.022	

Table 14.8 Cathodic current demands (mean and final)

Parameter	*Value*	*Units*
Area of pipeline (50 joint section)	1460	m^2
Factory coated	95.1	%
Field joint coated	4.9	%
Mean current demand		
Factory coated	0.10	A
Field joint coated	0.05	A
Total	0.15	A
Final current demand		
Factory coated	0.13	A
Field joint coated	0.07	A
Total	0.20	A

The corresponding CP design parameters are set out in Table 14.6. In this case, we are going arbitrarily to start our design exercise by looking to see if we can space anodes 50 pipe joints (610 m) apart. We then produce design calculations based on 50-joint pipeline modules.

We see that we are dealing with a buried line with an internal fluid temperature of 40°C. From Table 14.1, we see that DNV recommends a (bare steel) design current density of 0.03 A/m^2 in sediments for this temperature. This is uplifted by the design factor ($k = 1.5$) to get 0.045 A/m^2.

The coating breakdown parameters for this design are set out in Table 14.7. Combining these with the relative areas for field joint and factory-applied pipe coating and the 0.045 A/ m^2 current density gives the cathodic current demands shown in Table 14.8. The results indicate a requirement of 0.15 A for the mean, and 0.2 A for the final, current demand for each 50-pipe joint module.

It is also worth mentioning here that, if we had been carrying out this CP design exercise in (say) 2002, and we had been using EN 12474, our 40-year coating breakdown figures would have been 6%–8% (average) and 10%–16% (final). Taking the upper end of the coating breakdown figures, the mean and final current demands for each 50-pipe joint would have been very much higher: 2.5 and 4.9 A, respectively.

14.3.2.4 Anode design – mean current

We will assume that a decision has been taken to use zinc anodes for this pipeline project. Since we have modularised the design such that each anode protects 50 pipe joints (610 m), we need each anode to supply a mean current of 0.15 A. The minimum nett weight of anode alloy (W) needed is found by inserting the relevant values into the equation we met in Chapter 3.

$$W = \frac{8760 T I_m}{CU}$$

where
T=40 years
I_m=0.15 A per anode
C=580 Ah/kg (Table 14.4)
U=0.80 (for a bracelet anode).

This gives the result that each anode must contain a minimum of 113 kg nett weight of zinc sacrificial anode.

At this point, it is worth doing a quick reality check to see how long the anode would need to be to contain a minimum of 113 kg of anode alloy. Zinc has a density of ~7100 kg/ m^3, meaning that the required volume of alloy is 0.016 m^3. We can approximate the cross-sectional area of a cylindrical anode to its thickness times its mean circumference. The thickness matches that of the weight coating (0.04 m), and for this estimate, we can take its mean circumference (~$\pi \times 0.8$ m). This gives us a cross-sectional area of ~0.1 m^2. Thus, the minimum required anode length would be only about 0.16 m. This would be far too short to be practically castable by any anode foundry. Our anode will have to be longer than would be dictated solely by the minimum weight requirement of 113 kg.

14.3.2.5 Anode design – final current

We now need to estimate the length of anode needed to deliver the final current of 0.20 A when buried in seabed mud (ρ=1.5 Ωm). We can use equation 14.2 (which is a simple rearrangement of the equation we used in Chapter 3) to work out the maximum allowable anode resistance (R_a).

$$R_a = (E_c - E_a) / I_a \qquad (14.2)$$

where
I_a=anode output (0.20 A)

E_c=protected potential of steel pipeline (−0.9 V)
E_a=operating potential of anode (−0.98 V, see Table 14.4)
R_a=anode resistance (Ω)

Using the design figures, we find that the maximum tolerable value of R_a is 0.40 Ω. Substituting this value, and the seabed resistivity (ρ=1.5 Ωm), into a rearranged version of the McCoy formula (14.3), gives a minimum anode surface area of 1.4 m^2.

$$\sqrt{A} = \left(\frac{0.315\rho}{R_a}\right) \quad (14.3)$$

Again, for a first-pass calculation, we can ignore the gaps between the anode half-shells. We can then calculate the minimum length the anode would have to be. On this basis, with the anode consumed to $(1-U)\times$its original volume, its end-of-life diameter would be:

- Pipeline OD (0.762 m) plus,
- 2×coating thickness (0.007 m) plus,
- $2\times(1-U)\times$original anode thickness (0.016),

amounting to 0.785 m. This equates to a circumference of 2.47 m, meaning that the minimum surface area requires a minimum anode length of (1.4 m^2÷2.47 m=0.57 m). Thus, the anode design is dominated by the need to satisfy the end-of-life current output requirement. It will comfortably exceed the minimum mass requirement. An anode length of 0.57 m might be castable, but would be rather short for an 0.8 m diameter bracelet.

This first-pass design calculation points towards increasing the modular length of pipeline so that we can make use of an anode design with a length that would be more convenient to cast (say 0.8–1.0 m). On this basis, we might consider increasing our modular pipeline length from our arbitrarily selected value of 50 joints to (say) 75 joints. The mean and instantaneous current output calculations indicate that this might make sense.

However, you may have noticed that, thus far, we have not yet considered what the code requires in respect of anode spacing. DNV mirrors ISO-15589-2 in calling for anode spacings >300 m to be... *justified by attenuation calculations or other mathematical modelling*. It follows that our first-pass allocation of one anode for every 50 pipe joints, and our thinking that we might go beyond that, requires justification that the extended spacing is acceptable. This brings us to the topic of anode spacing.

14.4 ANODE SPACING

14.4.1 Early practice

Zinc anodes have been in use from the 1950s onwards on trunk lines [12] in the Gulf of Mexico. However, prior to the 1960s, it was common practice for one operator to isolate in-field flowlines from platforms [20] and protect them by welding magnesium anodes to each end.

From the mid-1960s, it was established practice to protect pipelines using bracelet anodes and to electrically insulate the pipelines from the jackets. However, as Britton [29] has remarked, the use of bracelet anode design was, in many ways, the worst CP option since it combined low output with low anode utilisation factors. Bracelets were universally adopted because they enabled the CP to be pre-installed on the pipeline.

The anode spacing is the quotient of the line length divided by the number of anodes. For early pipelines, the CP designs resulted in relatively small spacings. Mackay [4] summarised the North Sea experience up to the early 1970s, telling us that pipelines were typically 0.762 m diameter (30-inch) and were protected by zinc anode bracelets (~225 to ~450 kg) and located at an approximate spacing of 152 m (~12 pipe joints). The typical anode spacing of 100–200 m was still in place by the end of the 1970s, by which time over 2000 km of pipelines had been laid in the North Sea [5]. Shortly later, ffrench-Mullen and Jacob [14] added more detail on spacing. They provided examples of anode spacings for nine North Sea trunk lines laid prior to 1984. Of these, eight had spacings of less than ~188 m (15 pipe joints). One had a spacing of ~495 m (40 pipe joints), showing that at least one operator had a more realistic expectation of coating performance than was fashionable at the time.

The early codes, which followed the industry practice of the time, were commensurately ungenerous with their advice on anode spacing. Table 14.9 attempts to summarise the evolution of anode spacing advice given in the codes. As can be seen, there was nothing in the 20th- century codes to encourage spacing intervals of more than 200 m.

Table 14.9 Code advice on pipeline anode spacing

Year	*Code*	*Anode spacing advice offered*
1973	BS CP 1021	Section 6.3.3.2 stated "*...galvanic anodes may be attached to the submarine pipeline at suitable intervals along its length*". However, no guidance was provided as to what constituted "suitable"
1981	DNV-TNA-703	Section 3.2 provides the general, but unhelpful, recommendation: "*The anodes should be distributed over the steel surface to achieve a uniform current distribution*"
1986	DNV-RP-B401	Although this RP covered pipelines, it gave no advice on anode spacing
1993	DNV-RP-B401	Section 7.10.5 stated "*The distance between successive pipeline anodes should not normally exceed 150 m*"
1997	NORSOK M-503	Section 5.7 stated "*...Anode spacing should not exceed 200 m. Amount of anodes shall be increased by a factor of 2 for the first 500 m from platforms and subsea installations*"
2001	EN 12474	Offers a spacing methodology based on attenuation calculations but provides no guidance on suitable attenuation constants to use in that methodology
2003 & 2016	DNV-RP-F103	Invokes the advice in ISO 15589-2 that spacings <300 m do not require any justification. It then adds "*...According to this RP, short pipelines (<10–30 km approximately depending on the design and operating conditions) may be protected by installing anodes on subsea structures located at the pipeline termination(s)...*" The RP then describes how such longer spacing can be calculated. This advice has persisted through to the current edition (2016) of DNV-RP-F103
2004 & 2012	ISO 15589-2	Stated that "*...Anode spacing exceeding 300 m shall be justified by attenuation calculations or other mathematical modelling*" Replicates the attenuation calculation methodology of EN 12474; but, likewise, without any guidance on the crucial attenuation constant. The 2012 edition is similar
2007 & 2018	NORSOK M-503	Broadly follows ISO 15589-2 but offers a methodology similar to DNV-RP-F103 for longer pipeline lengths

There were various reasons that these spacings were so modest.

In the first place, early expectations of subsea pipeline coating performances were exceedingly pessimistic. In the late 20th century, a pipeline CP designer might reasonably assume that end-of-life (30 years) coating breakdown would exceed 20%. However, developments in coating systems, and experience gained with actual performance, meant that by the early 21st century it had become more reasonable to base designs on <1% coating breakdown. In addition, designers have almost universally selected a target protection potential of −0.9 V for a pipeline CP design. This has been based on the (reasonable) expectation that there would be an MIC threat to any pipelines that might become buried in seabed mud, and the (unreasonable) belief that −0.9 V was required for protection against MIC. This meant that, irrespective of whether the designers opted for zinc or aluminium alloy anodes, the available driving voltage for CP, with a value of E_c set at −0.9 V, will have been no more than 150 mV.

Furthermore, quite a few pipelines were laid on the seabed rather than being intentionally buried. For such lines, there was an expectation that some parts would silt over, but no knowledge as to which parts. CP designers inevitably responded to this uncertainty by assuming the worst. There has been more than one pipeline for which the CP designers had assumed that the entire length of the pipeline would be on the seabed, and requiring cathodic current associated with fully aerated seawater; but that all of the anodes would be silted over, thereby having their outputs limited as a result of exposure to the higher resistivity mud environment. All of these factors contributed to a culture of placing anodes at short intervals along the pipeline. Such overdesign was understandable since the CP only amounted to 0.5%–1% of the cost of any submarine pipeline project.

14.4.2 Extending the spacing

14.4.2.1 The crucial resistance

When we designed a CP system for simple subsea structures in Chapter 3, the only resistance we considered was that of the anode to the environment. There was no need to consider the internal electrical resistance of the structure itself. This is because the internal resistance of massive structures (jackets, ships, floating installations, etc.) is negligible. The dominant resistance in the design is the anode to electrolyte resistance (R_a). However, for pipelines, the maximum length that can be protected from a single anode is controlled to a very large extent by the metallic resistance of the pipeline itself.

As seen in Table 14.9, DNV modified its previous position on anode spacing in 2003 when it advised that spacings between anodes could be very much greater than the few hundreds of metres advised in the earlier codes. In principle, this would enable quite long pipelines (up to 30 km) to be protected using anodes installed on terminal structures. This has considerable practical (and cost) advantages. This is particularly so for lines laid from a reel barge since it removes the need to install anodes whilst the pipelay is underway.

Nevertheless to benefit from this implied permission to extend the spacing beyond 300 m, we need to be able to estimate how far along a pipeline a single anode can throw its protection. This is not simple. We will consider three approaches.

1. The first is a crude "worst-case" check, but which provides a useful pass/fail result,
2. the Norsok method, which may be regarded as a refinement of the worst-case check, and
3. attenuation calculations, which attempt to describe the change of potential along the pipeline.

It should also be added that the latter two approaches are amenable to refinement by computer modelling. We return to this in Chapter 17.

14.4.2.2 A worst-case approach

What I am about to tell you now is obviously incorrect. However, it serves the point of illustrating how we might carry out a first-pass anode spacing assessment. I can also justify it on the basis that it has been used by others, including Statoil [**32**]. So please bear with me.

We start with our 610 m anode spacing. A single anode placed midway along a 50-joint section of pipeline is required to provide 0.2 A, to protect that section at the end of the design life. We now make the rash, worst-case, assumption that the entirety of the end-of-life CP current enters the line at a single massive coating defect at one extremity of the pipeline section. All of this 0.2 A CP current then flows back to the anode through 305 m of pipeline. This current generates a voltage drop in overcoming the longitudinal resistance of the pipe.

We need to estimate what this voltage drop might be. We start by calculating the longitudinal resistance of the pipeline (R_{pipe}) using the relationship in equation 14.4.

$$R_{\text{pipe}} = \frac{\rho_m L}{A_x} \tag{14.4}$$

where

R_{pipe} = electrical resistance (Ω) along pipeline wall
L = pipeline section length (m)
ρ_m = metal resistivity (in the absence of project-specific data)
$\rho_m = 0.2 \times 10^{-6}$ Ωm (carbon steel)
$\rho_m = 0.8 \times 10^{-6}$ Ωm (martensitic stainless steel)
$\rho_m = 1.0 \times 10^{-6}$ Ωm (duplex stainless steel)
A_x = pipe-wall cross-sectional area:

$$\pi\left(\left(\tfrac{1}{2}D\right)^2 - \left(\tfrac{1}{2}d\right)^2\right)$$

where D and d are the outer and inner pipe diameters (m), respectively.

In this example, the outer diameter (D) is 0.762 m, and the inner diameter (d) is D minus twice the final wall thickness, where the final wall thickness assumes that all of the (internal) corrosion allowance has been consumed. Thus, d is 0.738 m. From these values, we arrive at the cross-sectional area of the pipe (A_x) of 0.028 m^2. From equation 14.4, the longitudinal resistance (R_{pipe}) of the 305 m carbon steel pipe section is given by:

$$R_{\text{pipe}} = \frac{0.2 \times 10^{-6} \times 305}{0.028} = 2.18 \times 10^{-3}\ \Omega$$

By Ohm's law, the voltage drop generated by 0.2 A flowing against this pipe-wall resistance amounts to <0.5 mV. This is trivial, so we can ignore it. Thus, we arrive at the unsurprising result that there is no problem applying CP to this well-coated pipeline by placing anodes at 610 m intervals. Indeed, the small voltage drops tell us that, even using the extreme worst-case assumption that the entire CP current enters the line at its extremity, we could extend our spacing very considerably beyond 50 pipe joints.

14.4.2.3 Norsok method

The same approach to determining the length of pipeline that can be protected by a single anode is provided by both DNV-RP-F103 and NORSOK M-503. Although the appearances of the equations from these two codes are slightly different, they are constructed using essentially the same methodology.

The method leads to an estimate of how far a given voltage can drive the current along a length of resistive pipeline. In doing this, it has to take account of the fact that the longer the length of line, the more current will be picked up. Here both codes introduce the same over-simplification. They assume that the cathodic current is picked up uniformly along the pipeline and, therefore, that it is acceptable to treat the protection current as if it all enters the line at a single position half-way along its protected length.

By way of illustration, formula 14.5 replicates the NORSOK M-503 formula, although we have simplified some of the symbols used.

$$L_{max} = \frac{-R_a i_f \pi D F_{CF} + \sqrt{(R_a i_f \pi D F_{CF})^2 + \dfrac{\rho i_f D F_{CF}}{d(D-d)} \Delta E}}{\dfrac{\rho i_f D F_{CF}}{2d(D-d)}} \tag{14.5}$$

where:

L_{max} = maximum distance between anodes, or anode array connections (m)
R_a = (final) resistance of anode, or anode array (Ω)
i_f = (final) current density for bare steel (A/m^2)
D = pipeline outer diameter (m)
d = pipeline inner diameter (m)
ρ = pipeline metal resistivity (Ωm)
ΔE = driving voltage ($E_a - E_c$) (V)
F_{CF} = (final) coating breakdown factor (factors for pipeline and field joint coatings combined pro-rata to their areas)

Although we have described this calculation as "simple", formula 14.5 still manages to look pretty formidable. The formula in 14.5 is more complicated than it needs to be. For one thing, it rolls the anode current outputs into the mix. In that respect, the formula in DNV-RP-F103 is simpler because it relates to pipelines connected to massive terminal structures with their own adequate CP. That permits the DNV formula to assume that the anodes behave as a zero-impedance current source.

14.4.2.4 Potential attenuation

14.4.2.4.1 The problem

When we talk about potential attenuation, we are simply looking at the reducing ability of an anode to protect the pipeline as the distance from the point of anode connection increases. We seek a method of determining the nature and rate of this attenuation so that we can determine how far apart we are permitted to space our anodes.

We can start by considering a notional situation of a single bracelet anode installed midway along a very lengthy (essentially "infinite") coated subsea pipeline as illustrated in Figure 14.1. Beyond a certain distance (x) from the anode, the pipeline receives no CP current, so its potential is the corrosion potential (E_{corr}). As we discussed in Chapters 5 and 6, E_{corr} is not a fixed parameter since it varies over time in a complex manner that depends on a multitude of metallurgical and environmental factors. Nevertheless, in most steady-state corrosion situations in seawater, E_{corr} eventually settles down to about −0.65 V. That figure is good enough for us to use in this evaluation.

As we start to reduce x and move closer to the anode (by moving to the left in Figure 14.1), we start to pick up CP current from the anode. This current causes the potential to shift in a negative direction from E_{corr}. At a distance x from the anode, the value of the potential is

E_x. In keeping with the terminology of equation 14.5, we refer to the shift in potential (i.e. the polarisation) at this point as ΔE_x where:

$$\Delta E_x = E_x - E_{corr} = 0\,\text{V} \tag{14.6}$$

where $x=\infty$, there is no protection: $E_x=E_{corr}$ and $\Delta E_x=0$.

You may recall that in Chapter 5, we introduced the electrochemistry term overpotential (η). This is the same as ΔE_x.

Staying with Figure 14.1, and continuing our approach towards the anode, we reach a point where E_x is equal to our designated protection potential (E_{prot}). Depending on which code, we could take this as −0.9 V (ISO 15589-2) or −0.8 V (DNV-RP-F103). If we take E_{prot} as −0.8 V, ΔE_x is equal to −0.15 V. At this point, the value of x is equal to half the length of pipeline that can be protected. Obviously, the anode also protects to an equal distance in the other direction along the line, but we have omitted this from the diagram for simplicity.

$$\Delta E_x = E_{prot} - E_{corr} = -0.8\,\text{V} - (-0.65\,\text{V}) = -0.15\,\text{V} \tag{14.7}$$

where $x=L/2$ (where L=the protected length)

Finally, we assume that the potential of the pipeline at the anode connection is very close to the potential of the operating anode (E_a). If this is −1.05 V, then ΔE_x becomes −0.4 V.

$$\Delta E_x = E_a - E_{corr} = -1.05\,\text{V} - (-0.65\,\text{V}) = -0.4\,\text{V} \tag{14.8}$$

where $x=0$.

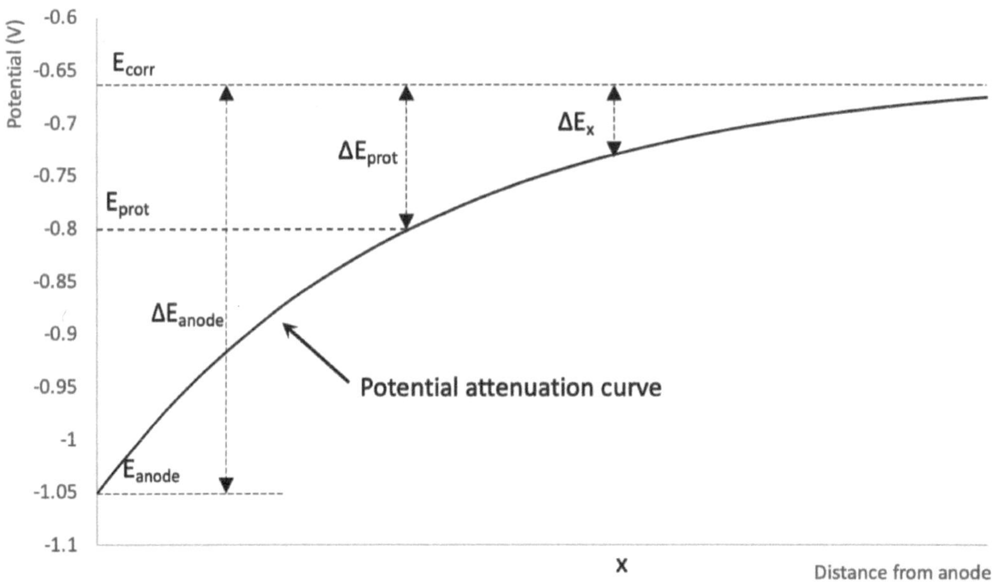

Figure 14.1 Attenuation of potential along an "infinite" pipeline.

Although we have drawn the potential attenuation as a curve in Figure 14.1, at this point we do not know anything about the shape of this curve. Indeed, we have yet to determine if it is actually a curve or a straight line. To make progress on this, we can start by developing a simple model of how the attenuation comes about.

14.4.2.4.2 Mathematical interpretation

CP current does not all enter a pipeline uniformly. It will enter the line at multiple locations along its length. Indeed, viewed globally, the coating may be regarded as slightly conductive entity permitting current to be picked up more-or-less uniformly along the pipeline length. This coating conduction may be because the coating has sustained physical damage or is slightly permeable to ions. The details and relative contributions of these conduction mechanisms need not concern us.

All we need do is make our first assumption: namely that the resistance the coating interposes to the flow of CP current onto the pipe is uniform over its surface. This situation is illustrated in Figure 14.2 which shows an idealised image of a pipeline with a coating exhibiting electrolytic conductivity. CP current enters the pipeline from the environment having passed through the coating. This current then flows through the pipe wall back to the anode. Remember that here we are discussing conventional (positive to negative) current flow. As we know, the electrons are flowing in the opposite direction.

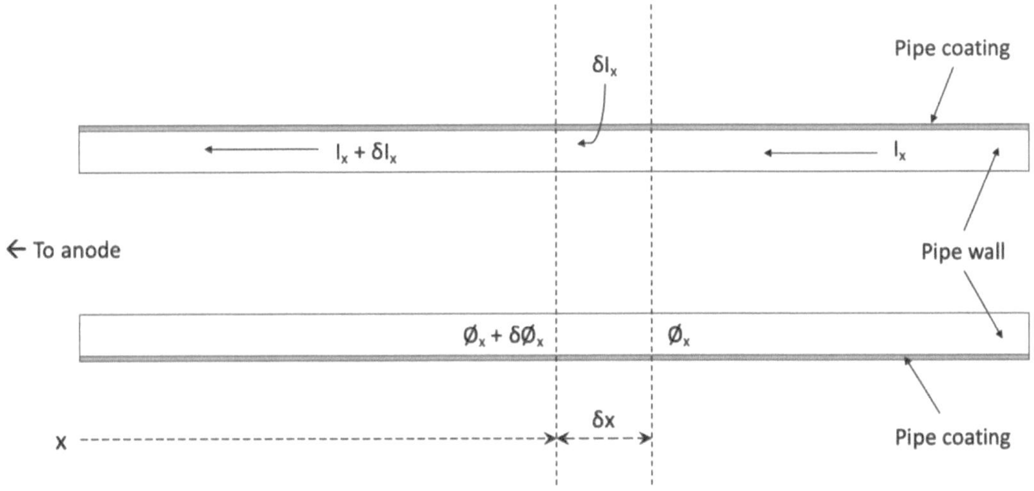

Figure 14.2 Model for pipeline potential attenuation calculation.

The current already flowing in the pipe at a distance x from the anode connection is I_x. This is the total amount of current that has been picked by the pipeline further away from x. In the very small interval of pipe length δx, there is a very small increment of current picked up (δI_x) such that the current continuing towards the anode is now $I_x+\delta I_x$. This change in current at x is given by equation 14.9.

$$\frac{dI_x}{dx} = -I_{cp(x)} \tag{14.9}$$

where

I_x=current flowing in the pipe wall at x

$I_{cp(x)}$=cathodic current entering the pipeline at x

The electrical potential in the pipe wall, also referred to as the coulomb potential, at point x is $\varnothing_x$. This potential is created by the flow of the current against the resistance of the pipeline.

$$\frac{d\varnothing_x}{dx} = -R_L I_x \tag{14.10}$$

where
R_L=longitudinal resistance of the pipeline per unit length (Ω/m)

As we discussed in Chapter 2, when we introduced the concept of "potential" (E), we are referring to the difference between the (coulomb) potential in the metal at a point x ($\varnothing_x$) and the (coulomb) potential in the solution ($\varnothing_{\text{solution}}$) at a point adjacent to x.

We now have to make a second assumption: namely that $\varnothing_{\text{solution}}$ is uniform along the length of the pipeline. From this, it follows that the rate of change of E_x along the pipeline is simply a measure of the rate of change of $\varnothing_x$. This assumption permits us to rewrite equation 14.10 as 14.11.

$$\frac{dE_x}{dx} = -R_L I_x \tag{14.11}$$

where
E_x=the pipeline potential at x
since $E_x = \varnothing_x - \varnothing_{\text{solution}}$

We can also add a further slight modification. Instead of considering the potential at any point x (E_x), we are looking at the shift in potential from its freely corroding value (ΔE_x). This enables us to rewrite equation14.11 as 14.12.

$$\frac{d(\Delta E_x)}{dx} = -R_L I_x \tag{14.12}$$

We can now combine equations 14.9 and 14.12 to arrive at equation 14.13:

$$\frac{d^2(\Delta E_x)}{dx^2} = -R_L I_{cp(x)} \tag{14.13}$$

14.4.2.4.3 Finding solutions

Equation 14.13 is a second-order differential equation. It tells us that we are dealing with a rate of change, of a rate of change, of ΔE_x. Now, if like me, your eyes glaze over, either because you have long since forgotten how to solve a differential equation, or you never actually knew in the first place, then fear not. We do not need to know how to solve it because that work has already been done. In many CP texts, this has been credited to either Morgan [15], Uhlig [13] or both, on the basis of the work published in the original editions of their seminal books published in 1959 and 1963 respectively. In fact, the origins of the solutions pre-date either publication.

It seems that Morgan may have been the first to publish the attenuation concept in relation to onshore pipelines. However, some of the earliest work in this area was carried out with reference, not to buried pipelines, but to buried telephone cables. The lead sheaths of these cables were protected by cathodic protection or, as it was referred to at the time, "forced current drainage". Hence, what appears to have been the first publication in the corrosion literature, originated from Pope of Bell Telephone Laboratories in New York [1] in 1946. He, in turn, drew on earlier work by Sunde, also from Bell Laboratories.

The solutions to equation 14.13, given in CP standards EN 12474 and ISO 15589-2, are arrived at by making the further assumption that, for small values of $I_{cp(x)}$, the value of ΔE_x is linearly proportional to the transverse resistance of the coated pipe:

$$\Delta E_x = I_{cp(x)} R_t \tag{14.14}$$

where R_t is the transverse (or leakage) resistance for a unit length of pipeline (Ωm)

We can now make use of equation 14.14 to rewrite equation 14.13 as follows.

$$\frac{d^2(\Delta E_x)}{dx^2} = -\left(\frac{R_L}{R_t}\right)\Delta E_x \tag{14.15}$$

if we set the above limiting values:

- $\Delta E_x = 0$ at $x = \infty$,
- $\Delta E_x = E_a - E_{\mathrm{corr}}$ at $x = 0$,

this second-order differential equation has the standard solution [**1,3,22**], shown in equation 14.16.

$$\Delta E_x = \Delta E_a \exp\left\{-\left(\sqrt{\frac{R_L}{R_t}}\right)x\right\} \tag{14.16}$$

In reality, we are not interested in protecting pipelines of infinite length. We are concerned with pipelines of finite length with anodes, or anode sleds, connected at fixed distances apart. The object of any attenuation study then becomes to determine how far apart the two anodes may be whilst ensuring that the midpoint on the pipeline is fully protected. The situation is more akin to Figure 14.3, which shows the polarisation along a section of pipeline brought about by two anodes.

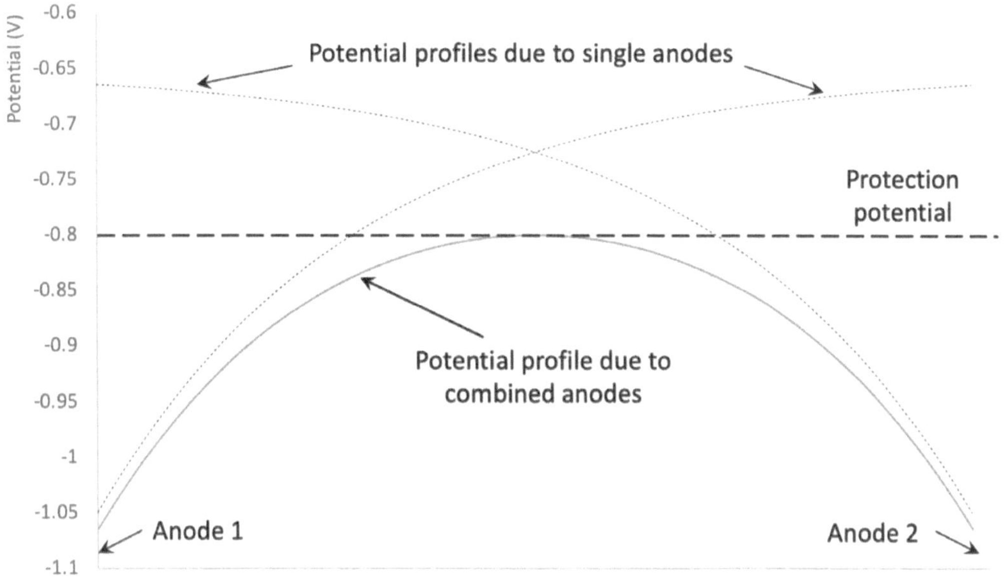

Figure 14.3 Potential between two anodes.

In this illustration, the spacing between the anodes has been adjusted to just bring about full protection at the midpoint. This occurs at an anode separation distance (a) such that, at the midpoint between the two (where $x = a/2$), the overpotential is just enough to achieve protection. At this point, there is no current flowing in the pipeline, so:

$$\frac{d(\Delta E_x)}{dx} = 0 \tag{14.17}$$

where $x = a/2$.

If we make a further (conservative) assumption that the beneficial effect of each anode beyond the midpoint between them can be discounted, we can solve equation 14.15 with this new boundary condition. The boundary condition $\Delta E_x = \Delta E_{\text{anode}}$ at $x = 0$ remains unchanged. Applying these boundary conditions to the solution of equation 14.15 yields the results:

$$\Delta E_x = \Delta E_{\text{prot}} \cosh\left\{\left(\sqrt{\frac{R_L}{R_t}}\right)(x - a/2)\right\} \tag{14.18}$$

$$\Delta E_a = \Delta E_{\text{prot}} \cosh\left\{-\left(\sqrt{\frac{R_L}{R_t}}\right)a/2\right\} \tag{14.19}$$

14.4.2.4.4 The attenuation "constant"

It is all too easy to be beguiled into believing that these complicated, yet elegant, equations must be correct. After all, their derivations are to be found in a number of authoritative texts, and versions have been published in normative annexes of ISO standards. However, we need to apply our critical faculties rather than just accepting them as the basis of yet another CP design spreadsheet. As Ashworth points out [22]... *this analysis is far from exact.*

The potential attenuation equations which we set out above were originally developed for telephone cables and then applied in the context of onshore pipelines. The latter, unlike pipelines in seawater, have traditionally been protected by ICCP systems with anode groundbeds located tens of kilometres apart. Indeed, Ashworth also states that... *the question of attenuation is not significant in the case of sacrificial anode protection systems where the individual source outputs are low.*

Even for onshore pipelines, it is very difficult actually to find any correspondence between the potentials measured in an over-the-line survey and the shapes of the potential profile in Figure 14.3. I have been involved in close interval potential surveys (CIPS) on >1500 km of onshore oil and gas pipelines. I can endorse Ashworth's view that, at best, the analysis... *demonstrates that attenuation is likely to follow an exponential decay and that it will be less severe for larger diameter pipe than for smaller.* However, beyond that generalisation, it is very difficult to see how the attenuation equations can assist a pipeline CP designer.

The reason for this difficulty emerges when we examine arguments, or independent variables, of the hyperbolic cosine functions in equations 14.18 and 14.19. A key parameter is the attenuation constant (α).

$$\alpha = \sqrt{\frac{R_L}{R_t}} \tag{14.20}$$

This has units of m^{-1} such that the arguments of the hyperbolic cosine functions, $\alpha(a/2)$ and $\alpha(x - a/2)$, are dimensionless. The way this function works is that the smaller the value of the attenuation constant, the greater will be the length of pipeline protected by an anode.

There are two ways of reducing the attenuation constant: make R_L smaller or make R_t larger. This explains Ashworth's comment that attenuation is less severe for larger diameter pipe. The cross-sectional area of the metal will be greater for a larger diameter, thicker-walled pipeline. So, the resistance per unit length of pipe will be lower. Similarly, we can readily see that, if the quality of the coating is improved, such that it increases the electrolytic resistance between the pipeline and the seawater, R_t will increase; and so too will the length of pipeline that we can protect from each anode. Both of these effects are intuitively obvious. It explains, for example, why it is usually futile to try to apply CP to thin-walled uncoated tubing.

We have already seen that R_L can be calculated quite readily. All we need to do is set the pipeline length (L) equal to unity in equation 14.4. Our problem comes when we try to assign a value to R_t. In this respect, ISO 15589-2 relates the leakage resistance to the insulation resistance of the pipeline coating.

$$R_t = \frac{R_c}{\pi D} \tag{14.21}$$

where:

R_c=pipe-to-seawater insulation resistance (Ωm^2)
D=pipeline outer diameter (m)

Beyond that, the ISO is not that helpful since it gives no guidance on how to source a value for R_c. It advises that a... *value for the insulation resistance... should be selected based on practical experience and consider the following; type of coating; exposure conditions such as to seawater or seabed sediments; design life of the pipeline and anticipated progressive reduction in coating resistance over the design life; pipeline installation method and projected extent of coating damage.*

This all sounds well and good. Unfortunately, however, it leaves users of the standard completely in the dark as to either what value of R_c we should start with or how that value will be changed according to the circumstances that it lists. As a matter of experience, trying to measure the electrolytic resistance of a wetted coating is problematic enough. Trying to predict the way it will change over future exposure is simply unrealistic. Instead of suggesting that we calculate R_c on the basis of *practical experience*, the ISO might reasonably suggest that we take a wild guess! If you got a result that was accurate within a couple of orders of magnitude, it would be something of a triumph. Of course, you would not be able to celebrate that triumph because you would never know what the "accurate" figure should be.

14.4.2.4.5 Further problems with the other attenuation assumptions

If the practical impossibility of divining a reliable value for the attenuation constant were not bad enough, things actually get worse when we start to drill further down into some other aspects of these attenuation equations. The reason for this is not so hard to understand when we re-examine some of the other implicit assumptions:

1. the natural corrosion potential is known and does not vary with time or changing exposure circumstances;
2. the anode potential, likewise, is a fixed parameter;
3. the pipeline coating degrades more-or-less uniformly along its length;
4. the potential in the solution ($\varnothing_{solution}$) is uniform along the length of the pipeline;
5. the current density flowing onto the pipeline (at areas of coating breakdown) is linearly proportional to ΔE_x.

However, none of these additional assumptions is correct.

As we explained in Chapter 5 and 7, the natural corrosion potential (E_{corr}) can never be a precisely defined value (assumption 1). Similarly, the operating potential of the anode (assumption 2) will vary to some extent (see Chapter 10). However, the variations in these parameters are not terribly important. In keeping with everything else in CP design, we can simply use conservative values.

Pipeline coatings do not degrade, or suffer accidental damage, uniformly (assumption 3). More to the point, attenuation calculations are only relevant for the end-of-life coating condition. At this stage, not only is the conductivity of the coating at its maximum, but so too is the variability of that conductivity along the line.

Assumption 4, that the potential in the solution is uniform along the pipeline length is simply incorrect. Strictly speaking, it can only apply at an infinite distance from the anode. A corollary is that attempts to characterise the overpotential as simply being the difference between the coulomb potential in the pipe metal and the (assumed) coulomb potential in the environment must be regarded as suspect. In this respect, it is interesting to note the views of Britton and Baxter [44] who refer to... *proven attenuation models.* They say that it is noteworthy that their... *approach ignores the drain point potential spike which is caused largely by the anode sled induced potential gradients and may be ignored for the purposes of ultimate distance (anode spacing) calculations.* Since these authors have considerable experience in designing pipeline CP retrofits, it is reasonable to give this opinion some weight. However, their claim that the attenuation models are "proven" overstates the reality.

The proposition that ΔE_x is linearly proportional to the current density (assumption 5) can be traced back to the Butler–Volmer equation which we met in Chapter 5. An implication of that equation, first published by Stern and Geary [2], is that at small overpotentials (typically less than ~20 mV) the potential versus current response for an electrode experiencing steady-state corrosion is approximately linear. That is, the electrode (almost) obeys Ohm's law. However, the Stern–Geary approximation breaks down at applied overpotentials in excess of ~20 mV. For a cathodically protected pipeline, the applied ΔE_x will be between 150 and 400 mV, and the potential vs current density relationship is not linear. Generally speaking, electrochemical systems do not obey Ohm's law.

Taken as a whole, therefore, the flaws in the key assumptions underpinning the attenuation equations are sufficiently profound that they actually invalidate the equations. The assumptions had to be made in order to produce a standard form second-order differential equation for which an analytical solution is readily available. However, if the assumptions are wrong, we have no option but to assume that the resulting attenuation equation is likewise unreliable.

14.4.2.4.6 Other attenuation models

These limitations have been recognised for some time and have prompted further thought and publication, but not further experimentation or field measurement. For example, Pierson et al. [27] have addressed the (plainly incorrect) assumption that $\varnothing_{solution}$ is uniform along the length of the pipeline. They introduced the influence of the resistance in the electrolyte which, as we have seen in Chapter 10, is essentially the anode resistance (R_a). This gave rise to a more sophisticated looking attenuation equation, albeit that this had to be subsequently revised by Lysogorski and Hartt [31] to eliminate an error in the original version.

Nevertheless, this equation still makes the assumption that overpotential is linearly related to current density. This same assumption also remains in another rework of the attenuation equation developed by La Fontaine and Gibson [33], and which is endorsed by at least one company promoting CP retrofits for offshore pipelines [44].

Other workers have followed different paths for developing attenuation equations. For example, Kozelkova et al. [24] proposed an exponential type equation based on a "decrement factor". The paper contained the intriguing comment that... *when plotted the curve fits closely with the collective mental images of the authors.* Comforting as that may be, they produced no hard data to support their views.

Whilst these, and other, refinements add to the sophistication of the equations used to define the potential attenuation profile, none addresses the core uncertainties. In particular, the value of any attenuation profile produced can be no more reliable than the guess made at the order-of-magnitude value of R_c.

14.4.2.5 Recommendation

In considering which, if any, spacing analysis to use in a CP design, the first thing we need to understand is that there is no such thing as a "correct" answer, just as there is no such thing as a "correct" CP design. The Norsok method has the advantage that it can be implemented using the parameters that are already available to the CP designer. Indeed, it is essentially a methodology that takes the specified design assumptions about coating breakdown, cathodic current demand and protection potential, and manipulates them into an estimate of the length of pipeline that can be protected.

The methodology will be non-conservative since it ignores the reality that if the coating breakdown is uniform (as supposed), the pipeline adjacent to the anode will take disproportionately more current than more remote parts of the line. However, this concern can be offset by the knowledge that other design parameters are intentionally conservative.

The attenuation approach, whilst mathematically more alluring, suffers the disadvantage of requiring the CP designer to conjure up a figure of the attenuation constant, for which the design codes offer no advice. At best, it can be used to run a range of trial-and-error "what if" scenarios.

My advice: use Norsok.

14.5 ELECTRICAL ISOLATION: OFFSHORE

Question: Should we isolate pipelines from offshore structures?
Answer: Probably not.

14.5.1 Early offshore practice

An offshore pipeline does not exist in isolation. It serves to connect an offshore installation with another installation either offshore or onshore. These terminal installations will usually be steel and will be provided with their own CP systems. Traditionally, the structures and the pipelines of an offshore development have been designed by separate companies, as have their CP systems. This contributed to a desire, on the part of both entities, to establish electrical isolation between a pipeline and its terminal structures. This desire was justified on the grounds that the CP designs of the pipeline and the structure would be radically different. As we saw in Chapter 13, the structure was usually uncoated, and its target CP design potential was set at −0.8 V. Most pipeline CP designs, on the other hand, set a target protection potential of −0.9 V. However, since pipelines are coated, it was recognised that they would actually spend much of their service lives at close to the anode potential (≃1.0 V). Any pipeline CP designer might, therefore, be concerned that the bracelet anodes affixed to the pipeline might be consumed prematurely as a result of delivering current to the less negative structure. The suppliers of pipeline isolating joints, or insulating flange kits, had no reason to counter that concern.

For these reasons, many of the early offshore pipelines, particularly in the North Sea, were fitted with isolation joints at the top of their riser sections. This also meant that precautions had to be taken to ensure that the riser itself remained electrically isolated from the structure. This includes, for example, supporting the riser in rubber-lined clamps.

There was little in the way of helpful information in the early CP codes. BS CP1021 (1973) referred to isolating joints sometimes being required as part of the safety precautions at oil terminal jetties, but made no reference to their use, or non-use, on offshore installations. NACE RP0176, issued in 1976 for the CP of structures, made the point that it was *necessary or useful* to have information on electrical isolation from foreign pipelines or structures, in order to design a platform's CP system. However, it offered no advice on whether or not isolation should be installed. The 1981 DNV document (TNA 703) made no recommendation in respect of isolating joints offshore.

Given this lack of guidance, it is not surprising that pipeline isolation was by no means universal practice in the early days of offshore development. Some operators had a standard practice of isolating, others did not. This was exemplified by the case of the 0.76 m diameter oil line from Marathon's Brae "A" platform to BP's Forties "C" in the North Sea. This had isolation fitted at Brae, but not at Forties.

An interesting snapshot of practice on riser isolation in the UK sector of the North Sea was obtained in a survey carried out by the Marine Technology Directorate (MTD) in 1988 [**18**]. The results are summarised in Table 14.10. It covers 44 offshore installations to which pipeline risers are attached. Despite the territorial limitation of this survey, the results are interesting. There seems to have been a 50/50 split between operators favouring isolation between sacrificially protected structures and sacrificially protected pipelines. Over the subsequent years, that balance has shifted in favour of bonding. The prevalence of bonding of pipelines into platforms protected by ICCP seems likely to reflect an appreciation of the problems of stray current interference (to which we return in Section 14.5.4).

Table 14.10 UK riser isolation practice 1988

	Riser – structure connection	
Structure CP system	*Insulated*	*Bonded (or presumed bonded)*
Sacrificial anode	16	16
Impressed current or hybrid	1	10
None (concrete structure)	1	0

From [18].

It is worth a comment on the case of the pipelines connecting to concrete structures. Whatever the pros and cons of isolating a cathodically protected steel pipeline from a cathodically protected steel structure (and they are mostly "cons"), there is a strong case for isolating a pipeline from concrete reinforcement. If the pipeline were not isolated, it would inevitably be in direct contact topsides with equipment that was also earthed at some point to the concrete reinforcing steel. The reason that steel reinforcement in concrete does not normally require supplementary corrosion protection is that the high pH of the pore solution in the concrete matrix maintains the steel in a passive state. As we saw in Chapter 5, passive metals exhibit relatively noble (i.e. less negative) potentials than corroding or cathodically protected steel. Thus, the likely consequence of omitting isolation would be to draw current from the pipeline anodes in a fruitless attempt to polarise the passive reinforcement.

14.5.2 Recent codes

The 2003 and 2010 editions of DNV-RP-F103 make no mention of isolating pipelines from structures. The 2016 edition also has little to say. However, in one of its guidance notes, it states... *In case* (of) *interaction leading to premature consumption of pipeline anodes at interfaces to other CP systems is to be fully prevented, the CP systems will have to be electrically insulated by use of monolithic insulating joints.*

The 2004 edition of ISO 15889-2 recommends that... *offshore pipelines protected by galvanic anode systems should be electrically isolated from other pipelines and structures that are protected by impressed current systems.* Furthermore, it requires that... *offshore pipelines shall be isolated from other unprotected or less protected structures which could drain current from the CP system.* It then adds that... *if isolation is not practical or stray current problems are suspected, electrical continuity should be observed.* The 2012 ediition retains this somewhat equivocal advice but also makes the point that isolation... *should be considered...* at landfall terminals.

Rippon [35], representing the NACE task group TG 169 *Control of External Corrosion of Steel Pipelines in Natural Waters*, pointed out that ISO 15589-9 (2004) only gave outline guidance on where electrical isolation should be used. Partly, in response to this lack of steer, the task group issued its own publication the following year [38,40]. Unfortunately, the TG 169 publication falls a little way short of expectation. It briefly describes the types of situation where isolation might often be installed:

- at change of ownership, for example where a third-party pipeline riser connects to a platform,
- at pipeline landfalls,
- at interfaces between ICCP and sacrificial anode systems,
- at interfaces between coated pipelines and uncoated structures,
- at changes in pipeline materials.

However, it offers no hard advice on whether or not isolation should actually be installed. The character of the information it provides is exemplified by its coverage of bare structures tied to coated pipelines... *electrical pipeline isolation between galvanic anode CP systems installed on platforms and pipelines is often omitted. With good CP design, both platform and pipeline are typically adequately protected without impairing the lives of their individual CP systems. However, offshore pipelines are typically isolated from other unprotected or less protected structures, which can drain current from the pipeline CP system.*

This is useful information, but not necessarily something a CP designer can use. To the best of my knowledge, there has been no recent survey of operators to determine if there is any consensus on whether isolation is required offshore. What can be said for certain, however, is that many old isolating joints have been bonded across, either directly or via resistors. On the other hand, nobody has ever felt the need to carry out the very expensive exercise of cutting into an installed offshore pipeline and inserting an isolation joint.

To consider whether or not isolation should be installed on a new pipeline, or if existing isolation should be shorted out, we need to consider what the isolation is intended to achieve, and what might be its potential pitfalls. This brings us to the topic of "interference". This term has developed two quite distinct meanings within offshore CP: current drain, and stray current interference.

14.5.3 Current drain

The first calculation in the CP design exercise in Section 14.3 was to determine the lifetime current requirement of the pipeline. CP designers would like to focus only on the structure they have been commissioned to protect and not to have to attend to the current demands of somebody else's kit. Isolation aids this. If we take the typical case of a coated pipeline tied into an uncoated platform, then isolation of the pipeline from the platform enables the teams designing the CP for the pipeline and the platform, which may well be from different design houses, to pursue their work independently.

Vendors of isolating joints and insulating flanges are not reticent in promoting that benefit. Even so, pipeline engineers are generally unenthusiastic. They like to keep their "metal tubes" as unencumbered as possible. Many view a welded-in monobloc isolation joints, and more particularly flanges, as potential vulnerabilities. If isolation is installed at all offshore, it is more likely to be a monolithic joint, welded into the line, than a flange.

Even then, electrical safety engineers are likewise ill-disposed towards any form of isolation, preferring all metalwork to be bonded to a common earth to eliminate spark hazards. There are many carefully installed isolation flanges that have been shorted by electrical departments, frequently without having troubled to inform the responsible corrosion engineer.

It is difficult to support isolation purely on the basis that it simplifies the CP design process. Omitting isolation inevitably means that there will be some unintended current flow between the connected entities, but this need not be a practical problem.

We have already mentioned the Brae "A" to Forties "C" pipeline which was isolated at Brae but not at Forties. I acted as the operator's representative on the post-commissioning CP survey[1] in 1984. Sailing south from Brae the CP profile was unrelentingly flat. For the most part, the line was close to the potential of the Al-Zn-Hg bracelet anodes: typically, more negative than −1.1 V. Field gradient measurements at selected anodes, which were spaced at only 144 m, indicated very little current being drawn from them. Indeed, the measurements suggested that differential aeration effects were in play on the anodes. The lower parts of the bracelets, which were in the seabed, were delivering a trickle of current to the upper parts of the bracelets which were in open seawater. At that early stage in the coating's life, no current flow onto the extremely well-insulated pipeline could be detected.

However, about 5 km out from Forties, the mind-numbingly dull CP characteristics of the pipeline started to change. Field gradient measurements at the anodes showed that they were starting to deliver current. These indications increased steadily as the survey approached Forties "C". Since there was no indication that the quality of the coating was changing, this observation could only be interpreted as the Forties platform drawing CP current from the pipeline anodes. This was in no way surprising. Although Forties "C" was fully coated below the waterline, in keeping with its original owner's practice at the time, it was protected by zinc anodes. Hence there would have been a difference in potential in the region of 0.05–0.1 V between the structure and the newly installed pipeline. This potential difference readily explained this flow of current.

This was interesting, but it was not a problem. Even if we discount the point that the pipeline CP system was very conservatively designed, in common with pipelines generally at the time, we had no cause to be concerned about this current drain. It represents a "swings-and-roundabout" situation. One way of looking at it was not to regard the platform as taking current from the pipeline, it was simply borrowing it. Later in the pipeline's life, if the predicted coating breakdown were actually to manifest itself, the pipeline could benefit from the same electrical continuity to abstract current from the anodes on the platform.

[1] We deal with pipeline CP surveys in more detail in Chapter 18.

Nevertheless, the understanding that anodes on well-coated pipelines will experience some degree of accelerated consumption, as a result of being directly connected to less polarised terminal facilities, has prompted the practice of doubling up the anode provision on the last 500 m of non-isolated pipelines. This practice dates back to the early 1980s. For example, Sund et al. [10] informed us that insulating joints were considered for the Statpipe system in the North Sea, but were not actually installed. Instead, the designers doubled the anode provision within 500 m of the structures linked to the pipeline.

This realisation that CP current drain between structures and pipelines was something of a two-way street, combined with a growing accumulation of CP survey data showing that the magnitudes of such interactions are modest, has led operators to dispense with isolation between platforms and pipelines, including by bonding across existing isolation. For example, Thomason et al. [20] have reported Conoco's experience in the Arabian Gulf. Prior to 1984, pipelines were isolated from platforms, but this isolation had... *a tendency to break down*. From then on, the company progressively shorted isolation and dispensed with it on subsequent projects. This decision evidently worked out well. The authors made the point that shorting isolation between platforms and pipelines can be beneficial since stand-off anodes in open seawater have much greater current outputs than bracelets in seabed mud. This proved useful in one instance where a pipeline coating was severely damaged, and surveys showed that the anodes on the platform were exerting a beneficial effect on the pipeline out to a distance of 2.3 km.

Other operators were slower to dispense with isolation. For example, it remained the norm for ExxonMobil into the 1990s. According to Surkein et al. [21]... *as part of the normal corrosion control programme for offshore structures, insulating flanges are installed to electrically isolate subsea pipelines from offshore platforms*. However, the authors then went on to claim, for one North Sea installation at least... *computer modelling proved that with various changes to CP design, elimination would be acceptable*. This paper by another major operator doubtless contributed to the overall move of the industry away from pipeline isolation.

14.5.4 Stray current interference

Stray current interference, also called stray current electrolysis, caused by isolation is almost always associated with ICCP systems, and we described how it might occur in Chapter 11. In the case of isolated pipelines, the problem arises if the line picks up current from a host structure's ICCP system, and that current then returns to the host by exiting the pipeline and entering the seawater. In such situations, it is obviously safer if the pipeline is earthed to the host. The current simply passes through a direct metallic path to the negative connection point and does not cause any stray current electrolysis.

Just such a stray current interference problem was reported by Cochrane [17] on the Ninian Southern platform. This was fitted with hybrid CP comprising an ICCP system with 12 seabed anodes design to supply 70% of the 5800 A total current requirement. The remaining 30% was to be supplied by Al-In-Zn sacrificial anodes. Problems were first found in the mid-1980s. The export risers... *exhibited extreme overprotection (–2000 mV) and severe pitting on the electrically isolated riser caused by interference current "leaving" the riser at weak points in the coating and "entering" nearby jacket members*. As a result, all risers were electrically bonded to the structure. The same action was taken on the northern platform which had a very similar hybrid CP system design. In the latter case, some of the seabed ICCP anodes were never activated. This resulted in accelerated consumption of the platform's sacrificial anodes and, in due course, the need for a platform CP retrofit.

Electrical isolation, leading to stray current electrolytic attack, can cause damage relatively quickly on a pipeline. On that basis alone, there is a strong case for dispensing with electrical isolation offshore. However, the situation is not so clear cut when it comes to offshore pipelines when they arrive at their onshore terminals.

14.6 PIPELINE LANDFALLS

14.6.1 Some problems

Varanus island sits 70 km off the northern coast of Western Australia. At 1:40 pm on 3rd June 2008, a 16-year old, 0.3 m diameter sales gas line (SGL) from the island's processing facility ruptured at the beach crossing. The gas release led to an explosion and fire. This caused an adjacent 0.3 m diameter gas import pipeline also to rupture. Happily, there were no casualties among the 166 personnel on the island at the time. The damage to the pipelines was repaired fairly rapidly. Sixty percent of the plant capacity was restored within 3 months, and it was fully operational in 6 months. However, Western Australia's domestic gas supply was reduced by about 30% for 2 months, and the supply to industrial users, including mines and smelters, was reduced by 45%. This reduction is estimated to have caused a loss to the state's economy of around A$3 billion.

For 4 years afterwards, there were various threats of litigation by parties seeking to recover their share of the lost billions. This probably explains the lack of corrosion literature relating to the root causes of this incident. Operators are not keen to draw attention to failures at the best of times, and when litigation is in the air, their desire to disclose is supressed even further. There is, however, a NOPSA[2] report [**43**] in the public domain. This fact alone tells us that, from the Australian perspective at least, the beach section is part of the offshore pipeline. Accordingly, it falls under the purview of this book.

With regards to the failure itself, the report identified three contributory causes:

1. The anticorrosion coating (4.5 mm asphalt enamel) was badly degraded at the beach crossing section due to... *damage and/or disbondment from the pipeline.*
2. The CP was ineffective in the wet-dry transition zone. The line only had bracelet anodes for the offshore section. No CP provision was identified for the onshore section.
3. The inspection and monitoring of the beach section of the pipeline had been ineffective.

Perhaps more worryingly, a 2009 report for the government of Western Australia [45] concluded that... *Apache's understanding of the cathodic protection system on the 12-inch SGL was confused and confusing.* (We will have more to say about this confusion in Chapter 18.)

It has to be said that the Varanus island incident was not unique. In August 1989, a Shell-owned pipeline routed along the foreshore at Bromborough in NW England ruptured causing hot oil to pollute the Mersey Estuary. The report of the UK's Pipeline Inspectorate [**16**] highlighted a number of failings. Perhaps the most telling was the fact that the hot oil product temperatures had been allowed to increase (to ~80°C) without consideration of its damaging effect on the coal tar enamel coating. The report also highlighted a lack of inspection and maintenance of the magnesium anode CP system.

[2] NOPSA National Offshore Petroleum Safety Authority (now NOPSEMA National Offshore Petroleum Safety and Environmental Management Authority).

14.6.2 Isolation

Question: Should we isolate offshore pipelines from onshore terminals?
Answer: Probably.

In practice, an isolation joint is often installed close to the shoreline so that the onshore and offshore CP systems can be designed and managed separately [51].

This usually makes sense because an onshore facility is likely to include various buried piping services, ground-contacting reinforced concrete and tank bases, all of which will probably be tied into a common site earthing system. The facility will, in all probability, be protected by an ICCP system, but the facility hardware in the vicinity of the landfall may be polarised to potentials less negative than the pipeline. In the absence of isolation, offshore pipeline anodes within a few kilometres of the landfall will, therefore, undergo enhanced dissolution – possibly for a substantial part of the pipeline's life.

Fitting, and maintaining, an isolation joint obviates this possibility. Furthermore, if the pipeline anodes do eventually become depleted, and pipeline life extension is required, a resistive bond can be installed across the isolation joint, and the onshore ICCP system can be adjusted to feed some CP current to the offshore pipeline.

I have only been involved in one CP project where isolation was intentionally omitted at the onshore terminal. The pipeline engineers' view that an isolation joint was a potential weakness held sway. As a belt-and-braces response, the offshore pipeline project team connected a bank of magnesium anodes at the terminal fence. A similar approach of using magnesium anodes was reported by Manian [26] for the 2.2 km onshore section of a loading line; the offshore 7 km of which was protected by aluminium alloy bracelets. In that case, although no isolation was installed at the shore crossing, it was installed at the terminal. Unsurprisingly, since the line was new and its coating in good condition, it was observed that the offshore anodes were fully protecting the 2.2 km onshore section even before the magnesium anodes were fitted.

14.7 HOT PIPELINES AND RISERS

14.7.1 Ekofisk alpha

We started this chapter by noting the historical lack of external corrosion problems encountered on cathodically protected subsea pipelines. On the other hand, the same blanket statement cannot be made in respect of risers. In 1975, a 0.254 m (10-inch) hot oil riser on the Ekofisk "A" platform in the North Sea ruptured causing a fire. Although the fire itself did not cause any casualties, an accident with one of the rescue capsules during the ensuing evacuation of the platform resulted in three fatalities and two serious injuries.

The subsequent investigation showed that the riser failure had occurred above the waterline. As such, since CP was not involved, the incident itself falls outside the scope of this book. However, as was inevitable with such a serious incident, there followed a substantial research effort targeted at reducing the probability of similar events in the future. This work focussed on the external corrosion protection of carbon steel pipes transporting hydrocarbons at elevated temperatures. A good part of this work included the influence of elevated temperature on the CP requirements [8,9,11,13].

14.7.2 CP criteria

14.7.2.1 Protection potential

We discussed the protection potential for steel in seawater in Chapter 6. There we saw from a study commissioned by the UK Department of Energy [8] *that... adequate cathodic protection can be provided to either heated or unheated steel by polarisation to a potential of -0.8 V.* The report itself demonstrates that, for temperatures up to 92°C, the limit of the testing, a potential around −0.775 V produces entirely satisfactory levels of protection.

Fischer et al. [9] conducted similar experiments in Norway at around the same time. He was also able to conclude that... *a potential more negative than −760 mV will be adequate as a cathodic protection potential for hot steel (25°C to 50°C) in cold oxygen-rich seawater.*

The above experiments demonstrate unequivocally that −0.8 V is a safe, indeed a conservative, target protection potential for heated steel in open seawater. There is no obvious reason why this should not also apply to heated steel in seabed sediments, although – to the best of my knowledge – this has not been tested by experiment. Nevertheless, despite the lack of supporting evidence, note (f) in Table 1 of ISO-15589-1 calls for a design protection potential if −0.9 V where the temperature exceeds 60°C. It seems that this is another example of conservatism in CP.

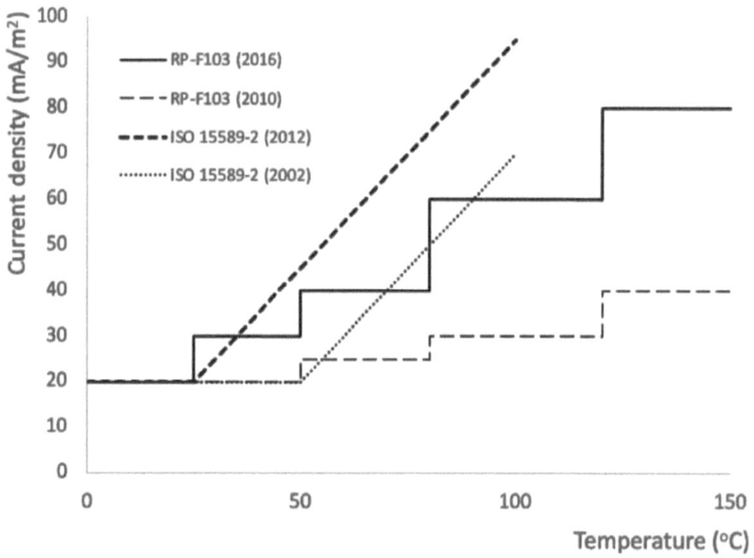

Figure 14.4 ISO and DNV advice on current densities for buried pipelines.

14.7.2.2 Protection current density

In addition to selecting an appropriate protection potential, the pipeline CP designer also has to ensure the provision of sufficient current. Figure 14.4 summarises the ISO 15589-2 (2012) and DNV-RP-F103 (2016) recommendations for the design current densities for buried subsea pipelines as a function of temperature. The figure also shows the advice given in the previous versions of these two codes. To interpret the ISO graph, it is necessary to know the external surface temperature of the pipeline. However, as a matter of experience, in seabed sediments, this is close to the internal fluid temperature, which is the basis of the

DNV graph. As can be seen, both codes recommend increasing the current density as the temperature increases, and both codes have substantially uplifted this temperature effect since their earlier editions.

14.7.2.3 Coating breakdown

As a general rule, the permeability of coatings to seawater increases as the temperature rises, as does their susceptibility to mechanical damage. However, the pipeline coatings listed in Tables 14.2 and 14.3 derive from DNV-RP-F103, Table A-1 of which gives their upper service temperature limits (e.g. 80°C for 3LPE and 110°C for 3LPP). We have to assume that the design breakdown figures given in this recommended practice document apply up to the maximum permissible operating temperature. In truth, if the code were to underestimate the rate of coating breakdown at elevated temperature, this would be more than adequately covered by the conservatism in the advised design current densities at elevated temperature; and further compensated by the design factor (k), if used.

14.7.2.4 Bracelet anode performance

We discussed fall-off in the performance of sacrificial anodes, in particular the fall-off in capacity at elevated temperature, in Chapter 10. We also saw that, as is customary, the codes veer comfortably on the side of caution when providing their advice.

14.7.3 Flow assurance

14.7.3.1 Keeping the product flowing

The pipeline engineer's wife who dismissed pipelines as "just metal tubes" was presumably unaware of the engineering discipline now known as "flow assurance". This would not be surprising since the term had only been in the industry for about 10 years at the time. This multidisciplinary endeavour, which has spawned many more publications since its inception than has CP, seeks to ensure that multiphase hydrocarbon fluids flow efficiently in pipeline systems. A crucial objective is to avoid the pipeline becoming blocked with any of a number of solids: gas hydrates,[3] asphaltenes, waxes and naphthenates. These solids might deposit from the transported fluid if its temperature were allowed to drop. It follows that a major contributor to flow assurance involves maintaining the temperature of the product above some critical value. This, in turn, sometimes requires that pipelines, particularly production flowlines, are thermally insulated, or even heated, to counter the cooling effect of the seawater.

14.7.3.2 Insulation

A good record has been enjoyed by insulating coatings used subsea. However, the range of polymeric materials which possess the required combination of thermal insulation properties and the other physical attributes needed for subsea pipeline coatings is fairly limited. The candidate coating systems include polypropylene and synthetic rubbers such as neoprene. From the pipeline CP design point of view, these systems are adequately covered by

[3] Gas hydrates, or more strictly "methane hydrates", are solid ice-like materials that can form in water-containing natural gas streams at low temperature and high pressure. They may be thought of as comprising methane molecules trapped in the interstices of an ice-like structure. Such chemical entities, in which one molecule is physically trapped within the structure of another, are termed clathrate compounds.

the (generous) coating breakdown figures advised by the codes. Providing these materials are used within their prescribed temperature limits, their code-based breakdown performances can simply be factored into the current demand calculations. There is no need for too much head scratching on the part of the CP designer.

However, these materials have not always been without their problems. For example, an interesting case arose concerning a 5.4 km 6-inch carbon steel flowline installed in the North Sea in 1987 [25]. The pipeline was thermally insulated with EPDM (ethylene propylene diene monomer) rubber. This insulation had been applied directly to the pipeline, without any anticorrosion coating. Potential measurements, taken at the time of pipe laying, gave readings in the range of −650 to −750 mV. This was surprising since it was known that well-coated pipelines usually adopted the anode potential (~1050 mV) soon after laying. The line was connected to the well and the platform the following year, and was gravel-dumped for stabilisation. An ROV survey after the gravel dump reported that the anodes at the end of the line, which were not covered in gravel, were severely wasted. A further potential survey in 1989 showed that the potential of the line now varied between −860 mV at the platform tie-in to −760 mV at the wellhead. Thus, although there had been some further polarisation since it was laid, the line had failed to reach the operator's specified target protection potential of −900 mV.

A desk-based review of the problem was influenced by anecdotal reports of problems encountered when applying the cast polyurethane field joint coating. Apparently, because of the speed of the pipelay, the joints were not allowed to cool before entering the seawater, leading to the cracking of some of the field joint castings. In the event, a CP retrofit was designed to make up the anticipated shortfall between the as-designed current output from the original bracelet anodes and the revised anticipated current demand assuming all of the field joint coatings (amounting to about 3.5% of the pipeline's surface area) had failed. This retrofit was installed early in 1991.

Surprisingly, potential measurements made prior to connecting the retrofit anodes showed potentials as positive as −525 mV. This is more than 100 mV more anodic than the natural corrosion potential for steel in seawater. It could only mean that accelerated corrosion of the flowline was in progress. Connecting the retrofit anodes went some way to mitigating this. However, a further CP survey in May 1991 showed that the retrofitted anodes were delivering considerably more current than anticipated, yet the pipeline was achieving only modest polarisation. Evidently, something very strange indeed was happening.

A spare insulated pipe spool was available, and samples were cut and taken to a laboratory for analysis. Electrical measurements confirmed that the EPDM coating was, unexpectedly, a very poor electrical insulator. Whereas the polymer was expected to exhibit a bulk resistivity of around 10^{14} Ωm, the measured value was 10 Ωm. Moreover, this fell to 2 Ωm at a pressure of 1.6 MPa, equivalent to the water depth of the flowline. In addition, electrochemical polarisation experiments confirmed that the coating was electrochemically active.

The explanation for this anomalous behaviour was not in the EPDM rubber itself, but in the carbon black used as a reinforcing pigment. (The tyres of your motor car are the colour they are because they too are reinforced with carbon black.) In this case, the coating contained about 40%. Thus, instead of the pipeline anodes being required to supply the very small current needed to protect the carbon steel at small areas of coating breakdown, they effectively found themselves coupled to a very large area of carbon cathode. The operator prudently changed pipeline insulation specification for future projects to ensure that carbon-loaded EPDM was thenceforth only applied over a non-conducting anticorrosion coating such as polypropylene.

However, in this case, not only was the original CP system wholly ineffective, more current than was expected was being drawn from the retrofit anodes. Even then, these were only just managing to protect the line. This, combined with a decision to extend the operating life of the flowline, forced a decision to install a second high current output CP retrofit.

14.7.3.3 Direct electrical heating

Although insulation can be vital in maintaining the transported fluids at elevated temperature, thereby providing flow assurance, its benefit is limited if there is a stoppage in production. At best, it can slow the rate of cooling. But, if the shut-in is sufficiently long, the pipeline and its contents will cool to seabed temperature. In some systems, this might cause the formation of solids which would plug the line. If this were to occur, it might prove impossible to restart the flow.

For this reason, a small number of production flowlines are fitted with direct electrical heating (DEH), the first having been installed in the North Sea on six 9-inch 13Cr Åsgard flowlines, ranging from 6.5 to 8 km in length in 2000 [41]. Up to 2005, further systems were installed by the same North Sea operator on other 13Cr flowlines: Huldra (8-inch 16 km) and Kristin six 10-inch, 6.0–6.7 km); and on the 316L-lined carbon steel Urd flowline (12½-inch 9 km). However, the impetus for DEH seems to have waned after 2005. By 2013, only a few more flowlines had been added to the DEH inventory [48].

The principles of the technique have been described in electrical engineering publications [42]. A high current AC supply (50 Hz, single phase) is connected across the length of flowline to be heated. The passage of the current through the pipe wall causes resistive heating. As pointed out by Harvey et al. [48], the power needed for maintaining temperature in a typical flowline might be in the order of 120 W/m; and to heat a cold line, and its contents, sufficiently to enable flow to resume might need 300 W/m.

The magnitude of the current and the voltage needed to produce this heating effect depends on the longitudinal resistance of the pipe. As we have seen above, when discussing the attenuation of potential, this depends on the diameter and wall thickness of the pipe. The early systems were designed to deliver currents up to 1520 A and required substantial AC voltages of up to 24 kV. Aside from requiring very substantial electrical conductors (up to 1000 mm^2 cross-section), this power demand means that DEH systems can only be available at platforms with a substantial surplus of available power.

In an ideal world, all of the DEH current would be contained within the pipe wall where it is required for heating. However, this would mean that no anodes could be fitted to the flowline, and the total responsibility for corrosion control would then reside with the coating. This was rightly deemed unacceptable [41] because the performance of the coating would have to be perfect. The alternative is to accept that the line needs to be fitted with sacrificial anodes for corrosion protection. These anodes will earth the line, meaning that the DEH system has to be designed on the basis that there will be current losses to the seawater.

The resulting arrangement for the application of DEH is shown in Figure 14.5. This involves the power cables being connected to the pipeline at two so-called current transfer zones. Roughly 60% of the current flows through the pipe wall. The remaining ~40% flows through the seawater as a result of the line being earthed by the sacrificial anodes which are installed at the current transfer zones and at intervals along the pipeline itself. This arrangement obviously prompts two related questions. The first is: does the passage of a large fraction of this high AC current across the anode-seawater surface damage the anode? The second is: does it pose a threat to the pipeline?

The codes for subsea pipeline CP offer no advice regarding any modifications of the CP design parameters where DEH is applied. The only information ISO 15589-2 provides with respect to alternating currents is that... *interference on pipelines, although rare on offshore pipelines, can cause safety and corrosion issues if not mitigated effectively*. It does not consider DEH. DNV-RP-F103 and Norsok M-503 are also silent on the matter.

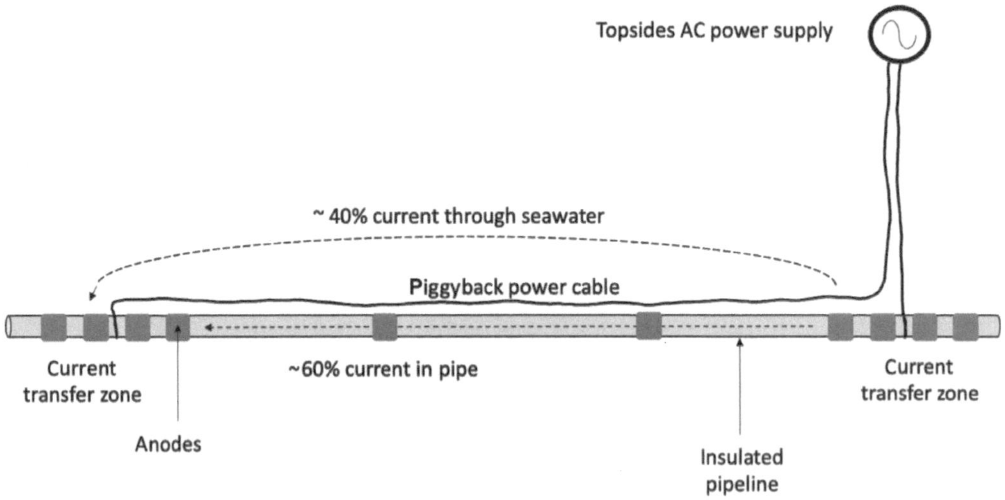

Figure 14.5 Schematic of DEH.

The issue of AC corrosion, and its interaction with CP, has been very extensively studied; but, as yet, imperfectly quantified in the context of on-land pipelines where AC pick-up arises as a result of buried pipeline running along rights-of-way shared by high voltage power transmission lines. On the other hand, AC current pick-up and discharge has not been a problem for subsea pipelines, even though some run parallel to underwater AC power cables.

It has long been known that an AC current poses a substantially lower corrosion threat than an (anodic) DC current of similar magnitude. However, the AC current flowing between the anodes and the seawater is very large (hundreds of amps). One might reasonably be concerned that, for 50% of the AC cycle, this high current is anodic. On this basis, we might anticipate dramatic anode consumption and even damage to the pipeline itself.

Fortunately, the effect turns out not to be too dramatic. The reason for this lies in the nature of the electrochemical interface between the metal (aluminium anode) and the electrolyte (seawater). Thus far, we have only considered this interface as behaving like a resistor: for example, when we discussed polarisation in Chapters 5 and 7, and where we considered attenuation equations earlier in this chapter. In fact, the electrochemical behaviour of the interface is too complex to describe simply in terms of resistance. Not only does the interface offer a route for charge transfer (i.e. electrical conductance), it is also the site of electrical charge separation. This charge separation imbues the interface with capacitive behaviour, referred to in electrochemistry as the double-layer capacitance.

Thus, the interface is actually an electrolytic capacitor, albeit a "leaky" one. This means that most of the AC current flow into the seawater crosses the interface without influencing the anodic or cathodic processes taking place. The process has been studied by the Norwegian research organisation SINTEF. The details are not in the public domain, but it is reported [37] that there is no increased corrosion of sacrificial anodes due to AC currents up to 100 A/m^2. They further found that, where steel is protected by sacrificial anodes, the tolerable current density on the steel is 300 A/m^2. However, following further investigations which indicated corrosion damage to carbon steel at AC current densities of 250 A/m^2 [46], this figure was subsequently revised down to 100 A/m^2 [47].

The use of DEH remains limited as far as subsea pipelines are concerned. The only operator to have disclosed an involvement is the Norwegian company Statoil (now Equinor). The implications for CP are that the extended anode spacings implied by ISO 15589-2 for well-coated pipelines do not apply to lines with DEH since the requirements for earthing the AC current trump the simple CP needs.

14.7.4 Seawater cooling

14.7.4.1 Pipelines

As a generalisation, the flow assurance engineer would prefer to run pipelines hot. The pipeline materials engineer, on the other hand, would prefer to run them cool in order to mitigate the internal corrosion threats. This might permit the selection of carbon steel ahead of a more expensive CRA. On some projects, where flow assurance permits, instead of thermally insulating the pipeline, the seawater is allowed to cool it. Once the fluid contents temperature falls to a value at which the predicted corrosion rate of carbon steel becomes tolerable, the pipeline material changes from CRA to carbon steel.

There have been instances of this cooling being achieved either by constructing the first part of the pipeline in duplex stainless steel (DSS) before transitioning to carbon steel for the rest. The DSS has typically been coated with fusion bonded epoxy paint and provided with CP designed to the same protection potential as the carbon steel to which it connects. The alternative of permitting a more noble protection potential for the DSS section, whilst permitting greater anode spacing, would require the installation of an isolation joint subsea which, whilst not unheard of, is viewed as problematic by pipeline engineers. Either way, the CP design, in particular the cathodic current density and the expectation of anode performance, is modularised in keeping with the change of the internal fluid temperature along the line.

14.7.4.2 Subsea coolers

It is convenient also to mention subsea coolers here. I realise that I am stretching a point by including these in a chapter on pipelines because these are piping. Nevertheless, this is a subtlety of no relevance to a sacrificial anode.

Many topsides hydrocarbon processing facilities are required to cool hot crude prior to further processing and export. This cooling can be achieved either by air coolers or direct heat exchange with seawater. The advantage of installing a subsea cooler is that it frees up the topside's space and weight that would otherwise be needed for conventional shell-and-tube or plate-type seawater exchangers or a bank of air coolers.

The piping in a subsea cooler is always a CRA (typically duplex stainless steel), and the factory-applied coating is usually fusion bonded epoxy. The cooler assembly is fixed into a steel support frame which is usually painted with an approved subsea coating system (e.g. System 7 per NORSOK M-501 – see Chapter 9). As a matter of convenience, the anodes are fixed on the frame rather than on the piping itself.

In principle, therefore, the CP design should take account of the potential attenuation down the entire lengths of the tubes since these are only assured to be in contact with the frames at their ends. Depending on the design of the cooler, there might be a very considerable length of tubing – a fact not changed by its convoluted layout. However, since the usual arrangement is to use aluminium alloy anodes on a well-coated frame (about –1000 mV), and the CRA tubes only require a modest polarisation target (typically –500 mV), attenuation is unlikely to be an issue.

14.8 PIPELINE RETROFITS

14.8.1 Why retrofit?

14.8.1.1 Something went wrong

We have already mentioned a case of something going wrong with the overall pipeline corrosion protection system when we recounted the problem of the carbon-loaded EPDM insulation on a flowline. That was something of a rarity. A more frequent problem has arisen as a result of anodes becoming detached during the pipelay. Such tales are mostly anecdotal, but some instances have made their way into the public domain literature. For example, Smith [19] recounts a case in the early 1980s in which a 12-inch gas line was laid from a platform to the shore 39 km away. In the event, only one of the five anodes allocated to the first kilometre of the lines remained attached after the pipelay. The anodes on the remaining 38 km of line which, unlike the first kilometre, was concrete weight coated, were fine.

14.8.1.2 Life extension

Mishaps such as mentioned above have not been particularly common and became very rare indeed as the technology of offshore pipelaying matured. Most pipeline CP retrofits are needed not because something went wrong, but because something went right. Hydrocarbon production remained viable longer than intended. So, the pipelines were required to operate beyond their originally intended design lives.

In 2017, Britton pointed out [50] that ~55,000 km of offshore pipeline had been installed prior to 1991 (>25 years earlier), of which 17,000 km had been installed prior to 1981 (>35 years earlier). Although some of those lines had been decommissioned, the majority were still in service and destined to remain so for the foreseeable future. In some instances, this had obliged the owners to consider retrofitting supplemental CP.

Please note that I have said "consider retrofitting". Just because a pipeline is approaching, or has just passed, the end of its design life does not mean that the pipeline owner immediately needs to reach for the cheque book! There is plenty of scope for more analysis before doing that.

14.8.2 When to retrofit?

14.8.2.1 What you cannot see...

If a pipeline is lying on the seabed, it is a straightforward matter to both determine the potential profile along its length (see Chapter 18) and, crucially, to make a visual assessment of the extent to which the bracelet anodes have been consumed. If the line is (say) more than 20 years old, is fully protected (potentials more negative than −800 mV) and a substantial portion of each anode (say >40%) remains in evidence, then it does not take much study to conclude that nothing will need be done for some time to come. That is not to say, of course, that we can forget about the line. It will be prudent, as well as a statutory requirement, to keep the condition of the CP system under surveillance.

The position becomes a little trickier where the pipeline, or substantial lengths of it, are buried in the seabed. We can still attempt to determine the potential profile along the line, even if that involves some loss of precision compared to pipelines that are accessible on the seabed. However, short of excavating the line at the location of each anode, it is not possible to determine the remaining life of the CP system.

In one exercise I was commissioned by a major operator, responsible for an ageing population of offshore pipelines, to assist with the evaluation of its pipeline inventory. The operator was mindful of the possible future need to install retrofit CP on some lines. It was also aware that, since most of the lines were buried, a substantial amount of investigative work would need to be carried out as a precursor to designing or commissioning any such retrofits. The crucial question was: Which pipelines should be investigated first?

A simple answer might have been: start with the oldest pipeline. However, experience suggests that would not always provide the most sensible answer. A more thoughtful study was needed.

14.8.2.2 Lost in the Iron Mountain®

You might think, as did I, that the obvious place to start would be a desk study in which all of the historical CP and coating information for each line was correlated and examined in the context of the CP inspection and monitoring data. Unfortunately, that approach adopts a view that an operator, or more particularly its personnel, have access to such historical data. The reality is less utopian.

We are in the 21st century where information technology rules. Most of the pipelines that are in use today were installed in the last century. Furthermore, many that are now old enough to be candidates for retrofitting were designed before the omnipresence of computers at every engineer's desk. The CP designs calculations were usually performed using a pencil. The first version of Microsoft Excel® for Windows did not make an appearance until November 1987. To younger readers, that might seem like a very long time ago; but, as we have observed, a lot of operational pipelines are way older.

Prior to the early-1970s, most CP design reports were typed on a wax stencil, and the requisite number of copies then run off. Xerox soon consigned wax stencils to rubbish skip, but the document production remained the domain of the typewriter. Most organisations moved quickly to word processors from the early 1980s, although the first version of Microsoft Word® did not turn up until 1984. Even then, the electronic storage medium for documents was going through its own tortuous evolution. When was the last time you saw a 5¼-inch floppy disk?

Ultimately, the definitive record of CP engineering was committed to paper: a lot of paper. Then arrived the new millennium and the cult of the paperless office. In the case of my client, many tonnes of paper had disappeared into a mystical place called Iron Mountain. With it, presumably, had travelled those few sheets of paper which would have contained all of the CP design data I needed for the pipelines they had asked me to study. Try as I may, I was unable even to find anyone within the organisation who had any idea about how to find the documents, let alone release them from their entombment.

There was nothing to be gained from questioning the operator's pipeline and corrosion engineers, most of whom were younger than the pipelines they were charged with managing. Several times in the recent past I have found myself assisting offshore operators on pipeline CP issues. In each case, I have asked the obvious question: is the line isolated; and in each case, the pipeline group had to refer my query to the offshore inspection personnel. It reminded me that the 1988 information summarised in Table 14.10 was largely compiled by one of the original MTD publication's [18] authors simply chatting to the incumbent CP engineers. Good luck to anyone who fancies repeating the exercise in the 21st century!

So, in the absence of crucial CP design information on the buried pipelines, I went back to basics and constructed reasonable designs using the codes current at the time. In general, the calculations showed that, by modern standards, the anodes were very substantially over designed on the basis of weight. The life-limiting factor would be the ability of the anodes' instantaneous outputs to keep up with the predicted progressively increasing cathodic current demand. The latter parameter was re-estimated on the assumption that the coating breakdown would more realistically follow the lower, but still conservative, predictions in ISO-15889-2 (2012).

The calculations for various scenarios were straightforwardly implemented on a spreadsheet. As is usually the case with such exercises, the result was a plethora of graphs resulting from the various "what if" analyses that were run. One example is given in Figure 14.6 for an asphalt enamel and concrete coated 20-inch gas trunk line that had been installed, unburied, in 1985. The evaluation was carried out in 2015 which was, technically, at the end of the CP system's design life.

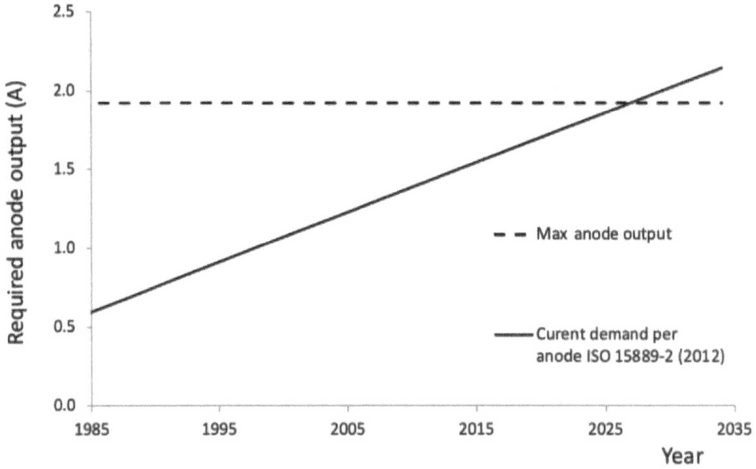

Figure 14.6 CP design re-evaluation for a 20-inch gas pipeline.

The original CP had been designed on the basis of the then-current DNV code (TNA 703). In this case, the bracelet anode dimensions and spacing (109.8 m) were known. This permitted a reconfirmation of the current output. This was then compared with the expected increasing cathodic current demand. Figure 14.6 illustrates the outcome for the line in the unburied condition and assuming the maximum seawater cathodic current density given in the ISO. As can be seen, in 2015 there was a reasonable expectation that the anodes would last for more than a decade going forward.

Of course, it would be hubristic to assume that the outcome of a spreadsheet-based desk study alone could permit the operator to leave the CP system on the pipeline unattended for (say) the next decade. However, when applied to the entire inventory of pipelines, this exercise permitted a risk-based prioritisation of the pipelines for more detailed investigation and analysis.

14.8.3 Retrofit strategies

14.8.3.1 Basic cases

Even allowing for the historical overdesign of early CP systems, some pipelines will eventually need a retrofit if required to remain in service for long enough after their original design lives. Drawing on experience with Gulf of Mexico pipeline retrofits, Britton [30] has stated that the overall cost element is made up as follows:

- Design: 2%–5%
- Materials 20%–25%
- Installation: 75%–80%

Clearly, one of the prime objects of the design work, which includes the problematic issue of determining the retrofit current demand, is to minimise the offshore installation costs. Where the retrofit current is to be provided using sacrificial anode sleds, the intention of the design will be to minimise the number of sleds that need to be deployed offshore.

Each pipeline retrofit scenario is, in many ways, unique. Nevertheless, there are some broad-brush generalisations that can be made.

The first of these generalisations is that, although the original CP system was, almost certainly, based on bracelet anodes, the imperative to minimise installation costs means that the retrofit will not be. Where there is an opportunity to provide CP current from terminal installations, for example by bridging across isolation flanges on host platforms or at the shoreline, then this opportunity is likely to be taken.

Things get a bit trickier when we try to determine the spacing between retrofit anode arrays. We have already seen how problematic it is to try to conjure up an attenuation curve for a pipeline. We find ourselves applying non-scientific intuition to non-scientific rules-of-thumb. In the absence of an attenuation calculation method that does not involve making wild guesses about the conductivity of the coating, we could do worse than to follow Britton's generalisation [30] that short in-field pipelines (<8 km) can be shorted to the jackets at each end. Long lines (>16 km) will require multiple anode sleds. Although not stated, the paper implies that lines between 8 and 16 km in length will require a single anode sled at the midpoint.

From the perspective of designing a retrofit anode sled for a pipeline, the basic principles are essentially the same as designing one for a fixed structure. The main difference arises from the fact that sleds for fixed structures are usually deployed within the exclusion zone around the structure. Pipeline retrofit sleds, on the other hand, may need to be designed to avoid the risk of damage due to trawling.

14.8.3.2 Connecting anode sleds to pipelines

Any retrofit will only be successful if a sound connection can be made to the pipeline. Unlike retrofits on structures, pipeline retrofits require the removal of a patch of coating ~100 mm wide. This can be achieved by high-pressure water jetting or underwater hydraulic wheels [23]. Various connection clamp options have been developed [44] which can be more easily installed by either divers or ROVs (see Figure 13.14 in Chapter 13). Alternatively, friction welding, which we have discussed in the context of retrofits on structures, has also been used successfully on pipelines [23].

14.9 CRA AND FLEXIBLE PIPELINES

As mentioned, CRA is used for pipeline construction where carbon steel is incompatible with the corrosivity of transported fluids. The CRAs used for pipeline construction are the 22% Cr and 25% Cr duplex, and 13% Cr weldable supermartensitic, stainless steels. We introduced these alloys, and their cathodic protection, in Chapter 8. As indicated in the codes, these materials can be protected at less negative potentials than required for carbon steel. Indeed, the known susceptibility of some of these materials to hydrogen embrittlement might suggest that a less negative potential would be essential.

It is, of course, possible to design a CP system to achieve a less negative potential. However, this requires the CRA pipeline to be electrically isolated from any carbon steel structures. This has been done. For example [34], subsea isolation has been installed on 13% Cr flowlines which were then protected using potential limiting diodes, using the methodology we described in Chapter 10.

However, as we have seen, electrical isolation can introduce stray current electrolysis threats. We have also mentioned that both pipeline and electrical safety engineers prefer to omit the isolation. This might mean polarising to potentials achieved using conventional Al-Zn or Zn-alloy sacrificial anodes. This places the onus for pipeline integrity on ensuring that the metallurgical condition of the CRA is sound and that it is not subject to excessive strain.

Finally, we need to point out that flexible pipelines have gained some prominence in offshore hydrocarbon production. These are usually constructed with various layers of high tensile steel armouring and have end fittings made of CRA or HSLA steel. It is relatively straightforward to protect the end fittings by attaching small anodes, typically: ~2 kg of Al-Zn-In or, more usually, zinc. In so much as a design is required for these, it can be carried out in accordance with a code such as DNV-RP-B401. However, there is no realistic prospect of anodes on the termination protecting anything more than a very small area of armouring that might become exposed as a result of mechanical damage to the outer polymeric sheath. This is a situation where corrosion protection has to rely almost entirely on the coating.

REFERENCES

1. R. Pope, *Attenuation of forced drainage effects on long uniform structures.* Corrosion **2**, 307 (1946) republished in Materials Performance **20**(12), 29 (1981).
2. M. Stern and A.L. Geary, *Electrochemical polarization: a theoretical analysis of the shape of the polarization curve.* Journal of the Electrochemical Society **104**, 56 (1957).
3. H.H. Uhlig, *Derivation of potential change along a cathodically protected pipeline*, in Appendix to *Corrosion and Corrosion Control*, 2nd edition. Wiley and Sons (New York) 1971, pp. 396–401.
4. W.B. Mackay, *North sea offshore cathodic protection* Paper, *1957* Offshore Technology Conference, Houston, TX, 1974.
5. R. Strømmen, *Current and potential distribution on cathodically protected submarine pipelines.* Transactions of the Institute of Marine Engineers **91** NSC Conference 49 (1979).
6. R. Mollan and S. Eliassen, *The effect of cathodic protection in gaps and crevices on offshore pipelines and risers.* Paper E1 4th International Conference *Internal and External Protection of Pipes* BHRA, Cranfield, 1981.
7. R.N. Duncan and G.A. Haines, *Forty years of successful cathodic protection in the Arabian Gulf.* Materials Performance **21** (1), 9 (1982).
8. G.P. Rothwell, P.E. Francis and K.F. Hale, *The effects of heat transfer on the external corrosion of submarine pipelines and risers,* Summary report OT-R-8292 Society for Underwater Technology for UK Department of Energy (London) 1982.
9. K.P. Fischer et al., *Hot risers in the North Sea: A parametric study of CP and corrosion characteristics of hot steel in cold seawater*, Paper no. OTC 4566 Offshore Technology Conference, 1983.
10. S. Sund et al., *Corrosion protection design parameters for Statpipe pipeline system*, Proceedings of UK CORROSION, 1983.
11. R. Mollen and S. Eliassen, *Design criteria examined for cathodic protection of hot submarine pipelines.* Oil and Gas Journal, 52–56 February 20 (1984).
12. J. Cochran and F. Mayes, *The design of cathodic protection systems of offshore pipelines in varied environments, Parts 1and 2* Paper no. 333 CORROSION/84.
13. N. Nilsen and B. Espelid, *Corrosion behaviour and cathodic protection of stainless steels for offshore hot risers,* Paper no. 320 CORROSION/1985.
14. T. ffrench-Mullen and W.R. Jacob, *Pipelines undersea*, chapter 12, in *Cathodic Protection: Theory and Practice*, Eds. V. Ashworth and C.J.L. Booker. Ellis Horwood (Chichester) 1986, pp. 214–225.

15. J.H. Morgan, p. 201, Chapter 6 in *Cathodic Protection*, 2nd edition. NACE (Houston, TX) 1987
16. D.A. Southgate, *Investigation report of the hot oil pipeline failure at Bromborough on Saturday 19th August 1989*, HMSO Publication ISBN 0 11 413703X, 1990.
17. W.J. Cochrane et al., *A computerized CP retrofit design of the Ninian Northern platform*, Paper no. 375 CORROSION/1990.
18. *Design and operational guidance on cathodic protection of offshore structures, subsea installations and pipelines.* Marine Technology Directorate (London) 1990 ISBN 1 870553 04 7.
19. S.N. Smith, *Analysis of cathodic protection on an underprotected offshore pipeline.* Materials Performance **32** (4), 23 (1993).
20. W.H. Thomason et al., *Shorting pipeline and jacket cathodic protection systems.* Materials Performance **32** (9), 21 (1993).
21. M.B. Surkein et al., *Cathodic protection design analysis and review of electrical isolation requirements.* Materials Performance **32** (5), 25 (1993).
22. V. Ashworth, *Principles of cathodic protection*, Section 10.1, pp. 10:23–10:27 in *Corrosion and Corrosion Control*, Eds. L.L. Shreir, R.A. Jarman and G.T. Burstein, 3rd edition. Butterworth-Heinemann (Oxford) 1994.
23. R.H. Winters and A.C. Holk, *Cathodic protection retrofit of an offshore pipeline.* Materials Performance **36** (9), 15 (1997).
24. I. Kozelkova, M.G. Rogers and R. Nuttall, *An acceptable approach to attenuation of cathodic protection along subsea flowline,* Paper no. 14 Proceedings of UK CORROSION, 1998.
25. W.R. Jacob and C.G. Googan, *High current CP retrofits for a submarine flowline*, Proceedings of Conference *Corrosion and the Environment*, Bath 15–17 April 1998 NACE (Houston, TX) 1998.
26. L. Manian, *Galvanic protection of an offshore loading pipeline in the Baltic Sea.* Materials Performance **39** (5), 24 (2000).
27. P. Pierson et al., *A new equation for potential attenuation and anode current output projection for cathodically polarized marine pipelines and risers.* Corrosion **56** (4), 350 (2000).
28. M. Surkein et al., *Corrosion protection program for high temperature subsea pipeline*, Paper no. 01500 CORROSION/2001.
29. J. Britton, *Cost saving offshore cathodic protection retrofit methods*, Proceedings of UK Corrosion, 2001.
30. J. Britton, *Offshore pipeline CP retrofit strategies*, Paper no. 02018 CORROSION/2002.
31. D.K. Lysogorski and W.H. Hartt, *A potential attenuation equation for design and analysis of pipeline cathodic protection systems with displaced anodes.* Corrosion **60** (9), 815 (2004).
32. S. Eliassen et al., *Design and installation aspects of cathodic protection for high temperature insulated pipelines: a case study based on the new ISO standard for cathodic protection of submarine pipelines,* Paper no. 04090 CORROSION/2004.
33. J.P. La Fontaine and G. Gibson, *Cathodic protection design of pipelines using an improved attenuation method,* Paper no. 04091 CORROSION/2004.
34. A. Sjaastad et al., *CP design of a super 13% Cr flowline*, Paper no. 04092 CORROSION/2004.
35. I.J. Rippon, *New ISO cathodic protection standard for offshore pipelines* Paper no. 04101 CORROSION/2004.
36. A. Tawns, *Reel-laying of rigid pipelines.* Materials Performance **43** (10), 24 (2004).
37. J.K. Lervik et al., *Design of anode corrosion protection system on electrically heated pipelines.* Vol. 2, p. 26, Proceedings of Fourteenth International Offshore and Polar Engineering Conference, Toulon, 2004.
38. NACE, Item 24228 Task Group TG 169, *Report on electrical isolation/continuity and coating issues for offshore pipeline cathodic protection systems*, 2005.
39. M. Roche, *External corrosion of pipelines: What risk?* Proceedings of 14th Middle East Oil and Gas Show (Re-published as SPE Paper 93600), 2005.
40. I.J. Rippon, *A new standard and state of the art report from TG169 cathodic protection in seawater*, Paper no. 06101 CORROSION/2006.

41. S.M. Hesjevik and S. Olsen, *Direct electric heating on subsea pipelines a challenge to corrosion protection*, Paper no. 06178 CORROSION/2006.
42. A. Nysveen et al., *Direct electrical heating of subsea pipelines: Technology development and operating experience.* IEEE Transactions on Industry Applications **43** (1), 118 (2007).
43. NOPSA, *Final report of the findings of the investigation into the pipe rupture and fire incident on 3 June 2008 at the facilities operated by Apache Energy Limited on Varanus Island*, 2008.
44. J. Britton and D. Baxter, *Extending the life of cathodic protection systems for offshore pipelines: some recently applied new technologies*, Paper no. 1016 EuroCorr, 2008.
45. K. Bills and D. Agostini, *Offshore petroleum safety regulation Varanus Island incident investigation*, 2009.
46. L. Sunde Liieby, S.M. Hesjevik and S. Olsen, *Effects from alternating current on cathodic protection of submarine pipelines*, Paper no. 11055 CORROSION/2011.
47. S.V. Hesjevik and S. Olsen, *Cathodic protection of submarine pipelines with direct electric heating*, Paper no. 2443 CORROSION/2013.
48. D. Harvey, G. Winning and X. Hu, *Direct electrical heating of subsea pipelines: What are the effects on cathodic protection?* ICorr London Branch Meeting: *Cathodic Protection*, London, 2013.
49. A. Pedersen, T. Sydberger and E. Skavås, *External corrosion control of submarine pipelines*, Paper no. 7430 CORROSION/2016.
50. J. Britton, *Ageing subsea pipelines external corrosion management*, Paper no. 9642 CORROSION/2017.
51. S. Song, T.G. Cowin and A.C. Nogueira, *Systematic design and management of an offshore cathodic protection system*, Paper no. 11321 CORROSION/2018.

Chapter 15

Ships and floating structures

A lesser-known Davy

If you were Professor of Chemistry at two Irish universities, a Fellow of the Royal Society, and had discovered acetylene, then you might expect history to record you as, at least, the top scientist in your family. This was not so for Edmund Davy, whose more illustrious cousin happened to be Sir Humphry, who was responsible for the first-ever application of anodes when he used them to protect the copper sheathing on the wooden hull of *HMS Samarang*. As we have explained, although the corrosion was controlled, the loss of the antifouling properties proved to be a bigger problem. Marine CP was close to being dead-in-the-water; almost before it had begun.

This is where Edmund Davy makes his brief, but crucial, appearance in our story. He had assisted his illustrious cousin with his experiments on protecting copper from corrosion with zinc or iron blocks. In addition to experimenting on copper, the cousins also carried out experiments to see what would happen if zinc blocks were attached to iron. Edmund had evidently learned from this. In 1829, after he had moved to Ireland, he demonstrated that zinc blocks could successfully protect the ironwork of mooring buoys [29]. Thanks to Edmund, marine CP was still hanging on.

Nevertheless, even before *HMS Samarang*, now without anodes, saw her first active service in the Anglo-Chinese Opium War (1839–1842), the end was in sight for sail-powered wooden-hulled warships. Steam-powered iron-hulled vessels were being built from the 1830s. Early vessels were driven by paddle wheels. Screw propellers, usually in bronze, started displacing these from the 1840s. By the middle of the century, any nation seeking to project naval power around the world was employing steam-powered wooden vessels clad with up to 100 mm of iron armour. Well before the century was out, the wood had disappeared. The use of the term "iron-clad" lapsed from the nautical lexicon. Hulls were now being constructed in iron, and in due course, steel.

In this chapter, we are only concerned with the external hulls of ships, and other floating structures, together with external protuberances such as mooring cables and tethers. We deal with internal surfaces, such as ballast tanks, sea intakes and seawater piping in the next chapter.

15.1 SHIPS' HULLS

15.1.1 Early days

Drawing on his own extensive experience in US Naval research, and with access to some reports that did not make it into the public domain, Peterson has described what is known of the early history of marine CP [11]. His useful account acknowledges the scarcity of

DOI: 10.1201/9781003216070-15

documented material (at least in the English language) covering the first half of the 20th century.

The practice of fixing zinc blocks to the hull was established in some navies before World War II. Peterson cites a 1938 US Navy instruction calling for zinc "wasters" to be fitted on all vessels operating in salt water. It seems that the British and Canadian Royal Navies were also following this practice. This may have been a recognition of the propensity of galvanic interactions between bronze propellers and steel hulls. However, for a long time, it probably owed more to tradition than to any understanding of CP. Apparently, it was not uncommon for a shipyard, having taken the trouble to attach the zinc blocks to the hull, then to paint them[1]!

The CP of naval ships then moved onto a sounder footing after World War II. In June 1949, the Royal Canadian Navy fitted magnesium anodes to the bilge keels of HMCS New Liskeard. She was dry-docked in May 1950 and Barnard reported [1] that... *no corrosion was observed on the underwater hull except in a few re-entrant places*. This enabled the conclusion that CP offered... *the possibility of lengthening the period between dry-dockings of active ships from one year to two*. On the basis of these encouraging findings, Barnard was instrumental in drafting the Canadian Navy's first specification for the CP of ocean-going ships [2]. Other navies soon followed.

15.1.2 CP design

15.1.2.1 Differences between fixed structures and ships

A steel plate in contact with seawater is unaware of whether it is part of a fixed or a floating object. So, we could have included ships in Chapter 13 along with the various other steel structures. The reason that we did not do this is that corrosion, and its control by CP, depends upon the nature of the metal, the nature of the environment and, crucially, the circumstances of exposure of the metal to the environment. A ship resembles a fixed offshore structure, in that it is made of steel and exposed to seawater; but the circumstances of that exposure are crucially different.

CP is required to protect fixed structures for extended periods, sometimes in excess of 40 years. The object is to mitigate corrosion losses that would diminish the load-bearing capacity and to control corrosion fatigue.

In the case of ships, however, the primary impetus for applying CP was to help keep them smooth [15]. The rusting that occurred on hulls painted with traditional oil-based or bituminous paints caused roughening which increased the hydrodynamic drag. This interfered with ships' speed and very markedly increased the fuel costs. The thinning of a hull was regarded as a less important consequence of corrosion.

15.1.2.2 Current demand

15.1.2.2.1 Bare steel

Irrespective of whether the ship is to be fitted with an ICCP system or sacrificial anodes, the first stage in the design exercise should be the same: determining the cathodic current demand. This is conceptually the same as for coated fixed structures. The codes recommend dividing the hull into zones for this purpose. Such zones might comprise the forward, midships and aft hull sections, and the rudder. The surface area of each zone is calculated from the drawings and usually assumes conservatively that the vessel will always be fully laden.

[1] Some shipyards still paint anodes. The difference is that, nowadays, they apply grease to the anodes first. When the ship is launched, the grease and the paint are quickly washed off and the anodes start to work.

A design current density is then allocated to each zone. As in Chapter 3, we could conjure up a current density for bare steel and a reasonable figure for the coating breakdown factor, and then combine the two. Indeed, some codes offer bare steel current densities and breakdown factors for hull coatings. Some also provide current density advice for ships based on geographical locations. This has some relevance for vessels with a predetermined trading pattern such as harbour tugs, ferries or coastal traders. However, major vessels such as container ships, bulk carriers and tankers can find themselves trading anywhere between polar waters and the tropics. There seems little point in selecting design current densities based on geography.

The key factor determining the current demand is the speed of the vessel. As we saw in Chapter 5, this is readily explained in terms of electrode kinetics. Increasing the seawater velocity increases the rate of mass transport of dissolved oxygen to the steel surface, thereby increasing the limiting cathodic current for oxygen reduction (see Figure 5.9).

However, even with this insight into the relationship between velocity and cathodic current density, we are a long way from being able to link a bare steel current density to a vessel's speed. The relevant laboratory experiments have generally been carried out using clean, polished specimens. As discussed in Chapter 7, the development of calcareous deposits has a profound effect on the shape of the polarisation curve. It also affects the hydrodynamic flow over the specimen.

The advice given in the codes is, perhaps understandably, not that helpful. For example, EN 16222 gives guidance values on current densities for bare steel: 100 mA/m^2 for up to 1 m/s, 250 mA/m^2 for between 1 and 3 m/s and (at least) 500 mA/m^2 for speeds above to 3 m/s. This seems plausible until we realise that 3 m/s is <6 knots,[2] which is well below the cruising speed of even a ponderous tub. We are given no advice at all for the cruising speed of a typical cargo vessel which is typically around 15 knots. More relevant figures are offered by ISO 20313 which suggests a current density of up to 350 mA/m^2 for a ship moving at up to 20 knots (10.3 m/s), and up to 500 mA/m^2 for faster vessels.

15.1.2.2.2 Coated hulls

There is no point spending too much time on the bare steel current density versus speed question. Ships' underwater hulls are almost universally painted with an anticorrosion system topped with at least one coat of antifouling. Furthermore, unlike coated fixed offshore structures, hull coatings are repaired or renewed when the vessel is routinely dry-docked. Shipowners regard the coating as the primary corrosion barrier, with CP providing a supplemental role. Unfortunately, whereas broad-brush bare steel current densities can be inferred from laboratory test data, the rate of accumulation of coating damage remains a matter of guesswork.

As shown in Table 15.1, EN 16222 and ISO 20313 offer guidance on current densities based on ships' speed and dry-docking interval. It is seen that increasing the speed, or accepting greater coating damage by extending the dry-docking interval, leads to increased current density recommendations. As with so much of the code advice on coating performance, this advice is a distillation of guesswork amplified by committee-based conservatism. DNV-CG-0288, for example, offers less-than-precise current density advice... *Average current densities needed to obtain full cathodic protection of well coated ships' hulls will usually be about 10 mA/m^2 or more. In special cases, e.g. on ice breakers, up to about 60 mA/m^2 may be needed. The current density demand will... be different at different locations of the hull and its accessories such as propeller, rudder, and sea chests.*

Other workers claim that even higher current densities are needed for ice breakers. For example, a research group at Helsinki university [18] has argued that *60 mA/m^2* is appropriate

[2] A knot is a nautical mile per hour. As near as makes no difference, it is 2 km/hour.

for general sea-going vessels; but in the case of vessels trading in ice, the... *total power of the protection system must be at least five to ten times that of ordinary operating ships.* Unfortunately, however, the paper does not provide any evidence to support this opinion.

Table 15.1 Ships' hull current density recommendations

	CP design current density (mA/m²)				
Dry-dock interval	<1.5 years	<3 years	<3 years	<5 years	<5 years
Code	EN16222	EN16222	ISO 20313	EN16222	ISO 20313
Vessel trading in seawater					
Speed not specified	15–25	26–45		46–75	
<20 knots			11–18		17–28
>20 knots			18–25		28–40
Vessel trading in ice					
Speed not specified	>25	>45	35–55	>75	60–90

It is worth noting that the figures in the 21st-century codes (Table 15.1) have not changed much since the 1970s. Capper and Willis [6] advised a global figure or 35 mA/m² for ships. In a published discussion on the paper, this figure was endorsed by Trotman who pointed out that, nevertheless, many ship owners opted for 60 mA/m², and the US Navy went as high as 100 mA/m². These figures indicate how little faith ship owners of the time had in the hull anticorrosive coating systems, even though they regarded the paint as the primary corrosion mitigation.

In the early days, there was a debate about whether or not ships in tropical waters would need more or less current than ships in cooler latitudes. There was a belief that, because warmer waters had lower resistivities, more current was required. In 1986, Foster and Moores reported a study of the ICCP log from a Canadian naval frigate [12]. This sailed south from the North West Pacific with the hull consuming about 40–60 mA/m² whilst under way. By the time the vessel had spent a couple of days in the warmer mid-Pacific, this had dropped to 10–25 mA/m². As observed on fixed structures, this lowering of cathodic current density is due to the beneficial effect of higher temperatures on calcareous deposit formation.

15.1.3 Propellers and shafts

15.1.3.1 Materials

Propellers are commonly made from copper alloys, such as nickel aluminium bronze (NAB), which has good corrosion resistance to seawater. Other alloys, including cast stainless steel, are sometimes encountered, particularly on high-speed vessels. As noted in Chapter 8, stainless steels may be susceptible to crevice corrosion in seawater. Inevitably, the packed gland where the shaft passes through the hull gives rise to a crevice. This situation has been examined by Kain and Dunoff [23] who found that... *cathodic protection using zinc sacrificial anodes was almost totally effective in eliminating crevice corrosion... on the static test shafts.*

To some extent, the question of whether a propeller needs CP in its own right is secondary.

The propeller shaft emerges from the reduction gearbox, which is oil filled, and passes through the hull via pressure-lubricated, and therefore insulated, bearings and a stern gland. This means that, when rotating, the propeller should not be earthed to the hull or its CP system. However, when not rotating, there is a reasonable expectation that, somewhere on the shaft, there may be metal-to-metal contact within a bearing. Under these circumstances, the propeller would draw galvanic corrosion current, either from the hull itself or, it would be hoped, from the hull's CP system [17].

15.1.3.2 Bonding

Because of this uncertainty as to whether or not the hull and propeller are connected, the pragmatic approach is to ensure that they are positively connected at all times. This is referred to as *earthing*, or *grounding*, the propeller shaft to the hull. This shaft earthing requires the hull's CP system to satisfy the current demand of both the hull itself and the propeller. According to Lenard [24], propeller shaft grounding was first introduced in 1961. The methodology has changed little since. The secure electrical connection is achieved by an arrangement of conducting brushes and a slip ring on the shaft as seen in Figure 15.1.

The outcome of this shaft grounding is that between 20% and 60% of the hull CP current is needed for the propellers. Occasional use has been made of coated propellers, but this is unusual. Instead, the current drain of the propeller is regarded as a given, and anodes (ICCP or sacrificial) are located preferentially near to the stern to satisfy this current demand. This is illustrated in the case of a small steel craft in Figure 15.2.

15.1.3.3 Current demand

ISO 20313 recommends a design current density of >500 mA/m^2 for propellers, although it does not advise on how much greater. Although the standard does not consider different propeller materials, we know from Chapter 8 that the same design current density figures apply to all alloys.

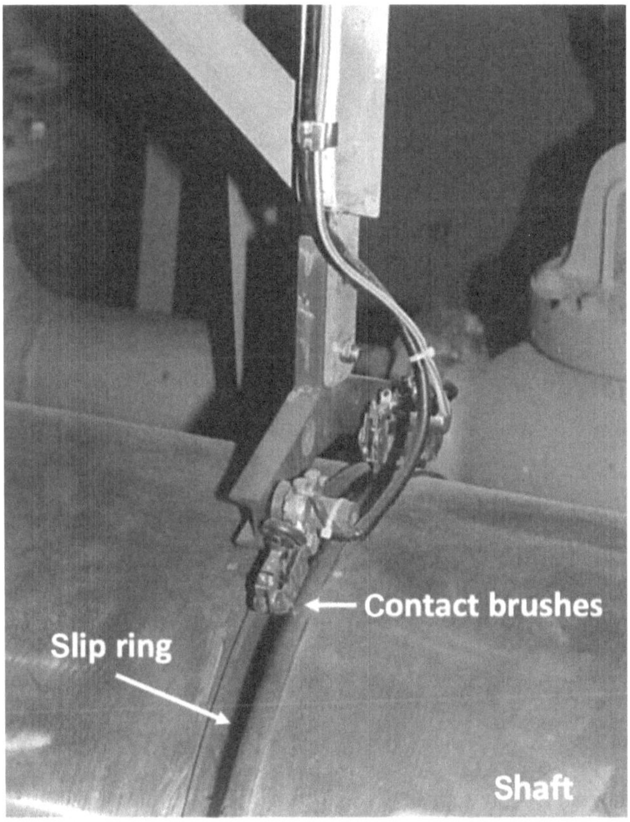

Figure 15.1 Propeller shaft bonding arrangement.

Figure 15.2 Anodes around the propeller. (Courtesy of Aberdeen Foundries.)

15.1.4 Rudders

15.1.4.1 Bonding

For the same reason as propellers, a ship's rudder should not be in electrical contact with its hull. The rudderstock emerges from oil-filled turning gear and passes through lubricated bearings. Unlike the propeller, it is sometimes possible to mount some sacrificial anodes on the rudder itself (as seen in Figure 15.2). However, the current demand per unit area of rudder is likely to be much greater than that of the hull since it experiences higher water velocities and is more prone to coating damage. Furthermore, the presence of corroding anodes on the rudder would impact adversely on its efficiency and could generate unpleasant or damaging levels of vibration.

The most effective way to protect the rudder is, therefore, also to bond it electrically to the hull and make use of the current output of the hull anodes. This is usually achieved by attaching a flexible cable between the exposed top of the rudderstock and a convenient earthing point. The arrangement is illustrated in Figure 15.3.

15.1.4.2 A cautionary tale

A few years ago, I was called in to examine a problem of severe corrosion damage to a ship's rudder. The owner was perplexed because the ship was relatively new and was fitted with an ICCP system, and there was no evidence of corrosion damage on the hull itself. It did not take long to find the source of the problem. The shipyard had been so diligent in its painting of the vessel, and it had also painted the connection lug intended to earth the rudderstock to the hull Figure 15.4. The paint acted as an insulator with the result that the rudder sat in the wash of the propeller without the benefit of any CP.

15.1.5 Sacrificial anode systems

15.1.5.1 Development

Although the efficacy of CP on ships' steel hulls was first demonstrated using magnesium anodes, these were soon supplanted by zinc. The reason for this was the magnesium was too active for the bituminous underwater hull coatings favoured at the time. Its use resulted in cathodic blistering and cathodic disbondment. Much of the 1950s work on zinc anodes was carried out by the US Navy in its quest for a dependable alloy formulation. Aluminium alloys started to gain popularity after the 1960s but have not supplanted zinc in the shipping industry to the extent that it has on offshore structures. There are probably a couple of reasons for this. The first is that the use of zinc is seen, to some extent, as traditional. The second is that zinc comes into its own if the vessel spends a good deal of time trading in lower salinity waters such as river estuaries.

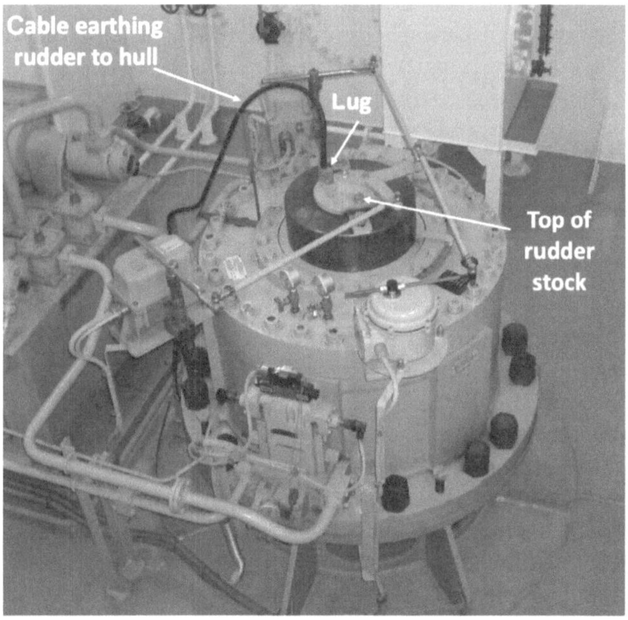

Figure 15.3 Rudder bonding arrangement.

Figure 15.4 Painted rudder earthing connection lug!

15.1.5.2 Design

In an ideal world, designing a sacrificial anode CP system for a ship should mimic the process of designing one for a fixed coated structure (Chapter 3). Unfortunately, the world falls somewhere short of ideal. In practice, shipbuilders tend to design ships and leave the positioning of the anodes to the draughtsman who, in turn, repeats what has been done for previous new-builds. The anodes are then either replaced in total at each dry docking, or else the owner opts to change just those that appear to be substantially consumed.

Although this approach to CP design, and CP rehabilitation at each dry-docking, might seem a trifle arbitrary, this probably does not matter that much. I have never heard of a steel ship being lost at sea simply because the underwater hull sprang a corrosion leak. On the other hand, I have been involved in several dry-dockings at which a class surveyor, having found 25% reduction in hull plate thickness, has required the damaged steel to be cut out and replaced.

15.1.6 Impressed current

15.1.6.1 Early days

We introduced the ICCP option in Chapter 3. We then outlined systems in Chapter 11 where we discovered that one of Thomas Edison's less successful inventions was a ship ICCP system based on trailing anodes. Although he patented his invention in 1890, he did not have sufficiently robust insulation and connections to produce reliable systems.

However, that was not quite the end of the line for trailing anodes. Howard Rogers [4] described a consumable anode ICCP system in which the anode was an aluminium wire, up to 50 m long, that was trailed behind the ship. Its advantage was that, being "remote" from the hull, it provided good current distribution and was not damaging to coatings. More aluminium wire was simply paid out to replace the material that dissolved. A significant disadvantage was that the anode had to be reeled back in when the ship was manoeuvring or in port.

15.1.6.2 Present day systems

Trailing wires were going out of use by 1964 [3], by which time the use of hull-mounted low-profile anodes, supplied by a DC electrical conductor penetrating the hull (see Figure 15.5) was the norm.

Ship designers are likely to use an established design that has worked previously on similar sized vessels. Innovations arrive slowly. The items of hardware – power supplies, references and cabling – have changed comparatively little over the years. However, as we will see in Chapter 17, ship CP systems, particularly on military vessels, have been the focus of various modelling programmes aimed primarily at improving current distribution.

Shipboard ICCP anodes are now usually MMO (see Figure 15.6) whereas, previously, vessels used platinised titanium (see Figure 11.2) or lead-silver. Zinc (the smaller electrode in Figure 15.6) remains the standard choice for the reference.

There is also a trend for power supplies to be automatic. Similarly, there is an increasing tendency for automatic acquisition and storage of the main CP system data: output voltage, output current and hull potential. Nevertheless, the chore of manually logging these parameters on each watch still persists on many vessels. As a matter of experience, it does not seem to make much difference whether the data are recorded on paper or electronically. Either way, they are generally ignored.

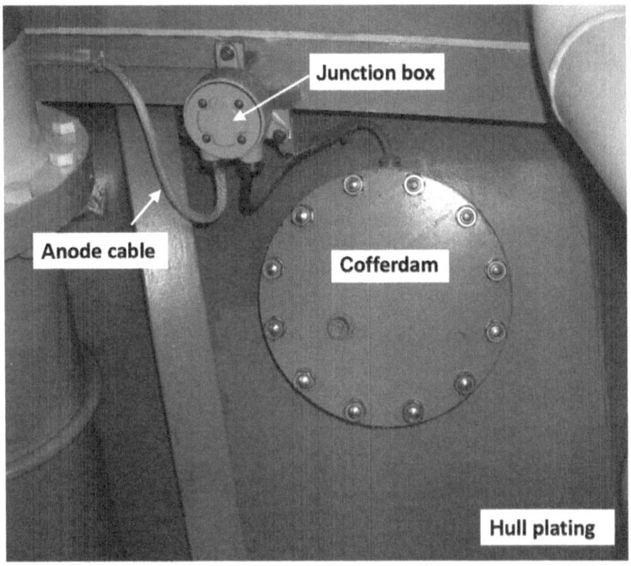

Figure 15.5 Cofferdam for hull penetration of anode cable.

Figure 15.6 MMO anode and zinc reference. (Courtesy OES Group Ltd.)

15.1.6.3 Fitting-out

15.1.6.3.1 Temporary protection

After launching, a ship can spend a considerable time afloat whilst the fitting-out work is completed; a process that can easily take over a year. Corrosion at areas of hull paint damage, and at unpainted locations where the hull stood on wooden blocks during construction, will inevitably be subject to some corrosion. Furthermore, the propeller and its shaft will usually have been installed prior to the launch. If the shaft settles on its bearings, or if the hull earthing is connected, then this introduces the additional threat of galvanically accelerated corrosion of the hull.

For these reasons, it is established practice to provide temporary CP to the hull during fit-out. For example, ISO 20313 warns that conditions in fitting-out berths are often severely corrosive. It recommends, but does not require, temporary sacrificial anodes. The design element of temporary systems is minimal. The methodology usually follows habitual shipyard practice. Typically, magnesium anodes are suspended on cables attached to temporary brackets welded to the deck. The choice of magnesium accords with the location of most fitting-out berths in brackish water.

15.1.6.3.2 Welding

When a ship, or any other floating installation, is being fitted-out whilst afloat alongside a quay, it is important to avoid stray current electrolysis as a result of DC welding operations. The hazard and its remedies are described in EN 50162 [21]. Where a DC welding power supply is located on the quay then, unless the earth return cable is securely bonded from the ship back to the welding power supply, there is a risk of high currents tracking back to the quay via the seawater. This will cause pronounced electrolytic corrosion damage to the underwater hull where the current discharges at areas of coating damage. As the EN document points out, the problem is eliminated if the welding stations are move onboard the ship.

15.1.6.4 Laying-up

Trading ships will inevitably spend time moored or at anchor, often this is whilst waiting for the next charter. The only impact this has on the CP system is that there is a lower current demand. The dissolution rates of sacrificial anodes will slow down and the output of automatically controlled ICCP systems will be reduced.

However, it is occasionally required to put vessels out of service for longer periods, or even indefinitely. This occurs when vessels are mothballed or, for example, repurposed as shore-side entertainment venues.

For mothballed vessels, where maintenance will be limited, ICCP systems will not normally be powered up. Similarly, sacrificial anodes will reach the end of their lives but will not be replaced due to the absence of dry-docking. Lloyd [19] described the US Navy's 1980s practice for mothballed vessels, which it termed its "inactive fleet". For the most part, sufficient electrical power was available on board each vessel to permit the rigging up of "temporary" ICCP systems which used platinised titanium anodes deployed on cantilever supports over the side. This option is only practicable because the vessels are static and laid up in sheltered water. As with an active ship, operation of an ICCP system on mothballed vessels requires a structured campaign of monitoring, adjustment and maintenance.

Where a vessel is converted into a fixed near-shore facility, then its CP can be treated as that of any other fixed structure, albeit with a longer (even indefinite) intended design life. An example is provided by the aircraft carrier *USS Lexington* which has been converted into a museum. Davis [22] has described its ICCP system which is notable as being one of the rare instances in marine CP where the anodes (MMO) were intentionally buried beneath the mud line.

15.1.6.5 Alongside berths

15.1.6.5.1 CP issues

We encountered the possibility of electrical interactions between structures in Chapters 11, 13 and 14. If a cathodically protected host structure is bonded to an unprotected, or less well protected "foreign" structure, then the unprotected foreign structure will drain CP current intended for the host. Conversely, if the foreign structure is not bonded to the host, but is influenced by the electrical field generated by its CP system, then the foreign structure may suffer severe electrolytic corrosion damage.

Where a small vessel protected by sacrificial anodes berths alongside an unprotected jetty, then there will be no electrical interaction effect unless the vessel makes electrical contact with the jetty. The effect will then be enhanced consumption of the anodes. We discuss the relevance of this in relation to leisure craft in Section 15.1.7.3. For trading vessels, however, this is unlikely to present anything in the way of a problem.

The issues become more complex when the vessel, the jetty or both are protected by ICCP. Various scenarios could arise. For example, in Chapter 13, we described seabed sleds for protecting harbour piling. It is easy enough to picture a vessel berthing between the remote sleds and the jetty. The vessel's hull would then intersect the harbour ICCP system, picking up current from the sleds on one side and discharging it from the side adjacent to the jetty piles. Conversely, a ship's ICCP system might generate more intense electrical fields in the water adjacent to the jetty than the jetty's own CP system. Under these circumstances, the ship's CP might damage the harbour. ISO 20313 looks only at the latter possibility of the shipboard ICCP system causing stray current electrolysis to a harbour. For example, it recommends that... *if a ship is laid up, or is berthed for long or repeated periods alongside quays or jetties, it is recommended that interaction testing should be carried out to demonstrate that adjacent structures are not adversely affected.* However, it offers no definition of *long or repeated periods.*

The option of turning the ship's ICCP off whilst it is alongside is reasonable, although it is not required nor is it always practiced. If it is turned off then, obviously, it is important to remember to switch it back on when sailing. On balance, as advised by Morgan [15], the most sensible CP option is to electrically bond a ship to the jetty's ICCP system when berthed. It is not practicable to switch off the harbour system whenever a vessel was alongside. Apart from the difficulties of depowering large systems, it would mean that a busy harbour was never protected.

15.1.6.5.2 Spark hazards

When it comes to ships loading and unloading flammable cargoes, the avoidance of electrical sparks outweighs matters of corrosion mitigation. In the past, IMO required electrical bonding between ships and jetties but has now relaxed that requirement. Instead, it calls for special precautions where ships are loading/offloading flammable material. It defers to the International Safety Guide for Oil Tankers and Terminals (ISGOTT) [27].

15.1.7 Non-ferrous hulls

15.1.7.1 Aluminium

15.1.7.1.1 High-strength alloys

High-strength aluminium alloys are employed for hulls where a combination of strength and lightness is required. Examples include fast catamaran-type ferries and high-speed military vessels. Although aluminium alloys exhibit good resistance to general corrosion in seawater, they are prone to pitting corrosion. Generally speaking, the alloying additions that confer strength also bring the disadvantage of increasing the pitting susceptibility. The corrosion control of aluminium-hulled vessels relies for the most part on protective paint systems. However, like all metallic hulls, CP is required to protect the areas where the hull coating is damaged or degraded. In that respect, the practice of applying CP to a painted aluminium hull is broadly the same as applying it to their steel counterparts. Suitably sized aluminium alloy sacrificial anodes will polarise the structural alloy to potentials more negative than the pitting potential. ICCP is also suitable providing it is appreciated that aluminium is amphoteric. It dissolves in both acids and alkalis. This means that it is particularly important to avoid excessively negative potentials due to the risk of cathodic hydroxyl generation. It is partly for this reason that EN 12473 specifies a maximum negative potential limit of –1.1 V for aluminium alloys.

15.1.7.1.2 Mixed metals

An interesting CP situation arises in hulls constructed of more than one metal. Examples include high-speed hydrofoil craft, as are used by some navies. These vessels have high-strength aluminium alloy hulls. However, these alloys are not strong enough to take the high loads passed through the hydrofoil struts, so these are usually manufactured in a high-strength stainless steel.

Aluminium-hulled vessels employing high-strength 15.5 PH stainless steel retractable hydrofoil struts provide an interesting example of providing CP to a mixed-metal system. The 15.5 PH (UNS S15500) is a precipitation hardened martensitic alloy. Its yield strength depends on heat treatment but could be as high as 1170 MPa. As we discussed in Chapter 8, the resistance of stainless steels to pitting and crevice corrosion in seawater depends on a parameter known as the pitting resistance equivalent number (PRE_n), the value of which depends on the concentrations of certain alloying elements, notably chromium and molybdenum. The 15.5 PH has a PRE_n of ~15, which is well below the value (~40) needed for unprotected service in seawater. Accordingly, like the aluminium hull, CP is also required for the hydrofoil struts.

15.1.7.1.3 Schottky barrier diodes

Unfortunately, 15.5 PH is also susceptible to hydrogen embrittlement. According to Dees [16], of the Boeing Corporation, it is not possible to pick a single target value for the potential that will guarantee protection to both the aluminium hull and the 15.5 PH struts, whilst also avoiding the threat of embrittling the latter. The solution arrived at was to electrically isolate the struts from the hull, fit sacrificial anodes on the hull and restrict the flow of CP current from the hull anodes to the struts. Early attempts to limit the current involved installing a resistive wire as a link between the hull and the struts. This proved unsatisfactory. Although the CP current to the struts was reduced, it was not practicable to control the current sufficiently to ensure protection of the struts whilst avoiding their being polarised to potentials at which hydrogen evolution can occur, and embrittlement becomes a credible threat.

In order to ensure adequate protection, without over-polarisation, the struts were connected to the hull via Schottky diodes. Their principle of operation was as we have described in Chapter 10.

15.1.7.2 Copper and Cu-Ni hulls

There is a small, niche market for boats for which the hulls are either constructed in a copper-nickel alloy (usually 90/10 Cu-Ni), or else constructed in a more conventional material (steel or GRP) and then clad with Cu-Ni sheathing. Examples include fast fire-fighting vessels. There is also still a small market for copper sheathing in the pleasure boat industry.

For these vessels, the roll of the cuprous material on the hull is to provide antifouling properties. For this to work, the copper has to leach from the surface to inhibit the development of hard fouling such as barnacles. Since CP would inhibit this copper dissolution, and negate the antifouling benefit, it should not be applied to these vessels. The Cu-Ni or copper is permitted to corrode. Pircher et al. [10] provided a review in the mid-1980s. They reported that for a fast vessel (24 knots) the maximum rate of Cu-Ni loss was found to be 0.08 mm/year. This was in the zone affected by propeller turbulence. Thus, 2 mm of Cu-Ni cladding should last 20 years on a ship.

Cu-Ni merits consideration for GRP pleasure boat hulls. However, it is a risky coating for a steel hull. Any impact damage poses the risk of galvanically coupling a large area of Cu-Ni cathode to a small area of expose steel. This could lead to rapid electrolytic penetration of the steel hull.

15.1.7.3 Pleasure craft

15.1.7.3.1 Small anodes

Millions of small, and not-so-small, sea-going boat hulls are constructed traditionally in wood, and nowadays more frequently in GRP. At first sight, it is perhaps surprising to find that many of these non-metallic craft are fitted with one or more zinc anodes. Of course, the function of the anode is not to protect the non-corroding and non-conducting hull itself. Its role is to control the corrosion of skin fittings such as water inlets and outlets. These are frequently made of admiralty brass.[3] The hull-mounted anodes also protect (painted) steel or cast-iron components such as propeller shaft brackets and keels. A typical small boat hull anode is illustrated in Figure 15.7. For hydrodynamic reasons, it should always be installed with the thicker part of the anode at the forward end.

All of the skin fittings and other protected components should be electrically connected to the anode by cabling inside the hull. Sometimes this cabling is integrated with the boat's electrical earthing system.

Other components, however, cannot be earthed to a hull anode in this way. For a boat with an inboard engine, the propeller shaft will be isolated from the earthed engine block for the same reason that it is isolated from the hull on a large ship. For this reason, boat chandlers sell small shaft-mounted zinc anodes, as also illustrated in Figure 15.7.

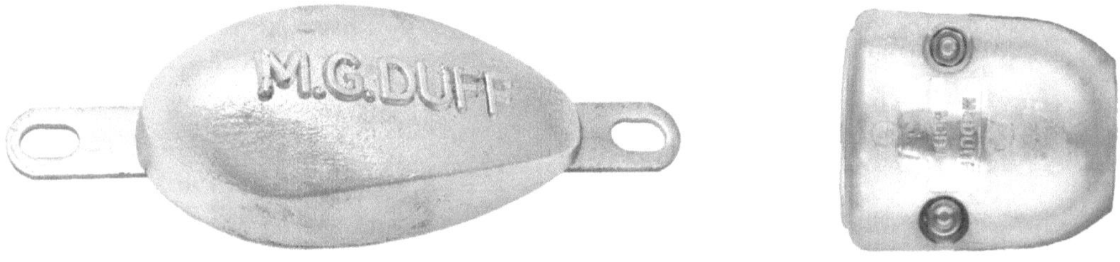

Hull anode (2kg) Shaft anode (0.67 kg)

Figure 15.7 Small craft hull and shaft anodes. (Courtesy of MG Duff.)

15.1.7.3.2 Galvanic isolators

A small pleasure craft is likely to spend the majority of its time tied up in a marina. Most marinas provide mains shore-power. It is not uncommon for boat owners to leave shore-power cables permanently installed for the purpose of keeping batteries topped up, and domestic equipment, such as fridges and dehumidifiers, operating. This introduces a vulnerability.

[3] Admiralty brass (UNS C44300) is an alloy of copper (~71%), zinc (~28%) and tin (~1%). It also contains a small amount (<0.06%) of arsenic. It has good corrosion resistance in seawater but is not immune to attack.

It is possible for all the boats that are connected into the marina's shore-power to share a common earth. This earthing is also likely to be bonded to the marina's steel piling.

This means, for example, that a pleasure craft equipped with a small hull anode, as illustrated in Figure 15.7, might be electrically coupled to unprotected steel piles and even, perhaps, one or more badly maintained rusty steel boats. This could drain and deplete the anode, without the boat owner being aware.

If this is a concern the boat owner might be well advised to install a galvanic isolator in the earth cable. These proprietary devices permit the passage of AC current, but block any potentially damaging stray DC.

15.2 FLOATING INSTALLATIONS

15.2.1 Drill ships and semi-submersibles

15.2.1.1 Some history

As the hydrocarbon production industry moved into deeper waters the first wells were drilled by self-propelled mono-hulled drill ships, the first example of which was *CUSS 1* which was delivered in 1956. Its unappealing name derived from the consortium of oil companies that commissioned it: Continental, Union, Superior and Shell. From the standpoint of hull corrosion protection, drill ships differed little from other similarly sized vessels. They were expected to dry-dock regularly, so long-term CP performance was not a high priority.

In the event, monohulled vessels failed to offer the levels of dynamic stability needed for drilling in deep waters, and a range of semi-submersible vessels emerged. Semi-submersible vessels have now been the primary choice for offshore drilling for half a century, with the first vessels being introduced into the North Sea in 1973. They have also been adapted for service as construction barges, firefighting vessels and floating accommodation units, for which the name "flotels" has been coined. Some have also been converted to offshore production facilities.

Whereas drill ship hulls were universally painted, owners of semi-submersibles have opted for varying degrees of coating. Jensen and Abrahamsen [8] recorded 15 semi-submersibles deployed in the North Sea in the mid-1970s. The proportion of the underwater surface that was intentionally coated varied from 6% to 62% in the case of the 12 vessels protected by sacrificial anodes. The three vessels protected by ICCP systems had coating coverages of 3%, 5% and (surprisingly) 100%. Based on divers' reports of anode consumption, they recommended a CP design current density of 40 mA/m^2 for painted semi-submersible hulls.

This figure seems to be on the high side. It is more than double the 2019 DNV-RP-B101 recommendation for the hull of an FPSO. In contrast to ships, semi-submersibles are not normally fitted with their own propulsion systems, so the design of the CP systems does not have the complication of satisfying the high current densities arising from high water flow rates. It seems reasonable to conclude that the industry's expectation of hull coating performance has increased since Jensen and Abrahamsen's 1982 paper.

Although hull coatings have improved, the number of dry-docks capable of accommodating these vessels remains limited. This means that there has always been an incentive to reduce the frequency of hull inspections and, where practicable, to carry them out using divers in sheltered waters. This was preferable to transferring the unit to a dry dock, perhaps many thousands of kilometres away. Furthermore, the imperative to maximise hydrocarbon production at the time often weighed against taking units off station for inspection and repair.

In 1999, I was involved in the case of a semi-submersible drilling vessel which had been launched with a sacrificial anode CP system designed for 5 years, but which had stayed on station west of Shetland for 14 years. A hull CP potential survey had revealed potentials in the range −733 to −915 mV (mean: −783 mV). Although these results confirmed that the anodes were well on the way to being fully consumed, and a CP retrofit was in order, the fact that the system had worked for nearly three times the intended life emphasises the conservatism in the codes of the time.

15.2.1.2 A recent example

The conservatism of the codes was further emphasised when I subsequently examined an older semi-submersible. The vessel had been launched in 1977. In this case, the columns had been protected by coal tar epoxy (CTE) paint, whereas the pontoons and their sponsons (lateral members) had remained uncoated. The vessel had been dry-docked in 1989 and new sacrificial anodes fitted. No remediation of the columns' CTE coating had been deemed necessary. Although the 1989 CP system had a design life of 10 years, the unit ended up staying on station for more than double that time before being relocated to a sheltered shallow-water facility. There it was to be refurbished prior to a further deployment intended to last 15 years.

In preparing the CP design for the next 15 years, the coating breakdown predictions offered by the relevant codes could have been employed. However, as we have discussed, these codes intentionally err towards conservatism. This is illustrated in Table 15.2 in which the coating breakdown "predictions" from DNV-RP-B401 and EN 13173 were retrospectively applied to the following years:

- 1977 The initial breakdown at the start of the service life.
- 1989 The start of the service life for the replacement CP system.
- 1999 The originally anticipated end-of-life for the replacement CP system.
- 2011 Time of the inspection and CP upgrade design.
- 2014 Intended offshore redeployment.
- 2029 Notional end of service life.

As can be seen, retrospectively applying the EN or DNV guidelines at the time the 2011 CP design work was carried out would have led to a design based on a coating breakdown of between 53% and 80%. However, these same codes would have predicted breakdown in excess of 40% by 2011, at which time it was possible actually to inspect the CTE coating at first hand. As can be seen from Figure 15.8, the coating remained in very good condition. In this picture, the only features are barnacles. No steelwork is visibly exposed. A realistic breakdown estimate would have been <2%. However, in keeping with the conservatism that permeates all CP design and evaluation work, a figure of 10% was adopted; and even this was increased to 15% for the purpose of the design.

Table 15.2 Coating breakdown predictions – semi-submersible columns

Year *Source*	*1977*	*1989*	*1999*	*2011*	*2014*	*2029*
DNV-RP B401 (2010)	2.0%	16.4%	28.4%		46.4%	64.4%
EN 13173 (2001)	1.0%–2.0%	13.0%–20%	23.0%–35%		38.0%–57.5%	53.0%–80%
1989 CP design		5.0%	30.0%			
Observation				<10%		
2011 CP design					15.0%	30.0%

As we move to a greener world, it is reasonable to assume that we have now passed the peak for the employment of semi-submersibles in hydrocarbon production. Nevertheless, some will still be around for many years to come. It is also a safe bet that, irrespective of the design lives advised to the CP design contractors, many will end up being required to stay on station for longer than originally intended. Also, the end of offshore hydrocarbon production will probably not see the end of the semi-submersible. Even today, some are being reconfigured and redeployed for the construction of offshore wind farms. There is also an example of one being employed for launching space vehicles.

Figure 15.8 CTE coating (after 34 years) and anode (after 20 years).

15.2.2 FPSOs

15.2.2.1 Hulls

FPSO vessels have been developed to exploit offshore hydrocarbon reservoirs where the water is too deep for fixed platforms, or the locations are too remote for economic construction of a pipeline. The first FPSO was installed in the Spanish Mediterranean in 1977. The early vessels were converted bulk oil carriers, but purpose-built vessels are now the norm. There are presently (2021) over 160 operating around the world.

An FPSO is, to all intents and purposes, a ship. It is single-hulled and enters service fully painted. Like semi-submersibles, FPSOs are towed onto station and moored in place. This means that the CP designer is spared the need to deal with the effects of the propeller and does not consider speed or trading pattern. However, as emphasised by DNV-RP-B101, an FPSO will normally stay on station for 15 years or more whereas a trading ship will dry-dock regularly (typically every 5 years). RP-B101 offers design current densities for coating lives of between 10 and 40 years. For this reason, the CP design of FPSO hulls draws more on the experience of coated fixed offshore structures than that of ships.

DNV-RB-B101 sits firmly on the fence regarding the question of whether FPSOs should be protected by sacrificial anodes or ICCP systems. It points out that the choice between the two... *needs to be considered in relation to a range of advantages and/or limitations of the system in respect to installation, operation and maintenance.* Other workers, however, have not been shy about expressing a preference. A 2004 paper [25], for example, firmly concludes that a sacrificial anode system, based on Al-Zn-In anodes, is... *superior for to a hull-mounted ICCP system.* No doubt, a purveyor of ICCP would challenge the reasons put forward to justify that conclusion. Nevertheless, the provenance of the paper (ExxonMobil) lends weight to its conclusion.

This view has also been supported by other workers. For example, a couple of years later, Johnsen and Nilsen [26] concluded that... *corrosion protection based on a combination of protective coating and an ICCP system is not the preferred solution for FPSO's... planned to operate offshore without any coating maintenance for 10 to 15 years. This is based on a possibility of a "self-accelerating" coating damage due to low protection potentials.* However, the argument is not wholly convincing. The proposition advanced is that where CP is applied at very negative potentials (controlled potentiostatically in these tests at –1500 and –2000 mV), the coating will suffer disbonding (see Chapter 9). This will be focussed adjacent to the edge of the anode shield where the steel potentials will be the most negative. This disbonding will then increase the cathodic current demand of the hull, and the ICCP system output will increase to compensate. This in turn will further stimulate disbonding of the hull coating progressively away from the shield edge. And so, the "self-accelerating" cycle continues.

This scenario is, of course, what does indeed happen on real coated structures protected by ICCP. However, providing the coating is correctly chosen and applied, and the ICCP system is operated responsibly, not at –2000 mV, the extent and rate of the cathodic disbonding of the coating should not pose a practical problem within the nominal on-station life of an FPSO. For example, the same authors did not detect any cathodic disbondment of the epoxy coating in tests they conducted at –1000 mV.

Properly designed ICCP systems are fine for FPSOs, providing a competent person is charged with overseeing the monitoring, inspection and maintenance of the system. Many of the FPSO in service are protected by ICCP.

That is not to say, of course, that ICCP systems have been faultless. There have been a number of FPSO CP retrofits. However, some of these have been because FPSOs fitted with sacrificial anodes for a short-term deployment have then been required to remain on station for extended periods. Interestingly, these retrofits have often borrowed from the technology developed for fixed structures. For example, Franco et al. [28] report the case of an FPSO launched in 1996 with sacrificial anodes sufficient for a 5-year life. In 2002, the future life on station was assessed as being a further 15 years. Evidently, a retrofit was required. The option employed was to deploy two 400 A ICCP sleds, using buoyant MMO anodes, of the type discussed in Chapter 13, on the seabed. Their paper goes on to report that one of these sleds failed about 8 months after installation. The "most likely" cause of this was deemed to be a lightning strike on the FPSO.

15.2.3 Tension leg platforms

The Hutton tension leg platform (TLP), installed in the North Sea in 1984, was the world's first positively buoyant, rigidly tethered hydrocarbon production platform. The first in the Gulf of Mexico was the Auger platform installed a decade later. In total, approximately a dozen such installations had been installed worldwide for hydrocarbon production by 2020. There has since been some interest in employing the TLP concept for offshore wind power generation.

The TLP concept is intended to provide an economic method of drilling and producing in water depths ranging from ~300 to ~1500 m. By arranging for the buoyant structure to be connected under sustained tension to its seabed anchors, it is possible to minimise movement compared with, for example, a moored semi-submersible.

Because weight control is critical for a TLP, the fully immersed surfaces of Hutton's hull were designed with zero corrosion allowance. This decision required that corrosion would need to be effectively controlled from day one [13]. Furthermore, the design life (20 years) and the industry's pessimistic expectation of paint coating performance, which we discussed in Chapter 13, meant that the hull CP system was designed on the basis of an uncoated hull. This pointed to the need for a CP system with an output capacity of 4200 A.

The same weight restrictions that barred a steel corrosion allowance also forced the CP decision from a sacrificial anode system, which would have weighed over 300 tonnes, to an ICCP system with an all-up weight of about 9 tonnes; although 240 sacrificial anodes were also installed at what was judged to be critical locations. The resulting ICCP system was, arguably, the most comprehensively engineered and intensively monitored ever installed on a floating structure. It comprised 60 Pt/Ti anodes; each fed from a dedicated power module. Each ICCP anode was mounted on a 60 mm thick GRP dielectric shield, the dimensions of which (4.7 m×1.2 m) had been determined by one of the early computer modelling projects [14]. The current to each anode was continuously monitored, as was the hull potential using 24 zinc reference electrodes.

Interestingly, although the estimated polarisation current demand for the bare steel hull was over 4000 A, a review of the first year's monitoring data showed full and consistent polarisation being achieved with 214 A from the ICCP system and an (estimated) contribution of 75 A from the 240 strategically positioned sacrificial anodes. This much lower cathodic current density, equating to ~14 mA/m^2, was attributed to the benefit of the residual, thin (40 μm) temporary holding primer applied to the stock steel plate delivered to the fabrication yard.

It is interesting to note that the measured performance of "temporary" primer in reducing the cathodic current demand falls in the same ball-park as the estimated performance of the intentionally more durable hull coating systems applied to trading ships (Table 15.1).

15.2.4 Moorings

15.2.4.1 Tethers and tendons

The protection of the high-strength low-alloy tendons used to tether the Hutton TLP, described above, to the seabed anchors presented something of a challenge at the time. The designers set a negative potential limit of −1050 mV to avoid the risk of hydrogen embrittlement. There is no published evidence that they considered using the hull ICCP system to deliver current to the tendons. It is, therefore, a matter of conjecture whether the hull ICCP system could have been designed and operated to achieve the dual requirements of a potential less negative than −1050 mV at the top of the tendon, and that the attenuation down the tendon would not compromise a −800 mV target at its base.

Instead, the designers opted for a flame-sprayed aluminium (FSA) coating [9]. Due to its porous nature, the FSA was sealed with a vinyl copolymer. The selection was made after a review of the (limited) long-term field experience of FSA and some laboratory investigations. The latter focussed on the ability of the FSA, which exhibits some measure of sacrificial anode behaviour, to provide protection to the steel exposed at artificially created defects in the coating. The results were encouraging, as were the reports from other workers, that the FSA coating returned the fatigue characteristics of the steel to its in-air value. In the event, the decision to go with FSA was sound, and the tendons performed well through their operating lives of nearly 20 years.

15.2.4.2 Chains

If a ship is trading successfully, it spends comparatively little time at anchor. Generally speaking, therefore, ships neither seek to provide corrosion protection for their chains nor do their hull CP systems factor in any current drain to the anchor chain. Conversely, other marine structures, ranging from semi-submersibles and FPSOs to cardinal or marker buoys, spend most of their operational lives with their mooring chains deployed. This requires a decision about attempting to protect the chain, and about the amount of CP current a chain will take from the cathodically protected floating unit.

The US Navy had an interest in protecting mooring chains for a while during the late 1960s. Drisko [5] reports that riser chains attached to cathodically protected mooring buoys were observed to be in better condition than those securing unprotected buoys. This prompted some fieldwork to determine the practicality of installing sacrificial anodes on mooring chains. Initial attempts using magnesium anodes were soon abandoned in favour of experimenting with zinc. The approach that was adopted was to cast 220 kg trapezoidal zinc blocks onto some individual 0.89 m long 63 mm diameter links. These blocks were then installed at intervals along the mooring chain. Stab potential measurements along sections of chain laid on the seabed indicated that the benefit of the anodes only extended for ~6 m (~seven links) due to the effect of the contact resistances between each link. The researchers came up with a modification of the design which involved weaving a galvanised steel cable through the chain, and brazing or clamping it to every seventh link, to improve the electrical conductivity. This had the desired effect of extending the throw of the CP from each anode.

Evidently, the US Navy was sufficiently encouraged by these results to conduct further trials. The same author provided an update 10 years later [7]. Unfortunately, it is difficult to judge the efficacy of the system from the data presented. An example inspection data sheet is provided which gives potentials obtained by divers in the range −908 to −919 mV. On this basis, one would expect the links to be well protected. However, measurements showed locations where more than 10% of the 63 mm chain diameter had been lost. The data do not tell us how long the test had been running. However, if we assume a decade (the interval between the two publications), we are forced to conclude that corrosion has been progressing at ~0.3 mm/year. You may recall from Chapter 1 that this is much higher than the typical free corrosion rate for steel naturally exposed in seawater. It seems that the dominant degradation mechanism for chain links is wear rather than corrosion. Accordingly, applying CP to mooring chains is pointless.

This view seems to be endorsed by, amongst others, DNV-RP-B401. This does not suggest that mooring chains should be cathodically protected, but it advises that the CP design of the host structure should assume that the first 30 m of chain drains CP current. Based on the US Navy work, this figure – as so much else in RP-B401 – is conservative.

15.3 JACK-UP RIGS

Like semi-submersibles, jack-up rigs are examples of MODUs (mobile offshore drilling units). The early "offshore" hydrocarbon wells were drilled in the shallow waters of the Mississippi delta using simple barges with drilling derricks mounted on them. Similar barges are still used for drilling wells in shallow waters and swamp areas such as the Niger delta. However, the progress of oil production into deeper, more exposed, waters meant that simple floating barges could not provide a sufficiently stable base for drilling operations. One solution was the jack-up drilling rig, the first example of which entered service in the Gulf of Mexico in 1954.

The essential characteristic of all such jack-ups is that they are towed to the drill site and then they lower their legs using a rack-and-pinion system. Once the feet of the legs, referred to as spud cans, are fixed in the sea bed, this rack-and-pinion system elevates the hull a suitable distance above the surface of the sea. Early jack-ups had up to ten tubular legs. This was the case for the Sea Gem which made the first British offshore gas discovery in 1965, before tragically capsizing in 1967 with the loss of 13 lives. Subsequent designs have mainly opted for either three or four lattice-type legs.

When raised on its legs, the air-gap ensures that the drilling platform is unaffected by waves. This means that a jack-up is less prone to weather disruption than a drill ship or semi-submersible. As of 2021, more offshore wells have been drilled using a jack-up rig than by any other method. In recent times, a number of these units have been adapted for offshore wind farm installation work.

In the late 1980s, it emerged that there was a cracking problem in the high-strength steel legs of some jack-ups. The problem was first noticed in February 1988 when surveys found cracks in the heat-affected zones of welds connecting the high-strength legs to the spud cans. Further problems were found in rigs of the same design in March, June and August of that year. It eventually became clear, however, that the problem was not restricted to a single class of rig. Within a year of the first reported case, similar cracking had also been found in four other types of jack-up rig. In general, the legs and spud cans were painted. The cracking was found in areas where the paint had either deteriorated or else had previously been removed in order to carry out underwater weld inspection.

These observations prompted a substantial research effort into elucidating the cause or causes of the problem and evaluating remedial options. The causes were soon identified as hydrogen embrittlement resulting from the corrosion engineering, and classification society, requirement to apply CP to the legs and spud cans.

Coming up with a solution was a little more difficult. There was evidently a requirement to limit the potential of the high-strength steel legs to a value at which the corrosion rate was acceptably low, but was not so negative as to incur the penalty of hydrogen embrittlement. The latter requirement necessitated limiting the potential to no more negative than –825 mV. An HSE report [20] included a review by one of its authors (Robin Jacob) of the possible options. Based on the Boeing design for high-speed hydrofoils, which we have mentioned above, and his experience of applying CP to duplex stainless steel flow lines, he suggested connecting anode through Schottky barrier diodes.

Unlike the case of the hydrofoils, where the diodes and wiring could be housed within the hull, this application would involve diodes physically installed between the anodes and the structure, as described in Chapter 10. Nevertheless, the anode manufacturing and installation issues were mastered, and several rigs were fitted with such potential limited systems in the early 1990s. When these rigs were subsequently inspected, it was observed that the twin objectives of controlling corrosion and preventing cracking had been achieved. In addition, the voltage-drop due to the diodes had also limited anode consumption. There was also anecdotal evidence that the rigs' coatings were in good condition. It was conjectured that the intentional limiting of the potential would have provided the additional benefit of reducing any cathodic disbondment of the paint (see Chapter 9) around sites of damage.

REFERENCES

1. K.N. Barnard, *Cathodic protection of an active ship in sea water.* Corrosion **7** (4), 114 (1951).
2. E.C. Reichard and T.J. Lennox Jr, *Shipboard evaluation of zinc galvanic anodes showing the effect of iron, aluminum and cadmium on anode performance.* Corrosion **13** (6), 410t (1957).

3. R.A. Lowe, *The protection of ships' hulls*, Proceedings of Symposium *Recent Advances in Cathodic Protection*. Marston Excelsior Ltd (Wolverhampton) 1964.
4. T. Howard Rogers, Chapter 12 *Cathodic protection*, in *Marine Corrosion* Newnes (London) 1968, p. 201.
5. R.W. Drisko, *Cathodically protecting mooring chains in seawater*. Materials Performance and Protection **9** (7), 20 (1970).
6. H. Capper and A.D. Willis, *Anode design for ships hull protection*, Proceedings of Symposium *Cathodic Protection* London, Marston Excelsior Ltd (Wolverhampton) 1975.
7. R.W. Drisko, J.F. Jenkins and J.F. Wadsworth, *Cathodic protection and inspection of fleet moorings*. Materials Performance **19** (7), 42 (1980).
8. F. Jensen and E. Abrahamsen, *Cathodic protection of mobile offshore units, seven years experience from the North Sea*. Materials Performance **20** (12), 23 (1981); S. Eliassen (Discussion on this paper) ibid. **21**(7), 30 (1982).
9. M.T. Cooper, W.H. Thomason and J.D.C. Vardon, *Sprayed aluminium coatings for the corrosion control of the Conoco Hutton tension leg components*, Proceedings of UK Corrosion, 1983.
10. M. Pircher, B. Ruhland and G. Sussek, *Use of copper nickel cladding on ship and boat hulls*, Copper Development Association Publication TN 36, 1985.
11. M.H. Peterson, *A short history of marine cathodic protection*, Paper no. 285 CORROSION/86.
12. T. Foster and J.G. Moores, *Effects of temperature and salinity on the cathodic protection demand for ships' hulls*, Paper no. 294 CORROSION/86.
13. J.D.C. Vardon, A.D. Willis and G. Payne, *Commissioning and early life operation of the Hutton tension leg platform cathodic protection system*, Paper no. OTC 5270 Offshore Technology Conference, 1986.
14. M.A. Warne, *Application of numerical analysis techniques*, chapter 3, in *Cathodic Protection: Theory and Practice*, Eds. V. Ashworth and C.J.L. Booker. Ellis Horwood (Chichester) 1986, pp. 38–67.
15. J.H. Morgan, *Ships' hulls*, chapter 7B, in *Cathodic Protection*, 2nd edition. NACE (Houston, TX) 1987, pp. 291–329.
16. D.D. Dees, *Combined cathodic protection of 5000 series aluminum alloys and 15-5 PH stainless steel in seawater*, Paper no. 73 CORROSION/87.
17. T. Foster, *Coated propellers: reducing the current demand of ships*. Materials Performance **38** (3), 21 (1989).
18. O. Forsén et al., *Some aspects on the designing of cathodic protection systems of ice-going vessels*, Proceedings of EuroCorr, 1989.
19. D.B. Lloyd, *Cathodic protection activities in the inactive fleet*, Paper no. 536 CORROSION/91.
20. K. Abernethy et al., *Hydrogen cracking of legs and spudcans on jack-up drilling rigs a summary of results of an investigation*, HSE Offshore Technology Report OTH 91 351, 1993 (ISBN 0-7176-06147).
21. EN 50162, *Protection against corrosion by stray current from direct current systems*, 1994.
22. J.G. Davis, *Cathodic protection for immersed surfaces on museum aircraft carrier U.S.S. Lexington*. Materials Performance **33** (4), 12 (1994).
23. R.M. Kain and I. Dunoff, *Influence of packing material on the corrosion resistance of stainless steel boat shafting and other materials in seawater*, Paper no. 00639 CORROSION/2000.
24. D.R. Lenard, *The effects of shaft grounding on the corrosion of underwater hulls*, Paper no. 03231 CORROSION/2003.
25. M.B. Surekein, J.P. La Fontaine and L. Brattas, *A comparison of impressed current and galvanic anode cathodic protection design for an FPSO hull*, Paper no. 04095 CORROSION/2004.
26. R. Johnsen and A.C. Nilsen, *Cathodic protection of FPSOs: Current density requirements at potentials more negative than -1000mV Ag/AgCl*, Paper no. 06292 CORROSION/2006.
27. The International Chamber of Shipping (ICS) and The Oil Companies International Marine Forum (OCIMF), *International Safety Guide for Oil Tankers and Terminals (ISGOTT)*, 5th edition. Witherby Publishing (London) 2006 (ISBN 10-1856092917).
28. R.J. Franco, et al., *Aft anode sled failure at a floating production unit hull*, Paper no. 07075 CORROSION/2007.
29. P.A. Schweitzer, *Fundamentals of Corrosion*, p. 298. CRC Press (Boca Raton, FL) 2010.

Chapter 16

Internal CP

16.1 "FULLY SEALED" SYSTEMS

16.1.1 Corrosion threats

16.1.1.1 Corrosion by dissolved oxygen

Although a sealed volume of seawater will be denied future replenishment, it will contain dissolved oxygen when it first enters the enclosed space. This raises the question: What happens to that oxygen? We can explore this using a simple calculation based on the sealed internals of a wind farm monopile foundation. The original concept was that such foundations would be fully sealed and, in accordance with the then-current DNV guidelines, CP would *not normally* be required. However, the phrase "not normally" implies that there might still occasionally be a requirement for a specific project. This situation invites a calculation based on the Laws of Chemical Equivalence. The numbers are rounded for convenience.

1. Consider a monopile foundation 7.0 m is diameter containing a column of seawater 20 m high. We also assume that there is a 6 m column of air between the surface of the water and the air-tight deck.
2. The volume of seawater is $\pi \times 3.5^2 \times 20 = 770\,m^3$, and that of the trapped air is $\pi \times 3.5^2 \times 6 = 231\,m^3$
3. If we assume (conservatively) that the initial oxygen concentration is 10 mg/L (0.01 kg/m^3) in the trapped volume of seawater, then this contains 7.7 kg of dissolved oxygen.
4. The 231 m^3 of air has a density of 1.225 kg/m^3 and an oxygen concentration of 21%. This implies an oxygen mass of 59.4 kg in the air space.
5. We now assume that, over time, all of the oxygen in the air will dissolve in the seawater, replacing the oxygen consumed by corrosion. Thus, a total of 7.7 + 59.4 = 67.1 kg of oxygen will be available to corrode the steel below the water line.
6. We use the simple equation that we set out in Chapter 1:

   $$2Fe + 2H_2O + O_2 \rightarrow 2Fe(OH)_2$$

   and we conservatively ignore the further consumption of oxygen required to oxidise ferrous (Fe^{2+}) ions to the ferric (Fe^{3+}) species found in solid rust.
7. This tells us that 1 kg.mol of oxygen (2 × 16 kg = 32 kg) dissolves 2 kg.mol of iron (2 × 56 kg = 112 kg), where 16 and 56 are the (approximate) atomic masses of oxygen and iron respectively.
8. Thus, the available oxygen can dissolve 67.1 kg × 112/32 = 235 kg of iron (or steel).

DOI: 10.1201/9781003216070-16

9. Since the density of steel is typically about 7.8×10^3 kg/m^3, the volume of metal lost is $235/7.8 \times 10^3 = 0.03$ m^3.
10. The submerged internal surface area of the monopile is $\pi \times 7.0 \times 20 = 440$ m^2.

It follows that the thickness lost is 0.03 m^3/440 m^2 = 68 µm. This figure is <0.1% of the wall thickness of a typical monopile (~70 mm).

Thus, this example calculation endorses the conclusion that for this fully sealed monopile neither CP nor any other supplementary corrosion control would be required.

16.1.1.2 What really happens to the oxygen?

In fact, the conclusion that CP is not needed is correct, but the calculation itself is deeply flawed because it is based on two assumptions:

a. all of the dissolved oxygen participates in the corrosion reaction, and
b. corrosion ceases completely once the dissolved oxygen has been cathodically reduced.

Both of these assumptions are incorrect.

To see the problem with the first assumption, we need to understand a little more about why the dissolved oxygen concentration of seawater is (usually) 6–12 mg/L. In a simple model, this concentration is governed by Henry's law which tells us that the amount of gas that dissolves in a liquid is proportional to the partial pressure of the gas in the atmosphere above it. The proportionality constant depends on numerous factors, but principally on the salinity and the temperature. Since the partial pressure of oxygen in the atmosphere at sea level stays within a narrow band (~2×10^4 Pa) over the entire planet, and oceanic salinity is also fairly constant (~35‰), the main variable is the seawater temperature. Gases are more soluble in colder waters. As a result, we find the highest dissolved oxygen concentrations (~12 mg/L) in polar regions and the lowest (~6 mg/L) in tropical waters.

However, this simple Henry's law picture does not tell the whole story. The dissolved oxygen is not static. Open seawater contains both animals and plants. The animals, ranging in size from big fish to microscopic zooplankton, consume the dissolved oxygen. The photosynthetic plants, ranging in size from giant kelp to phytoplankton, regenerate it. Overall, there are innumerable complex biological interactions, assisted by the natural wave and tidal circulation, that maintain the oxygen levels.

When a volume of natural seawater is confined, as in a sealed monopile, this natural balance is destroyed. The absence of sunlight causes photosynthesis to stop and the phytoplankton dies. Oxygen is still consumed by the aerobic micro-organisms but is no longer replenished. The oxygen levels drop and the aerobic organisms likewise perish. The level of dissolved oxygen continues to reduce as it is consumed chemically by reacting with the decaying biomatter. This causes the sealed seawater to become anaerobic.

In practice, anaerobic conditions are developed relatively quickly. For example, Lee et al. [29] showed that, where seawater was sealed in glass vessels and denied light access or oxygen replenishment, the level of dissolved oxygen fell below the detection level of their equipment (0.1 mg/L) within 2 days. What this means for a monopile foundation is that the dissolved oxygen molecules, already present or partitioning from the trapped air above the water, are likely to be reduced biochemically within the bulk seawater, before they can diffuse to the steel wall to partake in a corrosion reaction. In other words, even our estimate of a metal loss of only ~0.068 mm turns out to be very conservative.

16.1.1.3 What happens then?

The answer is: not much. Corrosion does not cease, but it only proceeds very slowly. In the absence of dissolved oxygen, the only available cathodic process to drive the corrosion reaction is the reduction of hydrogen ions to hydrogen gas. As we saw in Chapter 1, this process in seawater is unlikely to produce a corrosion rate that exceeds 0.01 mm/year, or 1 mm/century. This is unlikely to pose an integrity threat to any flooded and sealed component of a structure.

16.1.1.4 What about MIC?

I have posed an intentionally misleading question here. Beyond doubt, there will be some MIC. The relevant question would be: Is MIC a credible threat? The answer to this question is: probably not.

For generations, corrosion engineers have been schooled into interpreting the presence of de-aerated seawater, or saline mud, as inevitably generating an MIC problem. There is some logic. As we saw in Chapter 1, the microbes principally implicated in MIC in marine environments are the sulphate-reducing prokaryotes (SRP). These organisms are inactive in aerated seawater. However, the oxygen does not kill them. They persist as inactive spores at very low concentrations, estimated to be as low as 1 spore per litre.

However, if a sealed volume of seawater, such as in a monopile foundation, is denied oxygen replenishment and becomes anaerobic, the spores will reactivate and the SRP will start to metabolise. Initially, the conditions will be very conducive to microbial reproduction. From the point of view of the microbes, the natural sulphate content of seawater (>2600 mg/L) provides an abundant energy source. The decaying biological material also provides a source of low molecular weight organic compounds that the organisms need for the synthesis of cellular material. Once activity commences, the reproduction rate will be very high, with the SRP population doubling in as little as 20 minutes. So, even if there had been fewer than a million spores in the monopile foundation when it was sealed, there would be numbers that bust your computer within a matter of days. Obviously, given the involvement of these organisms with MIC, this might be serious.

In reality, however, this reproduction rate could not be sustained for long. Doubling the bug population every 20 minutes would result in the impossible situation of the volume of microbes physically exceeding the volume of the monopile within a day! Obviously, this does not happen. The activity of the bugs ceases well before this; either because they are suppressed by the build-up of their own metabolic by-products, they run out of nutrients or a combination of the two. Ultimately, in a ***fully sealed system***, all life (microbial or otherwise) will exhaust its available energy supply.

It follows that, although MIC might initiate in a fully sealed system, it will not be sustained for long enough to constitute a problem.

16.1.2 Is CP needed?

DNV-RP-B401 tells us that...*Closed and sealed flooded compartments do not normally need CP*. This advice is based on the fact that, providing there is no access of oxygen to the contained volume of seawater, the corrosion rate will be too low to be of any practical concern. Similarly, we saw in Chapter 12 that deoxygenating the seawater, and thereby dramatically reducing the corrosion rate, returns the fatigue characteristics of steel to its in-air value. The DNV advice is, therefore, sound but only if the system remains fully sealed.

16.2 LEAKING SYSTEMS

16.2.1 Monopile foundations

In reality, there are very few fully sealed seawater systems. This has been emphasised by offshore wind farm monopile foundations. We discussed the construction of these in Chapter 13. The original design intent was that the internal space should be fully sealed. However, due to an error in the early editions of the DNV standard (DNV-OS-J101), the grouted connection between the monopile and its transition piece proved to be insufficiently robust to support the loads imposed on it by the turbine tower, aerofoil and the generator. The correction of this problem required expensive engineering interventions.

Unsurprisingly, this caused argument as to who should bear the cost caused by the error in the DNV standard. In May 2012, these arguments resulted in the instigation of legal proceedings concerning Scotland's first offshore wind farm: Robin Rigg. The litigants were the Danish company (MT Højgaard A/S) which had designed, constructed and installed the foundations in 2008/2009, and the German-owned wind farm operator (E.ON and various partners). In essence, Højgaard's view was that it could not be blamed for the problem since it had complied with DNV-OS-J101 as was required by the contract. Unsurprisingly, E.ON took the opposite view. At the risk of oversimplifying some complex legal reasoning: E.ON, or rather its lawyers, argued that Højgaard's duty of care outweighed any obligation to adhere rigidly to standards.

The case was first heard in the English Technology and Construction Court, which found in favour of E.ON in April 2014. That, however, was not the end of this landmark legal case. Højgaard took the matter to the Court of Appeal and was rewarded in April 2015 with a judgement [32] overturning the Technology and Construction Court's decision. Not to be outdone, E.ON then escalated the case all the way to the UK Supreme Court. This overturned the Appeal Court ruling in August 2017 [41]. The litigation process then had no further to go. The incontestable result was that MT Højgaard A/S found itself on the wrong end of a financial drubbing, a squabble[1] of lawyers found themselves comfortably enriched, and the guilty party (DNV) remained conspicuously uninvolved.

Well before the protracted Robin Rigg litigation was under way, it was becoming clear to the offshore wind industry that the error in DNV-OS-J101 had implications for the internal corrosion control of monopiles. The assumption that the internal environment would remain oxygen-free could no longer be relied upon. Across numerous European wind farms, the corrective engineering interventions, and associated increased inspection activity, meant that the normally closed hatches in the air-tight deck would now have to be opened periodically. In addition, the unanticipated movement of the transition pieces relative to the monopiles often placed unexpected loads on the steel J-tubes conducting the power cables through seals in the monopile walls. Inspection revealed that many of these seals were damaged, permitting inflow of oxygenated seawater.

16.2.2 Corrosion implications

I was at the annual Eurocorr conference in Stockholm in 2011. As with all such events, the competing draws of multiple parallel technical sessions meant that most presentations were made to small audiences. Often, authors would be happy to get above a dozen or so attendees at their presentations. I expected the same modest attendance as I made my way to the designated room for a presentation by FORCE Technology on the subject of corrosion in wind

[1] The English language has no collective noun for lawyers. A "squabble" has been suggested by others.

farm foundations [27]. I was wrong. The room was rammed: standing room only. One thing was abundantly clear: corrosion on the inside of wind farm monopiles was the new hot topic.

The paper dealt with... *the unexpected discovery of corrosion...*, and it was illustrated with pictures of corroded steel surfaces in the headspaces of monopiles that had been designed on the basis that they would be oxygen-free and, as a consequence, corrosion-free. Those pictures certainly held the attention of the overcrowded attendees. Where corrosion had... *proven to be worse than anticipated in the design...* the paper offered its advice on control methods. CP featured strongly in that discussion, but the point was made that existing standards did not cater for the design of systems for such applications.

To my mind, this interesting and seminal paper missed addressing one point. It proceeded directly from the observation of unexpected corrosion (in the void space above the water column) to a consideration of the appropriate control measures for the water column itself. It omitted the crucial step of analysing whether or not the corrosion actually needed to be controlled. This is a point to which we will return below.

For now, however, all we need to consider is that the internal CP of offshore monopile foundations probably engendered more investigation and discussion over the following decade than any other aspect of CP. So, it is to this interesting topic that we now turn.

16.2.3 Internal CP of wind turbine foundations

16.2.3.1 A touch of schadenfreude?

The English language has absorbed the German word "schadenfreude", loosely translated as a feeling of enjoyment that comes from seeing or hearing about someone else's problems. It is hard to imagine CP suppliers were untouched by this sentiment when news emerged of the endemic (and potentially very costly) issue of leaking monopiles. An unexpected new market for their products and services was opening up.

This was obviously good news for the CP industry, which has subsequently demonstrated no shortage of enthusiasm in offering CP systems, both sacrificial and impressed, together with monitoring and modelling, to the offshore wind industry. Not all of that commercial activity has turned out to be fruitful.

16.2.3.2 Code guidance

Unlike CP applications for the external surfaces of jacket structures, pipelines or ships, until recently there were no standards available for designing internal CP systems for structures containing seawater. What little information the standards provides was, at best, sketchy. EN 12499 (2003), which was current at the time, states that its scope encompassed... *every structure containing an electrolytic solution that it is technically possible to cathodically protect.* However, in reality, it was of no use for monopiles, and it did not address seawater-filled structures. Neither EN 12499 nor any other then-current standard provided current density recommendations for internal spaces. The more recently issued EN 17243 (2020), which does refer to seawater, is little more informative. The only advice it offers is that... *offshore wind turbine monopiles require special attention in relation to cathodic protection design and anode selection.* Beyond that, the reader is simply referred to three 2015 conference papers [33–35].

DNV-RP-0416 states that internal... *surfaces of the submerged zone shall be protected by either CP or corrosion allowance, with or without coating in combination.* But other than advising to use zinc instead of aluminium anodes, it offers nothing in the way of usable CP design guidance for monopile internals. ISO 24656 echoes the advice not to use aluminium but is otherwise similarly non-prescriptive with respect to internal CP.

16.2.3.2.1 Hydrogen gas

That is not to say that the CP industry was ignorant of all of the issues involved. For example, even the otherwise limited EN 12499 (2003) draws attention to the hazards of the build-up of hydrogen gas. As we have discussed in Chapter 12, it was well known that hydrogen is liberated on cathodically protected steel surfaces. It was also known, but perhaps not so widely, that aluminium alloy sacrificial anodes also liberate hydrogen when they dissolve. As noted in Chapter 10, measurements of hydrogen gas evolution were formerly used for estimating the electrochemical capacity of early aluminium anodes.

The concern about hydrogen in confined spaces was not simply a theoretical issue. Fairhurst [24] had previously emphasised that... *gaseous hydrogen is an extremely dangerous material.* He noted the case of an offshore structure where the headspace of a flooded platform leg was connected to a sizeable caisson. Both the leg and the caisson were sealed from the external atmosphere. Al-Zn-In anodes had been installed in the flooded leg. The accumulation of hydrogen generated over a 14-month period resulted in an explosive hydrogen-air mixture in the confined space. This was ignited during welding work on the external surface of the caisson. The resulting explosion blew off the leg cover plate resulting in a number of fatalities.[2]

Accordingly, it was established early [34,38,39] that any application of CP within a monopile would require the installation of ventilation in the head spaces. Thus, the original precept of maintaining an air-tight environment was reversed as a response to the failure to maintain that air-tight environment.

16.2.3.2.2 Acidification

Known unknowns (and forgotten knowns)

Donald H. Rumsfeld served two US presidents as Secretary of State for Defense. He once famously said... *There are known unknowns. That is to say, there are things that we know we don't know. But there are also unknown unknowns. There are things we don't know we don't know.* No doubt, Mr Rumsfeld's bleak remarks apply to a lot of human endeavours. As we shall see, there is also another category to add when it comes to corrosion: *forgotten knowns.*

Seawater is a "buffer solution". It has a capacity to resist changes in pH value. We encountered one of the mechanisms by which it does this in Chapter 7. The cathodically generating OH^- (hydroxyl) ions on the metal surface favour a local increase in the pH. This causes the chemistry of the seawater to change in a manner that tends to counteract that increase. The excess hydroxyl is consumed by precipitating magnesium hydroxide (equation 16.1) and by shifting the HCO_3^-/CO_3^{2-} (bicarbonate/carbonate) equilibrium to the right (equation 16.2). This, in turn, leads to the precipitation of calcium carbonate (equation 16.3). As we saw, these processes together result in the beneficial deposition of calcareous deposits.

$$Mg^{2+} + 2OH^- \rightarrow Mg(OH)_2 \tag{16.1}$$

$$HCO_3^- + OH^- \rightarrow CO_3^{2-} + H_2O \tag{16.2}$$

$$Ca^{2+} + CO_3^{2-} \rightarrow CaCO_3 \tag{16.3}$$

[2] Fairhurst did not provide further details, and very little documentation on this incident exists in the public domain. I was informed by ADMA-OPCO in the UAE in the mid-1990s that the incident had occurred on one of its facilities. Three workers lost their lives.

Where we are applying CP externally, the volume of seawater is, for all practical purposes, infinite. It follows that, even if the seawater immediately adjacent to the cathodically protected surface becomes depleted in calcium and magnesium, this local shortfall is soon made good by the transport of these ions from the "infinite" bulk of the seawater.

Similarly, the designer of an external CP system never needs to consider what happens to the dissolution products of the anode. Any soluble species will diffuse away. For example, workers looking for potentially polluting anode dissolution products in areas such as harbours have had difficulty detecting levels of aluminium that are distinguishable from the normal background level in seawater.

However, once the movement of the seawater is constrained, as is the case in internal CP, the effect of the anodic dissolution products on the local chemistry of the seawater needs to be considered.

In the case of aluminium anodes, the cations produced by the anode's dissolution reaction (equation 16.4) are rapidly hydrolysed to form insoluble aluminium oxide. This liberates hydrogen ions (equation 16.5), thereby lowering the pH at the anode surface. The buffering nature of seawater also tends to counteract this reduction in pH. Viewed simplistically, this is achieved by shifting the $CO_2/HCO_3^-/CO_3^{2-}$ equilibria towards the left (equations 16.6 and 16.7).

$$2Al \rightarrow 2Al^{3+} + 6e^- \quad (16.4)$$

$$2Al^{3+} + 3H_2O \rightarrow Al_2O_3 + 6H^+ \quad (16.5)$$

$$HCO_3^- \leftarrow CO_3^{2-} + H^+ \quad (16.6)$$

$$CO_2 + H_2O \leftarrow HCO_3^- + H^+ \quad (16.7)$$

Again, where the CP is applied externally, the depleted carbonate and bicarbonate are rapidly replenished from the bulk seawater, and the effects of hydrolysis of the aluminium remain unnoticed. Where CP is applied internally, on the other hand, and buffering capacity of the seawater is not replenished, the hydrolysis of aluminium ions becomes relevant.

There is nothing new in any of this basic chemistry. Furthermore, it has been discussed in the CP literature, albeit a long time ago. For example, in a paper presented in London to the 1st International Congress on Metallic Corrosion in 1961, Carson reported that a... *decrease in pH can be expected when seawater is electrolysed with aluminium... anodes* [1]. However, during the half-century between the 1961 Congress and the rush to get CP installed inside wind farm monopiles, Carson's paper gathered dust.

In the event, the early trials to apply CP inside wind farm monopiles involved suspending strings of Al-Zn-In anodes in the internal seawater columns. No doubt, there was a reasonable expectation that polarisation would be readily achieved. However, as subsequently pointed out by Tavares et al. [34], problems arise if... *secondary effects are not considered.* It soon emerged that applying CP in confined spaces was not so simple. The pH value of the seawater column reduced, from typical open sea values of above 8 to about 5. This may not sound much, but we should remember that pH is a logarithmic scale. A drop of 3 pH units represents a thousand-fold increase in acidity. Under these conditions, protective calcareous deposits will not develop. The result of the preliminary trials was that, although some measure of protection was achieved, the anodes struggled to achieve and maintain the desired levels of polarisation.

This is why, as noted, both the DNV and ISO codes advise against using aluminium anodes. The hydrolysis of zinc cations does not generate such a low pH.

This pH reduction was... *unexpected at the time* [39]. However, to extend Donald Rumsfeld's quotation about *known knowns* and *unknown knowns*: there are also *forgotten knowns*. That is to say, things we really should know, but don't.

16.2.3.3 Lessons learned

16.2.3.3.1 Sacrificial anodes

The early setbacks in applying CP to the monopile internals could not halt the growing momentum to achieve satisfactory CP inside monopiles. The combination of a burgeoning offshore wind farm industry, convinced that it was facing a serious corrosion problem, and a CP industry keen to market its equipment and services, guaranteed that there would be continuing development.

Detailed studies [31,32,35.36] were carried out on the corrosion inside monopiles (with and without applied CP) and on the changes in seawater composition arising from CP. This work provided useful design information.

It is fair to say that the knowledge now exists to enable CP experts to design suitable systems for the inside of monopiles, albeit that this information is not captured by codes such as DNV-RP-0416 or ISO 24656. Indeed, CP systems are now being installed in new monopiles in preference to the original strategy of maintaining a sealed and air-tight environment. The acidification problem can be mitigated by replacing aluminium anodes with zinc. Alternatively, or additionally, holes are intentionally cut into the monopile wall to allow seawater exchange, thereby overcoming both the hydrolysis problem and providing a source of replacement seawater that is naturally close to saturation with calcium carbonate; thereby encouraging the beneficial formation of calcareous deposits.

Other problems with early trial retrofit installations have also been ironed out. For example, casting anodes onto lengths of steel wire cables has largely been abandoned. Although this seemed a convenient way to deploy a string of anodes within a monopile, it had its drawbacks. Aside from casting issues in the anode foundries, the individual thin strands of the support/conductor cables were vulnerable to corrosion above the waterline. More dependable deployment methodologies were developed in which the dead-weight load of the anodes is supported by polymer rope, and the current is carried by insulated copper cables.

16.2.3.3.2 ICCP

Krebs has reported a successful application of ICCP inside 38 wind farm monopiles [42]. In this instance, Al-Zn-In anodes had previously been protecting the internal surfaces adequately. However, due to concern over acidification, a decision was taken to remove the sacrificial anodes and replace them with ICCP systems. These featured a rare example of the use of magnetite anodes. Part of the justification for the use of ICCP over sacrificial anodes in this case was the fact that this wind farm was in the Baltic Sea which has a lower salinity (~1%) than typical open seawater (~3.5%). On this basis, the resistivity of the seawater at this site would be expected to be higher than the ~0.3 Ωm value typically used for the North Sea. However, the resistivity value quoted by Krebs (110 Ωm) seems improbable.

The reason for selecting magnetite ICCP anodes over the more universally adopted MMO material was interesting. The authors stated that this was due to concern over chlorine generation. We saw in Chapter 11 that chlorine gas can be liberated on the surface of ICCP anodes. They reasoned that MMO anodes, which were originally developed for electrochemical chlorine production, would liberate more chlorine than magnetite. This is probably so since magnetite is more efficient at liberating oxygen. However, given the low salinity implied by his resistivity figure, it is hard to see how chlorine generation would be an issue for either anode type.

16.2.3.4 But is CP needed?

Many of the interesting papers emerging from the endeavour to provide internal CP to monopile foundations take care not to disclose which particular wind farm they refer to. An exception is a pair of 2017 papers relating specifically to the Lynn and Inner Dowsing (LID) wind farm off the east coast of England. These provide a useful description of the work, in particular the trial installations and monitoring, that cumulated in the design of the evidently successful internal CP systems [38]. There was also a detailed analysis of the acidification issue [39]. However, that is not the reason that I am citing these informative papers here. I mention them because they provide a backdrop to a more informative, but unpublished, backstory.

The LID wind farm foundations were also installed by MT Højgaard A/S. At the same time that it was embroiled in the Robin Rigg wind farm case, it was also fending off a smaller, but none the less substantial, legal claim from the LID's owner Centrica in the London Court of International Arbitration (LCIA). However, unlike the judgements handed down by the Appeal and Supreme courts, LCIA arbitration is a private process. The LCIA has strict confidentiality obligations which mean that the outcome is not made public. I must, therefore, be circumspect about what I disclose here.

The essence of this case was that, as a result of the problems with the grouted connections between monopiles and transition pieces, fully sealed monopile environments could not be guaranteed. This uncertainty over the sealing undermined the original corrosion engineering design philosophy which was based on maintaining an oxygen-free environment. Centrica reacted to this uncertainty by commissioning a specialist CP company: first to advise on remedial measures, and then to develop an internal CP system. As mentioned, a description of the CP trial work [38] upon which the final sacrificial anode design was based, and the detailed investigation of the acidification problem [39], have been published. Centrica then sought to recover the cost of this work from Højgaard. This cumulated in the LCIA arbitration. Højgaard defended the claim on the basis that, in its opinion, the remedial CP work was unnecessary. The LCIA found in Højgaard's favour.

As I acted for Højgaard, I am also bound by the LCIA's confidentiality protocols, so am unable to discuss the details of this case. Nevertheless, we can draw some generalisations based on information and opinions that are in the public domain.

The first point to be made takes us right back to the beginning of Chapter 1 where I recounted a very much earlier phone call from a lawyer asking me if seawater was corrosive to steel. The simple answer was: yes. However, a more relevant question would have been: Does it matter?

We have already made the point that the laws of thermodynamics tell us that corrosion is, ultimately, inevitable. The role of the corrosion engineer is to manage corrosion so that equipment and structures can achieve their design lives safely and economically. Sometimes managing corrosion requires an intervention, such as changing the material or the design, chemically treating the environment, applying a protective coating or applying CP. However, there is no natural law that says corrosion must always be controlled. Sometimes, no intervention is warranted. There is no justification for spending more on controlling corrosion than the cost of the corrosion damage itself.

In the case of wind farm monopiles, it is useful to revisit the 2011 FORCE Technology paper [25] and to consider some subsequent publications on the topic. The 2011 paper provides a description of the reasons why the assumption of a fully sealed monopile cannot be relied upon. It then provides a useful review of the types of corrosion issues that could possibly arise. The paper makes frequent use of the words "may" and "might" in this context.

It is noteworthy that the paper does not state that the likely occurrence of the corrosion phenomena it discusses will necessarily pose an integrity threat. Indeed, its evaluation shows that most of the putative corrosion mechanisms introduced by loss of sealing are unlikely to be critical. It certainly does not advise that remedial measures such as CP must necessarily be implemented. Its overall thrust is to highlight the previously unforeseen need for inspection in monopiles that were originally intended to be sealed and left undisturbed.

However, it seems that some wind farm custodians reacted more forcibly to the fact that some corrosion, above that originally anticipated in the design, could occur. For example, one wind farm operator expressed the view that for offshore wind farms... *it is essential to avoid corrosion to ensure an acceptable lifetime* [35]. My own view is that it is "essential" to assess whether the anticipated additional corrosion poses an integrity threat within the design life of the structure. Even if it does, it is unlikely to be practicable to "avoid corrosion". Corrosion mitigation would be the only achievable objective.

The issue of wall thinning due to corrosion does not arise. Even if monopile internals were fully aerated, the cumulative metal lost due to corrosion over a design life of (say) 20 years is unlikely to exceed 2 mm. This is inconsequential in terms of the dead-load capacity of the foundation, the wall thickness of which would have been determined by the need to take the piling loads.

The issue of potential concern for monopile foundations is fatigue or, more correctly, corrosion fatigue. As we saw in Chapter 12, the fatigue behaviour of steel in fully deaerated seawater is essentially the same as it is in air. Hence, the fatigue design for monopiles has usually been based on (worst-case) predicted loadings and the use of "in-air" S-N curves. Inevitably, the ingress of oxygenated seawater into a monopile invalidates the original "in-air" fatigue design assumption. It is this consideration that prompted many wind farm owners to retrofit internal CP in existing monopiles and to design new monopiles with CP installed from the beginning. In the latter case, rather than taking the optimistic route of trying to ensure an air-tight environment through life, the monopile design intentionally includes portals for seawater exchange. This means that seawater acidification issues no longer need to be addressed.

I have been involved in a couple of cases where it has been necessary to rework the original design fatigue calculations using "free corrosion" instead of "in-air" S-N curves. However, the opportunity was also taken to revise the intentionally worst-case predicted fatigue loadings to account for actual (measured) values obtained using strain gauges. The outcomes of these analyses were that, although the switch to "free corrosion" fatigue curves led to a reduced fatigue life compared to the "in air" results, the predicted fatigue life, taking in to account the actual measured loadings, still remained in excess of the design life of the wind farm.

Although the vast majority of offshore wind farms presently employ monopile foundations, these foundation designs are not all the same since water depths, soil conditions and the loads imposed by the turbine tower vary from wind farm to wind farm; and even within a single farm. It would, therefore, be unwise to make any broad-brush comments on whether or not CP is required for monopile internals. Nevertheless, it has been demonstrated, as a matter of law, that there has been at least one instance where CP had been installed unnecessarily.

In some cases, the wind farm owner confronted with evidence of unexpected corrosion has not asked the relevant question: Does it matter? Instead, the question posed has been: How do we stop it? Furthermore, there have been cases where this second question has been directed at CP equipment supply companies rather than a corrosion expert. As a matter of experience, if a CP supplier is asked to come up with a recommendation to control corrosion, the odds of it coming up with a solution, other than CP, are pretty slim.

16.2.4 Water ballast tanks

16.2.4.1 Ballast water and its management

Continuing under the heading of internal spaces that are not fully sealed brings us to the topic of water ballast tanks (WBTs). The move from wooden to steel-hulled cargo vessels saw a change from the use of rocks to seawater for ballast. There is a very wide variation in the way water ballast is managed in ships and other floating structures. For example, some naval vessels replace consumed fuel with seawater to maintain trim. Fuel tanks are intentionally maintained in an air-free condition. So, as noted above [29], the seawater pumped into in fuel-over-water tanks rapidly becomes fully deoxygenated, rendering MIC the dominant corrosion threat.

Elsewhere, the range of ballast tank usage is very varied. A semi-submersible, for example, will usually have some tanks designated as "permanent ballast". Once at its working location, these tanks are filled with seawater to take the unit to its working draught. They are then left filled for as long as the unit is on station. This could be a matter of months for a drilling unit, or years for unit converted to petroleum production. Although these tanks are nominally vented via their sounding pipes, the air refreshment is very restricted; and anaerobic conditions are to be expected.

The way in which water ballast is deployed on trading ships is also varied. Previously, single skin bulk crude carriers transported oil from loading terminals to refineries and returned with seawater ballast carried in the cargo tanks. However, single skin vessels have now largely been purged from the inventory of world shipping, not least because of the pollution risk they posed in the event of a collision or grounding. For example, after the single-hulled *Exxon Valdez* ran aground in Alaska's Prince William Sound in 1989, causing catastrophic environmental damage, legislation was passed banning the use of single-hulled tankers from carrying oil in US waters. The same ban also applies in Europe.

The majority of cargo ships are now double-skinned, with the spaces between the outer hull and the inner cargo holds or tanks used as ballast water tankage. Within any ship, the range of ballast tank usage varies considerably in response to the weight of cargo carried and the sea conditions. Some tanks are kept almost constantly full, some are filled and emptied on a regular basis, and some are usually empty, only being filled to provide extra stability in storm conditions. Furthermore, since 2017 IMO rules have been in force aimed at supressing the transport of invasive marine species. Vessels that trade internationally are required to install and operate a ballast water treatment system. They must implement, and document, a water ballast management plan. This includes performing ballast water exchange in designated locations (200 miles offshore), minimising ballast operations in environmentally sensitive areas and regularly cleaning ballast tanks to remove sediments.

16.2.4.2 Corrosion management

16.2.4.2.1 General

The corrosion of ships' WBTs has been a perennial problem in the shipping industry. It is also a problem that has spawned its fair share of litigation over the years. At first sight, WBTs and monopile foundations might appear similar since they are steel structures which hold seawater in a nearly static condition with limited aeration. However, there are important differences.

- Whereas monopiles are not coated internally, ship owners have shown an ambivalent attitude to WBT coating. In the past, some owners chose not to coat WBTs. For example, the MIC damage in Figure 1.8 occurred in a bulk carrier in which this WBT was left uncoated.
- Generally speaking, however, WBT's are usually coated, but the standard of the coating work has varied very considerably.
- WBT's are not normally fatigue critical. So, the integrity threat is plate thinning. The amount of plate thinning that can be tolerated depends on the classification society rules. Typically, a 25% reduction is tolerated. WBTs are usually surveyed (but not *inspected*) on a 5-year cycle. It is not unknown for surveyors to miss local areas of plate thinning. This is not necessarily due to indolence on their part. A WBT survey often takes place with the tank bottoms incompletely drained, leading the surveyor to take random steel thickness measurements on the shell side and vertical bulkheads. Since these vertical surfaces are less prone to MIC, the survey outcome might be misleadingly favourable.
- The ownership of cargo vessels will often change several times during its life, creating a disincentive to invest in corrosion mitigation.

16.2.4.2.2 Cathodic protection

The only viable option for the CP of ships' WBTs is sacrificial anodes. ICCP is wholly impracticable. Beyond that consensus, however, the attitude of ship owners to ballast tank CP is ambivalent. I have crawled through numerous WBTs. Where I have seen anodes (and in many tanks I have seen none), there has been little evidence of a CP design. The anode placement has often owed more to convenience than to addressing the corrosion threat.

Many ship owners, or ship managers, have taken the view that, since the anodes can only be effective for the part of the time that the tanks are in ballast, they have limited relevance.

The area of the ballast tank that is most vulnerable to corrosion, particularly MIC, is the floor plating between the stiffeners. This is also the area where the coating is most vulnerable to damage. It is an area where sediments settle, and which rarely drains fully, even when the tank is fully deballasted. Anodes installed for convenience on bulkheads or stingers will only provide protection when the tanks are filled. It is likely that an anode will spend around 50% of its operational life providing no protection whatsoever. This is illustrated in Figure 16.1 which shows an anode mounted on a stringer in a deballasted WBT still containing sludge on the base plate.

It is perhaps for this reason that IMO rules pay little attention to the option of providing CP in WBTs. Instead, it requires tanks to be provided with "hard coatings", by which IMO means two-pack epoxy type paints.

The main contribution of IMO rules is to restrict the maximum height at which aluminium anodes can be installed in WBTs that might also contain flammable vapours (for example, WBTs sharing a bulkhead with a tank carrying hydrocarbons). The concern is that an anode may detach and fall onto rusty steelwork creating an incendive spark. This concern is mirrored in EN 17243 which sets a maximum height above the tank floor of:

$$H = 28 / W$$

where:

H = installed height (m)
W = anode weight (kg)
No such ignition threat arises from zinc.

Figure 16.1 WBT anode (50% effective?).

16.3 STEEL STRUCTURES CONTAINING AERATED SEAWATER

16.3.1 Sea chests

A sea chest is a sizeable space within the ships' hull that is open to the external seawater. It provides an intake reservoir from which seawater can be drawn into the piping that feeds the ballast, engine cooling or fire-fighting systems. Sea chests are usually protected by removable gratings and contain baffles so that the enclosed volume of seawater remains relatively undisturbed by the ship's movement.

The sea chest steelwork is painted to the same specification as the hull. It also draws some protective current from the hull's CP system. However, irrespective of whether the hull is protected by ICCP or sacrificial anodes, sea chests will also be provided with their own sacrificial anodes behind the gratings. Thus, the corrosion protection of the sea chest and its grating normally pose little in the way of a problem.

A possible exception to this arises in the case of box coolers. These are heat exchangers employed for engine cooling on some smaller vessels such as tugs and support boats. Instead of a conventional tube-and-shell or plate type exchanger, the heat transfer tubes are placed within the sea chest itself. This reduces the internal space required and cuts down on the need for seawater pumps and piping.

It is important for efficient heat transfer that the cooler tubes do not become fouled. For this reason, they are usually fabricated in a copper-nickel alloy which has a low natural corrosion rate in seawater. This slow corrosion liberates copper ions, thereby inhibiting the attachment and growth of fouling organisms. This means that the cooling tubes must be kept isolated from the hull and sea chest CP system. Otherwise, the release of the copper ions would be prevented and the exchanger would soon be fouled by a combination of calcareous and biological deposits.

This arrangement poses some interesting CP challenges. For example, there is a risk of stray current corrosion damage of the electrically isolated Cu-Ni tubes if they pick up hull CP current and then discharge it back to the sea chest steelwork. As we have seen, stray current electrolysis problems can arise where a metallic conductor intersects the electrical field generated by a CP system.

John et al. [30] have described an interesting solution to this problem which involves an automatically adjustable resistive link between the tubes and the cathodically protected hull. The value of the resistance is adjusted to minimise cathodic polarisation of the Cu-Ni and to minimise the drain on the hull CP system. It also permits periodic disconnection of the tubes from the hull to allow the antifouling properties of the Cu-Ni to develop.

16.3.2 Seawater intakes

16.3.2.1 Multi-metal systems

We have already encountered multi-metal systems in the open seawater environment. For example, an offshore production manifold will often comprise a carbon steel structure and CRA piping. It is straightforward to define the CP requirements for this situation. The total entity is treated as if it were carbon steel. However, where a volume of seawater is contained within a multi-metal system, matters can become more complicated. We consider two types of such situations: shore-side seawater intakes and offshore seawater lift caissons.

16.3.2.2 Shore-side seawater intakes

Various industrial enterprises are located near the shore in order to access seawater for process or cooling purposes. Examples include desalination plants, refineries and power stations. All possess large capacity conduits to bring the seawater to the heart of the facility, plus some form of preliminary screening to prevent the entry of debris or sea creatures.

Reinforced concrete has often been the material of choice for constructing the conduits. In principle, given proper design and control over such parameters as the concrete composition and curing conditions, and the depth of cover, the steel reinforcement should not normally require any supplementary protection. In reality, problems with achieving the necessary concrete quality, or cover depth to the rebar, have resulted in corrosion issues necessitating internal CP.

For example, Simon [20] describes the case of a 400 m long reinforced concrete tunnel (3.6 m × 3.6 m) carrying seawater to a nuclear power plant. The reinforcement was showing considerable distress after 20–25 years. In itself, this did not pose a particularly difficult challenge to the CP design. However, difficulties arose because the tunnel could only be drained, and CP installation work carried out, during the relatively short periods during which the plant was offline for refuelling. This time constraint forced the design towards sacrificial anodes rather than ICCP, with the system calculated to provide a 10-year life. Embedded reference half-cells were also installed. These showed that polarisation to potentials more negative than −0.8 V was being achieved. However, had the situation not been so, nothing could have realistically been done to remedy the situation until the next refuelling.

In addition, seawater intake screens, designed to remove debris on a continual basis from the incoming seawater, either have to be manufactured in materials that are fully resistant to ambient seawater (e.g. 25Cr SDSS), or a lesser grade material supplemented by CP. Norris [40] has described the application of CP to a type 316 stainless steel screen. In this case, the screen was connected to the ICCP system that had already been installed to protect steel reinforcement in the concrete seawater conduit. A key challenge was to ensure that all parts of the continuously moving screen were in metallic contact at all times when immersed. Following discussions with the screen manufacturers, this was achieved through metal-to-metal contact at the sprockets. Based on the commissioning data, the installation was successful, demonstrating that, in this case at least, the choice of CP in conjunction with a less expensive alloy was cost effective.

As an interesting aside, Norris also commented on the commissioning data for the system, pointing out that the screens and distribution bays were commissioned at 20% design current (i.e. 30 mA/m^2) and that this... *demonstrates that a high initial current density value as specified by BS EN 13174: 2012 was not required.*

A little caution is needed here. ISO 13174, in keeping with other CP standards, is intentionally conservative in its current density recommendations. However, that conservatism does not mean that it advises values that are five times higher than is required. The important point to note is that ISO 13174 relates to carbon steel. As we mentioned in Chapter 8, and as will re-emerge below when we consider CP inside stainless steel piping systems, cathodic current densities on stainless steels will be low, providing they are not biofilmed. Thus, all Norris's observation of low current densities on the stainless steel screen tells us is that it was not biofilmed when the CP commissioning data were collected.

16.3.2.3 Seawater lift caissons

Seawater is employed for various purposes on offshore hydrocarbon production facilities. It is used for process cooling, for reservoir injection to enhance oil recovery and for fire-fighting. The seawater is obtained using lift pumps located some metres below LAT. These pump the seawater up a pipe, referred to as a riser. Both the lift pump and its riser are made of a CRA such as 25 Cr SDSS.

The pump and its riser would be vulnerable to mechanical damage and fatigue if left exposed to the waves. Accordingly, they are installed in a protective carbon steel pipe termed a "caisson". Typically, a caisson would have a diameter of about 1 m, and a wall thickness around 10 mm. The annulus between the carbon steel caisson and the CRA pump and riser is freely flooding with aerated seawater. Although there would not normally be any intentional contact between the carbon steel and the CRA within the caisson itself, the two will inevitably be earthed to the structure. Hence, the seawater lift caisson presents us with a galvanic corrosion threat.

I visited a North Sea "Condeep" platform in 1987 with the operator's chief metallurgist. Coincidentally, whilst we were there a seawater lift caisson fractured at a circumferential weld. The lower portion ended up on the seabed. The now-unsupported seawater lift pump and riser had to be pulled to prevent fatigue failure as they had become exposed to the waves.

We took a little time out from our intended work to consider this occurrence. The cause of the failure was obvious enough. The pump casing and the riser were fabricated in stainless steel, and its protective caisson was in carbon steel. The annulus was filled with aerated seawater, and the two were inevitably in electrical contact since the entire topside structure was a common earth. Thus, the arrangement constituted a classic galvanic corrosion cell. Even worse, a decision had been made to coat the inside of the casing pipe sections. The coating was applied everywhere except, obviously, where the pipe joints were welded together. Thus, not only had the design created a classic galvanic corrosion cell, an attempted remedial action had involved a classic blunder of corrosion engineering: painting the anode in a galvanic corrosion cell. This had the effect of focussing and intensifying the corrosion at the uncoated circumferential welds of the caisson.

The chief metallurgist, who had had no hand in the original corrosion engineering design of the facility, was miffed that such a basic error could get through the system. Nevertheless, the problem was relatively easy to remedy. Steps were set in place to get the remaining stainless steel pumps and risers pulled and painted. (In a galvanic couple, it is always the cathode that should be painted.) In addition, in so far as space permitted, anodes were to be attached to the riser. The role of the anodes would not be to protect the stainless steel but to polarise the total carbon steel – stainless steel galvanic couple to −0.8 V or more negative.

As far as I am aware, those remedial actions were instigated by that operator. It also replicated these actions across all of its assets. Furthermore, back in the 1980s, even if operators were guarded about corrosion failures, the information was disseminated informally around other North Sea corrosion engineers. In due course, the information appeared in the public domain. Palmer [13] provides more information regarding the widespread nature of this corrosion problem in the North Sea up to that point. Older platforms employed relatively small pumps in large uncoated caissons. Under these circumstances, the galvanic effect was not large. Caissons had lasted for up to 20 years without failure. There was then a tendency for platforms to be installed with larger pumps in smaller caissons. This reduced the failure times to between 6 and 10 years, with caisson corrosion rates of up to 2 mm/year adjacent to the pump and its strainer. It seems likely that this increased caisson corrosion rate prompted the intuitively reasonable, but entirely incorrect, response of internally coating the caisson. This was practiced increasingly from the mid-1980s. However, this served only to intensify the galvanic attack at areas where the paint suffered mechanical damage during pump installation. Local corrosion rates of up to 7 mm/year were experienced, with through-wall penetration occurring in 2–3 years. This was followed by fatigue failure in 3–7 years.

The solution to the problem was easy enough to determine. For example, Miller and Tate [12] gave details of a CP design based on anodes installed in the annular space. Interestingly, their design was based on the stainless steel pump casing and riser being coated. Furthermore, the design also assumed that the pump and riser would be pulled and repainted on a 3-year cycle. It would be interesting to know if this painting was actually done, but at least the classic need to paint the cathode in a galvanic cell had been formally re-stated.

It seems, however, that seawater lift caissons are another example of a topic that falls under our category of "forgotten knowns". Subsequent publications have once again illustrated the ability of the corrosion industry to forget a lesson it had once learned. For example, about 15 years after the problem was "sorted" in the North Sea, Wigen and Osvoll reported [25] a recurrence. During a diver inspection, a seawater lift caisson was found to be perforated in several areas. This was despite the platform only having been operational for 4 years. Preliminary investigations, in which a reference half-cell was lowered down the annulus between the 6Mo stainless steel riser and the inside of the caisson, made it... *obvious that no cathodic protection system is installed to prevent corrosion on the caisson, and further the recorded potential values prove that there was full galvanic effect between the stainless steel seawater pump and the carbon steel caisson.*

The paper describes how a CP retrofit system was designed with the object of preventing the caisson perforations from growing any bigger. In this case, the remedial CP design, which was carried out with the assistance of boundary element modelling (BEM), which we discuss in the next chapter, involved attaching anodes to the riser that were small enough to fit in the annulus between the riser and the caisson. From the CP perspective, the anode design and modelling are interesting. We may assume that it did what was required, although the scope of the paper did not include coverage of how the retrofit anodes subsequently performed.

A more interesting observation is that a well-known problem, that is well-understood, actually reoccurred in an industry that sees itself as corrosion-aware. Some insight into this is provided by Heselmans et al. [26] who suggested that, although the problem was known, it did not elicit much interest because the pumps themselves have not been corroding, so the galvanic corrosion was not regarded as a problem. They also included the interesting observation that the... *only way of protecting the inlet area is installing anodes on the pump itself. However, this is not desirable because the pumps and motor are relatively sensitive equipment which should not be impacted by any destabilising weights such as depleting anodes. For that reason, pump manufacturers may void warranty if anodes are installed.*

In addition, Wyatt et al. [37] refer to the... *view often held by pump manufacturers that there is no need to apply coating to corrosion resistant material.* In that 2016 paper, the authors were reporting the remediation of another occurrence of a seawater lift pump caisson corrosion failure. It seems that was another case of a *forgotten known* being forgotten again.

16.4 SEAWATER PIPING SYSTEMS

16.4.1 Unlined carbon steel

Using unlined carbon steel in a piping system handling seawater is usually asking for trouble. It only works in piping where the seawater only flows occasionally, or else it has been chemically treated to reduce its corrosivity. An example of the first application would be the occasional use of steel in ships' ballast tank piping. The second example is seawater used for injection into hydrocarbon reservoirs. In that case, the seawater is deoxygenated by a combination of mechanical deaeration and chemical oxygen scavengers. Even then, problems with deaerators, or lapses in chemical treatment, have caused a good number of corrosion failures of flowlines.

Internal CP of a carbon steel seawater pipeline only has a chance of being successful if it has a large diameter and is internally coated. The internal CP of unlined carbon steel seawater piping is unlikely to be practicable. Irrespective of whether a sacrificial anode or ICCP system is used, the attenuation of potential along the line would be dramatic.

This was illustrated by Carson [1]. Although the paper was primarily concerned with investigating changes in seawater chemistry as a result of applying CP, it incidentally included data on potential attenuation down small-bore carbon steel pipes. In one set of experiments, CP was applied to an uncoated steel tank using an aluminium anode.[3] Seawater flowed into and out of the tank via 19 mm nominal bore unlined steel pipes. Reference electrodes were positioned at selected intervals along the inlet and outlet lines. There are not enough relevant data points to construct a potential attenuation curve. Nevertheless, it is clear that a potential more negative than −0.8 V only persists for a very short distance (apparently <300 mm). Beyond about 600 mm along the pipe, there was no longer any observable influence of the header tank anode on the potential. Testing of this sort led to the establishment of a rule of thumb that CP could only be provided for a distance equal to no more than five to six diameters internally along a pipe [22].

For example, DNV-RP-B401 provides little design advice for structure internals. Nevertheless, it offers recommendations for calculating how much current should be allowed for in the case of open pile ends. It advises that... *the top internal surface shall be included for a distance of 5 times the diameter and shall be regarded as seawater exposed.* This figure seems to align with this rule of thumb.

16.4.2 Lined carbon steel

The application of a polymeric lining to a carbon steel pipe will inevitably increase the throw of CP current internally. There are some examples where this combination of lining plus CP has proved satisfactory.

[3] As we saw in Chapter 10, using an aluminium anode would have been very innovative in 1960.

An example of a successful application on a large scale is the La Rance tidal energy installation near St Malo in France. Opened in 1966, this was the world's first tidal power station, and it remained the largest until 2011. It comprises twenty-four 10 MW bulb turbines capable of delivering power on both the flood and ebb tides. Other than the vanes, which are titanium, all of the seawater-wetted structures are coated carbon steel. These have been protected since installation by ICCP systems. According to Féron,[4] the power plant and its corrosion protection were still working well in 2020.

There is less experience of internal CP on smaller diameter lined carbon steel seawater pipelines. Nevertheless, the experience of Zadco in the Arabian Gulf [28] merits consideration. The company uses an extensive network of 14″–18″ diameter carbon steel pipelines for seawater transport and distribution around its Zakum field facilities. The lines, most of which were installed in 1982/1983, are internally lined with epoxy phenolic paint. Within a couple of years, these lines were leaking due to internal corrosion. The immediate response was to install ~1000 zinc pencil anodes. This proved unsatisfactory, due to rapid anode consumption which, in turn, imposed increased maintenance costs.

Accordingly, the sacrificial anode system was replaced by an ICCP system in 1987. This comprised 1000 Pt/Ti anodes installed through 25 mm Threadolet fittings. Power is supplied by 23 transformer/rectifiers. There are also 56 explosion-proof anode junction boxes and 56 zinc reference electrodes. The design assumed that the internal coating would be 100% broken down and that the cathodic current density in the flowing Arabian Gulf seawater would be 200 mA/m^2.

The anode spacing is set at ten pipe diameters, meaning that each ICCP anode is only expected to protect for five pipe diameters in each direction. In 2012, Al-Jaberi [28] reported that the system had provided 25 years good service, albeit that it had required upgrades in 1994 and 2009/2010.

From a CP point of view, this case demonstrates that, if necessary, CP can be made to work inside carbon steel pipes carrying seawater. On the other hand, from a corrosion engineering perspective, the end result of having such a wired seawater piping system, with ICCP anodes installed every ten pipe diameters, seems to have been an elaborate and expensive way to satisfy the primary objective: namely distributing seawater around the facility. It seems that phenolic epoxy-lined pipe may not have been a particularly clever original choice. For example, another operator in the region has enjoyed an excellent performance from cement-lined steel pipes for the same service with no need for the complexities of ICCP.

Sometimes CP is ***an*** answer, but it is not always ***the*** answer.

16.4.3 Corrosion resistant alloys

16.4.3.1 Materials for handling seawater

Although carbon steel is cheap, available and workable, it is a poor choice for handling seawater in anything other than straight runs of large bore pipework. Unlined it is too susceptible to corrosion, whereas the prospects of providing a reliable lining for convoluted pipework are not encouraging. As noted above, internal CP can be provided but we end up in the anomalous situation of using an intense and expensive application of CP just to make use of an inexpensive material of construction.

The only sensible corrosion engineering options for handling seawater is either to use a non-metallic material such as GRP or to use a CRA. There have been some successful

[4] D. Féron, *Green Energy and Corrosion Protection*, EuroCorr Webinar 10 September 2020.

industrial applications of GRP for seawater handling offshore. It is, of course, totally immune to corrosion. However, it has not gained wide acceptance because of historical problems with its fabrication, in particular making leak-free joints, and its perceived vulnerability to mechanical damage. Corrosion engineers usually opt for a CRA.

16.4.3.2 CRAs: selection and vulnerabilities

In Chapter 8, we outlined CRAs for seawater service. Traditionally, copper-nickel alloys (typically 90/10 Cu-Ni) were employed for piping. These are generally reliable materials, and they remain much used. However, they are prone to erosion corrosion at high water velocities. Often, the only practicable way of designing around this is to reduce the design flow velocity by increasing the pipe diameter. This adds weight, both in terms of the piping itself, and the mass of contained seawater. These limitations have led to Cu-Ni alloys being replaced by stainless steels in many offshore and marine applications.

As we also discussed in Chapter 8, although stainless steels are not prone to general corrosion in seawater, they can suffer localised attack in the form of pitting or crevice corrosion under some circumstances. To avoid these degradation mechanisms in seawater, stainless steel alloys need to have a pitting resistance equivalent number (PRE_N) of at least 40. This high PRE_N requirement excludes the more commonly encountered industrial stainless steels, such as Type 316 (UNS S31603). It forces the selection towards the more highly alloyed materials such as 6 Mo superaustenitic stainless steels (e.g. UNS S31254) or 25 Cr SDSS (e.g. UNS S32760).

These stainless steels are now very widely used for piping raw and, more usually, chlorinated seawater. They give very good service, but they are not totally immune to failure. In practice, the higher the seawater temperature, the greater the risk of pitting or crevice corrosion. So, even though the general performance of these materials has been good, there have been occasional leaks where conditions have combined to produce a localised corrosion threat. The most common occurrences have been leaks caused by crevice corrosion in pipework flanges downstream of heat exchangers.

Generally speaking, such problems have not been widespread. In some cases, for example, they have been linked to seawater exiting compressed gas coolers at higher temperatures than anticipated in the original materials selection exercise. The resulting seawater leaks at a flange, whilst neither particularly common nor a major safety issue, are expensive inconveniences which plant operators could well do without. As a result, investigations have been carried out to see if there was merit in using CP to reduce the instances of leaks in CRA seawater piping.

16.4.3.3 Internal CP of stainless steel pipework

When a CRA is coupled to carbon steel, as on a subsea production manifold, it has to be polarised to the protection potential of the less noble metal: −0.8 V. However, when it comes to the internal protection of CRA piping systems, there should be no carbon steel present. The sole objective of CP is then to protect the CRA from localised corrosion. As a consequence, it is not necessary to install an anode every ten pipe diameters in CRA piping. The reason for this is that the levels of cathodic polarisation required to protect a CRA, and the cathodic current densities needed to achieve that polarisation, are markedly lower than for carbon steel.

This is explained by re-examining the E-log $|i|$ diagram for a CRA such as a stainless steel in seawater. As explained in Chapter 8, stainless steel is usually in a passive condition, and its corrosion potential (E_{corr}) falls in the passive region and is, therefore, very much more positive than it is for carbon steel. Similarly, the CP current density to return pitting stainless steel to its passive condition is very much lower than it is for the protection of carbon steel. This is illustrated in the Evans diagram in Figure 16.2. Since lower CP currents are required, the voltage drop generated by the flow of the CP current through the seawater in the pipe is much lower than would be the case for carbon steel. The potential attenuation is much less pronounced. So, anodes can be placed considerably further apart.

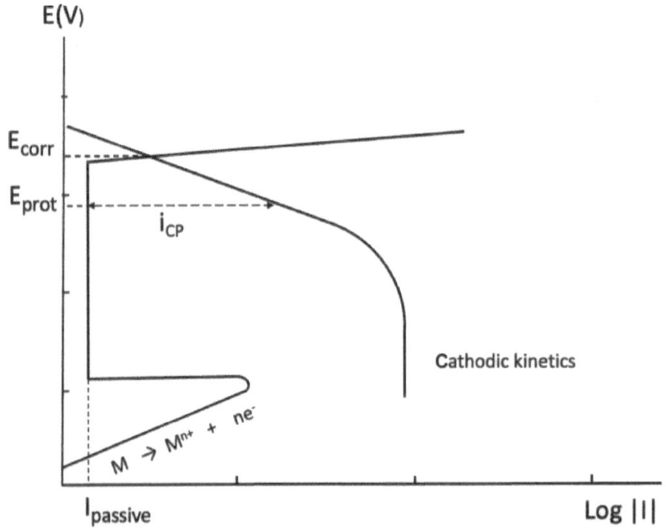

Figure 16.2 CP of a passive stainless steel in seawater.

16.4.3.4 Resistor-controlled cathodic protection

The Norwegian Foundation for Industrial and Technical Research (SINTEF) made use of these aspects of applying internal CP to stainless steel pipework carrying seawater. It developed a CP concept for using resistors to limit sacrificial anode current outputs to the lower values that are sufficient to protect CRAs. FORCE Technology, in conjunction with SINTEF, has implemented this concept in numerous seawater piping systems, using the designation "resistor-controlled cathodic protection" (RCP).

RCP was introduced in 1991 [21]. Published background dates back to 1992 when Gartland and Drugli [11] described the SINTEF work carried out on 6 Mo superaustenitic stainless steels. They proposed applying CP to 6 Mo stainless steel pipes in chlorinated seawater as a means to reduce the risk of local corrosion or to push the service temperature limit upwards. They reported that modelling indicated that when a sacrificial anode is coupled to the pipe via a suitable resistor, the potential of the pipe can be shifted into a range near 0 mV. At this potential, the risk of local corrosion initiation is much reduced, but the cathodic current density remains comparatively low at 0.1–1.0 mA/m^2.

The reason for the current densities being so low is that stainless steel is known to be a poor cathode in clean seawater. Gundersen et al. [10] have suggested that this is a consequence of charge transfer limitations across the oxide film. However, it has been known since the 1980s that, if stainless steel is immersed in natural seawater, after a few days it becomes much more electrochemically active. Its potential is ennobled and, importantly, a higher current density is then required to polarise it cathodically [7,8].

It was subsequently reported [14] that a suitable protection potential to prevent the initiation of pitting or crevice corrosion of either superaustenitic or 25 Cr SDSS in chlorinated seawater is 100 mV.[5] (Note: this is +100 mV, which is almost 1 V more positive than the values associated with the CP of carbon steels.) The CP could be applied with a conventional sacrificial anode (zinc or aluminium) with the output limited by either a resistor or a resistor in series with a diode.

The reference to diodes disappeared from subsequent publications from the same organisations [16–18,21,23]. These reported on the increasing use of the RCP technique for seawater cooling systems in (mainly offshore) petroleum production. Reportedly, over 1000 individual RCP anodes had been installed by 1999 [23]. It is, however, difficult to comment on RCP here because crucial details of the systems are not fully disclosed in the papers. For example, although the technique hinges on the use of resistors, there is no published information on resistor values or the basis on which they might be adjusted. Unsurprisingly, organisations involved seek to maintain a commercial edge.

Nevertheless, there is good evidence for the effectiveness of the technique. For example, 6Mo stainless steel was selected for the seawater piping on the Draugen platform. This selection was made on the basis that the seawater exiting the coolers would not be hotter than 30°C. Due to various start-up problems, however, temperature excursions of up to 70°C were experienced. This caused the initiation of localised weld and crevice corrosion, which resulted in leaks within 3–6 months. As a result, RCP was installed. At the time of reporting [21], the seawater system had operated for a further 2 years without leaks, even though occasional temperature excursions to ~70°C persisted.

In passing, it is worth mentioning that not all RCP applications have been an unqualified success. I am aware of a problem encountered in 2004 on another facility where a leak developed in the fitting installed for installation of an RCP anode. The welded flange fitting had been installed in an 8-inch 6 Mo seawater line and had developed a through-weld crack within a year. To be clear, this was not caused by the RCP. The welding of the flange had simply been botched, and fatigue had done the rest. It reminds us again to be aware of the Law of Unintended Consequences!

As with any insurance policy, it is not possible to quantify the cost benefit of installing RCP in seawater piping systems. In theory, a 6Mo or a 25 Cr SDSS system could be engineered such that the critical crevice temperature is not exceeded in any of the return piping. In practice, as exemplified on Draugen, sometimes things do not run entirely as intended. Any economic justification of RCP is likely to require comparing its costs with those which may, or may not, be incurred if the system is misoperated at some point in the future. Even then, RCP is unlikely to provide a remedy against all mishaps. For example, if chlorination fails for long enough to permit biofilms to form, then the likelihood of localised corrosion will increase, but the ability of RCP to protect will reduce.

RCP would be easier to justify if its installation permitted the selection of a lower grade, and therefore cheaper, material such as UNS S31603 stainless steel for the seawater piping. This potential application of the technique has been demonstrated in the laboratory [23]. However, to the best of my knowledge, no operator permits a stainless steel with PRE_N below 40 for seawater piping.

[5] These potentials are referenced to the saturated calomel electrode (SCE). I've ignored the difference between values on this scale and Ag|AgCl|seawater, which is <10 mV.

16.5 HEAT EXCHANGERS

16.5.1 A corrosion machine

A heat exchanger serves two purposes. Its primary function is to generate interesting examples of corrosion. Its secondary function is heating or cooling process fluids, but we need not dwell on that here. Back in the 1970s, when I was first studying corrosion, I was told by more than one of my lecturers that heat exchangers were ideal for teaching the subject. They provided examples of just about every form of corrosion known. Nothing in the intervening years has suggested to me that they were exaggerating. I have encountered general corrosion, pitting, crevice corrosion, galvanic corrosion and stress corrosion cracking in heat exchanges. The only thing I have not seen is MIC, but others claim that they have. It is not surprising, therefore, that there has been no shortage of attempts to manage corrosion in heat exchangers using CP. Some attempts have been more successful than others.

16.5.2 Early CP systems

Steam boilers, the bedrock of the industrial age, were arguably the first types of heat exchangers. These have almost always been constructed from iron or steel, and the primary defence against corrosion has, where practicable, been water treatment. Nevertheless, as long ago as 1905, Commander Elliot Cumberland filed a patent for the electrochemical protection of boilers using a method that we would today recognise as ICCP. He formed the Cumberland Engineering Company in 1912 to provide corrosion and scale control in industrial boilers. His protection technology was also applied to locomotives of the Chicago Railroad Company in 1924 [19].

Cumberland's technique spread from boilers to power station condensers. Prior to 1925, all ICCP in the UK power industry was referred to as "Cumberland protection" [3]. As in Cumberland's original invention, the anodes were made of soft iron. These were referred to as *wastage plates* although, according to Berkeley [6]... *the reason for the effectiveness of the wastage plates was not understood for some time.*

16.5.3 Seawater exchangers

In this book, our interest is in marine CP so the heat exchangers that interest us are seawater coolers located afloat, offshore or on the coast. In the latter case, coastal power stations make use of the abundant supply of seawater for condensing steam. Heaton [4] noted in 1975 that a power station condenser might have 40,000 tubes each around 25 m long, but that the role of CP was to prevent corrosion of the steel water boxes which were galvanically coupled to brass tube sheets and aluminium brass tubes. By that time, cantilevered Pt/Ti anodes were well established for this service, having started to be introduced into power stations in the early 1960s [2]. They were also well established for the protection of shipboard water boxes which, according to Willis (discussion in [4]), were normally coated to reduce the quantity of CP current needed.

Most seawater cooled heat exchangers are the classic tube-and-shell type, although plate type units are often found where space is something of a premium such as in ships' engine rooms. In the case of shell-and-tube exchangers, it is normal design practice for the seawater to flow on the tube side. One reason for this is that the insides of the tubes in an exchanger are easier to clean than the outsides. Thus, the tubes, tube plate and channels (or water boxes) will be in contact with flowing seawater.

Carbon steel would not normally be selected for tubes or tube plates, but lined carbon steel is not uncommon for the channels. Tube plates are usually brass or stainless steel. The tubes themselves are stainless steel or titanium. Thus, in many multi-metal seawater heat exchanger designs, galvanic corrosion issues are to be expected. For example, Nekoska and Hansen [9] reported rapid galvanic corrosion, with penetration rates of up to 25 mm/year, of Muntz metal[6] tube plates after a condenser had been re-tubed with titanium.

Given that a seawater heat exchanger will usually contain a mixture of metals, cathodic protection is the usual option for preventing attack on the less noble metal(s), which are usually the water box and tube sheet. As a rule, the materials selected for the tubes are expected to be fully resistant to seawater at the temperatures encountered. This generally forces the selection of titanium or highly alloyed stainless steels. Alternatively, copper-nickel alloys can be considered, providing the seawater flow rates are controlled.

In a materials selection exercise for an exchanger with seawater on the tube side, it is often the case that CP is provided to protect the water boxes and tube sheet from galvanic corrosion as a result of coupling to the tubes. Although the tubes do not require CP, they will take current. This has two implications. It means that the CP design must allow for this current drain and that the tube material must not be adversely affected by cathodic polarisation. For example, Nekoska and Hansen [9] also commented on some 1980s experience of ferritic stainless steel condenser tubes cracking as a result of hydrogen embrittlement. Since this vulnerability of ferritic stainless steels was known before the 1980s, this can be regarded as another instance of "forgotten knowns".

The topic of current drain-down tubes has been examined at various times. A long time ahead of most of the modelling work which features in the next chapter, Astley [5] produced a lengthy and detailed paper analysing the galvanic corrosion effects, or CP throw, in multi-metal tube type heat exchangers. The mathematics Astley presents is not for the faint-hearted, one of his equations occupies almost an entire page of the Corrosion Science journal. Nevertheless, what is particularly informative about this paper is that it compares mathematical predictions with actual measurements made in the laboratory. In particular, it compares the predicted and experimental attenuation of a Cu-30% Ni tube (11 mm diameter), with CP applied by means of a potentiostat, with a working electrode near the tube inlet. The rate of flow of the aerated seawater was 1.5 m/s.

Astley's results are reproduced in Figure 16.3, in which potentials are transposed to Ag|AgCl|seawater, and the distance recalibrated in terms of numbers of pipe diameters. BS CP 1021:1973, which was current at the time, recommended a protection potential of between −450 and −600 mV for copper alloys. On that basis, we see that the tubes are fully protected at the inlet, but that full protection is lost about ten diameters into the tube. Some polarisation is, however, detected at up to about 40 diameters. Obviously, these results are strictly only applicable to the tube material and diameter, and the seawater parameters, used in the study. It also relates only to clean tubes. Nevertheless, it emphasises the practical point that CP has limited penetration down Cu-Ni heat exchanger tubes. Conversely, in the absence of CP, the cumulative surface area of tube internals participating in a galvanic corrosion cell with less noble metals in the tube sheet and water box will be quite substantial.

Al-Hashem and Carew have also studied the issue of CP attenuation along heat exchanger tubes [15]. They used an experimental approach in which seawater was passed through Ni-Cu alloy (UNS N04400) alloy heat exchanger tubes (15 mm diameter, 6 m long). In this case, the CP was provided sacrificially using an iron anode located in a plastic header.

UNS N04400, also known as Monel® 400, has generally good resistance to corrosion in seawater but is susceptible to pitting and crevice corrosion in slow flowing or stagnant

[6] Muntz metal is a type of brass. It is an alloy of ~60% copper and 40% zinc.

seawater. The objective of applying CP to these tubes would be to polarise the alloy to a potential more negative than about –70 mV. This is the value found from other work to be sufficiently negative to prevent pitting in seawater (at ~30°C). Their results, which were based on exposure times of up to 100 hours, confirmed that for the tube length studied, the iron anodes could provide full protection.

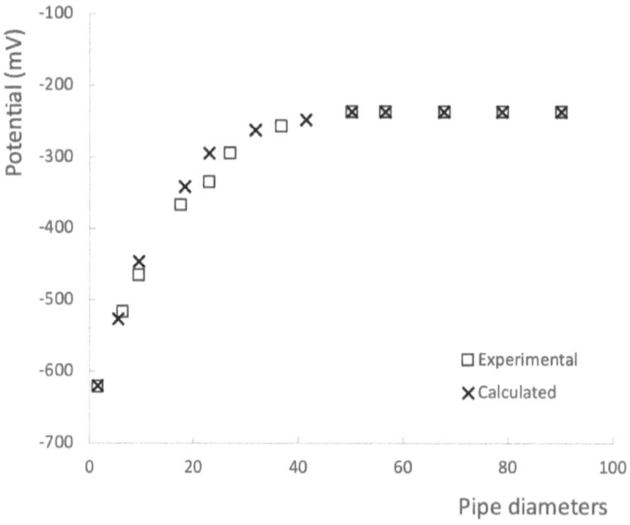

Figure 16.3 Potential attenuation down a Cu-30% Ni tube. (From Ref. [5].)

One of the reasons the iron could be so effective is because a current density of <10 mA/m^2 is all that is needed to achieve the modest polarisation needed to protect this nickel alloy. Thus, iron sacrificial anodes are capable of protecting the insides of these nickel alloy condenser tubes for essentially the same reason that RCP works in stainless steel piping.

Al-Hashem et al. subsequently extended their investigations, using a modified version of the same experimental set-up, to 70/30 Cu/Ni (UNS C71500) tubes [22]. In this case, the anodes were in the form of carbon steel inserts at the tube inlets. In this work, they did not nominate a protection potential for the alloy. As far as can be seen, they obtained protection (per codes such as BS CP 1021) near the steel anode inserts. However, beyond their first 1 m test segment (about 40 diameters) protection was lost. Thus, their results are consistent with those of Astley, confirming that it would be difficult to achieve cathodic protection of copper alloy exchanger tubes carrying seawater. They further showed that the situation is even more problematic if the seawater is polluted with sulfide.

REFERENCES

1. J.A.H. Carson, *An evaluation of an electrolytic process for the reduction of corrosion in a steel sea water piping system*, *Proceedings of 1st International Congress on Metallic Corrosion*, London, 1961.
2. W. Mathewman, *Protection of power station cooling systems*, Proceedings of Conference on *Recent Advances in Cathodic Protection*, IMI (Birmingham) 1964.
3. E. Burbridge, *A practical approach to power station cathodic protection*, Proceeding of Conference on *Cathodic Protection in Marine Environments*, Fawley Power Station, UK, Institute of Corrosion, 1975.

4. W.E. Heaton, *Cooling water systems protection*, Proceeding of Symposium on *Cathodic Protection*, Marston Excelsior Ltd (Wolverhampton) 1975.
5. D.J. Astley, *A method for calculating the extent of galvanic corrosion and cathodic protection in metal tubes assuming unidirectional current flow*. Corrosion Science **23** (8), 801 (1983).
6. K.G.C. Berkeley, *Anodes for cathodic protection: old and new*, Paper no. 48 CORROSION/84.
7. R. Johnsen and E. Bardal, *The effect of microbiological slime layers on stainless steel in natural sea water*, Paper no. 227 CORROSION/86.
8. S.C. Dexter and G.Y. Gao, *Effect of seawater biofilms on corrosion potential of stainless steels*, Paper no 377 CORROSION/87.
9. G. Nekoska and R.T. Hanson, *Effects of chlorination on cathodic protection current requirements*, Paper no. 286 CORROSION/89.
10. R. Gundersen et al., *The effect of sodium hypochorite on the electrochemical properties of stainless steels and on bacterial activity in sea water*, Paper no. 108 CORROSION/89.
11. P.O. Gartland and J.M. Drugli *Methods for evaluation and prevention of local and galvanic corrosion in chlorinated seawater pipelines* Paper no. 408 CORROSION/92.
12. W. Miller and S. Tate, *Cathodic protection and corrosion monitoring of offshore firewater and seawater lift pump caissons*, Proceedings of UK Corrosion, 1993.
13. J.C.W. Palmer, *Corrosion rate predictions and practical cathodic protection of seawater lift pump caissons*, Proceedings of UK Corrosion, 1994.
14. P.O. Gartland, R. Johnsen and J. Drugli, *The RCP method for prevention of local corrosion on stainless steel in saline water piping systems*, Proceedings of Conference on *Corrosion in Natural and Industrial Environments: Problems and Solutions*, Grado NACE, Italia, 1995.
15. A. Al-Hashem and J. Carew, *Effectiveness of cathodic protection of UNS N04400 heat exchanger tubes for seawater service. Br. Corros. J.* **30** (4), 312 (1995).
16. P.O. Gartland, R. Johnsen, S. Valen, T. Rogne and J.M. Drugli, *How to prevent galvanic corrosion in seawater piping systems*, Paper no. 496 CORROSION/96.
17. R. Johnsen, P.O. Gartland, S. Valen, and J.M. Drugli, *Internal cathodic protection of seawater piping by the RCP method*. Materials Performance **35** (7), 17 (1996).
18. E. Ommedal, R. Johnsen and I.V. Hollen, *Experiences with the seawater systems on Draugen: after three years in production*, Paper no. 420 CORROSION/97.
19. W. von Baeckmann, Chapter 1 *The history of corrosion protection*, in *Handbook of Cathodic Corrosion Protection*, 3rd edition, Eds. W. von Baekman, W. Schwenk and E. Prinz. English translation, Gulf (Houston, TX) 1997, p. 13.
20. P.D. Simon, A. Sudhakar and K.C. Garrity, *Cathodic protection of steel reinforced concrete circulating seawater conduits at a nuclear power generating plant*, Paper no. 237 CORROSION/1997.
21. R. Johnsen, S. Valen, P.O. Gartland and J.M. Drugli, *Internal cathodic protection of piping system by the RCP method: what is the experience?* Paper no. 421 CORROSION/97.
22. A. Al-Hashem, J.A. Carew and A. Al-Sayegh, *The sacrificial cathodic protection of UNS C71500 heat exchanger tubes in Arabian Gulf Sea Water*, Paper no. 414 CORROSION/97.
23. S. Valen, R. Johnsen, P.O. Gartland and J.M. Drugli, *Seawater piping systems designed with AISI 316 and RCP anodes*, Paper no. 321 CORROSION/99.
24. D. Fairhurst, *Offshore cathodic protection: What have we learnt?* Journal of Corrosion Science and Engineering **4**, paper 6 (2002).
25. S.M. Wigen and H. Osvoll, *Corrosion problems in seawater pump caissons. Practical solutions*, Paper no. 06105 CORROSION/2006.
26. J. Heselmans, N.W. Buijs and E. Isaac, *Sacrificial anodes for protection of seawater pump caissons against galvanic corrosion*, Paper no. 11056 CORROSION/2011.
27. L.R. Hilbert et al., *Inspection and monitoring of corrosion inside monopile foundations for offshore wind turbines*, EuroCorr, 2011.
28. M.S. Al-Jaberi, A.M. Al-Busaeedi and A. Ur-Rachman Hanif, *Impressed current cathodic protection system for internal corrosion protection of seawater plant piping*, Paper no. 1207 EuroCorr, 2012.
29. J.S. Lee et al., *Issues for storing alternative fuels*, Paper no. 1096 EuroCorr, 2012.

30. D.G. John et al. *Corrosion control issues for marine vessel copper-nickel boxcoolers*, Paper no. 3818 CORROSION/2014.
31. B.B. Jensen and F. Grønvold, *Corrosive environment inside offshore monopile structures and challenges in monitoring*, Paper no. 7365 EuroCorr, 2014.
32. England and Wales Court of Appeal: *MT Højgaard A/S v E.ON Climate And Renewables UK, Robin Rigg East Ltd and others*, EWCA Civil 407, April 2015.
33. B.B. Jensen, *Corrosion protection inside monopile wind turbine foundations: Commissioning and performance*, Paper no. 588 EuroCorr, 2015.
34. I. Tavares et al., *Internal cathodic protection of offshore wind turbine monopile foundations*, Paper no. 442 EuroCorr, 2015.
35. S.T. Briskeby, L. Børvik and S.M. Hesjevik, *Cathodic protection in closed compartments: pH effect and performance of anode materials*, Paper no. 5657 CORROSION, 2015.
36. B.B. Jensen, *Corrosion protection of offshore wind farms, protecting internal sides of foundations*, Paper no. 5762 CORROSION/2015.
37. B. Wyatt, et al., *Internal cathodic protection of offshore sea water pump caissons*, Paper no. 68002 EuroCorr 2016.
38. A. Delwiche and I. Tavares, *Retrofit strategy using aluminium alloy anode for internal sections of windturbine monopiles*, Paper no. 8955 CORROSION/2017.
39. A. Delwiche, P. Lydon and I. Tavares, *Concerns over utilizing aluminium alloy anode in sealed environments*, Paper no. 8956 CORROSION/2017.
40. J. Norris, *Cathodic protection of stainless steel 316L rotating screens on seawater intake structures*, Paper no. 9362 CORROSION/2017.
41. UK Supreme Court, Judgement *MT Højgaard A/S (Respondent) v E.ON Climate & Renewables UK Robin Rigg East Limited and another (Appellants)*, UKSC 59, August 2017.
42. T. Krebs, *ICCP system for internal protection of monopiles for offshore wind farms*, Paper no. 11120 CORROSION/2018.

Chapter 17

Modelling

Turning facts and assumptions into predictions

As I was starting to write this chapter, I heard an epidemiologist on the radio say "models are tools for turning facts and assumptions into predictions". I could never have picked a better opening sentence.

As coincidence would have it, I had recently been invited to comment on a draft of a proposed ISO standard. The confidentiality protocols of standards committees mean that I am not able to give details of that document here. Nevertheless, I feel it is important to comment on one of its clauses... *Verification by modelling that the current spread fulfils the protective criterion is recommended to be completed before any design is adopted.* No! Modelling can be a useful tool for developing a CP design; but it cannot provide "verification".

The only way a CP design is *verified* is by installing it, and then making the appropriate measurements. Unfortunately, there are numerous examples of papers issued by purveyors of modelling which either imply, or actually report, that their commercial modelling package was used to validate or verify a design [45,47,51,61,63].

As we will explore in this chapter, computers and modelling make an important contribution to CP. However, CP engineers have occasionally been guilty of over-reliance on the outcomes of computer modelling; and purveyors of modelling have occasionally been guilty of over-selling its benefits. As we shall see, this all-round over-enthusiasm has occasionally resulted in unnecessary expenditure.

We need to understand where the predictions that modelling produces can help us in either designing a CP system or appreciating how it is performing. We also need to appreciate how far we can rely on those predictions. With these objectives in mind, we can start by trying to answer the simple question: What do we mean by modelling?

17.1 WHAT IS A MODEL?

The origins of the English word "model" is a little obscure, but it is clearly linked to the Latin "modus": meaning a "measure". We can define a "model" as being an imitation of something, often on a reduced scale. This definition refers, of course, to physical models with which we are all familiar. It also applies to virtual models that are created in the more abstract language of mathematics.

Physical models have been around for a very long time. It is unlikely that any of the world's great man-made structures came into being without their having first been represented in wood, clay or, latterly, metal or plastic. Doubtless, there were once mock-ups of the pyramids. The explosion of computing capacity in the late 20th century expanded our definition of modelling into the virtual world. Nowadays, the design of a new chemical

DOI: 10.1201/9781003216070-17

plant, for example, no longer proceeds using plastic and wire models of vessels and piping. It is laid out virtually in three-dimensional cyberspace.

Irrespective of whether the model is created physically or virtually, the objective remains the same. It is to predict how things will turn out in the future. Both physical and mathematical modelling can be applied to a CP design. A physical model, as its name implies, requires the creation of a physical representation of the CP system, whereas the mathematical model involves describing it numerically in a computer programme. Using either approach, the model is then "run" in an attempt to investigate how the CP system will perform.

17.2 PHYSICAL MODELLING

In this chapter, we will start with physical modelling because it predates its mathematical counterpart, and because it is easier for us to grasp.

17.2.1 Full scale

To a limited degree, we have already encountered physical modelling in the work on polarisation in crevices, and under disbonded coatings, described in Chapter 9. In those experiments, a physical model of a crevice was constructed in the laboratory. The polarisation behaviour of the steel within the crevice was then investigated using small reference half-cells built into the model. In all practical respects, the model was a full-scale reproduction of the physical system. The crevice dimensions were typical of those anticipated in service, and full salinity seawater was used.

Elsewhere, there are very few documented examples of a full-scale physical model being used for a CP design or evaluation. Only one example of a one-to-one scale model comes to mind. Krupa et al. [42] reported an instance where concern had been raised about the ability of the external hull CP system on a military vessel to provide protection within the convoluted geometry of a sensitive nickel aluminium bronze (NAB) hull penetration. The question was answered by the simple, but perhaps pricey, expedient of constructing a full-size replica with reference electrodes located within the occluded areas. As a result, the authors were able to conclude that… *the physical model indicates that sufficient cathodic protection is available to prevent stress corrosion cracking of NAB weld metal and to protect the internal surfaces of the mating assembly.*

This instance, however, was something of an exception. Full-scale modelling is rarely if ever carried out to confirm any aspect of a CP design. Even military budgets do not permit the construction of full-scale ships' hulls, for example, for the purpose of refining the design of the CP system. Practicalities force the work to be carried out on a reduced scale. This scaling introduces some problems.

17.2.2 Reduced scale – reduced conductivity

By the 1980s, navies had moved from zinc anodes to ICCP systems to protect warships. This brought about an understandable desire to optimise the number and positioning of anodes and reference cells.

In the case of the US and British navies, this prompted experimental work using scale models of ships fitted with repositionable anodes and references. Workers on both sides of the Atlantic reasoned that the conductivity of their test solutions needed to be reduced pro-rata to the scale of their model. For this reason, tests were carried out in diluted seawater. For example, the British Navy [19,27] used an accurate, and doubtless costly, 1:60 scale model of a ship's hull fabricated in mild steel and coated with the Navy's customary hull paint. The model also included NAB propellers. In keeping with the 1:60 scaling, the seawater used in the test tank was diluted to a resistivity of 12.0 Ωm, rather than the 0.2 Ωm figure that would be representative of tropical waters. The early US Navy modelling also used precision scale models, with areas of paint selectively removed to simulate damage [22].

Convincing results were reported from this work. Early ICCP designs were improved by the inclusion of additional anodes and the allocation of the hull into independently controllable CP zones. For example, Parks et al. [28] reported on the case of the 9800 tonne-guided missile cruiser *USS Ticonderoga* which had been launched in 1981. It had originally been protected by a single zone ICCP system, but this had proved inadequate. They reported that corrosion rates of 25 mm/year had been encountered on selected areas of the rudder and struts. As a result of 1:96 scale modelling, it was decided to change the ICCP system to a twin zone arrangement, with anodes near the stern to satisfy the high-current demands of the propeller. The modified twin zone system was installed on Ticonderoga's sister ship the *USS Princeton* in 1987. It was found that this twin zone system showed considerable improvement over the single zone arrangement. The authors concluded that this... *confirms the validity of scale model testing as a tool for the design of shipboard ICCP systems.*

However, rereading those papers with hindsight, and a touch of cynicism, raises a number of questions. For example, it is not obvious to what extent the models improved the design, or the modifications were put forward on the basis of intuition; and the models then apparently endorsed that intuition. Judged by the absence of subsequent publications, it seems that the US Navy showed little further interest in physical scale modelling.

The British Navy, on the other hand, persisted for a while. In addition to modelling other shipboard components such as torpedo tube internals [25], it also sought to expand its scale modelling applications into the offshore petroleum sector. The technique was applied to offshore pipelines [29,32] and an offshore jacket [33]. However, these latter applications were not pursued within the CP industry.

Although these outcomes were regarded as encouraging, there was a recognition of an inherent problem. It was appreciated that diluting the seawater was relevant for modelling the flow of current around the hull. However, it inevitably distorted the electrochemistry taking place both on the anodes and, crucially, on the hull itself. Tighe-Ford [30] explained that this meant that the physical scale modelling was only relevant for the *initial state* of polarisation. Whilst this explanation was reasonable, it still leaves us in the dark when it comes to the behaviour of ships and structures after the initial polarisation.

In addition to the issue of electrochemistry, modelling and in-tank testing were always destined to be at a disadvantage compared to the emerging dominance of computational techniques. Warne summed up the position with physical modelling [14] when he said that such techniques... *can provide very useful data but, as they need physical representations of the geometries involved, they are difficult to set up, and tedious to apply.*

It is a moot point whether it was the appreciation of this problem with the basic electrochemistry, the experimental costs, or the emergence of computer modelling that sidelined the physical scale modelling approach. Whichever it was, publications in the corrosion literature on the subject tail-off after the early 1990s.

17.2.3 Reduced scale – full conductivity

Compared to the above military-funded modelling, in which both the physical size of the model and the solution conductivity are reduced in step, there have been very few examples of modelling involving physical scale representations immersed directly in seawater or the seabed. An exception is the 2017 publication by Hajigholami et al. [64]. In this case, a steel model of a platform jacket was constructed, not for the purpose of designing the full-sized structure, but for validating the outcome of numerical modelling of the CP system by the finite element method (FEM).

Unfortunately, the paper is scant on experimental details. For example, we are not told the scale of the model. However, we are told that it comprised four 60 mm diameter legs and twelve horizontal support tubes; four each at the base, half-height and the top. The images also indicate eight diagonal members. The model was unpainted and was fitted with an undisclosed number of 0.25 kg zinc stand-off anodes.

The model was installed in the Persian Gulf (resistivity 0.20 Ωm), and its potential was recorded at 20 key locations at daily intervals over 3 days. The process was then repeated in the higher resistivity (0.61 Ωm) Caspian Sea. FEM was carried out for the two environments, but the paper only shows a comparison between the mathematically predicted and physically modelled data for the Caspian exposure. Their calculated results for the same 20 locations, replotted as a scatter diagram, are shown in Figure 17.1. The dotted line in the figure indicates a 1 to 1 correspondence between predicted and measured potentials.

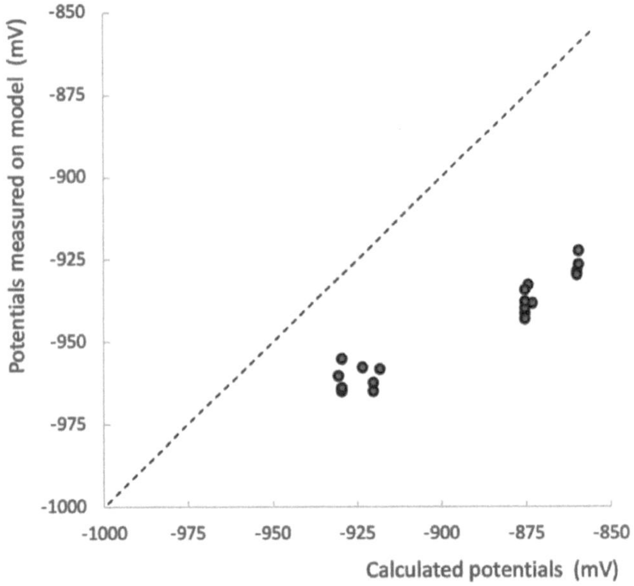

Figure 17.1 Comparison between measured and calculated potentials [64].

As can be seen, the potentials calculated by modelling are consistently less negative than the values measured on the physical model. The average discrepancy between the two is 52 mV. Given that, if a zinc anode sacrificial CP system works as intended, it will polarise a structure into a 200 mV potential range (−800 to −1000 mV), a discrepancy of 52 mV is not at all impressive. The prediction is even less convincing when we consider that all of the measured results fell within 43 mV (between −922 and −965 mV).

On that basis, it is difficult to support the authors' conclusion that the... *FEM simulation, therefore, was applied successfully for modelling the sacrificial anode system for steel jackets.* More relevantly, this work compares one modelling technique (physical) with another (computational). It does not compare either with results that might be obtained on a real structure.

We will return to FEM modelling below. But first, we need to pay our respects to some of the earlier applications of computer technology to the design of CP systems.

17.3 EARLY COMPUTER APPLICATIONS

The pioneering offshore CP systems that we described in Chapter 13 were designed without the benefit of computers. At best, those early CP engineers had the use of slide rules. Then, as the industry moved into the 1970s, the first four-function electronic calculators began to make their presence felt.

References to "computers", and to the even more radical word "modelling", started to appear in some CP publications in the early 1980s. For the most part, however, these did not deal with computer modelling as we would now interpret it. For example, Cochran [2] referred to a new "model" which, in reality, was the use of an advanced (for the time) calculator to carry out the type of routine CP design calculations that we went through in Chapter 3.

If you are too young to have lived through those early years of computer science, then you might care to seek out some of the early publications on the application of computers to CP. For example, those of us old enough to have programmed in BASIC might care to reminisce over a 1981 paper [4], which simply lists, line-by-line, the code for calculating the number of anodes for a length of pipeline.

However, even as that article was being put to press, the ascendency of the spreadsheet in engineering applications had begun. It started with Apple's *Visicalc*, launched in 1979. This was followed by *Lotus 1–2–3*, the go-to spreadsheet software of the 1980s, before that was in turn eclipsed by Microsoft's *Excel*. The contemporary literature contains examples of how these emerging spreadsheet applications could be applied to offshore CP designs (e.g. [34]).

Although once regarded as cutting edge, spreadsheets are now run-of-the-mill. As we saw in Chapter 3, they assist in producing straightforward code-based CP designs. They enable us to calculate, quickly and conveniently, how many anodes of a given type we will need to install in order to protect a structure in compliance with a selected code. We can also use their "what if" functionality to bring about some optimisation of that design. However, they are not modelling the type we are discussing in this chapter. Nisancioglu [31] has summed up, or rather he has dismissed, such applications as the... *use of elementary computer models for design and book keeping purposes*. It is not computer *modelling*.

17.4 COMPUTER MODELLING – THE BASICS

17.4.1 The convenience of the computer

Despite its many benefits, a spreadsheet constructed for CP calculations will not help us answer the question: where should we place the anodes? For that we need to go deeper into the realm of computer science and explore modelling. This helps us achieve the overall balance between the structure's cathodic current demand and the current that can be supplied by the anodes. It also addresses the geometrical question of where that current goes as it travels from the anodes, through the environment, onto the surface of the cathode.

It is far more convenient, and very much less costly, to sit at a computer than to construct, and then repeatedly modify, a physical model. The computer also does away with the need to construct a tank and procure a source of seawater with which to fill it. Once a model has been "built" in the computer it can be modified and rerun, as many times as the operator wishes, at the click of a mouse.

We should also mention here that the ambitions of CP engineers alone would not have generated the financial impetus for the development of the modelling software, which we will be discussing below. These modelling techniques have become commercially available because, in principle, they can be used to tackle a very wide spectrum of scientific and industrial problems. Examples include such varied endeavours as the study of electric and magnetic field, stress and heat flow analysis, fluid dynamics and electroplating. The expansion of these models into the CP arena only gained traction after they became well established in other fields of science and engineering.

17.4.2 Potential and current distribution

Most CP engineers are comfortable with creating and modifying designs using spreadsheets or other applications such as MathCad®. On the other hand, a few are likely to have acquired the computational skills needed to work in mathematical modelling. If you wish to get more deeply involved then I am afraid that you will have to look beyond this book for guidance. Our ambition here is to develop some insight into the processes within the modelling computer and, perhaps, within the minds of the individuals doing the modelling. So, the following is my layman's interpretation of what goes on.

At its essence, the objective of computer modelling is to develop a mathematical description of how a CP system performs on a given structure. If that mathematical description is valid and complete, then, in principle, the model can be run to explore the way in which changing any aspect of the design will affect the levels and distribution of the polarisation of the cathode. It can also evaluate, for example, the current drawn from the anodes and, thence, their rate of consumption.

For massive fixed or floating structures (Chapters 13 or 15), there are three aspects of a CP system that the model seeks to replicate:

1. the manner in which the current leaves the anode and enters the environment (seawater or seabed sediment);
2. the pathways the current takes as it flows through the environment to arrive at the surface of the structure (the cathode); and
3. the effect of that current as it flows onto the cathode surface.

As with the conventional CP design approach, modelling for fixed structures ignores the flow of electronic current within the metal of the structure itself. This is reasonable in the case of a massive hunk of steel such as an offshore structure of a ship's hull where the electrical resistance between any two points will be inconsequential.

However, as we saw in Chapter 14, this does not hold for pipelines. Here, both the CP design and any supportive modelling additionally need to account for the flow of CP current along the resistive pipe wall. This is of course amenable to modelling. Everything is.

Even if it had been available, modelling would have had little relevance to the early CP designs for the relatively small and simple structures employed for offshore hydrocarbon production up until the 1970s. By the same token, the hereditary practice of installing bracelet anodes, at intervals rarely exceeding a couple of hundred metres, obviated the need for any investigation of the current distribution in pipeline CP designs.

However, as structures became larger, more spatially complex, and more weight-sensitive, there developed an increasing interest in the way the CP current distributes itself over the surface. This interest was further stimulated as a result of CP surveys on some early prestigious offshore installations which, surprisingly, showed areas of local under-protection. This suggested that spatial issues, which had previously been ignored in CP designs, merited attention.

Perhaps inevitably, it came to be recognised that the 3-dimensional flow of current through the electrolyte needed to be considered when designing CP systems for complex structures.

Early modelling focussed on this aspect. Indeed, some of the pioneering workers in the field were not CP or corrosion specialists but electrical engineers with a professional interest in the behaviour of electrical and magnetic fields.

17.4.3 The Laplace equation

The origins of the computer modelling of CP systems lies in attempts to describe the distribution of the potential in the 3-dimensional space surrounding a protected structure. To this end, modellers have drawn on work carried out by the eminent mathematician, astronomer and physicist Pierre-Simon Laplace. One of his notable achievements, on a par with being one of the few intellectuals who kept his head attached to his neck during *La Terreur* following the French Revolution, was the publication of his second-order partial differential equation.

$$\nabla^2 E = \frac{\partial^2 E}{\partial x^2} + \frac{\partial^2 E}{\partial y^2} + \frac{\partial^2 E}{\partial z^2} = 0 \tag{17.1}$$

The Laplace equation, as equation 17.1 is now known, applies in numerous scientific and engineering fields. It was developed well before the advent of CP. Laplace was in his mid-70s by the time Sir Humphry Davy, assisted by the young Michael Faraday, was fixing anodes to *HMS Samarang*.

With CP in mind, equation 17.1 is written in terms of electrode potential (E) and the cartesian coordinates x, y and z. However, versions of the equation crop up where, instead of E, the field quantity could relate to electron probabilities in quantum mechanics, electrostatic and magnetic field strengths, stress distribution, heat flow, etc.

For our purposes, it is sufficient to picture the physical reality that the Laplace equation conveys. In a homogenous electrolyte such as well-mixed seawater, the current arriving at any point in space must exactly balance the current leaving that point. Viewed simplistically, it is a restatement of the Law of Electroneutrality. We can start by imagining the marine environment as comprising a 3-dimensional array of very small individual volumes, or elements, of seawater. The total electrolytic CP current entering each individual element from some of its adjacent elements must exactly balance the current flowing away to other adjacent elements. Each element has a resistance determined by its size, geometry and the resistivity of seawater, So, this flow of current between elements is associated with an array of potential differences. As Sander [13] has pointed out, providing the seawater is a uniform homogenous conductor, which is a reasonable assumption, then the distribution of the electrical potential through the seawater must obey Laplace's equation.

17.4.4 Solving Laplace

17.4.4.1 The need for a number-cruncher

I suspect that many CP engineers, like myself, might not be too familiar with solving second order differential equations. Happily, we do not need to be too hard on ourselves about that gap in our education. It turns out that Laplace's equation can only be solved analytically for a very small number of simple cases, none of which is relevant to us. For real CP systems, the electrolyte space is too complex for an analytical solution using the disciplines of calculus. Instead, the problem has to be attacked using brute force number-crunching. That's where computers come in. They can be set to work running remorselessly through a sequence of trial solutions, targeted at reaching an outcome that makes sense.

Using numerical computational techniques to solve partial differential equations, such as Laplace, dates back to the early days of electronic computers. Its origins can be traced to Alan Turing's work on mathematical biology in the early 1950s. Almost 30 years later, Strømmen [1] was among the first to suggest that numerical modelling techniques could be used in connection with CP, although he subsequently credited [3] R.N. Fleck's 1964 seminal dissertation on current distribution in electrochemical cells.

17.4.4.2 Defining the space

The first step in any numerical model is to assign the limits of the physical space being modelled. In marine CP, this physical space is the seawater and seabed mud. The dimensions of this conducting space provide the first problem to be overcome. In the code-based CP designs that we worked through in Chapter 3, CP was viewed as an exercise in delivering current from anodes to "remote" earth, and then returning that current from "remote" earth onto the surface of the structure. In this sense, "remote" earth can, quite literally, mean that the entire planet is involved in the electrolyte conduction path. This would be a challenge even for the most gargantuan number-crunching machine. In practice, however, the most dramatic changes in potential are observed close to the anode and immediately adjacent to the cathode. In modelling, the problem of defining "remote" earth is pragmatically overcome by allocating a volume of seawater (or seabed) sufficient to contain the flow of the vast majority of the CP current.

17.4.4.3 Defining the boundaries

For CP modelling, the physical boundary of the modelled space comprises the surfaces of the anodes and of the cathode. In addition to defining where the boundary exists in space, the modeller also has to assign "properties" to it. In this case, the "properties" of interest are the polarisation characteristics of the metals, anode and cathode, that form the boundary surfaces.

In mathematical terms, the boundary and its "properties" represent the limits of the finite integration of the Laplace equation. It will emerge that actually defining the "properties", by which we mean the potential-current relationships, of these boundary surfaces, is pivotal to the modelling exercise. For now, we simply need to appreciate that the boundary is allocated by the modeller. It may be a section of a structure, such as a critical node on a jacket, a section of pipeline or the stern of a ship. Alternatively, it could be an entire structure.

It is probably also worth mentioning here that most of the published work on modelling has placed more of an emphasis on creating the mesh, and then setting up and solving the multiple simultaneous equations, than it has on nailing down the boundary conditions. For example, Sander [16] applied a purely electrical engineering perspective to the case of the current and potential distribution on a steel pipe. With regard to the boundary condition, he stated that the... *solution requires information on the relationship between current density and electrode potential which in general will be non-linear; in the absence of reliable experimental data a linear relationship may be assumed.*

No! That might be an electrical engineer's viewpoint. However, from the electrochemical perspective, it is deeply flawed. The only thing that can be guaranteed by picking incorrect boundary conditions will be incorrect results. As the computer aficionados would say: "GIGO".[1] We will return to the issue of boundary "properties" in Section 17.5.4.

[1] GIGO: "Garbage in, garbage out". (Less polite versions are available.)

17.4.4.4 The finite element method (FEM)

The next step is to divide the volume of seawater, or seabed mud, being modelled into a mesh of very small discrete volumes, termed *finite elements.* These spatial elements are typically tetrahedra of differing sizes. If the elements are very small, the discrimination of the modelling will be high; but the cost in terms of computer memory and run-time for the number-crunching might be prohibitive. On the other hand, elements which are larger, whilst easing the computational load, will result in lower resolution outputs. A compromise is usually arrived at which involves adjusting the mesh size to reflect the (intuitively) anticipated intensity of potential variation. Typically, a smaller mesh size will be assigned close to a critical area such as a node, while the mesh elements are larger further away from the structure's surface where changes are expected to be more gradual.

It is no small task to build the finite element model in the computer. For any complex structure, the number of elements required can be enormous. For example, Bortels [46] has described an exercise on a ship's seawater ballast tank in which the steel surface was modelled as 300,000 surface elements. These surface elements, in turn, generated a mesh of about 3,000,000 three-dimensional finite elements. The herculean effort required to create such a mesh would be beyond the capability of mere humans. Earlier studies employed software referred to as a mesh generator. In later years the prevalence of computer-aided design (CAD) drawings means that the meshes can now be generated automatically from the CAD drawings.

Once the mesh has been created, and the boundary conditions decided upon, the computer can be set to work. The modelling processes themselves are proprietary. Nevertheless, we may surmise that it involves cycling through a series of iterations in which the (initial) potential difference between the cathode and the anode causes currents to flow from the anode (in accordance with its preprogrammed anodic polarisation response and its physical geometry), through the environment (in accordance with its description of resistivity and geometry in the model), and onto the cathode (in accordance with its preprogrammed geometry, dimensions and cathodic polarisation characteristics).

The computer "interprets" this current as shifting the potential within the array of environmental mesh elements, and over the cathode and anode surfaces. This results in a new distribution of calculated potentials throughout the domain. This new array of potentials then becomes the input for the next cycle of the iteration. It produces modified currents that, in turn, result in further modifications of the potential distribution. The iteration is repeated through enough cycles to arrive at a near steady-state picture of the distribution of potential through the volume of seawater and over the anode and cathode surfaces.

Thus, the modelling methodology differs from the DNV-RP-B401 cookbook type design approach we used in Chapter 3. The modelling attempts to mimic the mutual polarisation process. That is the real-world situation whereby the potentials of both the anode and the cathode move interactively as they progress towards a near steady-state value. Codes such as DNV-RP-B401 ignore this mutual polarisation of anodes and cathodes. It recommends constructing designs on the basis of invariable anode and cathode potentials.

17.4.4.5 The boundary element method (BEM)

It is now time to meet a particularly interesting, but unassuming, individual by the name of George Green. Born in 1793 near Nottingham, in the English Midlands, George was a child with a frail constitution. Nevertheless, he went to work in his father's flour mill and bakery at the ripe old age of five. His career in baking, which he detested, permitted him only 1 year of tuition. He attended the local village school between the age of eight and nine. Other than that, he was entirely self-taught. Notwithstanding his lack of a formal education, he developed an interest in science and mathematics. At the age of 35, he authored what was to become a seminal essay on electricity and magnetism. This essay contained what is now known as Green's theorem.

At the risk of over-simplification, it states that if the potential distribution of a closed surface satisfies certain conditions, the potential distribution in the electrolyte volume enclosed by that surface must satisfy the Laplace equation.

This is not an easy proposition to grasp. Do not be too hard on yourselves if, like me, you find Green's theorem too taxing. I am reliably informed that some people who actually make a living out of computer modelling don't actually "get it". On the other hand, if you want to dig deeper, references [9,15,48] are good starting points. But brace yourself for some heavyweight mathematics.

Fortunately, all we need to do is to take the core message from Green's theorem. That is that we do not necessarily have to solve Laplace's equation for the entire volume. If we can determine the distribution of potential on the boundary of the space (i.e. on the surface of our structure), then we could, if we wished, calculate the potential distribution throughout the entire volume of the seawater using one of Greens' equations. The latter point, however, is usually not that important to us. For the most part, in CP we are usually only interested in the potential of the metal surface. What matters to us is that, by applying Green's theorem, we can get a handle on the spatial distribution of potential over the surface of our structure using a drastically reduced number of elements. This is because we only need to define the 2-dimensional boundaries, not the 3-dimensional space. This is the basis of the BEM technique. It has considerable benefits in reducing the workload and computer processing capacity required.

Before moving on, we should round off our potted history of George Green. His essay did not have much influence at the time. He self-published and could only pay for a handful of copies to be printed. Most of these he gave away to (presumably bemused) family and friends. He made no money from his efforts.

Shortly afterwards, however, he inherited a tidy sum from his father. His flour milling and baking days were now behind him. Aged nearly 40, he fulfilled his long-held ambition to attend Cambridge University. He did well academically, but rumour has it that whilst there he succumbed to alcohol. He died a few years after graduating. Sometimes, a university education isn't quite as advantageous as it is made out to be.

17.4.4.6 FEM versus BEM

Both FEM and BEM software is commercially available for CP applications, and both have been used on prestigious, as well as trivial, projects. Depending on whom you talk to, FEM is superior to BEM [46], BEM is better suited to CP than FEM [9,12], or there is not much to choose between them [15]. I am not qualified to offer an opinion. However, I have spoken to purveyors of both systems. From what I can gather, both can be expected to give a reliable result, providing the boundary conditions are correct.

A corollary is that, if the boundary conditions are incorrect, neither modelling method has any hope of yielding a reliable result. We return to this issue below. But first, we need to record some other approaches to CP system modelling.

17.4.4.7 Other software approaches

17.4.4.7.1 Spreadsheets

Before moving on to examine the thorny topic of boundary conditions. We should mention that FEM and BEM are not the only two modelling methods that have been explored in connection with CP modelling. There were some early attempts to calculate the spatial distribution of potential by using spreadsheets. The thinking was that most engineers had personal computers and spreadsheet software, but a few had access to BEM or FEM. For example, in 2006 Ellor [41] described his attempt to use a spreadsheet to predict potential distribution on the hull of a warship. He argued, entirely without any supporting evidence, that... *a reasonable estimate of current distribution can be made on the basis of the principles of electrostatics.* He then compared the results of his calculation with the results of some physical scale modelling data produced by the US Navy, which we described in Section 17.2.2. All that can be concluded is that there was a greater discrepancy between his spreadsheet results and the observations recorded on the physical scale model than those produced by Hajigholami et al. [64] (see Figure 17.1).

The following year John et al. [43] also used spreadsheet-based software, originally designed for calculating the grounding resistances of complex structures, in an attempt to model the polarisation state of an offshore jacket. This had been installed in the North Sea more than 30 years earlier, and its sacrificial anodes were near the end of their lives. Accordingly, modelling was carried out in an attempt to optimise the anode retrofit that would inevitably be needed.

The construction of the spreadsheet is not detailed in the paper. However, it explains that the model looks at the interaction between every conductor and every other conductor and, hence, required a very large calculation matrix. However, even though they elected to use the same polarisation curve for the entire structure, the authors were candid enough to concede that the results obtained were not particularly convincing. Even after a time-consuming 70 iterations, the predicted potentials were still unstable. Furthermore, and perhaps unsurprisingly, their results showed poor correspondence with the available survey potential measurements.

Considering both Ellor and John et al., we can reasonably discard spreadsheets as an option for computer modelling of the spatial aspects of CP design.

Now, you may be wondering why I have taken the trouble to describe the John et al. paper since it was, ultimately, an unsuccessful spreadsheet exercise. There are two reasons. The first is that it serves as a useful indicator of just how problematic modelling can be. In that respect, it is commendable that the authors shared their results. It is generally helpful to science to learn about what doesn't work, as well as about what does.

The second reason is that this paper is one of the earliest examples in the NACE annual conference proceedings in which a multicoloured equipotential contour map of a structure is presented. Ironically, despite the authors' admitted inability to achieve results that matched reality, their diagram actually looks exceedingly convincing.

17.4.4.7.2 Finite difference method

Strømmen and Rødland [3] originally suggested the application of the finite difference method (FDM). They used this in determining the distribution of protection along a pipeline protected by a bracelet anode, or along a single structural member protected by a pair of trapezoidal anodes. Although interesting, the modelling was limited by the assumption that the current density on the steel was independent of the potential. More important, for all the theoretical analysis presented in the paper, there was no actual field data, or other experimental measurements, to support their contentions. A couple of years later Strømmen [6] wrote that FEM was more adaptable to complex geometry than FDM. He also mentioned that there had also been *recent interest* in BEM. In the subsequent decades, it has been FEM and BEM that have contested the CP modelling market.

FDM continued to be used to some extent in the 1980s. For example, Strommen [10] used it to re-examine the traditional anode resistance formulae that we discussed in Chapter 10.

However, by the time of the 1990 Marine Technology Directorate state of the art review of CP, it only merited a single paragraph in the chapter on modelling [26]. It has featured very little in the literature in recent times.

17.5 COMPUTER MODELLING – APPLICATIONS

17.5.1 Early days

In 1982 Munn [5], who had access to the computer resources of the US Navy, used FEM to predict the potential distribution generated by attaching a zinc anode to a steel plate in a small (0.2 m x 0.2 m) tray (0.03 m deep) filled with seawater. Unlike Strømmen and Rødland [3], who had used... *evenly distributed cathodic current of 54 mA/mm²* for their pipeline case, Munn used a *representative polarisation curve* in his analysis. Importantly, he recognised the importance of validating the prediction of the model using real data. Accordingly, he presented maps of lateral distribution of potential in the tray as predicted by the modelling and, crucially, as measured by moving a reference over a grid pattern. The agreement between the two maps, whilst not precise, was encouraging.

There was some progress made in developing numerical modelling during the 1980s. This can be traced through the papers championing FEM [8,14,18] or BEM [7,12,17,20,23,24]. For the most part, these papers gave a description of the technique, some being more mathematical than others. All of the papers also gave illustrations of how modelling might be useful, for example through somewhat contrived examples of predicted potential distributions on parts of real or hypothetical structures. However, unlike Munn, none of these early workers presented a paper in which modelling predictions were made and subsequently tested by field measurements.

The number and scope of these early publications were inevitably constrained by the availability of computers at the time. As we will see, modelling the actual behaviour of a CP system is not straightforward and is particularly demanding of computer power, which was not so available in the 1980s. As Hack pointed out in 1989 [21], one of the perceived disadvantages of computer modelling included... *the high amounts of computer resource needed.* This, in part, explains the persistence of physical scale modelling through the 1980s, and into the early 1990s.

17.5.2 Moore's law

Hack may not have been familiar with Moore's "Law". This was first advanced in 1965 by Gordon Moore, a cofounder of the Intel Corporation. Moore's law states that the number of components that could be fitted into an integrated circuit doubles every 18 months or so. The consequence of this "law" was that the *high amounts of computer resource* were becoming available even as Hack was offering his opinion.

Although the cognoscenti apparently pronounced Moore's law to be "dead" in 2019, it had enjoyed a spectacularly good run. Its outcome has been that the required *high amounts of computer resource* are now very readily available. This is something of a mixed blessing as far as CP is concerned. On the positive side, it means that the calculations that could not be carried out within an engineer's working life are now executed in seconds. This, in turn, means that the spatial problems associated with predicting where current will flow in complex geometries can now be tackled with confidence.

Conversely, we should be aware that there is also a negative side to modelling. Its availability, combined with the inevitable emergence of the commercial drive to sell it, has resulted in

CP engineers becoming inundated with its multicoloured graphical outputs. It is interesting to reflect on the proliferation of papers on this subject at corrosion and CP conferences in the period 2005–2021. The NACE corrosion conference, the largest annual corrosion event on the planet, in keeping with most other technical meetings switched from providing printed copies of papers to issuing them in electronic format in 1996. The printed papers were always monochrome. But, within the storage limits of the electronic medium (initially CD-ROM's), coloured illustrations and graphs started appearing in proceedings around 2001. Multicolour computer CP modelling outputs started to become prevalent from around 2004. This has given free rein to authors to populate the proceedings with multi-coloured equipotential or equi-current density contour outputs of their modelling. Of course, this is not a bad thing in itself. It is, after all, progress.

On the other hand, the complexity of the computerised analysis, combined with the beguiling polychromatic output, can engender a sense of authority. It is all too easy to forget that, behind all of this sophistication, the computer is just a machine for turning facts and assumptions into predictions. If the assumptions are wrong, or the "facts" are not quite as they should be, then the prediction becomes unreliable and even misleading.

17.5.3 When modelling gets it wrong

In so far as I am able, I will now share an example of this unreliability with you. It concerned an offshore wind farm installed in North West Europe. The foundations for the wind turbine towers were constructed using driven steel monopiles. As we discussed in Chapter 13, and as was established practice at the time, the monopiles were uncoated. External CP was provided by sacrificial anodes installed on the skirt of the transition piece.

However, due to the relatively small surface area of the transition piece skirt that would remain fully immersed at low tide, the anodes were clustered closely together. As we discussed in Chapter 10, this clustering of anodes had the effect of mutually supressing their outputs. In the event, and unsurprisingly with hindsight, the anodes struggled to polarise the monopiles. Early potential readings were not at all encouraging.

The wind farm owner, an organisation which itself possessed very considerable offshore CP experience, commissioned a corrosion consulting company, also very experienced in CP, to review the situation. The consultants quickly, and correctly, identified the core problem. The CP system had been designed, per contract, according to the then-current edition of DNV-RP-B401. Since this code had been developed for space-frame jacket-type structures, its recommendations for anode current calculations were implicitly based on anodes being liberally distributed over the structure. It did not then, nor does it now, provide any usable advice for monopile foundations with anodes clustered on the transition piece. The resulting CP system, the design of which had ironically been certified by DNV itself, failed to deliver the current that the RP-B401 guidelines had envisaged.

In addition to carrying out the retrospective analysis of the design, the consultants commissioned a computer modelling study. The study report was amply illustrated by colour-coded equipotential drawings. On the basis of these, the consultants arrived at a single unequivocal conclusion: the monopiles would never polarise to the target protection potential (−800 mV).

The issue was not corrosion, which would have been inconsequential over the design life. The concern was corrosion fatigue. As DNV has succinctly expressed it: *a wind turbine is basically a machine built to destroy its foundations.*[2] Fatigue loadings are the life-limiting factor for wind turbine monopile foundations. In this case, as was the industry norm, the

[2] J. Lichtenstein Eurocorr Webinar 9 September 2020.

foundation's fatigue design had been based on S-N curves for cathodically protected steel (refer to Chapter 12). It now seemed that the fatigue behaviour would follow "free corrosion" characteristics, thereby potentially compromising the design lives of the foundations.

The wind farm owner faced a serious problem. It was in possession of an authoritative report criticising the CP design, a firm conclusion from the modelling that the foundations would never reach their target protection potential, and it had a legitimate concern over the possibility of fatigue failure. Little surprise, therefore, that it commissioned the design of retrofit CP systems, with the object of re-establishing the cathodically protected fatigue behaviour.

In addition, while the retrofit design was being developed, the owner also pursued a programme of potential monitoring and surveying. This included the provision of fixed references, with data recording on selected monopiles; and a programme of dip cell potential surveys. (We discuss CP surveying and monitoring in the next chapter.) All of these actions were, of course, entirely sensible.

It will also not surprise you to learn that, even before the retrofit had been designed, the owner initiated legal proceedings against the foundation designers for recovery of the cost of the additional surveying, and the very much greater cost of the anticipated retrofit itself.

The owner's claim was made in the LCIA, which we came across in Chapter 16. Again, I acted for the foundation designer. Also, as before, LCIA rules limit what I can tell you about the case. Nevertheless, I can disclose that, after some years of legal jousting, the two sides agreed to settle just before the court sat. The details of the settlement remain confidential between the litigants and their lawyers. However, I can disclose that the wind farm owner settled for a sum that was very much less than its original claim.

Now, you may be asking yourself: why would a party with such an apparently strong case back down and settle for substantially less than the full sum it was claiming? The answer to this hinged on two facts.

The first was that the prediction made on the basis of the modelling turned out to be incorrect.

The comprehensive monitoring programme instigated by the owner demonstrated unequivocally that the monopiles did eventually polarise. More to the point, it was clear relatively quickly that the wind farm as a whole was in the process of polarising, and that the first of the monopiles to be installed had reached −800 mV or more negative. Thus, the conclusion drawn from the modelling that the monopiles would never polarise was self-evidently incorrect. We will take a look in Section 17.5.4 at why modelling can get it wrong.

The second fact undermining the owner's claim was that its own CP specialists were, or should have been, aware that every one of the monopiles was fully protected at a point in time when the retrofit project could have been shelved, and the majority of the expenditure avoided. Despite this, and with board approval, the owner went ahead and placed the order for the very costly retrofit installation work.

For what it's worth, we should record that the retrofit installation project went very well. The supplementary anode arrays had the effect of changing the condition of the monopiles from fully protected to, well, fully protected.

I am at a loss to explain why a group of competent and experienced CP specialists elected to keep faith with the modelling prediction, even when their own monitoring data quickly showed that the result was questionable, and then confirmed it to be incorrect.

Of course, this unhappy little legal saga does not mean that modelling has no value. The reality is to the contrary. However, it does demonstrate that modelling is not infallible. The task of the CP engineer is to understand where, and to what extent, modelling can be of assistance in either designing CP systems or providing valid information on how existing systems are operating. So, let us take a closer look at why we must exercise caution when it comes to modelling.

17.5.4 The boundary conditions

Once we have decided upon the size of the modelled domain There are two important issues to address when defining the boundary conditions. The first is the polarisation behaviour of the anode, and the second is the polarisation behaviour of the cathode.

17.5.4.1 The anodes

Generally speaking, we can be reasonably confident about the ability of a well-constructed model to produce a reliable picture of what is going on at the surface of sacrificial anodes in a CP system. We can rely on the fact that the anode alloys have been developed to be relatively non-polarisable. This means that their electrochemical characteristics have been intentionally tuned such that their potential does not vary much as the anodic current is drawn from them. This is reflected in the boundary conditions used for modelling. Figure 17.2 shows the polarisation characteristics of Al-Zn-In alloy anodes assumed by two groups of modellers: Marcassoli et al. [55], who use FEM, and Baynham and Froome [62], who use BEM. For comparison, the DNV recommendation is also included. Evidently, both modelling groups assume more negative operating potentials for the anodes than the (intentionally conservative) DNV guidelines.

In the case of impressed current anodes, which are nowadays usually MMO, the model should, in principle, account for the anodic kinetics of oxygen and chlorine evolution. In practice, because we do not need to worry about the operating potential of ICCP anodes, they are usually modelled as point sources of current.

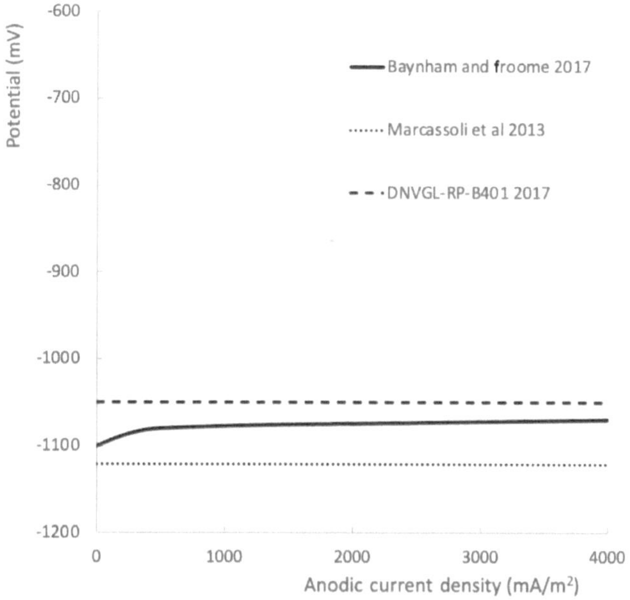

Figure 17.2 Some anodic polarisation curves used for modelling.

17.5.4.2 The cathode

As Decarlo pointed out [8] as long ago as 1983... *any technique used to predict current distribution on offshore structures must be able to describe the potential field within a complex structure and represent realistic cathodic polarisation behaviour.* Thus, with regard to the second requirement, the model has to be preprogrammed with the cathodic polarisation characteristics of the steel in the seawater, or seabed sediments. This turns out to be the major stumbling block for modelling.

Some of the proponents of modelling regard the polarisation as being a *property* of the metal in the seawater. Herein lies the problem. Polarisation is not a *property*. It is a *process*. Moreover, that process is crucially dependent on numerous environmental parameters: temperature, dissolved oxygen content, salinity and water flow. The process of polarisation also varies in a complex and, as yet, unpredictable manner depending on the nature of the biofilms that develop on the steel surface. Perhaps most crucially, it also depends critically on the cumulative previous polarisation history of that surface.

We introduced the vagaries of the polarisation of steel in Chapter 7. We can illustrate the difficulties involved in assigning a boundary condition for the cathode in numerical modelling by referring back to Figure 7.6, which shows both the scatter in experimental data, and the less-than-convincing correlation between laboratory (room temperature) data and field data from cooler seawater. Even in colder waters, it is now well established that mean current densities continue to decline over the years as calcareous deposits consolidate. For example, it has recently been stated[3] that cathodic current densities on a North Sea structure, which had been in service for 33 years, were in the range of 13–17 mA/m^2.

Despite this, all workers who have published CP modelling data for bare steel have been obliged to conjure up a single polarisation curve in order to force the modelling software to generate a result. Unfortunately, not all authors have disclosed the characteristics of the cathodic curve they used. However, some have, and their cathodic polarisation curves are presented in Figure 17.3. To aid comparison, these are replotted onto common axes.

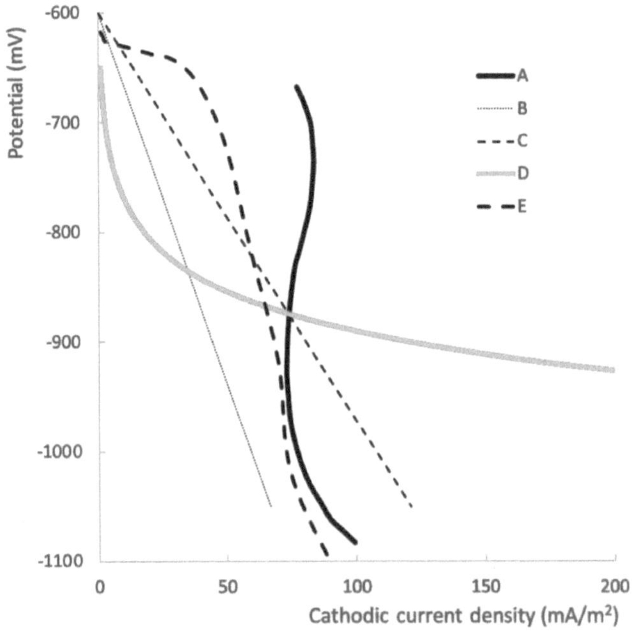

Figure 17.3 Some cathodic polarisation curves used for modelling.

[3] S.M. Wigen *Utilizing High Sensitive Field Gradient Sensor for Optimisation of CP Life Extension.* Joint Institute of Corrosion/Marine Corrosion Forum Webinar 09 October 2020.

Curve A was published by Baynham and Froome [62] in 2017. The linear relationships B and C were disclosed as part of confidential arbitration proceedings in 2015. Line B represents steel that has polarised well, and line C represents steel that has polarised poorly. Line D is the curve used by John et al. [43] in their 2007 spreadsheet which we discussed above. This curve, which uses theoretical electrode kinetic rather than experimental origins, has been transposed from a logarithmic to a linear current density scale. Curve E, published by Marcassoli et al. [53], has likewise been transposed from a logarithmic to a linear scale.

In principle, since these curves have been replotted onto the same scale, they ought to be comparable. The obvious fact is that they are not. Indeed, given the spread of polarisation data we found in Chapter 7, it is clear that there is actually no such entity as a single polarisation curve for steel in seawater. To reiterate a point that we have already made: this is because polarisation is a process, not a property. For steel in seawater, selecting any single polarisation curve to serve as the domain boundary for solving the Laplace equation is, to purloin a phrase from Winston Churchill, *a triumph of optimism over experience.*

The purveyors of modelling services are forced to select a polarisation curve or combination of curves. This has to be done in order to sell their service, but it does not necessarily mean that the curve is accurate. Indeed, the adjective "accurate" has no relevance to the polarisation characteristics of any metal in natural seawater. An oblique insight into the mindset of some modellers is revealed in a 2007 paper [44] which stated... *One can conclude that when the Potential Model is solved using the FEM, the technical problem is rather a mesh generation problem.*

I interpret this as indicating that the authors were more occupied in modelling the spatial distribution of the CP current flow than on the polarisation behaviour of the surfaces. This is interesting because the paper dealt with modelling the effect of an ICCP system on a fast catamaran. The twin hulls were constructed in painted aluminium alloy, and the propellers, which were earthed to the hulls, were NAB. Unfortunately, the authors failed properly to reference the source of information on the polarisation behaviour of either alloy.

The behaviour of the aluminium hull in seawater appears to have been based on the behaviour of a TSA coating. The behaviour of the NAB is based on a widely disseminated US Navy document by Hack [35]. This provides multiple laboratory polarisation curves (both potentiostatic and potentiodynamic) for this alloy in 30°C quiescent and flowing seawater. However, the modellers do not tell us how they manipulated these curves to generate their cathodic boundary conditions. It is clear from Hack's data that there is no single polarisation curve that encapsulates the cathodic behaviour of NAB over a normal range of seagoing ICCP applications. It is tempting to assume that *the Potential Model* was only "solved" to the extent that a polarisation curve was picked, seemingly at random, from the examples available. If this assumption is correct, then that was not a sound basis for then moving on to the problem of mesh generation.

17.5.5 What can computer modelling tell us?

17.5.5.1 Reasons to be careful

You might think that my trite answer to the question of what computer modelling can do for CP would be: *not a lot.* After all, I have offered some critical assessments of various examples of modelling work. However, to dismiss computer modelling because of its limitations would be to deny ourselves its benefits.

It is important to understand that modelling can indeed be useful, but it is not infallible. As we have seen, it does not validate a CP design, and it can be an unreliable predictor of the future. It also suffers from histories of having been over-relied upon by workers who lacked understanding, and of having been oversold by proponents who should have known better.

The deep underlying computer power involved in modelling, and its sophisticated graphical output, makes it all too easy for the CP engineer to be beguiled into believing that modelling must be accurate. More to the point, just because the predicted distribution of potential over a structure can be displayed with mind-boggling sophistication on the structure's "digital twin", does not mean that the information is accurate or, for that matter, that it is necessarily useful.

The trick is never to lose sight of the fact that the modelling is a tool for turning assumptions and facts into predictions. This means that the key to assessing how much reliance we can place on modelling depends entirely on the reliance we can place on our assumptions. It is always worth bearing in mind Prof. Sander's 1982 prescient comment [13] that... *too much must not be expected of modelling, even though it is apparently exact.*

17.5.5.2 Uncoated steel structures

In the days before modelling, some of the early problems of uneven current distribution on uncoated offshore structures could be traced to a failure to allocate anodes in relation to the local areas of steelwork to be protected. We discussed some examples in Chapter 13. In some cases, although the total sacrificial anode provision might have been appropriate for the structure, the placement of the anodes was poor. On occasion, this is thought to have been due to the problems associated with trying to visualise 3-dimensional structures when marking anode positions on 2-dimensional drawings. In other instances, entire areas of steelwork (e.g. well conductors) were apparently omitted from the design calculations. Rumour has it that the designers saw no need to protect the conductors so they ignored them, oblivious of the fact that they would take CP current irrespective of whether or not they needed it.

If modelling had been available to the CP designers, it is possible that it might have picked up on these errors, thereby avoiding the need for expensive retrofits. Conversely, it could be argued that all that was required was a little more diligence and CP understanding on the part of the designers.

The available evidence tells us that modelling is, as yet, unable to "predict" the future progress of polarisation of an uncoated steel structure. Whether or not it will be able to do so in the future remains to be seen. The rate and extent of polarisation depend intimately on the progress of calcareous deposit formation. This, in turn, is subject to the vagaries of the stochastic relationships between the electrochemical, chemical and microbiological processes involved. Expressed mathematically, instead of trying to solve second-order partial differential equations, we would need to be tackling problems of the third, or higher, order. Even if these more complex boundary conditions could be quantified, the computation would be challenging, to say the very least. As Warne [14] remarked: *considerable apprehension was engendered in that the time-dependency of polarization... would require iterative solutions of results which were themselves produced by the iterative process.*

I suspect that modelling will not be able to achieve reliable predictions of the future course of polarisation. More to the point, it is by no means clear why we should want it to. We must never lose sight of the comparatively limited objective of CP. Its role is to reduce an already low rate of corrosion (see Chapter 1) to a much lower rate, or else to permit the use of a more optimistic S-N curve for the fatigue design (see Chapter 12). CP is not a counsel of perfection. Thus, whilst many of the published multicoloured equipotential contour maps for cathodically protected bare steel structures are interesting, they lack practical relevance. In many cases, close examination shows (predicted) levels of polarisation which translate to corrosion rates ranging from too low to be of concern, to even lower still.

Although the petroleum production industry will remain active for the foreseeable future, the style of engineering is changing. The use of massive fixed uncoated structures, characteristic of the last quarter of the 20th century, has passed its zenith. Many of these are approaching decommissioning or have already been removed. It seems unlikely that we will see their like again. As operations move into more inaccessible and hostile locations, they are giving way to small-scale more intricate subsea installations serviced by FPSOs or directly from the shore. This subsea hardware is universally coated to the high standard dictated by codes such as NORSOK M-501 (see Chapter 9).

Whether or not massive uncoated steel structures will make a meaningful come-back in areas such as seabed mining or tidal energy is anybody's guess. Doubtless, uncoated steel will still remain the go-to option in other areas where it is already established: such as sheet harbour piling and, latterly, wind farm foundation monopiles. Both of these applications have attracted modelling activity. As we have already pointed out, not all of this has been successful. It is difficult to see that modelling has produced much value to these projects over and above that which could have been achieved using knowledgeable and experienced CP specialists.

If modelling is to justify itself in the CP of uncoated structures, then it will be in areas where relevant experience is lacking, and rule-of-thumb CP design practices have come to be recognised as inadequate or inefficient.

17.5.5.3 Coated structures

17.5.5.3.1 Old and new assumptions

In view of its difficulties in handling the polarisation of uncoated steel structures, a more rewarding future for CP modelling might lie in its application to coated steelwork. There is encouragement for the modellers here because, as noted, subsea engineering is making increasing use of complex coated equipment. Again, however, we need to be circumspect in our expectations of modelling.

Referring back to the unnamed epidemiologist we met at the start of this chapter, we must not lose sight of the fact that we are using our model to turn assumptions into predictions. With bare steel structures, our pivotal assumptions relate to the manner in which calcareous deposits form on the cathodically polarised surfaces. When we move to coated structures, the assumptions underpinning the modelling of this polarisation process become less of an issue because so little of the steel surface is exposed. However, what we are actually doing is replacing our assumptions about the future polarisation characteristics of the steel with our assumptions about the future rate and distribution of coating breakdown.

17.5.5.3.2 Worst case assumptions

Corrosion engineering in general, and the application of CP in particular, are exercises in future-proofing our assets. Since we do not know what the future holds, it often makes sense to assume the worst. This principle can be brought to bear on modelling. This was illustrated by Warne [14] in one of its very early practical CP applications. The case involved a "new type" of platform, which we now know was the Hutton TLP (see Chapter 15). The underwater TLP hull was painted, but only with an intentionally "short-lived" holding primer. The main burden of protecting the hull fell to its ICCP system. The designers had several important questions to answer:

1. Would the ICCP system avoid under-protection at critical node areas?
2. How big did the anode shields need to be?
3. What would be the anode-to-electrolyte resistance?

To answer these questions, FEM was carried out on the basis of worst-case conditions. For example, the ability of the design to avoid under-protection at nodes was evaluated without assuming any benefits from either the coating or the calcareous deposits. It was also examined under scenarios in which selected ICCP anodes were assumed to have failed. The modelling results confirmed that, even allowing for the extreme worst-case assumption, the ICCP system would indeed be capable of protecting these critical areas. In the context of this exercise, the question of whether the system's designed current capacity could be reduced, and still achieve the same result, had no relevance.

Similarly, the sizing of the anode shield assumed maximum output from the ICCP anode since that would generate the maximum electrical field strength around the anode. This led to an estimate of the minimum dielectric shield diameter that could suffice for any operating anode output.

The anode-to-electrolyte resistance was computed for the trial anode size and geometries, and the known seawater resistivity. Again, in this case, it was relevant to consider the worst-case maximum anode current outputs since the object was to ensure that the permissible output voltages of the ICCP power supplies would be sufficient.

No one would challenge the results or the relevance of the pioneering FEM modelling carried out for the Hutton TLP. It was evidently successful. The structure served on station for 28 years, substantially in excess of the original 20-year design expectation. The exercise illustrated Warne's assessment that... *modelling can provide comforting reassurance that all is well for a particular design.* On the other hand, the absence of a corrosion failure is always taken as demonstrating that the corrosion engineering design was correct. Nobody ever asks: was it over-designed or could the result have been achieved at a lower cost? It would be pointless now to pursue that question in respect of Hutton's ICCP design.

17.5.5.3.3 Code-based assumptions

As we discussed in Chapter 9, any CP design for a coated structure has to account for the deterioration of the coating over time. All things being equal, the cathodic current demand will increase as the coating ages. For this reason, CP design codes usually include guidelines on rates of coating breakdown. As we also discussed in Chapter 9, these guidelines are simply the code drafting committee's intentionally conservative best-guess at how a given generic coating type might be expected to perform over the long term.

The computer modeller is unlikely to have access to any more relevant information than the drafting committee. So, when a model produces a set of "time-stepping" outputs for a coated structure, it is essentially turning the assumed rate of coating breakdown into a prediction. Usually, this adherence to the progressive coating breakdown advised by the codes is accompanied by a decrease in size, and increase in resistance, of the sacrificial anodes. Modelling the anode and coating changes together generally results in the model predicting a slight overall attenuation of polarisation with time, accompanied by subtle changes in potential distribution.

From a purely theoretical standpoint, I am obliged to question a couple of the assumptions underpinning this time-stepping type of modelling process. Setting aside the pure guesswork involved in characterising the overall rate of coating breakdown, these modelling exercises make two further implicit assumptions. The first is that the coating damage is more or less uniformly distributed. In practice, this is never the case. The second challengeable assumption is that a single cathodic polarisation curve can be used to describe the steel surface, irrespective of the manner in which it becomes exposed to the seawater. This seems improbable. For example, one might expect different polarisation behaviour for steel under an area of coating that is undergoing cathodic disbondment, compared to steel where the coating is removed by mechanical damage.

Despite these challenges, the literature is not short of examples of this sort of "time-stepping" exercise for coated structures [39–40,48,49,51,56,58]. Unfortunately, but unsurprisingly, no one has yet published a modelling prediction for a coated structure, and then actually recorded what happened to that structure over the ensuing decades. It follows that any judgement we might make on the benefit of CP modelling on coated structures has to be based on intuition. It is difficult to see how this advances the case for modelling.

17.5.5.4 Sacrificial anodes

17.5.5.4.1 Resistance formulae

In principle, when we move from the cathode-side to the anode-side of the CP circuit, we move to an area that is more amenable to modelling. Whereas the cathode is intended to polarise from its natural corrosion potential (around −650 mV) to a protected potential (−800 to −1100 mV), the anode is intended to exhibit very little polarisation when it is working (see Figure 17.2). This means that the anode and its adjacent field can be modelled more conveniently, and more accurately, than the cathode and its field. It is, therefore, useful to explore what modelling can contribute to the anode side of our CP designs.

As we discussed in Chapter 10, in the absence of anything else to go on, the marine CP industry has adopted various earthing resistance formulae to enable designers to estimate anode output currents. These formulae were originally developed in electrical engineering, mainly for the purpose of designing safe earthing systems. Their accuracy has occasionally been challenged, but they have remained in use because they have been shown to err on the side of conservatism, and they yield workable designs.

In the early days of the North Sea development, there were several projects on which monitored anodes were installed (see Chapter 18). Attempts were then made to compare their predicted and measured current outputs. Strømmen [10] reviewed the state of this art in 1985. In addition, he applied the FDM methodology to produce comparisons between modelled and measured output.

Reading the paper 35 years later reminds us that nothing is straightforward in this game. For example, the measured outputs of the handful of examples of monitored stand-off anodes evidently produced unconvincing correlations with the theoretical values arrived at by the application of the Modified Dwight formula. However, we should not be too critical of these results. The basis of the anode output calculation was an Ohm's law equation of the type we met in Chapter 3.

$$R_a = \left(E_c - E_a\right) / I_a \tag{17.2}$$

where:

R_a = calculated resistance of monitored anode (Ω)
I_a = measured current output of monitored anode (A)
E_c = measured potential of the steel near the anode (V)
E_a = measured operation potential of the anode (V)

This analysis poses several difficulties. First, although the measurement of the total anode current (I_a) of a monitored anode could be expected to be reasonably reliable, the current density over the surface of the anode will vary. For example, anode edges and corners tend to dissolve more quickly. In addition, there would have been uncertainty about exactly where, and how, on the anode and cathode surfaces the potentials were measured; an activity which would have been carried out by divers. The relatively small natural variations in measured values of E_c and E_a, which were not recorded or analysed, would have added to the uncertainty in the calculation of R_a. Finally, R_a is directly proportional to the seawater resistivity which varies within the ambient temperature range. Those early trials did not record the local resistivity or temperature.

Taking these experimental difficulties together, therefore, it is perhaps not surprising that the few examples of anode resistances, back-calculated from measured outputs, did not agree particularly well with the calculated resistances. Indeed, given that the origins of Dwight were unrelated to CP, a closer agreement between prediction and reality would have been surprising.

The lack of agreement between resistances calculated from measured current outputs meant that the correlation with Strømmen's FDM results was likewise destined to be unpersuasive. Given the generally satisfactory nature of CP on large jacket-type structures subsequent to Strømmen's paper, there was presumably little incentive to delve too deeply into anode performance for some time.

Eventually, however, the ascendency of FEM and BEM coincided with renewed interest in modelling the anode side of the CP circuit. As mentioned, the rule-of-thumb anode resistance formulae that we discussed in Chapter 10, some of which we employed for our basic cookbook designs in Chapter 3, were all developed from electrostatic principles. They were required by the electrical engineering industry for the design of safety earthing systems: the object of the exercise being to construct a low resistance connection to ground or, as it is often termed, "remote earth". The electrical engineering origins of these formulae meant that they were the obvious choices for workers designing groundbeds for land-based CP. Their use subsequently diffused across into the offshore CP arena. As a matter of experience, they usually result in effective CP systems when used in cook-book designs.

There are two questions that we should be asking when we use formulae such as the modified Dwight.

$$R_a = \frac{\rho}{2\pi L}\left[\ln\left(\frac{4L}{r}\right)-1\right] \tag{17.3}$$

where:

ρ = seawater resistivity
L = anode length (m)
r = anode radius (m)

These questions are:

1. How accurate is this formula?
2. How relevant is it?

These questions have recently been examined by computer modelling.

17.5.5.4.2 Accuracy

We have already made the point that the modified Dwight formula, which is recommended by codes for long slender stand-off anodes, is an approximation. That is why we refer to it as a "formula", not an "equation". Dwight adapted it from an earlier more rigorous formula published by Hallén.

More recently, Baynham and Froome [65] have applied BEM modelling to a typical large trapezoidal stand-off anode (length 2 m, height 0.225 m, width tapering from 0.265 m at the base to 0.185 m at the top). Their results indicated that application of the modified Dwight formula to this anode shape over-predicted the resistance to earth by as much as 26%.

This does *not* mean that we should revisit our cookbook designs in Chapter 3 and rerun the calculations with a lower value for R_a. Firstly, this modelling result only applies to that specific anode shape and dimension. The paper also indicates that, as the anode length is reduced in relation to its radius, so too does the conservatism that is generated by the Dwight formula. Secondly, and more important, we might be at liberty to choose whether to design according to the DNV code or according to a methodology based on modelling. However, we should never cherry-pick design parameters from one methodology to use in the other. That is asking for trouble.

The result does, however, prompt us to question the relevance of carrying out CP design calculations on the basis of anodes transmitting their current to remote earth.

17.5.5.4.3 Where is "remote earth"?

As we mentioned in Chapters 3 and 10, the concept of the current leaving the anode and going all the way to "remote earth", and then coming all the way back, is difficult to grasp. The reason for this is that it is not what actually happens. Remember what we said in Chapter 3: the cookbook design methodology implied by codes such as DNV-RP-B401 makes no attempt to describe the electrochemistry underpinning CP. The methodology has only one objective. That is to get to a CP design that should work. With that objective, it is expedient to treat the anode current as if it all flows to remote earth. It makes the calculations simple, and, no less important, it is guaranteed to add conservatism to the result.

More realistically, if we mount an anode onto a steel structure then, as dictated by the laws of physics, the current will tend to follow the path of least resistance. In principle, therefore, this flow of current is a feature that lends itself reasonably well to modelling. This means that, if the steel is uncoated, the anode will behave as if it had a lower resistance than calculated on a remote earth basis. In effect, the steel cathode will short circuit some of the current that would, theoretically, flow to remote earth. Conversely, if the steel is painted, it will serve to block part of the current flux. This will have the same effect as increasing R_a. These effects have received some modelling attention (e.g. [65]). However, as with the observations on the imprecision of the Dwight formula, it is presently difficult to make use of the information when carrying out a code-based CP design.

17.5.5.4.4 Anode clustering

In Chapter 10, we discussed the suppression of current output that results from locating the anodes close to each other. The effect is essentially the same as increasing the resistance of the anodes. In Chapter 13, we illustrated this using the example of anodes on offshore wind farm monopile foundations. The designers had adopted a simple parallel resistor analogy. They assumed that the resistance of N anodes would be $1/N$ times the resistance of a single anode. In the event, it was found that clustering the anodes together on the skirt of the foundation's transition piece resulted in their combined resistance actually being much higher, and their combined current output substantially lower, than envisaged in the design. As we discussed, the outcome was that some of the monopile foundations took over a year to polarise fully.

Although this mutual anode interaction has long been appreciated in onshore CP [11], it has not featured very strongly in offshore CP codes. DNV-RP-0416 (2016) lacks both information and punctuation in stating... *Anode clustering shall be minimized interference effect of anode clustering must be considered in the design*. So, the reader is told that there is an issue with clustering but is left in the dark as to what to do about it. Also, prior to 2021, DNV-RP-B401 conveyed the misleading information that anode separations exceeding 0.5 m would not lead to a problem. It has since rescinded this advice, recommending the use of modelling to evaluate such interaction effects.

In Chapter 10, we considered developing a workaround based on formulae used for the design of multiple anode groundbeds on onshore CP. Such formulae have also been used for some time by experienced CP practitioners involved in the design of retrofit anode sleds.

However, the availability of numerical modelling gives us the possibility of an alternative bespoke approach to analysing the apparent increase in anode resistance caused by the mutual interaction of the anodes' electric fields. Modelling electric fields is, after all, what the Laplace equation is all about. For example, in 2013 Wigen et al. presented a paper [52] on the modelling of the CP of a jacket by remote sleds. Each sled comprised four platform-style anodes mounted as shown in Figure 13.11. The paper was not short of multicoloured images, including illustrations of the predicted current densities over the anodes' surfaces. Unfortunately, these are difficult to understand.

For example, when two sleds are positioned 5 m apart... *there is an obvious reduction in the anode output between the two anodes... facing infinite electrolyte volume, and the anodes facing the next sled*. This is what we would expect, although the fact that the effect is predicted to extend over a distance as great as 5 m is both interesting and surprising. In that case, however, the modelling did not drill down to the detail of the distribution of anodic current over the surface of any individual anode, since each is pictured as delivering a completely uniform current density.

A paper by Froome and Baynham at the same conference [54], on the other hand, did explore the detail of the variation of anodic current density within sled-mounted anode arrays. They defined the term *interference ratio* for an array of N anodes:

$$\text{Interference ratio} = \frac{\text{Resistance of a single anode}\left(R_a\right)}{\text{Resistance of anode array}\left(R_{\text{array}}\right)\times N} \tag{17.4}$$

They then applied numerical modelling to various arrangements of anodes, for which the electrical engineering literature contained no formulae for earthing resistance. By way of a notional, but not unrealistic, example, they examined the case of seven cylindrical anodes (2.1 m long×0.25 m diameter) deployed in two layers (three above four). These anodes were set at centre-line spacings of 0.6 m, both vertically and horizontally. The outcome of the modelling was that this anode disposition produced an interference ratio is 0.323. In other words, the sled would deliver only 32% of the current that would have been expected if the CP designers had calculated the output of a single anode using the modified Dwight formula (per the offshore CP codes), and then simply multiplied the answer by seven.

Further papers have followed in a similar vein. For example, Osvoll et al. [59] used numerical modelling to determine the interference ratio associated with other anode deployment arrangements encountered in offshore CP. This included anode sleds and anodes clamped around the circumference of a tubular member. Both of these arrangements have been, and still are, employed in platform retrofit exercises. The modelling results clearly demonstrated that the interference ratio reduced (got worse) as the anode spacing reduced.

The paper also addresses the effect of spacing between pipeline bracelet anodes. The latter has relevance because some CP designs in the past have called for bracelets that were too long to be cast conveniently. The designers circumvented the foundry problems by the plausible expedient of designing half-length bracelets and installing these as pairs. The modelling showed that this arrangement likewise needs to account for the interference ratio. In the example modelled of an 0.5 m diameter pipeline, it was concluded that the interference ratio was substantial unless the anodes were mounted more than 1.5 m apart.

In short, numerical modelling gives important and credible results when it comes to the anode side of the CP circuit. This is an area where the codes do not provide much useful information for the CP designer. In general, anode resistance formulae are published, and republished, in these documents with insufficient background on the limitations of their use. Numerical modelling offers a useful way forward, albeit that some anode current density versus potential relationship has to be known or inferred (see Figure 17.2).

17.5.5.5 Internal spaces and complex geometries

17.5.5.5.1 Opportunities for modelling

As noted, our cookbook-type designs for subsea structures adopt a scenario of anodes delivering current via remote earth. We have also recognised that this gives an intentionally conservative design approach. However, in part because of this intentional over-estimation of the anode-to-electrolyte resistance, we can allow ourselves the luxury of ignoring the generally much smaller cathode-to-electrolyte resistance. As a matter of experience, designing on the basis of these assumptions usually yields acceptable CP systems. Importantly, this approach is aided by the fact that we have an established set of anode resistance formulae at our disposal.

On the other hand, the concept of sending electrolytic current to remote earth makes less sense when we are trying to apply CP within internal spaces. It makes no sense at all if the internal spaces happen to be sealed from the external seawater. Despite this, it remains customary to design CP systems for internal spaces using classic resistance to remote earth formulae. An example is seen in Annex C of EN 17243, which offers a formula for potential distribution within a pipe. This adopts a methodology presented at an earlier Eurocorr conference [57] which advocates calculating the resistance of an anode located inside the pipework using the McCoy formula. There is a leap of faith here because the McCoy formula, which we met in Chapter 10, is based on anodes delivering current to remote earth.

Nevertheless, although the modelling details are open to challenge, it seems likely that modelling will contribute to the future design of CP for internal spaces. As might be expected, companies seeking to supply modelling services have pursued this opportunity. Again, as with other areas of modelling, the outcomes have thus far been mixed.

17.5.5.5.2 Water ballast tanks

Perhaps inadvertently, Bortels [46] illustrated some of the pitfalls when presenting the outcome of a FEM exercise on a (presumably fictional) ship's WBT. A sketch of the tank is provided, in addition to which we are told that its dimensions are 15 m×15 m×6 m, which is far more headroom than I ever encountered in my days crawling around in WBTs. We are also told that the WBT is... *made of steel and has an old coating with assumed uniform coating degradation.* We are also told that it is protected by 120 zinc anodes. Again, this is far more than I have ever seen in any WBT.

The result emerging from the modelling, which was illustrated by the inevitable multicoloured equipotential contour map, was that the tank was fully protected. It exhibited potentials in the range of –1000 mV (near the anodes) to –800 mV (at corners and sides of the tank).

As an exercise in modelling, this outcome may have been entirely satisfactory. However, from a practical corrosion engineering standpoint, it was less so. A number of issues arise. The first is the modellers' assumption that the WBT exhibited *uniform coating degradation*. As a matter of experience, WBT coatings' breakdown is anything but uniform. They degrade preferentially at features that are difficult to prepare and paint effectively: such as plate edges, cut-outs and weld seams. Furthermore, even on a fairly new WBT, the paint on the tank floor generally accumulates more mechanical damage than elsewhere.

The second issue is the modellers' assertion that the steel was... *requiring 100 mA/m² of protection current*. This implies exposure to fully aerated seawater, which makes no sense. Once charged with live seawater, ballast tanks are, to all intents and purposes, effectively sealed. As we discussed in Chapter 16, the environment becomes anaerobic within a matter of days. Any CP design based on fully aerated conditions would be excessive.

Finally, there is a problem that the sophistication of the modelling and its output could deceive the ship's designer into thinking that there was no corrosion issue to address in the WBT. Unfortunately, experience points to the contrary. WBT corrosion is a perennial problem on ships, and it is not necessarily a problem that can be mitigated by CP. The corrosion problems probably do not arise when cathodically protected tanks are completely full.

Corrosion is a greater threat when the tanks are "empty"; and any installed anodes are, therefore, inoperative (see Figure 16.1). Even nominally empty tanks retain water on the base plates, which keeps the internal volume humid enough to sustain corrosion of all of the salt-laden steelwork. Moreover, ageing tanks accumulate sufficient sludge on the floors to provide an ideal environment for micro-organisms. Even when the tanks are drained and nominally vented, conditions under this sludge remain anaerobic. MIC is a perennial problem in ships' WBTs. (We saw an example of such MIC in Figure 1-8).

It is largely because some of the main WBT corrosion threats are not mitigated by CP that IMO guidelines focus much more on the quality of WBT coatings (see Chapter 9).

17.5.5.5.3 Seawater lift pumps

As we discussed in Chapter 16, multi-metal seawater lift pumps fall under the heading "forgotten knowns". The entirely predictable galvanic corrosion problems were not always predicted in the 1970s, so the problem was "rediscovered" in the 1980s. It was "rediscovered" again in the 1990s, and yet again in the 2010s. The problem is clearly manageable: there are plenty of seawater lift pumps that operate satisfactorily without wrecking their carbon steel caissons. There are two complementary routes to mitigating the galvanic corrosion threat to an acceptable level. The first route is obeying the twin cardinal imperatives of managing galvanic corrosion:

- coat the cathode (the stainless steel pump and riser),
- do **not** coat the anode (the internal surface of the caisson).

The second route is to find a way of shoe-horning some anodes onto the pump assembly. This may seem counterintuitive if the pump and riser are fabricated in CRA. However, the pump and its riser are the only practicable places where anodes can be installed to protect the internal surface of the caisson, with which they are inevitably in electrical contact.

Unfortunately, pump manufacturers see no reason to coat their products since they have already incurred the cost of fabricating them in CRA. Similarly, operators have occasionally been guilty of not communicating that the coating is not needed to protect the pump. It is needed to protect the carbon steel that is galvanically connected to it. In addition, some operators have persistently failed to replaced anodes when pumps were pulled for servicing.

Nevertheless, as with all engineering mishaps, these galvanic corrosion failures have at least provided an opportunity for learning useful lessons. In this case, they are fertile ground for the application of computer modelling of the CP.

The restricted annular space between an uncoated stainless steel pump and its carbon steel caisson, which exacerbates the galvanic corrosion of the caisson, also renders the application of CP challenging. This is a case where the classic paradigm of anodes delivering current to remote earth, from whence it returns to protect a low resistance steel cathode, falls well short of describing what must be going on. This is an obvious opportunity for modellers to get involved.

In 1995 Miyasaka et al. [35], working for the research subsidiary of a pump manufacturer, described the use of BEM to evaluate the effect of zinc anodes, attached at two locations, for protecting a coated cast iron seawater pump casing and riser. The stainless steel impeller was intentionally isolated from the cast iron. They described how they obtained polarisation curves in the laboratory for cast iron as a function of flow rate and time (up to 20 days), and how they manipulated these so that they could be used as the boundary conditions. On a trial-and-error basis, they elected to combine the bare metal polarisation data with a 5% coating breakdown. Polarisation curves for the zinc anodes were also obtained but assumed to be largely independent of time. When they came to the modelling, they factored the current densities according to their assumed 5% coating breakdown factor. They then created a model of a 60° radial segment of the pump and ran the model on a mainframe computer. The processing took up to an hour for each run on the casing, and up to 35 minutes for the simpler geometry of the column.

Importantly, and unusually in the field of numerical modelling, these authors also carried out potential measurements along the inside of the casing and the column. The results showed very good agreement between the modelled and measured potentials.

Two years later, researchers from the same organisation described a similar exercise on an uncoated 316L stainless steel (UNS S31603) seawater pump casing and column [37]. The object of the paper was, in part, to demonstrate the capabilities of their BEM software. Based on the outcome, the authors claimed that... *even an engineer with no knowledge of corrosion will be able to operate this system easily and perform effective analyses.* There is an important point to be made here: an engineer *with no knowledge of corrosion* should not be meddling in the modelling of CP systems!

The art of modelling had developed considerably by the time the same predictable galvanic corrosion failures caused surprise by occurring yet again in the 2010s. A 2016 paper by Wyatt et al. [60] provides an informative and well-illustrated case study of unsuccessful galvanic corrosion mitigation. In this instance, the owners had at least tried to coat the SDSS pump and riser. The owners had also made some attempts to install anodes. Unfortunately, it appears that the quality of the coating was, at best, mediocre. This shortcoming was compounded by evident doubt as to whether the anodes had ever been properly connected.

More important, however, was the fact that the designers had committed the fundamental error of lining the steel caisson. In all likelihood, the lining had been carried out competently. However, the weld areas between caisson sections were left intentionally unlined. All that the lining had achieved was to reduce the exposed area of the carbon steel anode, thereby focussing the damaging effect of the galvanic corrosion current onto the circumferential welds of the caisson.

The problem remained undetected until underwater inspection of the caisson externals found the galvanic corrosion holes that had originated from the inside.

The paper llustrates the use of modelling in assisting the CP design for the modified, coated pump arrangement. As with all such modelling, it was an exercise in turning assumptions into predictions. In this case, the "assumptions" were principally the quality of the coating and its future breakdown characteristics, and the cathodic characteristics of the SDSS steel as its coating degraded. The "predictions" were the distribution of potential, both on the pump and riser and on the inside of the caisson. Now that the pump has been returned to service, it will prove very difficult to measure potentials in the annulus between pump and caisson. So, it is unlikely that the modelling predictions will be tested against reality.

17.6 GOING FORWARD

I have written this book because the need for marine CP is not going to disappear anytime soon. Furthermore, as subsea structures and equipment become more complex, there will be an increasing need to depart from the run-of-the-mill, code-based designs that we looked at in Chapter 3. This, in turn, will require some measure of inventiveness on the part of the CP designers. Doubtless, modelling will be a useful tool in supporting that inventive process. There is also little doubt that the purveyors of modelling will not be shy when it comes to promoting their services for this purpose.

What is important, is that the CP designers keep their feet on the ground. They must think hard about where modelling is useful, as opposed to simply illustrative, and about how much reliance to place on its results. It is always essential to remember that modelling produces predictions, not facts. It does not verify or validate any design.

A more measured assessment of where modelling belongs in the CP engineer's toolbox was expressed in 2012 by Baynham et al. [50] who... *demonstrated how simulation can be used to assist the optimisation...* of a CP design. The keyword here is "assist". In short, modelling helps the CP engineer in thinking through various CP design or evaluation tasks, but it cannot actually do that thinking.

REFERENCES

1. R. Strømmen, *Current and potential distribution on cathodically protected submarine pipelines*. NSC Conference Trans. I. Mar E. **91**, 49 (1979).
2. J. Cochran, *New mathematical models for designing offshore cathodic protection systems*, Paper no. OTC 3858 Offshore Technology Conference, 1980.
3. R. Strømmen and A. Rødland, *Computerized numerical techniques applied in design of offshore cathodic protection systems*, Paper no. 241 CORROSION/80.
4. P.E. Childress, *Computerized guidelines for galvanic anode installation in BASIC*. Materials Performance **20** (11), 39 (1981).
5. R.S. Munn, *A mathematical model for a galvanic cathodic protection system*. Materials Performance **21** (8), 29 (1982).
6. R.D. Strømmen, *Computer modeling of offshore cathodic protection systems utilized in CP monitoring,* OTC 4367 Offshore Technology Conference, 1982.
7. D.J. Danson and M.A. Warne, *Current density/voltage calculations using boundary element techniques,* Paper no. 211 CORROSION/83.

8. E.A. Decarlo, *Computer aided cathodic protection design technique for complex offshore structures.* Materials Performance **22** (7), 38 (1983).
9. J.W. Fu, *Calculation of cathodic protection potential and current distributions using an integral equation numerical method*, Paper no. 250 CORROSION/84.
10. R. Strømmen, *Evaluation of anode resistance formulas by computer analysis.* Materials Performance **24** (3), 9 (1985).
11. R.L. Benedict, *Effect of clustered anodes on deployment efficiency*, Pipe Line Industry, p. 43 April 1985.
12. R. Strommen et al., *Advances in offshore cathodic protection modelling using the boundary element method.* Materials Performance **26** (2), 23 (1987).
13. K.F. Sander, *The problem of predicting potential and current distribution*, Chapter 2, in *Cathodic Protection: Theory and Practice.* Eds. V. Ashworth and C.J.L. Booker. Ellis Horwood (Chichester) 1986, ISBN 0-85312-510-0.
14. M.A. Warne, *Application of numerical analysis techniques*, ibid Chapter 3.
15. J.W. Fu, *Numerical modeling methods for galvanic corrosion and cathodic protection analyses*, Paper no. 044 CORROSION/86.
16. K.F. Sander, *Computer modelling in the context of marine cathodic protection systems*, p. 177 proceedings of UK Corrosion Conference, 1986.
17. R.D. Strømmen, H. Osvoll and W. Keim, *Computer modeling and in-situ current density measurements prove a need for revision of offshore CP design criteria*, Paper no. 297 CORROSION/86.
18. W.M. Sackinger and B. Theuveny, *Method of calculation of cathodic protection current distribution for steel offshore structures in the arctic,* Paper no. 398 CORROSION/86.
19. D.J. Tighe-Ford et al., *The use of scale modelling for the evaluation of impressed current cathodic protection systems for marine structures*, Proceedings of UK Corrosion, 1986.
20. W. Keim, R. Strommen and J. Jelinek, *Computer modeling in offshore platform CP systems.* Materials Performance **27** (9), 25 (1988).
21. H.P. Hack, *Scale modeling for corrosion studies.* Materials Performance **28** (11), 72 (1989).
22. E.D. Thomas et al., *Scale modeling of impressed current cathodic protection systems*, Paper no. 274 CORROSION/89.
23. J.C.F. Telles et al., *Numerical simulation of a cathodically protected semi-submersible platform using the PROCAT system*, Paper no. 276 CORROSION/89.
24. P. Chauchot, B. Bigourdian and L. Lemoine, *Cathodic protection systems modelled by the PROCOR software*, Paper no. 401 CORROSION/89.
25. D.J. Tighe-Ford and S. Ramaswamy, *Modelling the impressed current cathodic protection of enclosed marine systems.* Proceedings of UK Corrosion, 1989.
26. MTD, Chapter 5 *Calculation and modelling for the design of cathodic protection systems*, in *Design and Operational Guidance on Cathodic Protection of Offshore Structures, Subsea Installations and Pipelines.* MTD Publication 90/102, 1990, ISBN 1-870553-04-7.
27. D.J. Tighe-Ford and J.N. McGrath, *The physical scale modelling of marine impressed current cathodic protection: A review*, Paper no. 308 CORROSIO/91.
28. A.R. Parks, E.D. Thomas and K.E. Lucas, *Physical scale modeling verification with shipboard trials.* Materials Performance **30** (5), 26 (1991).
29. P. Khambhaita and D.J. Tighe-Ford, *Design and analysis of remote surveys of cathodically protected seabed pipelines,* Proceedings of UK Corrosion, 1992.
30. D.J. Tighe-Ford, *Three types of marine cathodic protection: An evaluation for the offshore design engineer and operator*, Proceedings of UK Corrosion, 1992.
31. K. Nisancioglu, *Modelling for cathodic protection*, Chapter 2, in *Cathodic Protection: Theory and Practice.* Eds. V. Ashworth and C. Googan. Ellis Horwood (Chichester) 1993, ISBN 0-13-150038-4.
32. D.J. Tighe-Ford, P. Khambhaita and W.S. Cheung, *Remote surveys of cathodically protected seabed pipelines: A modelling study,* Paper no. 532 CORROSION/93.
33. D.J. Tighe-Ford and J.N. McGrath, *Design considerations for offshore oil rig cathodic protection.* Materials Performance **32** (4), 18 (1993).

34. A.J. Summerland and T.C. Osborne, *Spreadsheets for cathodic protection design of offshore, Pipelines* Proceedings of UK Corrosion, 1994.
35. M. Miyasaka et al., *Application of BEM to cathodic protection of a seawater pump*, Paper no. 288 CORROSION/95.
36. H.P. Hack, *Atlas of polarization diagrams for naval materials in seawater*, Carderock Division Naval Surface Warfare Center Report CARDIVNSWC-TR-94/44, April 1995.
37. M. Miyasaka et al., *Development of boundary element analysis system for corrosion protection design,* Paper no. 428 CORROSION/97.
38. R.A. Adey and J. Baynham, *Design and optimisation of cathodic protection systems using computer simulation*, Paper no. 00723 CORROSION/2000.
39. Y. Wang, *Boundary element evaluation of shipboard cathodic protection systems*, Paper no. 04102 CORROSION/2004.
40. H. Osvoll, A. Sjaastad and F. Duesso, *Evaluation of impressed current system on FPSO's by use of CP computer modelling,* Paper no. 04103 CORROSION/2004.
41. J.A. Ellor, *Electrostatic modeling of cathodic protection systems*, Paper no. 06108 CORROSION/2006.
42. M. Krupa et al., *Cathodic protection of highly complex and shielded components*, Paper no. 06109 CORROSION/2006.
43. G. John, D. Buxton and I. Cotton, *Polarisation modelling as part of galvanic anode upgrade of a hybrid offshore cathodic protection system*, Paper no. 07081 CORROSION/2007.
44. L. Bortels et al., *3D software simulations for cathodic protection in offshore and marine environments,* Paper no. 07085 CORROSION/2007.
45. Y. Wang and K.J. KarisAllen, *Validating impressed current protection numerical modelling results using physical scale modelling data*, Paper no. 08253 CORROSION/2008.
46. L. Bortels, *Expert 3D software simulations for cathodic protection in offshore and marine environments*, Paper no. 09516 CORROSION/2009.
47. L. Bortels et al., *Design validation of ICCP systems for offshore wind farms*, Paper no. 10390 CORROSION/2010.
48. J.M.W. Baynham et al., *Optimising a SACP system using simulation, to achieve uniform anode life and protection potential,* Paper no. 1057 EuroCorr, 2011.
49. L. Dong, M. Lu and Y. Du, *Study on the boundary conditions of coated steel in numerical simulation of cathodic protection systems*, Paper no. 11318 CORROSION/2011.
50. J.M.W. Baynham, T. Froome and R.A. Adey, *Jacket SACP system design and optimization using simulation,* Paper no. 0001282 CORROSION/2012.
51. C. Baeté et al., *CP system validation of offshore structures through modeling*, Paper no. 0001657 CORROSION/2012.
52. S.M. Wigen, H. Osvoll and M. Gouriou, *CP of offshore jacket by remote anode sleds and discussion about current drain to buried structures*, Paper no. 2121 CORROSION/2013.
53. P. Marcassoli et al., *Modeling of potential distribution of subsea pipeline under cathodic protection by finite element method*, Paper no. 2333 CORROSION/2013.
54. T. Froome and J.M.W. Baynham *Assessing interference between sacrificial anodes on anode sleds,* Paper no. 2344 CORROSION/2013.
55. P. Marcassoli et al., *Design of cathodic protection retrofitting of subsea pipelines assisted by finite element method (FEM) modelling*, Paper no. 7729 EuroCorr, 2014.
56. M.-J. Lee and C.-S. Lim, *ICCP system design on the hull of an ice breaker by computational analysis,* Paper no. 4004 CORROSION/2014.
57. H. Osvoll and J.C. Werenskiold, *CP design of internal components*, Paper no. 7747 EuroCorr, 2014.
58. C. Baeté, *ICCP retrofit challenges for an offshore jacket complex*, Paper no. 6012 CORROSION/2015.
59. H. Osvoll, G. Lauvstad and A. Bilsbak, *Anode interference has to be accounted for in a CP design*, Paper no. 6156 CORROSION/2015.
60. B. Wyatt et al., *Internal cathodic protection of offshore sea water pump caissons*, Paper no. 68002 EuroCorr, 2016.

61. C. Baeté et al., *CP design considerations for subsea pipeline system under GRP cover*, Paper no. 53182 EUROCORR/2016.
62. J. Baynham and T. Froome, *Assessment of effects of cavities and narrow channels on CP design in the marine environment*, Paper no. 94617 EuroCorr, 2017.
63. M.S. Hong et al., *Optimization of CP design with consideration of temperature variation for offshore structure*, Paper no. 9672 CORROSION/2017.
64. M. Hajigholami et al., *Modeling the cathodic protection system for a marine platform jacket.* Materials Performance **56** (4), 34 (2017).
65. J. Baynham and T. Froome, *Comparison of resistance to cathode with resistance to ground in the marine environment*, Paper no. 11352 CORROSION/2018.

Chapter 18

CP system management

Confused and confusing

In the 1970s, ICCP systems in the North Sea were regularly switched off because they interfered with TV reception [7]. I must confess that the image of offshore workers huddled together in the TV shack watching cartoons, while the structure corroded beneath them, brought a smile to my face. However, the problem was not necessarily very serious. As soon as competent and senior personnel became aware and explained just why it was not a smart idea to let the structure rot, the practice ceased.

On the other hand, the outcome of the investigation into the 2008 Varanus Island explosion (Chapter 14) was much more disturbing. The official report concluded that... *Apache's understanding of the cathodic protection system on the 12-inch SGL was confused and confusing.* It would be comforting to think that Varanus Island represented an isolated, never-to-be-repeated, type of failure. Somehow, I doubt it.

So far we have explained, at various levels of detail, how CP works. We have also examined how the codes can be used to produce run-of-the-mill CP designs. More importantly, we have also addressed some technical issues that are not so readily captured by the codes. It is now time to explore what needs to be done to be assured that our asset is, and will remain, adequately protected. This is a cradle-to-grave exercise, in which the appropriate levels of competence need initially to be directed at designing, constructing and installing the CP system. If it is an ICCP system, we then have to ensure that it is operated and maintained effectively throughout its life. Even sacrificial anode systems, which are occasionally and ill-advisedly described as "fit-and-forget", still need something of a watchful eye.

If we have got the design, construction and installation right, and we have established sensible operating and maintenance procedures for an ICCP system, or a sufficient inspection regime for sacrificial anodes, then all should be well.

Usually, that is the case. Occasionally, however, things do not go quite to plan. For example, the operational life might get extended beyond the original design intent. In that case, we then must consider what further actions we need to carry out to assure the integrity of the structure. Against this background, this chapter looks at the topics of surveying, inspecting and monitoring CP systems.

18.1 SURVEYING, INSPECTION AND MONITORING

I recall a meeting in the mid-1980s with an oil company's senior inspection engineer. I quickly formed the impression that the guy was unfulfilled in his job. He told me that his company's inspection philosophy was... *not to go looking for trouble.* At first, I assumed

DOI: 10.1201/97810032160702-18

that he was joking, but apparently not. He added that, when trouble did crop up, his company's mitigation philosophy was to... *throw money at it.*

Happily, things have moved on, with that operator, and more generally. The profile of inspection, and the related activities of surveying and monitoring, is now much more prominent among integrity-aware enterprises.

The terms surveying, inspection and monitoring are employed widely, and sometimes interchangeably, by CP practitioners. It would be nice if we could be clear about the differences between them. To this end, Table 18.1 presents my own view of the meanings of these terms, and how they complement each other in the context of CP management. However, we should bear in mind that, within the industry, there is some blurring of the boundaries between these definitions.

Table 18.1 CP surveying, inspection and monitoring

	CP management activity		
	Surveying	*Inspection*	*Monitoring*
Frequency	Occasional	Periodic	Frequent or continuous
Locations on the structure	Most	Many	Few
Preplanning to select locations	Not required	Important	Critical
Level of information	General	Detailed	Intensive

"Surveying" spans a wide range of activities: from taking a cursory look to check that anodes remain in place, to measuring the distribution of potential, or current density, over the structure or along a pipeline. However, a potential survey that is focussed on a specific area, for example a location of suspected under-protection, might reasonably be termed an "inspection".

In addition to localised measurements of potential or current density, "inspection" also covers the detailed examination of CP system components including: the physical condition of anodes (ICCP or sacrificial), fixed references, cabling or conduits, ICCP electrical hardware and instrumentation, or the extent of sacrificial anode consumption.

"Monitoring" refers principally to continuous measurement of the structure's potential or, occasionally, an anode's current output. Potential monitoring employs fixed references, installed at specific locations on the structure. The locations have to be preselected. In so far as is reasonably practicable, these locations are where protection is judged to be critical. Examples might include fatigue-critical welds or occluded areas where CP current might be restricted.

18.1.1 The need for measurement

The Dow Chemical Corporation has made some telling contributions to CP. For example, it was at the forefront of the development of aluminium alloy and MMO anodes. It is, therefore, fitting to grab a quotation from the Dow physical chemist J.J. Grebe who advised with commendable succinctness that... *if you cannot measure it, you cannot control it.* Numerous self-proclaimed management gurus have since trotted out reworded versions of the same advice.

Throughout this book, we have focussed on how to apply current to polarise our structure to a potential that assures protection, but is not so negative as to be wasteful or damaging. This only has relevance if we are able to measure that potential with confidence. So, we now take a closer look at the measurement of potential, together with other relevant parameters: current and current density.

18.2 MEASURING THE POTENTIAL

18.2.1 The story so far

We first encountered the concept of potential in Chapter 1. We followed this with a more in-depth examination in Chapter 2, where we discovered that the potential (E) is the difference between the electrical potential energy in the metal ($\varnothing_m$) and the electrical potential energy in the solution ($\varnothing_s$):

$$E = \varnothing_m - \varnothing_s \tag{18.1}$$

Although this potential difference across the metal-solution interface cannot actually be measured, we can make a relative measurement against an arbitrarily selected reference: the standard hydrogen electrode. In Chapter 2, we also defined some secondary references, which are more robust and more convenient for use outside the laboratory. This explained why the Ag|AgCl|seawater half-cell is the first-choice reference for marine CP.

We also learned that the only reliable way to determine whether or not a structure is protected is to measure its potential. With this in mind, Chapter 6 was devoted to determining what the appropriate protection potential for steel should be. At the end of the exercise, it emerged quite clearly that, for all practical purposes, −0.8 V does the job nicely. Importantly, this is irrespective of the nature of the environment. Then, in Chapter 8, we found that, although −0.8 V is appropriate for iron and steel in any environment, other target protection potentials are required for other metals that we use in marine construction.

There is one further point about which we should remind ourselves. The object of CP is to make the value of E more negative. From equation 18.1, we see that there are two ways that we can bring this about. We can make $\varnothing_m$ more negative, or we can make the solution potential ($\varnothing_s$) more positive. This is something to bear in mind when we examine some of the techniques for measuring potentials on subsea installations, particularly pipelines.

18.2.2 Alternative references

We now need to explore potential measurement in a little more detail. We can begin with reminding ourselves that, although the offshore CP codes and standards are written on the assumption that Ag|AgCl|seawater will be used, there are innumerable other references available. Table 18.2, which makes use of data published by Ashworth [11], summarises most of the references that find their way into the CP literature. The first column in the table gives the common name for the reference, many of which are imprecisely referred to as "electrodes" rather than "half-cells". The second column gives the more appropriate half-cell description. In this column, the phase boundary is indicated by a single vertical line ("|") in keeping with electrochemical terminology. The third column gives the potential difference between the half-cell and the standard hydrogen electrode. The fourth column gives the protection potential of steel (E_{prot}(steel)), as prescribed in most international standards, when reported against that reference.

18.2.2.1 Standard hydrogen electrode

We first came across the standard hydrogen electrode (SHE) in Chapter 2 when we discovered that, although potential itself is unmeasurable, we have little difficulty in measuring a potential difference. Thus, once electrochemists had agreed upon the SHE as the standard reference, and arbitrarily assigned it a potential of zero, things became fairly straightforward. It was simple enough to measure potential difference of other half-cells against SHE. Strictly speaking, of course, the SHE should be called the standard hydrogen *half-cell*. The actual electrode is the platinum. However, the memorable acronym "SHE" is so ingrained in electrochemistry that there is no point making a fuss about the imprecision of the terminology.

Table 18.2 Reference half-cells

		Potential (V)	
Common name	*Half-cell*	*vs SHE*	E_{prot} *(steel)*
Standard hydrogen electrode (SHE)	Pt|1.2 M HCl + H_2 (1 atm.)	0	−0.55
Saturated calomel electrode (SCE)	Hg|Hg_2Cl_2|KCl (saturated)	+0.24	−0.79
Silver chloride electrode	**Ag|AgCl|seawater**	**+0.25**	**−0.80**
Silver chloride 0.5 M KCl electrode	Ag|AgCl|KCl (0.5 M)	+0.24	−0.79
Copper sulfate electrode (CSE)	Cu|$CuSO_4$ (saturated)	+0.32	−0.87
Zinc	Zn|seawater	−0.75	+0.20

We have included the SHE in Table 18.2 for the sake of good order. For the most part, it has only academic relevance in the world of CP. Nevertheless, it has been traditional when publishing in electrochemical journals to express potentials on the SHE scale, irrespective of the nature of the reference actually used to make the measurement. Keep this in mind if you go delving into older literature. For example, many of the papers, which we discussed in Chapter 6, referred potentials to SHE. I converted all of these to Ag|AgCl|seawater.

The SHE is ~0.25 V more negative than the Ag|AgCl|seawater half-cell. This means that the protection potential for steel, which most codes state is −0.80 V on the Ag|AgCl|seawater scale, would be −0.55 V referred to SHE. The potentials are identical. It is just the numbers that differ.

18.2.2.2 Saturated calomel half-cell

I have included the saturated calomel half-cell, universally but imprecisely referred to as the saturated calomel electrode (SCE), because it is the most widely used half-cell in electrochemistry. However, it is generally too fragile for regular field applications, so its use is generally restricted to the laboratory. Nevertheless, it is sometimes used in field CP applications for recalibrating other more robust, but less precise, references.

The SCE is illustrated in Figure 18.1. It comprises a platinum wire inserted into a bead of mercury (Hg). The mercury, which is actually the electrode, is also in physical contact with a paste of mercurous chloride (Hg_2Cl_2) known as calomel, after the mineral from which it was originally derived.

Hg|Hg_2Cl_2 is an ion-selective electrode. For use as a reference, the calomel is in contact with a saturated potassium chloride (KCl) solution. Providing the reservoir of KCl solution is kept saturated, by ensuring that solid crystals are always present, the SCE is a very accurate reference. We obtain our potential readings by measuring the potential difference across the cell:

$$\mathrm{Hg} \mid \mathrm{Hg_2Cl_2} \mid \mathrm{KCl(saturated)} \parallel \mathrm{seawater} \mid \mathrm{steel}$$

This arrangement also involves a liquid junction. In electrochemical terminology, this is designated by the double vertical line "||" between the saturated KCl solution and the seawater. We will say a little more about liquid junctions in Section 18.2.3.4.

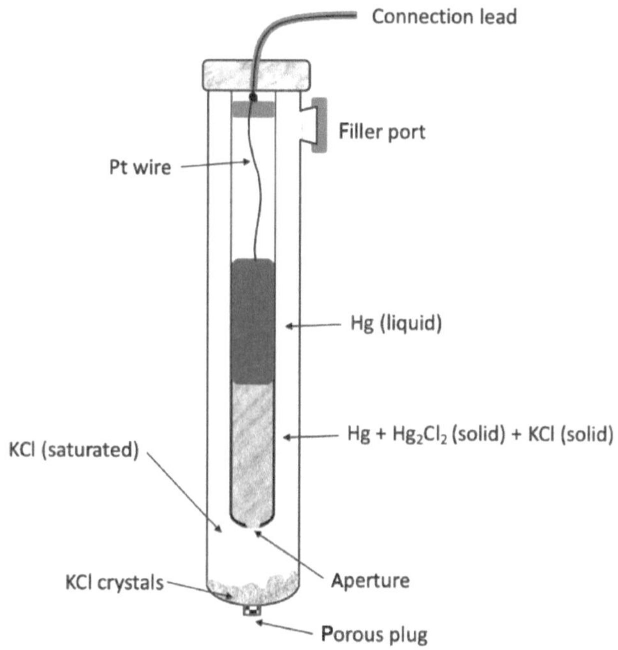

Figure 18.1 Saturated calomel half-cell.

18.2.2.3 Silver chloride electrode

To give it its full name, this is the "silver-silver chloride seawater half-cell". Table 18.2 shows this reference in bold text to emphasise its pre-eminence in marine CP. The electrode is a silver wire onto which a layer of insoluble silver chloride salt has been fused. This is an ion-selective electrode. Its potential depends on the soluble chloride content of the electrolyte. Since the chloride content of open seawater is nearly constant, the potential of the half-cell formed by dipping the Ag|AgCl electrode into seawater is also constant enough for all practical CP purposes. Thus, the −0.8 V target protection potential for steel applies in full salinity seawater anywhere. When we make a potential measurement of steel using the Ag|AgCl electrode, we are actually measuring the potential difference across the cell:

Ag | AgCl | seawater | steel

However, although these half-cells are simple and robust, they become unreliable in lower salinity waters, such as estuaries. Their potential is also prone to drifting if the electrode is immersed in natural seawater for prolonged periods.

18.2.2.4 Silver chloride (0.5 M KCl) electrode

When measuring potentials in waters that are not of normal salinity, it makes sense to use a reference in which the Ag|AgCl electrode is encapsulated in an environment of known chloride ion concentration. This fixes its potential. The chloride is invariably provided as a solution, or more usually a gel, of potassium chloride (KCl). Taking the example of a half-cell filled with 0.5 M KCl, we measure the potential difference across the cell:

Ag | AgCl | KCl(0.5 M) || seawater | steel

As with the SCE, this cell also includes a liquid junction ("||") in an electrochemical cell. In practice, the connection between the KCl solution and the seawater is provided by a porous plug made, for example, from sintered glass. Ag|AgCl|seawater, by contrast, does not have a liquid junction.

The chloride ion concentration in an 0.5 M KCl solution is fairly close to that of open seawater (~0.54 M). As can be seen from Table 18.2, using 0.5 M KCl, gives a reference potential that is about 10 mV more negative than Ag|AgCl|seawater. This means that the same target protection potential that appears as –800 mV using Ag|AgCl|seawater would be –790 mV if measured against Ag|AgCl|KCl (0.5 M). Both ISO 20313, which deals with the CP of ships, and ISO 13174, which covers harbours, advise the use of Ag|AgCl|KCl (0.5 M) rather than Ag|AgCl|seawater. This makes sense because harbours are frequently located in lower salinity estuaries, and ship CP surveys usually take place in or near harbours. Neither standard troubles the reader with the information that switching from seawater to 0.5 M KCl shifts the reference by about 10 mV.

The choice of KCl concentrations for half-cells is not limited to 0.5 M. A wide variety is available, each of which provides a different reference potential. This is illustrated in Figure 18.2, which shows how changing the concentration of chloride ions in the electrolyte around the reference electrode yields a different measurement of –0.8 V as measured against Ag|AgCl|seawater.[1]

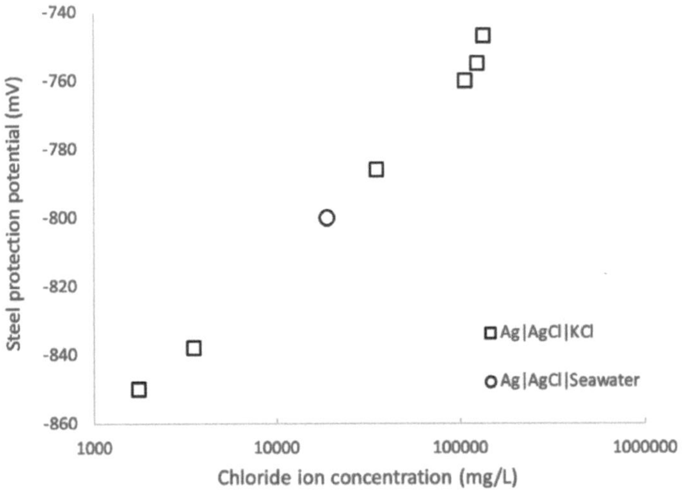

Figure 18.2 Protection potential for steel (vs Ag|AgCl).

18.2.2.5 Copper sulfate electrode (CSE)

Unlike the silver chloride and calomel references which use chloride ion-selective electrodes, the copper sulfate electrode exhibits a potential determined by the equilibrium between metallic copper and a saturated solution of Cu^{2+} ions. The CSE, more correctly the Cu|$CuSO_4$(satd.) half-cell, has been the universal reference for land-based CP potential measurement for about a century. It consists simply of a non-metallic tube containing a copper rod in a saturated solution of copper sulfate. Electrolytic contact with the electrolyte (usually soil) is achieved through a porous wooden plug at the base of the tube. The reference is easy to construct and maintain.

[1] The data for this graph were obtained from various sources, including *www.silvion.co.uk*.

As we discussed in Chapter 6, the pioneering work of Robert Kuhn showed that a suitable protection potential for buried steel and cast iron pipelines was −0.85 V measured against CSE.

Although CP was first applied to marine pipelines within a decade of Kuhn's work, there was little interest in attempting to measure CP potentials in seawater until a couple of decades after that. Since the CSE had proved itself so reliable for onshore CP, it was also inevitably used for early marine CP measurements. It is, however, unsuitable for seawater immersion. The half-cell and its wooden plug quickly become contaminated with ions from the seawater, causing a loss of accuracy. It was soon established that the Ag|AgCl|seawater half-cell was a more reliable option.

Nevertheless, examples of potentials referenced to CSE are to be found in the early marine CP literature. The potential of −0.85 V versus CSE is the same as −0.78 V versus Ag|AgCl|seawater. Interestingly, early workers on marine CP correctly adopted −0.78 V as the target potential for full protection. The experimental work analysed in Chapter 6 confirms that they were right to do so. It seems that the subsequent rounding of this parameter to −0.8 V was due to the conservatism of standards' drafting committees.

Before we move on, it is worth remarking that the abbreviations "SCE" and "CSE" can be easily confused. Speaking from experience, if you happen to get them mixed up, you will not be the first.

18.2.2.6 Zinc

All of the five references mentioned above are "true" reference half-cells. They comprise electrodes that are in equilibrium with the electrolytes in which they are immersed. Their behaviour is described by the Nernst equation. Providing the concentration of the relevant species and the temperature are known, then the potential difference with respect to the primary SHE standard will also be known.

By contrast, zinc is not a true reference. In an aqueous solution, it cannot enter an equilibrium state with zinc ions because it spontaneously corrodes. The measured potential is its corrosion potential which, as we have seen, is not a fixed parameter. However, as a matter of observation, when zinc is left to corrode in natural waters, including seawater, its corrosion potential settles down to an approximately constant value close to −1.0 V.[2] Furthermore, once this steady-state potential has been reached, it stays reasonably stable until all of the zinc has corroded away. A lump of zinc serves as a moderately precise, but particularly durable, reference. In this way, it complements electrodes such as Ag|AgCl which, although offering impressive accuracy, have limited durability.

ICCP systems on ships or fixed offshore structures require the measurement of the potential to control the current output. To this end, a robust reference, albeit of modest accuracy, will always be preferred to an inherently more accurate, but more delicate, alternative. For this reason, zinc is the generally preferred reference for marine ICCP systems.

As we explained in Chapter 11 it is possible to have the best of both worlds when it comes to selecting references for marine CP. A dual combination of Ag|AgCl and zinc, as advocated by Morgan [1], can be thought of as providing the twin benefits of accuracy and durability. Generally speaking, Ag|AgCl provides accurate measurements of the structure potential. This is particularly relevant during the commissioning and early operation of ICCP systems as they polarise the structures. The Ag|AgCl also has the ability to calibrate the zinc as it progresses towards its near steady-state corrosion potential. Reliance can then be shifted to the zinc in a timely manner, and before the Ag|AgCl has deteriorated to the point of unreliability.

[2] We have now returned to our convention of referring all potentials to the Ag|AgCl|seawater half-cell, unless otherwise stated.

18.2.3 Errors in potential measurement

18.2.3.1 Operatives

Britton [16] recounts an amusing instance where a contractor had performed a CP potential survey on an offshore platform and had reported all of the potential readings in units of ohms instead of volts. The mistake does leave us wondering about the level of competence being brought to bear by a CP contractor.[3] If the inspector did not understand what he was measuring, it is difficult to see how he would have had the ability to spot a genuine measurement problem.

A digital voltmeter (DVM) will always generate numbers, even when improperly connected. A human filter capable of recognising a problem with the measurement is vitally important in ensuring that spurious results are not recorded or, more crucially, acted upon. That human filter should be an operative trained and experienced to at least competence level 2 (Cathodic Protection Technician) as set out in ISO 15257. No doubt, advances in artificial intelligence will render that human replaceable in the not-too-distant future. But, for now, real people are all that we have.

For the purpose of this chapter, we will assume that the operative, human or otherwise, making CP measurements is competent.

18.2.3.2 Equipment and operatives

The German Nobel laureate Werner Heisenberg famously enunciated his uncertainty principle in 1927. It basically says that you cannot measure anything without disturbing it. Although the principle was formulated with specific relevance to the eerie world of quantum mechanics, it applies to a greater or a lesser extent to all physical measurements.

The measurement of potential is a case in point. In theory, all that is required is to connect a voltmeter across the cell formed by the metal-to-electrolyte and the reference half-cells. However, it is a practical impossibility to measure a voltage without drawing at least some current. As we know, passing current through an electrode causes polarisation.

This effect posed something of a headache in the pioneering days of electrochemistry when voltages had to be measured using current-consuming moving coil meters. Nowadays, however, high input impedance DVMs mean that we can measure the potential difference between a specimen and a reference whilst only drawing a miniscule current that is too low to produce any measurable polarisation. Technically, Heisenberg's uncertainty principle still applies, but CP engineers are justified in ignoring it.

18.2.3.3 Temperature

The potentials quoted for the equilibrium reference half-cells are based on a standard temperature of 25°C. Sadly, a great number of offshore installations do not enjoy such balmy conditions. Strictly speaking, where the temperature departs from 25°C, we should apply a temperature correction to our measurements. All reference half-cells based on an electrode equilibrium will have a temperature coefficient. In principle, these can be estimated from the Nernst equation which, as we saw in Chapter 4, includes temperature. In practice, however, the calculation is not straightforward. So, it is better to find temperature coefficients that have been determined under strict laboratory conditions. Typical temperature coefficients for the half-cells listed above will be in the range of −0.5 to −0.8 mV/°C. This means that the reference potential of a half-cell in cooler waters (say 10°C) could easily be 10 mV

[3] The incident reported by Britton occurred before there were any standard governing the competence of offshore CP personnel.

more positive than the value in Table 18.2. Thus, measuring a code-compliant potential of −800 mV in cool waters means that the potential actually being achieved would be somewhere around −810 mV. This conservative error is inconsequential.

18.2.3.4 Liquid-junction or diffusion potential

You may be wondering why some of the reference cells in Table 18.2 use a solution of potassium chloride (KCl), rather than the cheaper and more ubiquitous sodium chloride (NaCl) as the electrolyte. The reason stems from the rate at which potassium and chloride ions move in an aqueous solution. In the jargon of the electrochemist, they have similar mobilities. To understand why this is important, we need to understand a little about liquid junction potentials or, as they are also known, diffusion potentials.

18.2.3.4.1 Effect in reference half-cells

The encapsulation of an electrolyte in a half-cell means that, in order to create the cell across which the voltage can be measured, the half-cell electrolyte has to be brought into contact with the seawater. This is typically achieved by means of a porous sintered glass plug in the reference half-cell (see, for example, Figure 18.1). The zone where the KCl solution contacts the seawater is referred to as the liquid junction. As noted, this is indicated by the double vertical line in the depiction of the cell.

Where two electrolytes form a liquid junction, a potential difference can arise. We illustrate this in Figure 18.3 which depicts a simplified situation in which the half-cell chloride concentration is set equal to that of the seawater: ~19000 mg/L. For convenience, we also assume that the only salt present in seawater is sodium chloride (NaCl). Figure 18.3 indicates the half-cell KCl electrolyte on the left-hand side and the NaCl (seawater) on the right.

Ions will always tend to diffuse from a region of high concentration to a region of low concentration. Since the chloride ion concentration on each side of the junction in Figure 18.3 is the same, there will be no nett change of concentration of chloride on either side. On the other hand, there will be a nett diffusion of sodium (Na^+) ions into the half-cell electrolyte and a nett diffusion of potassium ions (K^+) into the seawater.

The rate at which individual ionic species diffuse is characterised by their ionic mobilities. We can leave the details of this to the electrochemistry enthusiasts. All that we need to be aware of is that the potassium ion has a ~38% higher mobility in water than the sodium ion. We can, therefore, picture potassium ions moving from left to right in Figure 18.3 faster than sodium ions move from right to left. This relative difference in ionic transport rate means that the diffusion of these species across the liquid junction produces some degree of charge separation. There will be a slight nett excess of positive charges accumulating on the seawater side. This charge separation across the liquid junction is the source of a potential difference.

In reality, this simplified situation will be complicated by the fact that the chloride concentration in the half-cell is unlikely to be identical to that in the seawater. In addition, the diffusion of other ions in the seawater, such as calcium, magnesium and sulphate, across the junction will also influence this potential difference.

All that matters to us, however, is the fact that liquid junction potential differences are not very large. It would be difficult to set up a CP measurement arrangement where it amounted to as much as 5 mV. The phenomenon of liquid junction potential differences has relevance in other fields of science, such as cell biology, but it has no impact in the world of the CP engineer. A measurement error of a few millivolts in a CP potential reading has no practical consequences in terms of the level of protection afforded.

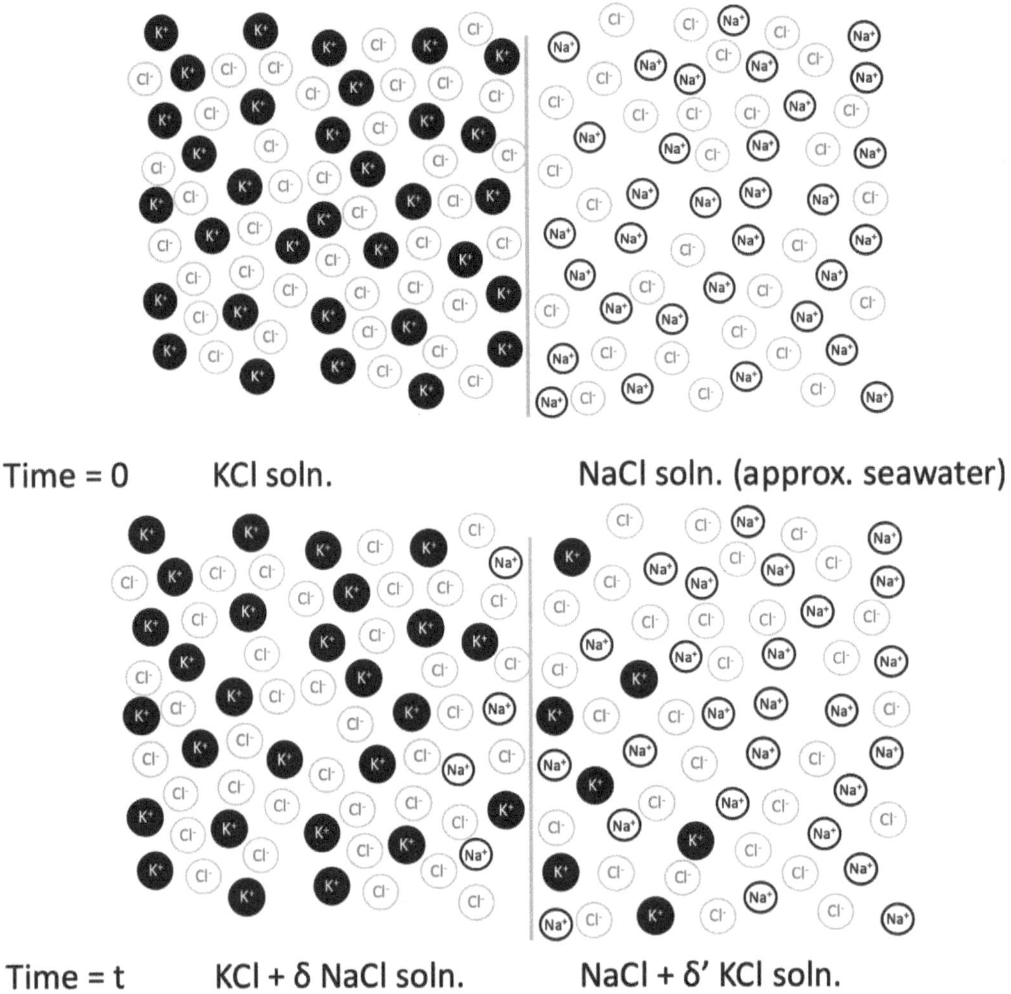

Figure 18.3 Liquid junction potential.

Nevertheless, this minor liquid junction potential error, combined with the above error due to the temperature dependence of the reference, highlights the frivolity of the habit of inspectors reporting, or modellers predicting, CP potentials to discriminations of ±0.1 mV.

18.2.3.4.2 Effect at the cathode surface

An additional diffusion potential arises due to the polarisation of the cathode itself. In Chapters 5 and 7, we discussed polarisation experiments from which the relationship between the applied potential and the nett rate of the anodic or cathodic current on the surface of the metal (usually steel) was plotted. It was implicit in these polarisation curves that the potential was the true potential at the metal-electrolyte interface.

However, we also learned in those chapters that the act of applying a cathodic current changed the chemical composition of the seawater at the interface, in particular it raised the local pH. Thus, polarisation creates concentration gradients at the steel surface, and the migration of ions in response to these gradients must also generate diffusion potentials.

The obvious question is: do these diffusion, or liquid junction potentials influence the potential value that we measure for the steel?

The short answer has to be: yes. Any potential measurement we make on a cathodically polarised surface in seawater is influenced by the fact that our measurement will also include a contribution arising from the diffusion potential at the surface.

Fortunately, there is also a longer answer. It is still yes, but it does not actually matter. As noted, liquid junction potential differences only amount to a few millivolts at the very most. More relevantly, the whole technology of CP has evolved based on steel-to-electrolyte potential measurements which have always included a contribution from these extraneous potential differences. In other words, when workers found that steel was fully protected in seawater at a potential of around –780 mV (Chapter 7), or when Uhlig found that the fatigue behaviour of steel returned to the in-air value at –760 mV (Chapter 12), those potential measurements included the effects of diffusion – just as will any measurement we might make today.

It follows that CP engineers have no need to be concerned about the liquid junction potential. This is just as well. Few have heard of it; fewer still have any idea what it is.

18.2.3.5 The IR problem

Although the "error" introduced into CP potential measurements as a result of ionic diffusion processes may safely be ignored, there is an additional source of error in potential measurements that cannot be dismissed. Figure 18.4 illustrates the situation. In this diagram, we have an anode which is delivering current onto the steel surface. We would like to measure the potential of the steel structure in order to find out whether or not this current is achieving full protection. Moreover, we might prefer to do this whilst sitting comfortably above the water.

All that is required is that we connect the positive terminal of our DVM to the structure and the negative terminal to an Ag|AgCl reference electrode. In this illustration, if we then lower the reference on its cable a little way into the water (case A), it will be somewhere in the region of the anode. We would not be surprised to find our meter reading around –1050 mV, or more negative. This confirms that the anode is working.

Reassuring as this is, what we actually need to know is the steel potential. We can investigate this by lowering the reference deeper into the water. As we do this, and we move away from the immediate influence of the anode, the readings on the meter will progressively shift to less negative values. If we pay out our cable until the reference is practically on the seabed (case B), about as far away from the anode as possible, we might see a reading of (say) –820 mV. That might tempt us into concluding that the entire structure is fully protected. However, we should be cautious.

Our depiction of our DVM simply registering the potential difference between the cathode half-cell and the reference half-cell is too simplistic. This electrochemical potential difference across the steel-seawater interface is not the only voltage we would be measuring. There is also the voltage that arises from the flow of the CP current (I) against the electrical resistance (R) of the seawater. By Ohm's law, the product of a current multiplied by a resistance is a voltage. This voltage is termed the "IR-error". The DVM measures a voltage that is the sum of the polarised potential of the steel plus this IR error. There is no way that the meter can distinguish between these two voltage components.

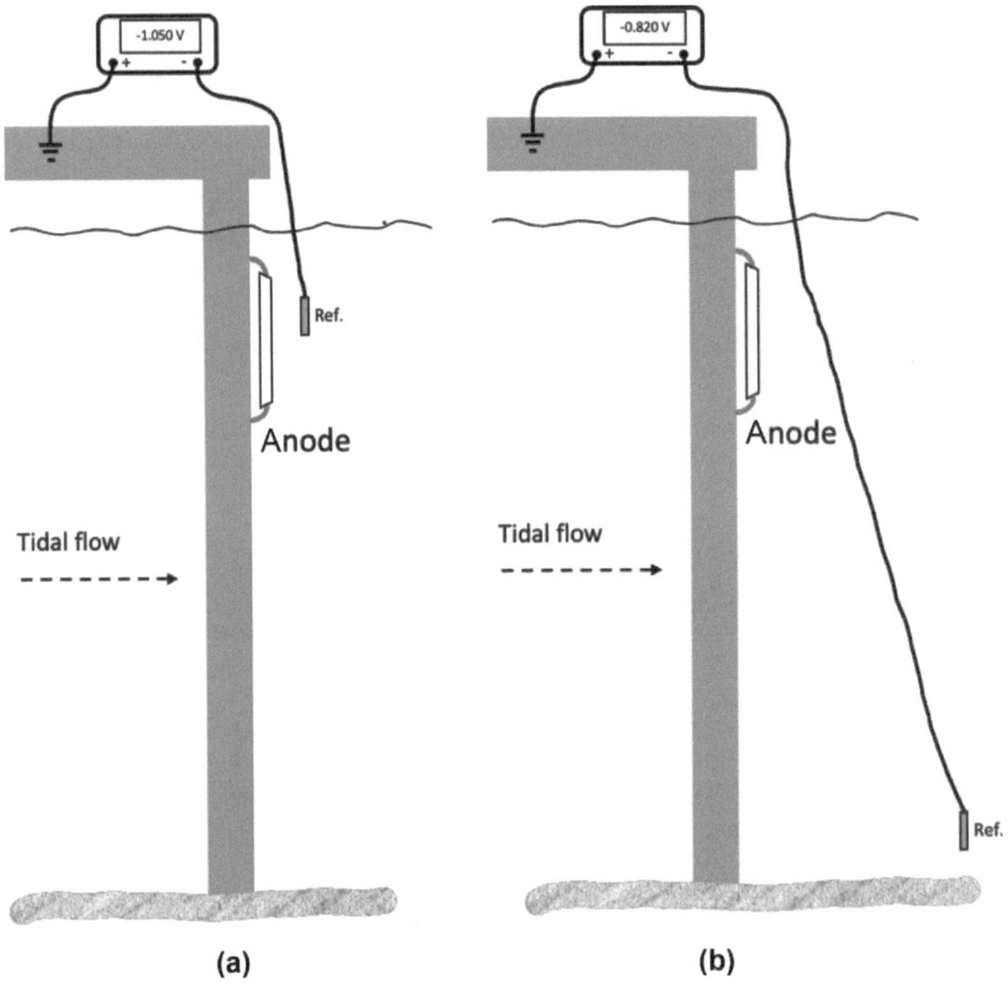

Figure 18.4 How the IR error arises.

It is very difficult to calculate the magnitude of the IR error, although we can be sure that the larger the current, the greater it will be. Indeed, this is one of the areas where computer modelling, which we discussed in the previous chapter, can be particularly useful.

On the other hand, we do know the polarity of the error. The reference half-cell will always be positioned somewhere in the electric field created by the current flowing from the anode to the cathode. This means that any attempted measurement of the cathode potential will actually return a value that is more negative by a voltage equal to the IR error. This holds irrespective of whether the anode is sacrificial, operating at a potential that is more negative than the cathode, or impressed current operating at a potential that is relatively positive. What counts is the direction of the current flow.

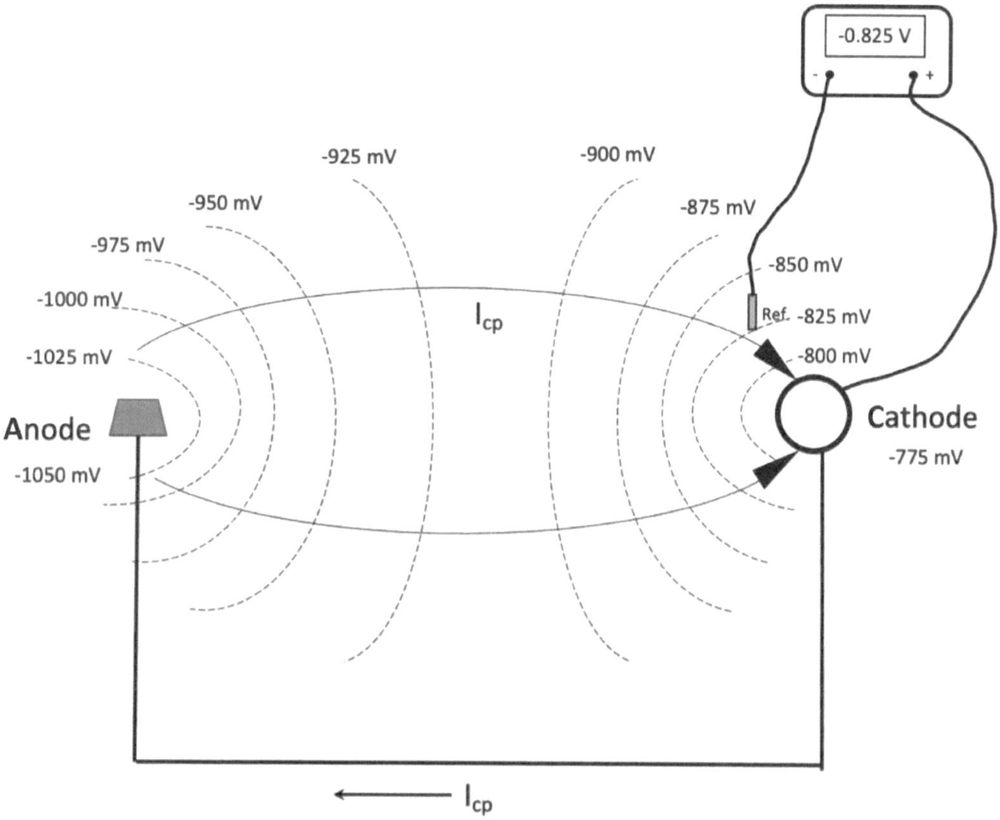

Figure 18.5 IR-error: positioning the reference.

The important consequence of this is that the polarity of the IR error will always be such that the measurement indicates a greater degree of protection than is actually being achieved on the structure. IR errors are always non-conservative.

18.2.3.6 IR-error mitigation

18.2.3.6.1 Positioning the reference

The first approach to reducing the IR error is to position the reference as close to the structure as is reasonably practicable. Figure 18.5 provides a highly stylised simplistic representation of the electric field surrounding an anode and cathode. As the CP current (I_{cp}) flows, it creates an electrical field gradient[4] in the space around the anode and the cathode. To illustrate the problem, we assume an anode potential of −1050 mV and a cathode polarised to −775 mV. I have represented the electrical field as a series of equipotential contours at 25 mV spacings. From this diagram, it is clear that the closer we can get the reference to the part of the structure on which we are trying to measure the potential, the more reliable our measurement will be. In our illustration, because the reference is positioned somewhat away

[4] I am, of course, indulging in chicken-and-egg reasoning here. It would be equally correct to say that the potential gradient is generated by the potential difference between the anode and cathode, and that the current then flows as dictated by the potential gradient.

from the cathode surface, the measured potential is –825 mV, which would be reported as complying with a requirement to achieve –800 mV, or more negative. In this illustration, however, the actual potential of the cathode is –775 mV which, whilst not presenting a corrosion or fatigue problem, is technically out of compliance.

18.2.3.6.2 Instant-off measurements

The alternative method of eliminating the IR error relies on the fact that, when we switch the current off, the purely ohmic IR component of the measured potential disappears almost instantaneously. The decay of polarisation of the steel, on the other hand, is subject to what we might call "electrochemical inertia". Once the polarising current is removed, it takes time for the structure to depolarise to its free corrosion potential.

The situation is illustrated in Figure 18.6. With the current applied, the measured potential will be IR volts more negative than the actual polarised potential of the steel surface. At the point at which the current is switched off, the IR voltage drop disappears, and the electrochemical polarisation begins its slower process of decay. At a point in time immediately after the switch-off, the potential measured remains, to a very good approximation, the true polarised potential of the steel. In practice, of course, the steel is not left to depolarise, as is indicated in Figure 18.6. Once the measurement is taken, usually on a fast-recording data-logging DVM, the current is switched back on to maintain the polarisation.

The "instant current-off" technique can only be applied to ICCP systems. Its origins lie onshore where ICCP is the norm, soil resistivities are high and, it follows, IR errors are larger. The technique is hardly ever used in offshore CP.

18.2.3.6.3 Below the seabed

A substantial part of the infrastructure we seek to protect is below the seabed. This includes, for example, piled foundations and buried submarine pipelines. This burial introduces a problem when it comes to positioning the reference close enough to obtain an accurate potential measurement. In principle, the reference could be mounted on an ROV-mounted probe and pressed through the sediment to a position close to the steel surface. Indeed, there has been some recent work in which this has been done. It seems unlikely, however, that this approach will find itself widely applied for general CP surveying in the near future.

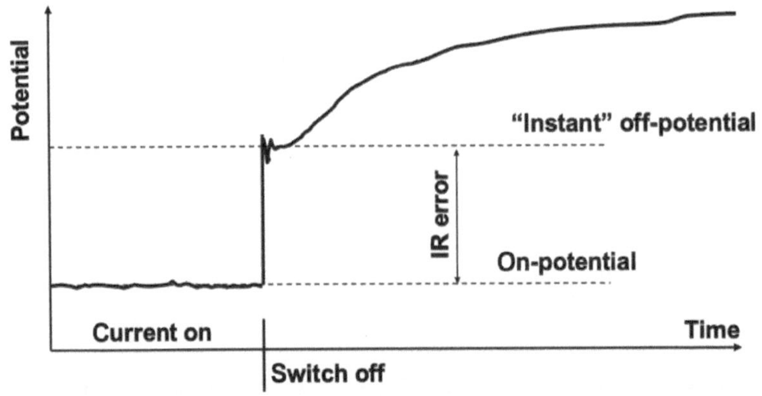

Figure 18.6 The instant-off method.

With regards to pile foundations, the present approach seems to be based on intuitive reasoning along the following lines. If the structure is found to be well protected down to seabed level, then it is also assumed to be adequately protected below the seabed, at least to a depth beyond which the consequences of corrosion or fatigue are no longer relevant. There is a logic to this in that although the seabed resistivity is higher than the open seawater, by a factor of about 5, the current density required to achieve protection reduces by a similar factor since dissolved oxygen diffusion is restricted. On this basis, the CP industry has generally been unworried by its inability to measure potentials below the seabed.

18.2.3.7 The effect of seawater flow

There are mechanisms by which the flow of seawater can, in theory at least, influence the potential. The first is a phenomenon known as the streaming potential, which arises due to shear stresses generated by the flow across the surface. The second arises from the tidal movement of seawater through the earth's magnetic field.

18.2.3.7.1 Streaming potentials

The streaming potential, or rather the streaming potential difference, is an electrochemical phenomenon that is relevant to branches of surface science such as colloid studies. As we have seen, there is some charge separation at any solid-electrolyte interface. This zone of charge separation is referred to as the electrochemical double layer. If an electrolyte moves laterally across this interface, the shear force it generates has the effect of sweeping some of the surplus ions in the double layer along with it. This sets up a lateral potential gradient over the surface.

Since fixed structures experience tidal flows of seawater across the surface, and ships obviously experience much more rapid flows, it is reasonable to ask the question: do streaming potentials influence CP or CP measurements?

The answer, for all practical purposes, is: no.

The value of the streaming potential difference across a surface falls dramatically as the liquid conductivity increases, and as the pressure differential decreases. In practice, seawater is too conducting, and lateral pressure differentials too mild, to pose a problem. The CP engineer has no need to be concerned about streaming potentials.

18.2.3.7.2 Tidal flow

From our school physics, we should recall that the movement of a conductor through a magnetic field produces a current in that conductor. This is the underlying principle of the generation of electricity. It follows that a body of seawater, which is a conductor, moving through the earth's magnetic field, will likewise generate a current. Boteler and Cookson [13] have calculated that a tidal flow of 1 knot (~0.5 m/s) produces an electric field of 26 mV/km. Technically, this has some relevance to the measurement of potentials along a pipeline, particularly near shore, or in other geographical locations, where tidal flows can be high. From the practical standpoint, however, we can ignore it.

18.2.4 Potential surveys – structures

18.2.4.1 Dip-cell surveys

We have already seen the essentials of a dip-cell (or "drop cell") survey in Figure 18.4. This is the simplest form of offshore CP potential survey to carry out. A weighted portable reference is lowered on its cable, and potentials are recorded at regular depth intervals (typically

1 m) as the cable is paid out. In the case of a vertical structure, such as a wind farm monopile foundation, the reference can be kept reasonably close to the steel surface by lowering it down a weighted, pre-installed guide wire. There is less merit in using a guide wire in the case of most space-frame structures where the vertical members slope outwards.

Dip-cell surveys require neither pre-planning nor the mobilisation of divers or remotely operated vehicles (ROVs). However, because of the separation of the dipped reference from the structure, potentials measured in this manner are likely to contain larger IR errors than those measured in diver or ROV surveys Furthermore, as explained above, these increased IR errors are always of a polarity such that the recorded values appear more negative than the actual steel-to-seawater potentials. Thus, dip-cell survey data will always be non-conservative to some extent.

Nevertheless, such surveys are useful for structures of simple geometry such as monopiles. They can also complement the fixed monitoring on installations protected by ICCP. This is because dip-cell surveys can be carried out without the need to switch the system off.

18.2.4.2 Diver and ROV surveys

18.2.4.2.1 Proximity measurements

Potential surveys in which divers or ROVs obtain measurements at specific locations on the structure will produce more accurate data than dip-cell surveys, but at the expense of deploying an offshore survey spread. For this reason, CP surveying of fixed offshore structures by divers or ROVs is integrated into an overall underwater inspection programme. The basics of the measurements need not differ from the dip-cell technique. In the case of the ROV survey illustrated in Figure 18.7, a recording DVM is located in the support vessel's ROV control cabin.

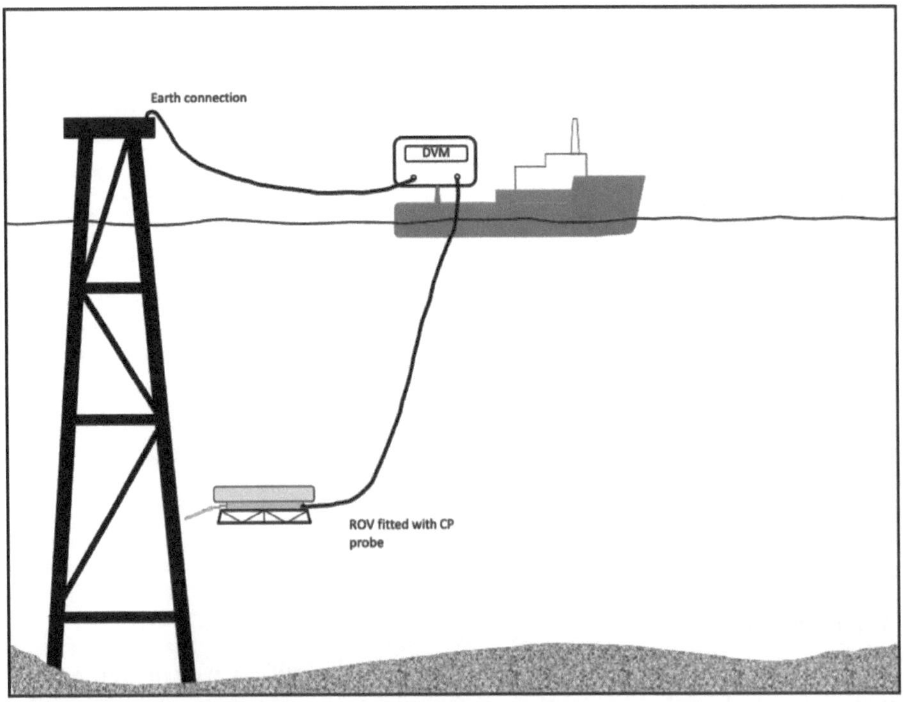

Figure 18.7 ROV survey.

The positive terminal is securely connected by cable to a convenient earthing point on the structure's topsides. The negative terminal of the DVM is connected, via a cable within the control umbilical, to a reference cell mounted on the ROV.

The ROV then manoeuvres the portable reference to a position much closer to the point of interest than would be possible using a dip-cell. The potential captured by the DVM is then recorded, often as a data stream in the survey video recording.

The survey process using divers is similar, except that it is a diver rather than an ROV that positions the reference. Obviously, communication protocols have to be established with the topsides support to ensure that potential data are recorded accurately.

18.2.4.2.2 Stab potential measurements

A technique that was originally developed for diver use, but which was subsequently adapted for ROV deployment, is the hand-held potential measurement device. A schematic of the instrument is shown in Figure 18.8. A manufacturer's exploded illustration of a proprietary unit, termed a Bathycorrometer®, is shown in Figure 18.9.

Using such a probe avoids the need for the diver to carry a cable connection to a topsides DVM, thereby removing the safety hazard that the cable might become snagged. These units, which have been in use since the 1970s, contain all that is needed to make a steel-to-seawater potential reading. The sharpened stainless steel tip makes a direct metallic contact with the structure. This tip is connected to the positive terminal of the internal DVM, the negative terminal of which is connected to the integrated reference half-cell. The latter is positioned close to the point of contact to minimise the IR error. The potential measurement is displayed on the instrument's diver-facing panel. These readings are either recorded by the diver (Figure 18.10), for example using a helmet-mounted camera, or communicated directly to the topsides support team.

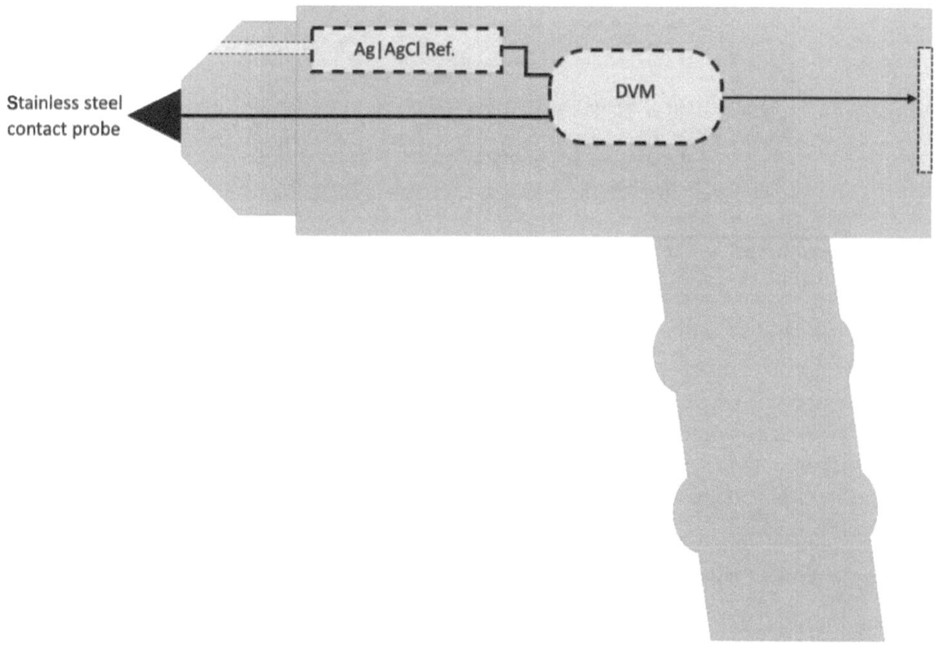

Figure 18.8 Diver-held potential measuring probe (schematic).

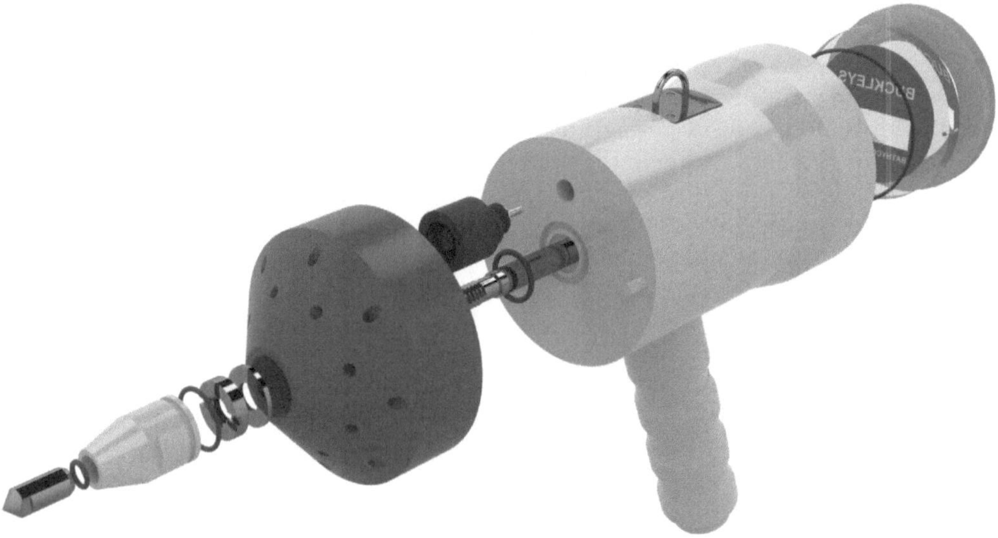

Figure 18.9 Bathycorrometer®. (Courtesy of Buckleys (URVAL) Ltd.)

Figure 18.10 Bathycorrometer® in use. (Courtesy of Buckleys (URVAL) Ltd.)

Some of you, mindful of my earlier comments about Heisenberg's uncertainty principle, might be wondering how much error is introduced into the potential measurement by stabbing the steel surface with a stainless steel probe. The answer is not much. The surface area of the stainless steel is very small, particularly in comparison with the massive area of the carbon steel structure. As far as anyone has been able to ascertain, its effect on the polarised potential of the steel is too small to be detected.

18.2.4.2.3 Issue with ICCP systems

Structures fitted with ICCP should have fixed monitoring installed as part of their control systems. This means that some information on the state of polarisation of the structure will always be available. However, a broader picture of the spread of the potential over the structure requires a survey. In principle, diver or ROV potential surveys are also relevant to structures protected by ICCP. We discussed the safety aspects of ICCP systems in Chapter 11. Providing reasonable safety precautions are adopted, divers can work safely without the need to switch the systems off. Nevertheless, old habits die hard. It remains customary offshore practice to switch ICCP systems off when underwater activities are in progress.

Unfortunately, switching the ICCP system off obviously invalidates the potential data subsequently obtained. The quality of the data diminishes the longer the structure is allowed to depolarise. Nevertheless, the literature contains reports of CP surveys having been carried out regularly on structures fitted with ICCP systems that had been intentionally switched off for the purpose. For example. Daley and Ingraham [15] describe underwater inspection work, including the taking of potential measurements, on structures in the Cook Inlet fitted with ICCP using remote sleds. All of the potential measurements were taken with the current turned off. It is hard to see the rationale underpinning this activity. All that can be reasonably stated is that, since the structures were found to be adequately protected with the current switched off, they spent most of their operational lives comprehensively over-protected.

18.2.4.2.4 Advantages and limitations

Diver or ROV potential surveys do not require the prior installation of any fixed equipment, nor do they require a great deal in the way of pre-planning. The CP surveys are generally carried out as part of a wider rolling programme of structural integrity inspection. However, because divers or ROVs are required, potential survey data are only obtained on a periodic basis.

Furthermore, surveys cannot be undertaken during adverse weather conditions. Similarly, surveying might be impractical during periods of high tidal flow. Since both storms and high tidal flows are antagonistic to CP, it follows that survey data obtained when conditions favour diver or ROV operation are likely to overstate the level of protection achieved throughout the year.

18.2.4.3 Survey frequency – sacrificial systems

18.2.4.3.1 Petroleum production structures

In 1979, Nagel and Willis [3] reported that the measurement... *of solution potentials around offshore steel structures on a regular basis is a relatively recent requirement. This has been made mandatory by Classification Societies who now require proof that satisfactory cathodic protection is being provided.* It is little surprise that the older CP codes occasionally used adjectives such as "regular" to describe the required frequency of surveying or inspection, but none was particularly prescriptive.

In the case of structures involved in petroleum production, the responsibility for determining the through-life inspection programme resides with the operator. In many jurisdictions, this requirement is reinforced by legislation that requires the operator to demonstrate the overall safety of their installations through the compilation of a safety case. Against this legislative background, the codes have been either silent on non-prescriptive in respect of the frequency of potential surveys.

Some information was provided in early editions of DNV-RP-B401. For example, the 1986 edition advised a potentials survey 3 months after installation in the case of an uncoated structure protected by sacrificial anodes, extending to 12 months for a coated structure. The logic behind this advice was not explained. As a matter of experience, some uncoated structures were found not to be fully protected within a year. Coated structures, on the other hand, would be expected to be fully polarised within a month. No advice was offered on the frequency of subsequent potential surveys.

The 2005 edition, whilst still highlighting the need for post-installation inspection, dropped any advice on time spans. This absence has remained in subsequent editions of DNV-RP-B401 and of NORSOK M-503. These are now only concerned with providing design guidance. Their scopes do not include through-life inspection or surveying.

NACE SP0176 is more detailed and illuminating. It recommends that a survey should be conducted after each CP system is energised. It does not say how soon after, but it does require that the survey includes *potential measurements in sufficient detail to demonstrate conclusively that protection has been attained.* This standard practice also provides advice on setting the frequency of potential surveys throughout life. It suggests that initially, or in the absence of supporting data, surveys... *should be performed at least annually.* Thereafter, it suggests that a risk-based approach be used to set the frequency, based on a 5-year interval unless more frequent inspection is required. It gives examples of the circumstances that might prompt more frequent inspection. For sacrificial systems, these include visual observations of enhanced anode consumption, the incorporation of additional steelwork into the CP system or the anodes approaching the end of their design life.

18.2.4.3.2 Offshore wind foundations

As we discussed in Chapter 13, and again in the context of modelling in Chapter 17, the introduction of near-shore wind farm structures was not without its problems when it came to CP. These problems have, in turn, prompted the development of codes detailing CP design and CP inspection and surveying.

DNV-RP-0416, published in 2016, recommends that the CP be applied to the structures as soon as possible after installation. It then states... *After maximum 365 days, a CP survey shall be performed to confirm that the structures... are adequately protected.* It goes on to advise that, because a wind farm will contain numerous very similar foundations, in the same general geographical location, with a commonality of CP provision... *it is recommended as a minimum to survey one structure for every 20 installed structures. However, the actual number of structures to be surveyed should reflect what in each case is deemed necessary in order to obtain the required representativeness of the survey.*

There is a case that surveying only 5% of the structures (1 in 20) is, in practice, a little thin. It would be meaningful if all of the structures were installed at the same time of year, and all had their anode connected very promptly after installation. However, in practice, wind farm monopile installation progresses through the year; and practical considerations, not least the weather, can influence the length of time between the foundation being installed and its anodes being attached. My own view is that a 20% sample is more appropriate for the first-year survey. This could then be built into a rolling 5-year programme as is typically employed for hydrocarbon structures.

As I write this chapter (May 2021), the new ISO 24656 *Cathodic protection of offshore wind structures* has been issued in draft form for public consultation. It gives recommendations regarding the frequency of potential surveys. The draft recommends that a baseline potential survey... *should be carried out after minimum 30 days of polarisation for coated structures and minimum 60 days for non-coated structures and before one year after the installation of the structure and connecting the galvanic system.* In terms of sampling, its advice is more exacting, and more detailed, than that of DNV-RP-0416. It also distinguishes between structures to which the anodes are welded, and those for which anodes are cable connected.

In the case of structures with welded anodes, and on larger wind farms comprising over 100 structures, it recommends 25% should be subject to the baseline surveys. On smaller farms, the percentage of structures sampled should be increased. It also suggests that the sampled foundations should not be selected at random. Rather, they should be picked to provide a spread covering the full spectrum of CP situations, such as minimum and maximum water depth. On the other hand, where anodes are cable connected, as is the case when anodes are mounted on the transition piece which, in turn, is continuity bonded to its monopile, it recommends that 100% of the structures are given a baseline survey.

The standard then goes on to advise that measurements... *should be repeated at the same representative locations at agreed time intervals throughout the life of the structure, so that trends in the protection level and hence CP system performance can be determined.* It further recommends that surveys are incorporated into a rolling programme such that surveys of all of the structures in the wind farm are measured over a period of 3–10 years.

18.2.4.3.3 Harbour structures

For sacrificial anode systems, ISO 13174 requires that a dip-cell survey be undertaken, within 3 months of installing the anodes. The readings should be at three water depths: at the mean water level (MWL), at the seabed and midway between the two. These should be repeated at regular horizontal spacings, typically 5 m, along the harbour structure. The standard then advises that following the 3-month survey... *further testing should be performed, typically between 9 and 12 months, prior to the end of any defect liability period, and then at intervals of 2 to 5 years subject to performance.*

A potential survey for ICCP systems is also required within 3 months of commissioning. This is to supplement the data from the fixed monitoring. The same vertical and horizontal spacings for the readings are advised. The standard also recommends that "IR-free" potentials are measured. In principle, the instant-off technique can be used. However, since harbour ICCP systems can be very extensive and interconnected, synchronised switching of CP sources to obtain instant-off potentials might be impracticable. Nominally "IR-free" readings can be obtained by positioning the dip-cell very close to the structure. Normally, this should not pose too much of a problem in the relatively quiescent waters of a harbour. ISO 13174 points out that ICCP system... *can pose a risk to divers and are normally switched off during diving operations in their vicinity. If this is impracticable, divers shall be informed that the system is energised so that the necessary actions can be taken to ensure their safety.* Again, this points to the use of dip cells with ICCP systems operating as being the most relevant survey option.

Providing the commissioning survey is satisfactory for an ICCP system, ISO 13174 does not set a timetable for further surveys. Instead, it focusses on functional checks of the hardware, plus measurements on the system performance. The functional tests are essentially observations that the DC current and voltage outputs remain as expected. Typically, these functional checks should be monthly in the first year and, subject to continuing satisfactory performance, quarterly thereafter. The performance assessment will be a matter of recording the potentials on the fixed monitoring devices. The standard recommends that this should be quarterly in the first year. Then, subject to continuing satisfactory performance, it can be relaxed to 6- or 12-monthly.

18.2.4.3.4 Ships and floating installations

18.2.4.3.4.1 EXTERNAL HULLS – SACRIFICIAL ANODE SYSTEMS

In the case of ships' hulls fitted only with sacrificial anodes, ISO 20313 recommends a dip-cell potential survey within 1 month of the anodes being fitted, a repeat survey before the end of the anode CP system guarantee period and a survey a year before the next dry-docking and anode replacement.

I find this advice intriguing. Although the ISO recommends these potential surveys on such vessels, my experience is that the recommendation is universally ignored. Once a vessel has set sail from its construction yard where the anodes were fitted, or from a dry-dock where anodes were replaced, there is no intent whatsoever of re-entering dry-dock again before absolutely necessary, for example to get the antifouling paint replaced or for a class survey. Even if the anodes were under-performing and a dip-cell survey found that full polarisation was not being achieved, it is very unlikely that the owner would react by speeding to the next available dry-dock for anode replacement. It would usually make more economic sense to keep trading, even if it meant accepting some corrosion and even replacing plating, if necessary, at the next docking.

There is the additional problem that a hull CP system will be working hardest when the ship is underway. However, that is not a good time to be dangling half-cells on cables over the side. Conversely, a dip cell survey might occupy crew members for an hour or so if the vessel is laid up at anchor; but the relevance of the results would not be obvious.

This is another example of the flip side of the aphorism that you can only control what you can measure. There is no point measuring something, in this case the potential of a ship's hull, if nobody is going to take any action as a result of that measurement.

18.2.4.3.4.2 EXTERNAL HULLS – ICCP SYSTEMS

For ships fitted with ICCP systems, ISO 20313 recommends the same regime of post-commissioning and pre-docking dip-cell surveys as for the sacrificial anode systems. There is a stronger case for dip-cell surveys on vessels with ICCP systems. If under-protection or over-protection is observed, then there is an opportunity to do something about it by adjusting the current output of the system. Nevertheless, as with hulls protected with sacrificial anodes, a survey would only really be relevant when the vessel was underway. Unfortunately, this is when it is generally impracticable to carry out dip-cell surveying.

It follows that dip-cell surveys are only likely to be relevant for hulls protected by ICCP, but which are normally tethered. For this reason, the technique has found occasional use on FPSOs fitted with ICCP systems.

18.2.4.3.4.3 BALLAST TANKS

ISO 20313 states that ballast tank system potential measurements can only be made using permanently installed reference electrodes. This information is as correct as it is pointless. As mentioned in Chapter 16, I have crawled through many unsavoury ships' ballast tanks. During those exploits, I sporadically encountered anodes or the remains thereof, but I have never seen either a zinc reference or its cabling. As we explained in Chapter 16, ballast tanks are only ever protected by sacrificial anodes, and often only half-heartedly so. As with sacrificially protected hulls, there is little point determining the potential when there is no practicable response that can be made on the basis of the information.

18.2.5 Potential surveys - pipelines

18.2.5.1 Background

The application of CP to fully immersed offshore pipelines has been remarkably successful. As we know, there have been some isolated issues. Examples include the mechanical detachment of bracelet anodes during pipelay or the application of conducting coatings to hot flowlines, which we discussed in Chapter 14. For the most part, however, the offshore inventory of hydrocarbon-carrying pipelines has been well protected. We can be confident about this because most pipelines are periodically inspected by in-line inspection (ILI) tools, often referred to as intelligent pigs. These devices can identify both internal and external pipe wall corrosion damage. There have been plenty of pipelines that have exhibited internal corrosion damage as a result of exposure to the transported fluids. Some have been damaged to the extent that they have had to be taken out of service. However, I have yet to see any report of external corrosion damage.[5] This is despite the fact that some of the pipeline inventory is now over 50 years old.

There is, however, a slight downside that rubs some of the gloss off of this success story. An ILI survey has a wall thickness discrimination of somewhere around ±0.1 mm. This means that if, for example, an eternal CP system had ceased to provide protection (say) a few years before the ILI survey was run, this fact would not necessarily be revealed. However, corrosion would be in progress and may have developed substantially by the time of the next run, typically 5 years in the future.

As we know, the only way to be sure that CP is effective is to measure the potential of the pipeline along its length. Nagel and Willis [3] appreciated this when they reported in 1979 that it was... *not yet mandatory that pipeline potentials be established.* However, this attitude was in the process of changing. By 1981, DNV had published its *Rules for Submarine Pipeline Systems.* This widely used document provided some limited instructions. Section 6.3.5.1 stated that potential measurements... *are to be carried out to ensure that the pipeline system is adequately polarised.* It added that this testing... *is to be carried out within one year after installation.* Beyond that, the document was vague about how these potentials should be measured. This is not surprising because it is far from straightforward to measure the potential profile along a pipeline, and this difficulty increases massively if it is buried.

DNV published a more informative paper in 1983 [8] in which the various inspection methodologies for offshore pipeline CP systems were evaluated. The two techniques for determining the potential were the trailing wire and close-remote reference surveys. These have not fundamentally changed since.

18.2.5.2 Trailing wire surveys

18.2.5.2.1 The process

Any survey in which a diver or an ROV uses a reference connected to a voltmeter on the structure's topsides, or on a support vessel (as in Figure 18.7), and then manoeuvres the reference around the structure could be called a trailing wire survey. Here, however, we are considering a longer length of pipeline where the survey vessel connects one terminal of its onboard DVM to the pipeline using a thin wire and then sails away above the pipeline.

The simplicity of the trailing wire survey, which is also referred to as a "towed-fish" survey, is seen in Figure 18.11. It is based upon the long-established over-the-line survey procedure for onshore pipelines. It has to start where the pipeline is taken above the waterline, either at an offshore platform or shore crossing. A cable is attached to the pipeline from a drum on the survey vessel. The reference is attached to a weighted "towed-fish", also deployed from the vessel. The vessel then progresses along the pipeline route, paying out the wire from the drum and towing the reference above the pipeline and, hopefully, as close as possible to it. Equipment on board records the potential as a function of distance, as measured either by GPS or the length of cable paid out.

[5] Written in May 2021.

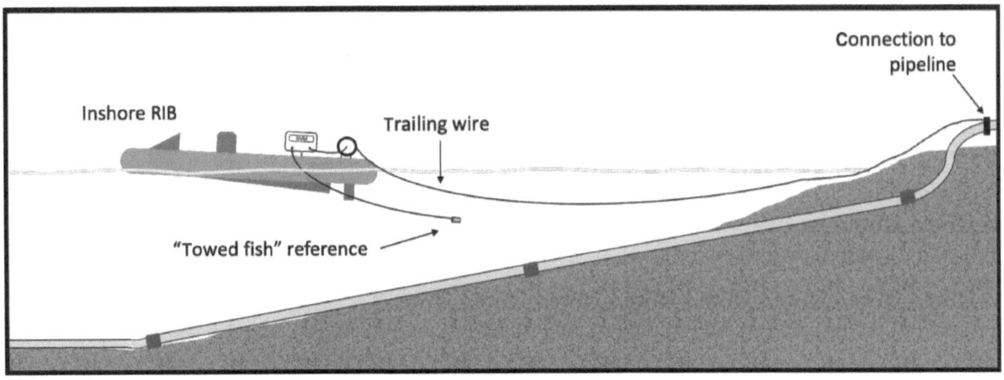

Figure 18.11 Trailing wire survey (inshore).

18.2.5.2.2 The limitations

Unfortunately, the limitations of the technique also emerge from consideration of Figure 18.11. We have already discussed the IR-error that arises if we are unable to position our reference immediately adjacent to the pipeline. No less of a problem is the fact that, even if the vessel has the ability to position its track directly above the pipeline, there is no guarantee that the reference cell on the towed fish will stay close. It will inevitably wander off-track as it falls under the influence of lateral currents. The magnitude of this displacement is almost certain to increase with increasing water depth. As illustrated in Figure 18.5, allowing our reference to drift away from the structure or pipeline will inevitably produce falsely re-assuring potential readings due to the IR error.

The principal drawbacks, which have been recognised for some time [9], are that it is too insensitive to detect local variations in the pipeline potential. There is also a history of problems with the insulation of the trailing wire and its propensity to break. Added to this is the environmental impact of any discarded wire on the sea floor.

At best, a trailing wire survey will give a general overview of the CP level of the pipeline. For a relatively new pipeline, with a coating in good condition and a code-driven abundance of anodes, a trailing wire survey will tell you that the pipeline is well polarised. However, in the case of an aged pipeline with substantial accrued coating damage, causing anodes to work harder to provide protection, the general overview that you will get might be misleadingly optimistic.

It is only relevant where it is not practicable to deploy a steerable device such as an ROV that can keep the reference close to the pipe. For this reason, it is the only available technique for shallow water surveying, such as shore approaches. Even then, it would be unwise to place too much reliance on the data produced. The survey data obtained on the beach itself, particularly if compared on a year-on-year basis, are probably more reliable and illuminating.

18.2.5.3 ROV surveys

An ROV offers the possibility of keeping the reference cell much closer to the pipeline than can be achieved with a towed fish. Typically, as illustrated in Figure 18.12, the reference is mounted on to the ROV and is connected so that data are transmitted via the control umbilical to the survey ship. In its essentials, the probe, which has been around since the 1970s [5], is not dissimilar to the diver-held unit shown in Figure 18.8. It usually has two reference half-cells instead of one. The second reference can be used for current density monitoring (see Section 18.4.1.2).

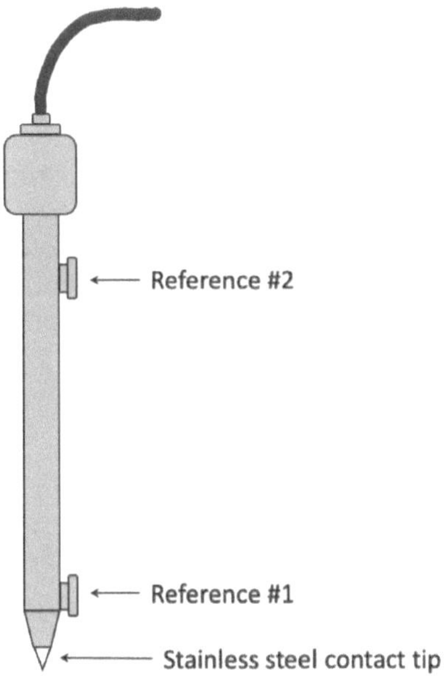

Figure 18.12 ROV-mounted probe.

Another limiting feature of the towed fish arrangement is the length and vulnerability of the trailing wire. ROV surveys overcome this by employing the close-remote reference technique, as is illustrated in Figure 18.13. Here, instead of measuring the potential difference between the reference and the pipeline, the technique measures the change in potential between two references. The "close" reference tracks along the pipeline. The "remote" is held far enough away from the line not to be influenced by its CP system.

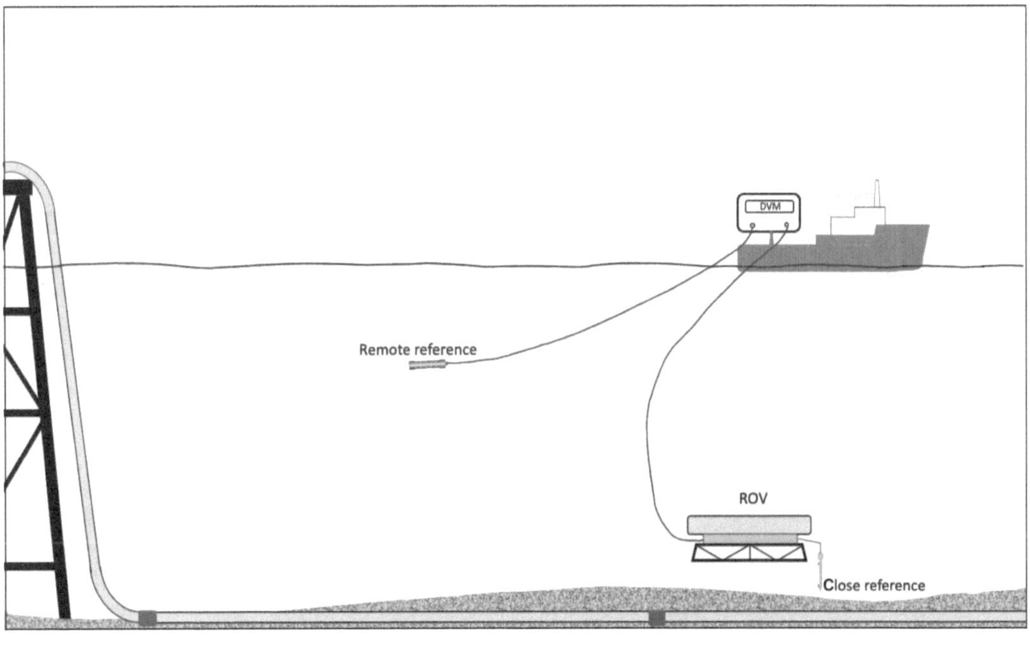

Figure 18.13 Close-remote survey.

The survey proceeds as follows.

- It commences at a point where a stab potential measurement can be obtained. As often as not, this will be at an exposed pipeline bracelet anode. Alternatively, it might be at a site where the pipeline is exposed by coating damage. Less desirably, it may be possible to drive the steel tip of the probe through the coating, for example at a flange, to make a direct contact.
- The pipe-to-seawater potential is recorded as the potential difference between the pipeline at the point of contact with the stainless steel probe tip and the lower reference half-cell. This is the pipe potential calibration reading (E_{cal}):

$$E_{cal} = E_{pipe} = E_{probe} - E_{close}$$

- At the same time, the potential difference (ΔE) between the close and remote reference cells is recorded.

$$\Delta E = E_{close} - E_{remote}$$

- The ROV then proceeds along the pipeline keeping the reference as close to the pipeline as possible. The survey speeds usually vary between 0.25 and 0.4 m/s [24].
- The survey then depends on the expectation that E_{remote} remains constant, and it is only the pipe-to-seawater potential difference that changes along the length of the line.
- There is a further assumption that the potential in the metal of the pipeline ($\varnothing_m$) is constant along its length. On that basis, the change in the potential (E) along the length of the line is dictated by the change in the solution potential ($\varnothing_s$). It follows that recorded changes in the value of ΔE directly mirror the changes in the pipe-to-seawater potential. So, the pipe potential at any point is given by adding the value of ΔE to the potential recorded at the calibration.

$$E_{pipe} = E_{cal} + \Delta E$$

- However, as we discussed in connection with the attenuation of potential along a pipeline $\varnothing_m$ does not remain constant along the length of the line. The assumption of a constant value for $\varnothing_m$ is almost correct for a pipeline that is not drawing much CP current, particularly if the anodes are closely spaced. However, for pipelines for which long lengths are protected either by ICCP or anodes at the terminal structures, the lateral changes in $\varnothing_m$ become significant.
- To compensate for any tendency for the potential of either the close or remote reference to drift, or for changes in the value of $\varnothing_m$, it is important that fresh stab potential readings are taken at regular intervals along the line, and a fresh reading of ΔE recorded. These readings are taken where an anode is accessible. Each stab potential measurement provides a recalibration for the value of potential inferred by tracking ΔE.

Where the line is buried, however, the reliability of the close-remote potential survey technique starts to come into question. Since it is critical that the value of $\varnothing_s$ is measured as close to the pipe as possible, burial interposes a fundamental problem; and the deeper the burial, the less reliable the measurement. Britton [24] has offered an experience-based suggestion that the maximum offset between the pipe and the reference should be 0.6 m. However, there is limited solid information available to either support or refute this. Arguably, this is an issue that computational modelling might usefully address. It requires modification of existing commercially available packages to take into account the joint lateral variation of both $\varnothing_s$ and $\varnothing_m$. At the time of writing, progress is being made in this area.

Despite these concerns over the efficacy of the close-remote survey technique, it has to be admitted that there is no obvious alternative survey technique available. This might point to the benefit of fixed monitoring in some cases (see Section 18.2.6.2).

18.2.5.4 Beach crossings and shore approaches

One of the findings of the investigations into the Varanus Island pipeline explosion (Chapter 14) was that there was evident confusion on the part of the operator as to whether the pipelines crossing the beach were the responsibility of the onshore or the offshore maintenance departments. Clearly, that is something that needs to be established by any pipeline operator. The widely adopted convention is that the transition from offshore to onshore takes place at the first flange, or isolation joint, above the high-water mark. This can sometimes be a considerable distance on the landward side of the shoreline.

The relevant codes are not fully definitive. ISO 15889-2:2012 sets its scope as being... *applicable to all types of seawater and seabed environments encountered in submerged conditions and on risers up to mean water level.* It makes no reference to shoreline crossings, nor does its appendix on *CP monitoring and surveys* contain any relevant information for the sea-land transition. On the other hand, the corresponding land-based pipeline CP standard ISO 15889-1:2015 applies to... *landfalls of offshore pipeline sections protected by on-shore based cathodic protection installations.* Thus, the onshore standard only applies if the pipeline is connected to the land-side CP system. There is a lacuna: neither Part 1 nor Part 2 of ISO 15889 holds for a section of shore approach lines which relies on its offshore CP system (usually bracelet anodes) to protect the onshore stretch.

Irrespective of where the responsibility for maintaining the shoreline crossing section of line resides, there will be a section of the "offshore" pipeline for which land-based CP survey and monitoring practices will apply. Various options are available. The benefits of these depend, to some extent, on whether or not the onshore and offshore sections are isolated.

18.2.5.5 Telluric currents

Since the early days of the telegraph in the mid-19th century, it has been known that spurious electrical currents flow through the surface of the earth. These soon acquired the catch-all name telluric currents, after the Latin "tellus" meaning "earth". Of the multiple natural sources of telluric currents, the phenomena of most relevance to CP are the geomagnetically induced currents. These arise from the solar wind, a stream of charged particles continuously ejected by the sun, interacting with the earth's magnetic field. This interaction, which is most pronounced during periods of intense solar activity, induces low frequency (1–25 Hz) AC currents in soil, seawater or any other conductor.

The interaction between telluric currents and pipelines has been much studied, particularly for onshore pipelines. The consensus is that they do not pose a serious corrosion threat, but that they can, and do, interfere with CP measurements. Their effect is much more prominent on onshore than offshore pipelines. They receive no more than a passing mention in any of the offshore pipeline CP codes, and there is comparatively little published offshore research.

An exception is the 1993 paper by Weldon et al. [10] who reported the results of an extensive campaign of trailing wire surveys in the North Sea in 1979. These were affected by seemingly anomalous readings. The problem was particularly pronounced on longer survey sections. Subsequent investigations indicated that the problem was due to telluric currents.

In 1981, they carried out a study on Total's twin Frigg to St Fergus pipelines. As part of that study, they deployed static electrode arrays at a compression platform (MCP-01) about half-way between Frigg and the onshore terminus at St Fergus in Scotland. They also installed a static array at St Fergus. The data from these arrays were recorded during the period the offshore trailing wire survey was carried out.

The potentials, measured on board the survey ship, were recorded together with both the distance along the line and the time. At the conclusion of the survey, the shipboard data were compared with the data from the static arrays. The latter showed some periods of intense telluric activity during the survey, and these periods coincided with fluctuations in the recorded survey potentials. Using mathematical techniques that they outlined, but did not fully describe, they were able to remove the spurious telluric effects from the pipeline potential profile. It has to be admitted that the paper now has mainly academic interest, not least because the trailing wire approach is now generally restricted to surveys of relatively short lengths of pipeline, particularly at shore approaches. The more usual close-remote survey techniques are less affected by telluric phenomena.

One of the few, if not the only, instance of pipeline CP systems being operated with telluric currents in mind was reported for pipelines carrying Algerian natural gas across the Strait of Messina which links Sicily with mainland Italy [6]. These lines which are comparatively short (~14 km), but which transit deep water (~360 m), were laid in the 1970s. They are protected by an ICCP system on each shore. These ICCP systems were intentionally operated to achieve more negative potentials than necessary for protection. The author explained that values of the off-potential were found to be... *between −1.0 and −1.07 V.* He *went* on to explain that this level of protection... *was maintained, not so much because of railway interference (negligible) so much as to counterbalance the effects of random telluric currents which cause potential changes (up to ±0.3 V) at the ends of the pipeline.* Unfortunately, he did not provide any details of how this value of potential fluctuation was determined.

18.2.5.6 Survey frequency

As with structures, the codes are generally non-prescriptive when it comes to how often a pipeline should be subject to a potentials survey. ISO 15889-2 (2012), for example, is silent on this. It simply requires that the design exercise also addresses the types of survey to be carried out and that this information has to be captured in an operation and maintenance plan.

In practice, it would be unusual to mobilise a costly offshore survey spread solely for the examination of the pipeline's CP. Rather, CP measurements on pipelines are surveyed externally as part of rolling programmes of wider inspections. Once there is a perceived need to fly an ROV down a pipeline, it makes sense also to fit it with relevant CP survey equipment.

18.2.6 Fixed potential monitoring

As we saw in Chapter 11, all ICCP systems should have fixed references. Without them, there would be no basis for controlling the system. This control may be either automatic, and continuous, or a manual activity carried out as part of the day-to-day operation of the system. In Chapter 11, we also discussed the relative advantages and disadvantages of Ag|AgCl and zinc references. The point was made that dual electrodes, combining a short-lived Ag|AgCl reference and a more durable, but less precise, zinc reference had practical advantages. An example of such a dual reference is given in Figure 18.17. In this case, the unit is also fitted with a current density monitoring plate, which we discuss in Section 18.4.1.1.

On the other hand, the permanent installation of fixed potential monitoring hardware is not easy to justify for structures protected by sacrificial anodes. The advantage of sacrificial anodes being self-adjusting is countered by the disadvantage of there being no way of intervening to adjust their outputs. This is another example of the flip side of Grebe's aphorism that you cannot control something that you cannot measure. If you cannot control it, then there is little reason to measure it.

18.2.6.1 Structures

Despite the impossibility of adjusting sacrificial anode outputs, various workers have expressed surprise that more use is not made of fixed monitoring on offshore structures. For example, in 2013 Narayan [20] pointed out that there were estimated to be ~7000 offshore structures involved in hydrocarbon production. All are cathodically protected but, surprisingly in his view, very few are provided with fixed monitoring.

The case for installing fixed potential monitoring for a sacrificial anode system can be advanced on several fronts. In some instances, the monitoring can be justified for research purposes. Elsewhere, it provides information on levels of polarisation without the expense of having to carry out a diver or ROV underwater inspection. This has particular benefits for structures in deep waters. In addition, permanent monitoring also has the advantage that data can be obtained during adverse weather and sea state conditions when surveying or inspection would be impossible.

Only surveying when the water is calm means limiting the information collected to periods when the cathodic current demand is at its lowest. Thus, the CP potential measurements are likely to convey a non-conservative picture of the level of protection. This issue was illustrated by Wyatt [12] who presented data from a monitored anode, installed in the Northern North Sea in the 1970s, to demonstrate a correlation between time-averaged wave height and both anode current output and structure potential. As we discussed in Chapter 7, such observations underpin the advice in emergent codes for an increase in design current densities for wind farm foundations [26].

Of more relevance here is that the exercise demonstrates that there is some benefit to be had from fixed potential monitoring on sacrificial systems, even if no system adjustments can be made in response to the data. For near-shore structures, such as wind farms, there is a case for installing monitoring equipment, including data loggers and transmitters, on a sample of foundations. The majority of the potential data in the early years of operations would be obtained from dip cell surveys or divers, both of which require costly offshore mobilisations. However, providing confidence in the fixed monitoring could be established through correlation with the survey data, the need for subsequent mobilisations for CP surveys could be substantially reduced.

18.2.6.2 Pipelines

A rare example of fixed potential monitoring of a pipeline protected by sacrificial anodes has been on a relatively short (<10 km) length section of 10″ oil pipeline installed by BP in the Beaufort Sea offshore Alaska [18]. The line was coated in dual-layer FBE and cathodically protected by closely spaced (~110 m) Al-Zn-In bracelet anodes. In this case, the decision to install fixed monitoring was justified by the environmental sensitivity of the area and the lack of experience generally with artic subsea pipelines.

This case was very much an exception to the general rule that sacrificially protected pipelines are not normally fitted with fixed potential monitoring. There are some obvious reasons for this. The first is that it is very difficult to pre-install any equipment on a pipeline since the activities involved in offshore pipelay would, as like as not, dislodge it. This means that the monitor has to be installed as a retrofit operation, thereby substantially increasing the cost of the exercise.

For a lengthy pipeline, it would be impractical to hardwire a monitor to an installation either on the beach or on a terminal structure. This means that monitors have to be self-contained and self-powered, and they have to include a workable technology for communicating the stored data. Although options for the data transfer exist, for example by using acoustic transponders, the exercise still requires the mobilisation of an offshore vessel. Alternatively, innovative photoelectric devices that can be powered up by lights on an ROV or an antonymous underwater vehicle (AUV) have been demonstrated [24]. However, the long-term durability of such equipment remains to be confirmed.

Perhaps a more profound disincentive to installing fixed monitoring on subsea pipelines is the difficulty in answering the question: how many monitors do I need?

The answer is simple. You only need one. But there is a problem. You only need one providing you know where on your many kilometres of pipeline you need to put it. Since pipelines are prone to integrity threats at unpredictable locations, it is difficult to make a convincing case for fixed monitoring. At best, a fixed pipeline monitor will only give an idea of how the CP is performing in that general vicinity.

18.3 CURRENT MEASUREMENT

18.3.1 ICCP systems

Although the potential is the sole protection criterion, measurement of the current provides useful information on how the ICCP system is working. The current output is simple enough to obtain since it is a metered system parameter. When recorded in conjunction with other system parameters such as DC output voltage, it can provide some useful information on the health of the CP system itself. In the case of coated structures, recording the progressive increase in the current needed to maintain polarisation can inform the owner about the rate of coating deterioration. Conversely, if the current remains stable, but a progressively increasing DC voltage is required then the CP circuit resistance is increasing. This may indicate an incipient problem with the ICCP anodes.

18.3.2 Sacrificial systems

18.3.2.1 Current clamp meters

Non-contact clamp-on current meters, working on the Hall effect, have been available for many years. Many propriety brands are available. However, just as the word "Hoover" has found its way into the English Dictionary as a synonym for vacuum cleaner, the "Swain clamp" or "Swain meter" has become entrenched in the CP lexicon. The name derives from the Swain Company, and its eponymous founder, a former NASA guidance systems engineer: William H. Swain. Importantly, for marine CP, the company has developed marinised versions of its equipment which can be used by divers or mounted on ROVs.

Where a retrofit CP system is installed based on sacrificial anode sleds, it is becoming a routine exercise to place a current clamp meter around the cable to confirm that it is connected to the structure, and to measure the current output [22].

It is also possible to use clamp meters to measure the currents flowing in the cores of standoff anodes, albeit this is an activity that has not been practiced a great deal. There are, however, some interesting examples of their use. Mateer [14] found that clamp meter readings on trapezoidal anodes broadly supported the use of the modified Dwight formula for calculating anode resistance. Interestingly, in one instance he observed a 3:1 variation between the individual currents passing through the two legs of a single anode. However, caution needs to be exercised when considering such an unusual result. The discrepancy between the two readings might be more to do with the difficulties in making the underwater measurement, in particular the need to "zero" the instrument to nullify the effects of extraneous magnetic fields. Indeed, some years after Mateer's paper, Lemieux and Hartt [19] carried out comparative testing of various methods of determining current outputs from anodes. Their assessment of the current clamp method fell well short of a ringing endorsement.

18.3.2.2 Monitored anodes

18.3.2.2.1 Structures

The overall current outputs of sacrificial anode systems, as a whole, are never measured. In the past, occasional attempts were made to measure the outputs of individual anodes. That work was motivated by research interests rather than having any direct relevance to the day-to-day management of CP. A typical rationale for installing some monitored anodes was to assess the reliability of the anode resistance formulae (see Chapter 10) for future CP designs.

Installing a monitored anode requires that, instead of welding it directly to the structure, it is mounted via electrically insulating bushes. Its only contact with the structure is via a resistive shunt, as indicated schematically in Figure 18.14. An installation on an offshore jacket is shown in Figure 18.15. In this arrangement, the current output is measured on a topsides DVM as the voltage drop across the shunt resistor. Monitored anodes were generally fitted on large oil production structures, mostly in the 1970s, when project budgets were elastic enough to accommodate a little research and development. Their use, although having continued into the 21st century [20], receives little support on cost-limited modern-day offshore CP projects.

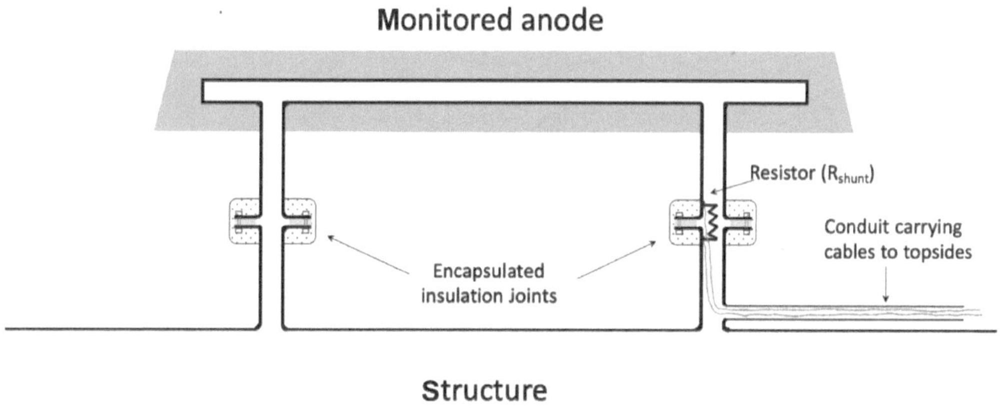

Figure 18.14 Arrangement for a monitored anode.

One source of scepticism over the use of monitored anodes was that even the small voltage drop developed across the shunt resistors would distort the result by suppressing the anode's output. The answer to this concern lay in converting the shunt resistor into a zero-resistance ammeter (ZRA) as explained by Backhouse [2]. This could be achieved by rewiring an operational amplifier or potentiostat (see Chapter 5) as indicated in Figure 18.16. Since the operational amplifier outputs current in a feedback loop to minimise the potential difference between its two inputs (+ and −), it must drive the same current as the anode current (I_{anode}) through the shunt resistor (R_{shunt}). So, I_{anode} is determined by Ohm's law from the output voltage of the operational amplifier, but there is essentially no voltage drop in the metallic circuit between the anode and the structure. Hence the term: "zero resistance".

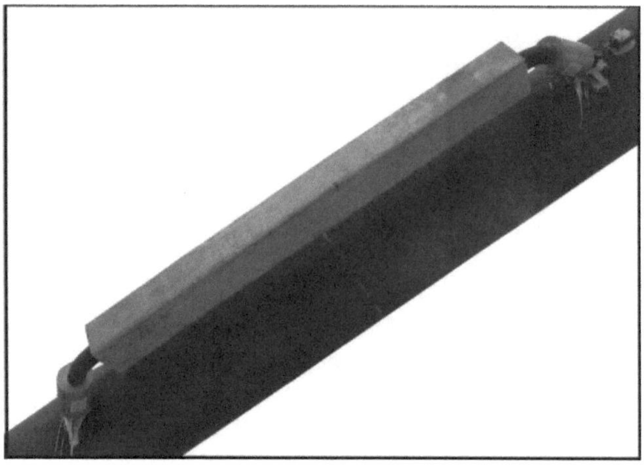

Figure 18.15 Monitored anode. (Photo courtesy of Deepwater Corrosion Services Inc.)

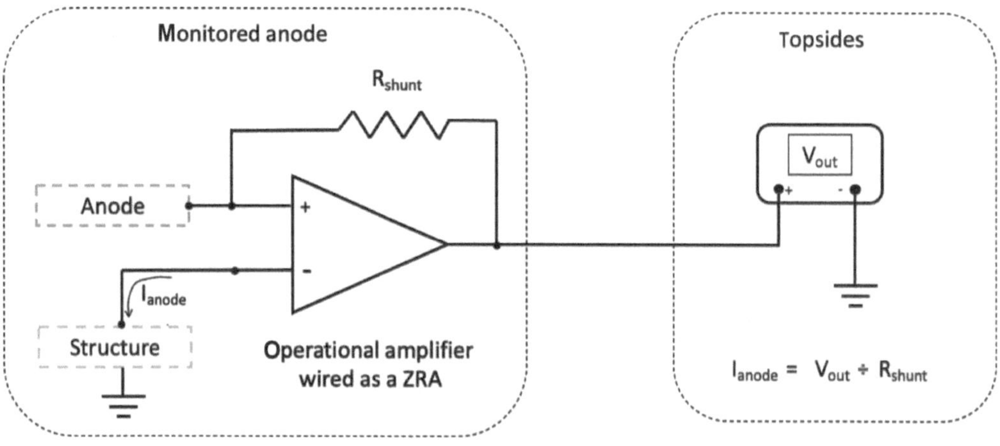

Figure 18.16 Zero resistance ammeter.

18.3.2.2.2 Pipelines

A rare example of a monitored ~47kg bracelet pipeline anode has been described by La Fontaine et al. [18]. This was on the short stretch of 10″ pipeline in the Beaufort Sea, which we have already mentioned (see Section 18.2.6.2). The arrangement was analogous to that used for stand-off platform anodes described above. Soon after installation, the data revealed an anode current output of 0.00001 A. Since the anode was exhibiting a potential of –1.087V, and the pipeline was evidently well polarised (–1.068V), it was concluded that the anode was operating normally. This meant that the pipeline was drawing very little current due to the protective efficiency of its double-layer FBE coating.

The authors did not comment upon the implication of this low anode current. You may care to insert the figures of a 47kg anode outputting a mean current of 0.00001 A into equation 3.4. If you do, you get a life in excess of 800,000years! The figure is, of course, meaningless because the anodes would have self-corroded long before then. Nevertheless, it can only be hoped that the FBE coating obliges by breaking down, as envisaged by the codes. Otherwise, the anodes will be condemned to a long and dull existence.

18.4 CURRENT DENSITY MEASUREMENT

18.4.1 Fixed monitoring

18.4.1.1 Current density plates and probes

18.4.1.1.1 Uncoated

In addition to monitored anodes, the pioneering days of large fixed offshore petroleum installations also saw a few examples of current density (CD) monitors. These comprised a wrap of sheet steel around a tubular member, but insulated from it. Alternatively, they are in the form of cylindrical probes [17]. An example of a proprietary corrugated type CD monitoring plate is shown in Figure 18.17. As can be seen, this example is attached to a robust potential monitoring unit. As with the monitored anodes, the plate or probes are connected to the structure via a current measuring shunt. There is little published in the way of long-term results.

Figure 18.17 CD plate and dual reference. (Photo courtesy of Deepwater Corrosion Services Inc.)

18.4.1.1.2 Coated

I have only come across one example of a coated CD coupon. It was on the same Beaufort Sea pipeline project that we have already mentioned. The coupon was coated with double-layer FBE. It is not particularly surprising that the cathodic current density on the coupon was too low to be detected.

18.4.1.2 Field gradients

In contrast to the very few permanently installed CD plates, there have been numerous publications on current density measurements using field gradient sensors. As we know, an electrical field gradient is generated around a structure that is under CP. This generates the IR error when it comes to determining the true steel-seawater potential. Conversely, this same field gradient can be exploited beneficially to estimate the cathodic current density on the steel.

This information is of more than just academic interest. For example, reliable current density measurements in some geographical locations have been useful in countering the generally conservative guidance figures in the codes. As we saw in Chapter 13, this enabled designers to produce less costly CP designs. More relevantly, very considerable cost savings may now be possible if life-extension CP retrofit designs can be based on the measured CDs, rather than conservative code-based guidelines.

An illustration of how field gradient, and thence CD, can be estimated is given in Figure 18.18. For convenience, the example shown relates to the anodic CD on a pipeline bracelet anode, but the principle is also used to estimate cathodic CD on the pipeline itself or on any structural component. The technique makes good use of the IR drop which we found to be something of nuisance when measuring potential. If we place the ROV-mounted probe, shown in Figure 18.12, onto a working anode, it will intersect the field gradient generated as the current leaves the anode and flows towards "remote earth" through the seawater. Since the distance between the two matched references is known, we can use the potential difference measured between them to tell us the field gradient. Moreover, if we know the resistivity of the seawater, then, by applying some basic geometry, we can convert this field gradient to a current density at the surface of the anode. From this information, and the known surface area of the anode, we can calculate a value for the anode current. By the application of Faraday's laws, and the estimated remaining anode mass, we can get a figure for its remaining life.

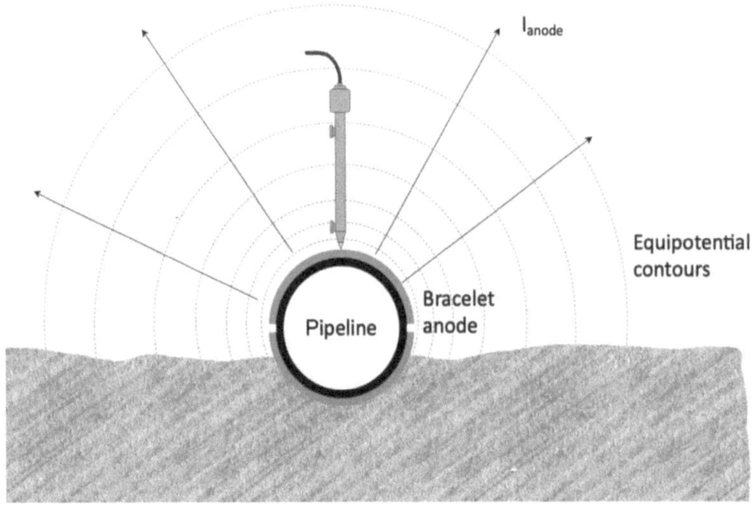

Figure 18.18 Field gradient estimation of CD.

18.4.2 Surveys

18.4.2.1 Pipelines

This measurement of field gradients, and its translation to current densities, might seem straightforward. However, you will spot some of the flaws in the basic analysis. The first problem is that Figure 18.18 is a two-dimensional sketch indicating current flowing away at right-angles to the pipeline axis. A little thought would remind us that the current flows in three-dimensional space, and that where it leaves the ends of the anode it will not follow the same radial flow. This means that any simple geometrical analysis will fail to describe the current pattern. So, it will produce an inaccurate estimate of the anode's output. This is the sort of problem that lends itself to computer modelling.

An alternative method for measuring field gradients has been developed by The Foundation for Scientific and Industrial Research, better known as SINTEF. This technique was then developed commercially by CorrOcean [4]. It used an ROV-mounted rotating T-sensor. Reference electrodes were fixed at the tips of the T's cross-bar. The T was then rotated about its main axis. The theory was that, as the T rotated, any bias arising from the manufacture of the two references would be cancelled out.

Presumably as a guard against disclosing exploitable technical information, the paper is thin on detail. For example, we are not told the nature of the reference electrodes or the spacing between them, or the rate of rotation of the sensor. The paper also states that the finite difference method (FDM) of numerical modelling was used to convert field gradient measurements into current densities. Again, however, no details are provided.

The paper shows a "schematic" potential profile of a pipeline but does not explain how potentials would be calculated using an ROV fitted with a rotating T-sensor rather than a stabbing dual-electrode type probe. The limited data reported for field trials made no mention of potential. In the event, the rotating T-sensor seems to have actually been used very little for pipeline CD surveys. The dual electrode (Figure 18.12) persisted as the preferred choice.

About 30 years later Force Technology, the business successor to CorrOcean, reintroduced the T-sensor concept in an updated form [21]. As with the original publication, crucial details of the equipment have been withheld. But, at least, we are now told that the references are platinum. The new incarnation has been named the **F**ield **G**radient **S**ensor ("FiGS"). The most obvious difference between the 20th- and 21st-century units is that the new version has twin rotating sensors so that it can record field gradients in three dimensions whereas the T-sensor device was limited to two. Arguably, however, the main difference between the two incarnations lies in the massive increase in the availability of computing power that, courtesy of Moore's law, can now be brought to bear.

Werenskiold and Osvoll [21] presented the result of a case study that provides an impressive demonstration of the sensitivity of FiGS. It involved a thermally insulated, partially buried and rock-dumped, pipeline on which the anodes were showing surprisingly high consumption rates.

The survey and its analysis determined that the pipeline anodes were delivering around 10 A in total, but that only about 0.05% of this (5 mA) was being taken by the pipeline itself. The remaining 99.95% of the current was being taken by the much less well-coated terminal structures. The survey also claimed to detect coating defects that were not visible to the naked eye, and one of which was buried. The paper went on to explain that potential measurements... *cannot detect these defects because the variation in potential across these defects is in the order of 0.03 mV at a measurement distance of 0.70 m above the pipeline.*

This statement indicates a divergence between the ambitions of the CP survey specialists, who inevitably seek to improve the detection abilities of their systems, and the pipeline integrity specialists, who seek to keep the line operating safely. To be able to pick up such miniscule coating defects is clearly a testament to the sensitivity of the technique. However, an integrity-minded CP engineer might take the view that coatings are expected to contain defects. That is, after all, why we apply CP to coated pipelines. From that perspective, a coating defect that is only disturbing the pipeline potential by a fraction of a millivolt is not exactly a show stopper. Indeed, a more pragmatic application of FiGS was reported a couple of years later [23]. This later paper described its contribution to determining whether or not CP retrofit was required to extend the operating life of aged lines which had suffered visible coating damage, and anode detachment, as a result of trawling.

18.4.2.2 Structures

The 1980s T-sensor was used to measure the field gradients, and thence the current densities, on North Sea structures. That work contributed to the adjustments in the mean design current densities in the codes of the time.

For structures associated with hydrocarbon production, the recent CP emphasis has switched from designs for new structures, to retrofit designs targeted at structures for which the operating lives have been extended. As we discussed in Chapter 13, installing current capacity as part of an offshore retrofit is very much more expensive than installing the same current capacity in a fabrication yard. So, there is an incentive to limit the mass of retrofit anode to no more than is needed to do the job.

Assuming that the structure is still fully polarised at the time the decision is taken to retrofit, all that the retrofit design needs to do is to ensure that the same current can continue to be provided through to the revised end-of-life date. It follows that an accurate knowledge of the actual cathodic current density on the structure could lead to considerable savings in the cost of any retrofit. There are two challenges that need to be overcome. The first is that cathodic current densities are much lower than anode current densities. So, the field gradients that need to be measured are much weaker. The second is that the field gradients, and cathodic current fluxes, on a 3-dimensional structure are vastly more complex than suggested, for example, in Figure 18.18. The first challenge is apparently addressed by developments in the sensitivity of the instrumentation. The second is tackled by computer modelling.

Lauvstad et al. [25] have published an example of such an exercise, again using the FiGS arrangement to measure field gradients, and then computer modelling to relate this to cathodic CDs on real structures. In this instance, it was reported that the mean CDs actually being taken by the three Ula field platforms in the North Sea was no more than 15 mA/m^2. Designing a CP life extension retrofit based on these mean figures would be considerably less expensive than using code-based values for new structures.

This calculated low cathodic current density is qualitatively supported by circumstantial evidence of prolonged anodes life on numerous structures. It seems that prolonged cathodic polarisation leads to continuing improvement in the protective quality of the calcareous deposit. Unfortunately, as with so many papers in this area, commercial sensitivity prevents the disclosure of many of the underlying details, so independent scrutiny of the results is difficult.

18.5 INTERACTION

18.5.1 Sacrificial systems

In Chapter 2, we made the point that current flowing from the seawater onto the surface of a structure is always protective. Current leaving the structure, and flowing into the seawater, by contrast, is always damaging. Currents flowing through the environment that are not associated with the CP design intent are referred to as "stray" or "interference" currents.

One of the perceived advantages of sacrificial anodes is that the voltages involved are sufficiently low that there is little or no stray current activity. For this reason, stray current testing is not normally applicable to sacrificial anode systems.

That is not to say, however, that the problem can never arise. For example, it was discovered in the 1990s that unbonded moveable components on wellhead xmas trees could pick-up and discharge damaging electrolytic current from the tree's sacrificial anodes. The problem was remedied by ensuring that all moveable components were supplied with bonding straps to ensure that they were earthed into the CP system.

18.5.2 ICCP systems

18.5.2.1 General

If stray current interference problems arise, they are more likely to be a result of the much higher system voltages developed in ICCP systems. We outlined the basic interference process in Chapter 11, and we encountered an example involving in-field pipelines in Chapter 13 and a case of damage to a ship's rudder in Chapter 15. For reasons such as these, stray current interaction testing is recommended for ICCP systems. The concern is not that the ICCP system will cause harm to its host structure, but that it might cause electrolytic damage to adjacent structures that are not metallically connected to the host ICCP system.

The classic way to assess stray current interference derives from land-based CP practice. It involves measuring the potential on the adjacent non-connected third-party structure or pipeline, often referred to as the "foreigner". The ICCP system is switched off and on, and the effect that this has on the measured potential of the adjacent structure is recorded. As we have explained, CP current does not follow lines of sight. For this reason, it may be difficult to gauge which location on the adjacent structure is most at risk of interference. A certain amount of trial-and-error positioning of the reference is usually beneficial when carrying out interference (or interaction) testing.

18.5.2.2 Harbours

The two areas of marine CP where ICCP is regularly employed are harbours and ships. In respect of harbours, ISO 13174 advises, but does not require, that interaction testing... *should be undertaken to demonstrate that adjacent structures are not adversely affected by the cathodic protection system.*

Notwithstanding this advice, it has to be borne in mind that large harbour facilities will have multiple, sometimes large capacity, ICCP power sources. In these cases, it may not be practicable to carry out the desired synchronised switching of the systems.

If switching is practicable, ISO 13174 provides the following guidance. Where the adjacent structure is fitted with sacrificial anodes, the effect of switching the ICCP system on and off should not cause the potential on the adjacent structure to swing outside the prescribed protection limits of between −0.8 and −1.1 V. The same limits also apply if the adjacent structure is also protected by its own ICCP system. The only point to make here is that stray current interference can work either way. So, it is best to carry out on/off switching on both systems to examine their mutual effects.

Where the adjacent structure is not cathodically protected, it is usually advised that switching the ICCP should not cause the free corrosion potential (E_{corr}) of the adjacent structure to change by more than ±20 mV. Although this figure in the standard derives from a rule-of-thumb used in onshore CP practice, it is easy enough to justify knowing what we have already learned in this book. In Chapter 1, we spent some time considering the long-term free corrosion rate of steel in seawater. It is not a fixed figure, but the evidence is that a figure of 0.1 mm/year is likely to be conservative. Furthermore, in Chapter 5, we learned from electrode kinetics that the Tafel slope for steel in seawater, and most other aqueous electrolytes, is very close to 60 mV per decade of current. This means that a 20 mV positive swing increases the anodic current by a factor of $10^{(20 \div 60)}$, which turns out to be roughly double. The implication may be that, if a structure deemed not to merit protection in its own right, then increasing the corrosion rate from around 0.1 to 0.2 mm/year does not have any practical implications.

Where the interaction testing shows that the potential of the adjacent system swings by more than the prescribed limits, the situation will need to be investigated and corrected. By convention in most territories, and by law in some, the onus is on the party introducing the new ICCP system to carry out this work. Whereas the interference testing itself can reasonably be carried out by personnel with the expertise equivalent to competence levels 2 or 3 (per ISO 15257), selecting the appropriate remedial actions requires the specialist knowledge of individuals equivalent to competence level 4.

Usually, interaction effects can be solved by the adjustment of the DC power supply outputs that is normally carried out as part of the commissioning of the new system. In some cases, however, further intervention will be needed. Where this situation arises, the remedial options include electrically bonding the two structures together, either directly or through a resistive link; or installing limited sacrificial anode protection on the third-party structure.

18.5.2.3 Ships and boats

Interaction testing is irrelevant for ships since they are normally at sea. There is, however, the possibility of interactions with jetty ICCP when they come alongside. We discussed this topic in Chapter 15, in particular the decision as to whether to isolate or bond when alongside at a jetty. Irrespective of what is done, however, the responsibility will lie with the ship's officer responsible for electrical systems. Neither CP management nor CP personnel are likely to be involved.

In Chapter 15, we also discussed the CP applied to small boats moored in marinas, and the installation of galvanic isolators in the mains earthing cable to prevent the adventitious connection of boats' anodes to unprotected craft or unprotected marina piling. In practice, there is no worthwhile CP monitoring that the small boat owner can carry out. All that can be done is replace the anodes, if need be, when the boat is brought ashore for other reasons.

REFERENCES

1. J. Morgan, *Monitoring and control of cathodic protection of offshore structures*, Paper no. 20 CORROSION/75.
2. G.H. Backhouse, *New developments in monitoring equipment for offshore cathodic protection systems*, Proceedings of Symposium on *Corrosion Protection Offshore*, CEFRACOR Paris, 1979.
3. R. Nagel and A. Willis, *New monitoring methods of cathodic protection systems*, Proceedings of Symposium on *Corrosion Protection Offshore*, CEFRACOR Paris, 1979.
4. R. Strømmen and A. Rødland, *Corrosion surveillance of submarine pipelines by electric field strength monitoring*. Materials Performance **20** (10), 47 (1981).
5. G.H. Backhouse, *Recent advances in cathodic protection surveys of subsea pipelines*, Paper no. 108 CORROSION/81.
6. F. Santagata, *The philosophy of corrosion protection of pipelines crossing the straits of messina, and its implementation*, Proceeding of UK Corrosion, 1982.
7. R.W. Wilson, *Corrosion control in the oceans*, Proceedings of UK Corrosion, 1982.
8. T. Sydberger, *Evaluation of inspection methods for offshore pipeline systems*. Materials Performance 22 (5), 56 (1983).
9. S. Eliassen, R. Mollan and T. Sydberger, *Potential surveys of submarine pipelines*, p. 135 Proceedings of UK Corrosion, 1983.
10. C.P. Weldon, A. Shultz and T.T.S. Ling, *Telluric current effects on cathodic protection potential measurement on subsea pipelines*. Materials Performance **22** (8), 43 (1983).
11. V. Ashworth, *Measurements for corrosion management*, in *Seminar on Surveying, Inspection and Corrosion Monitoring of Fixed Offshore Structures and Pipelines*. Global Corrosion Consultants Ltd. (London) 1985, pp. 22–38.
12. B.S. Wyatt, *Cathodic protection of fixed offshore structures*, Chapter 8, in *Cathodic Protection: Theory and Practice*. Eds V. Ashworth and C.J.L Booker. Ellis Horwood (Chichester) 1986, pp. 143–171.
13. D.H. Boteler and M.J. Cookson, *Telluric currents and their effects on pipelines in the Cook Strait region of New Zealand*. Materials Performance **25** (3), 27 (1986).
14. M.W. Mateer, *Overlooked data available from an offshore subsea survey*. Materials Performance **30** (7), 26 (1991).

15. J.C. Daley and D. Ingraham, *Underwater maintenance and inspection of CP systems (Cook Inlet, Alaska).* Materials Performance **35** (1), 23 (1996).
16. J. Britton, *Cathodic protection surveys of offshore platforms: A new approach*, Paper no. 729 CORROSION/98.
17. J.N. Britton, J.P. La Fontaine and G. Gibson, *Recent advances in offshore cathodic protection monitoring*, Paper no. 00672 CORROSION/2000.
18. J.P. La Fontaine, T.G. Cowin and J.O. Ennis, *Cathodic protection monitoring of subsea pipelines in the artic*, Paper no. 01502 CORROSION/2001.
19. E.J. Lemieux and W.H. Hartt, *Galvanic anode current and structure current demand determination methods for offshore structures*, Paper no. 03078 CORROSION/2003.
20. R. Narayan, *CP monitoring for fixed jacket offshore structu*res, Proceedings of Conference on *Offshore Cathodic Protection*, Institute of Corrosion (London Branch) 2013.
21. J.C. Werenskiold and H. Osvoll, *New tool for CP inspection*, Paper no. 7746 EuroCorr, 2014.
22. J. Vittonato and M.-A. Pellet, *Platform cathodic protection retrofit with anodes racks and subsea current measurement*, Paper no. 7615 CORROSION/2016.
23. G.Ø. Lauvstad et al., *Field gradient survey of offshore pipeline bundles affected by Trawling,* Paper no. 55072 EuroCorr, 2016.
24. J. Britton, *Ageing subsea pipelines external corrosion management*, Paper no. 9642 CORROSION/2017.
25. G.Ø. Lauvstad et al., *CP inspection of jacket based on high sensitive field gradient sensors,* Paper no. 11709 CORROSION/2018.
26. B.S. Wyatt, J. Preston and W.R. Jacob. *Cathodic protection of offshore renewable energy infrastructure: part 1.* Corrosion Management **156**, 19 (July/August 2020); Part 2, *Corrosion Management* **157**, 16 (September/October 2020).

Index